5500

ANTIFUNGAL COMPOUNDS

VOLUME 1
Discovery, Development, and Uses

Edited by

MALCOLM R. SIEGEL

Department of Plant Pathology
University of Kentucky
Lexington, Kentucky

HUGH D. SISLER

Department of Botany
University of Maryland
College Park, Maryland

Senior Editor of Volume 1:   MALCOLM R. SIEGEL

MARCEL DEKKER, INC.   New York and Basel

Library of Congress Cataloging in Publication Data

Main entry under title:

Antifungal compounds.

    Includes bibliographical references and indexes.
    CONTENTS: v. 1. Discovery, development, and uses.
    1. Fungicides. I. Siegel, Malcolm R. II. Sisler,
Hugh Delane, 1922-
SB951.3.A57   628.9   76-46827
ISBN 0-8247-6557-5

MARCEL DEKKER, INC.
270 Madison Avenue, New York, New York 10016

Current printing (last digit):
10 9 8 7 6 5 4 3 2 1

PRINTED IN THE UNITED STATES OF AMERICA

PREFACE

The need for fungicides is created by fungal activities detrimental to the welfare of mankind. As the bases for such destructive activities are identified, compounds are sought to control the fungi involved. Fungi had been identified as the basis of many problems a half century or more ago, but at that time very few fungicides had been found to control them. A great number of compounds have since been discovered, new methods of application have been developed, and new concepts on modes of action have evolved. At no time in the past has interest in fungicides been greater than during the current period of transition from the protective to the systemic fungicide.

A number of years have elapsed since a comprehensive coverage of fungicides appeared in two volumes edited by D. C. Torgeson (Fungicides: An Advanced Treatise, 1967). Since then, there have been marked advances in the development of systemic compounds, and a rising awareness of the toxicological hazards and the environmental impact of fungicides has evolved. With the advent of more selective fungicides, fungal resistance has emerged as a major problem. In this rapidly changing field, it is the purpose of this book and its companion volume to summarize and evaluate recent developments, to integrate these with significant developments of the past, and to attempt some projections into the future. We believe the various contributors to the volumes have achieved a reasonable measure of success for these goals.

The overall organization of the two books follows a biological—biochemical—ecological approach rather than one based on chemical class. However, individual authors have utilized the latter approach where it was most appropriate for their particular contribution.

This book, the first of the two volumes, focuses on the discovery, development, and use of fungicides and the problems associated therewith in plant pathology, medicine, wood preservation, and industry. The second volume considers fungicides according to their effects and their fate as they interact with biological, biochemical, and ecological systems.

We are grateful to the many authors whose efforts have made these volumes possible. We wish to thank Carolyn Siegel, Patricia Sisler, Libbie Jones, Debby Owen, and Jeanne Kelleher for generous assistance in the preparation of the manuscripts.

Malcolm R. Siegel
Hugh D. Sisler

CONTENTS

LIST OF CONTRIBUTORS

PAUL A. BACKMAN
Department of Botany and Microbiology, Auburn University, Auburn, Alabama

H. P. BURCHFIELD*
Division of Life Sciences, Gulf South Research Institute, New Iberia, Louisiana

ELROY A. CURL
Department of Botany and Microbiology, Auburn University, Auburn, Alabama

CHARLES J. DELP
Biochemicals Department, Research Division, E. I. du Pont de Nemours & Co., Inc., Wilmington, Delaware

JOSEPH W. ECKERT
Department of Plant Pathology, University of California, Riverside, California

DONALD C. ERWIN
Department of Plant Pathology, University of California, Riverside, California

E. EVANS†
Fisons Agrochemical Division, Chesterford Park Research Station, Saffron Walden, Essex, England

WILLIAM E. FRY
Plant Pathology Department, Cornell University, Ithaca, New York

---

*Present address: Research Associates, New Iberia, Louisiana
†Now deceased.

RENÉ J. LACOSTE[*]
International Division, Rohm and Haas Company, Philadelphia, Pennsylvania

MICHAEL P. LEVI
School of Forest Resources, North Carolina State University, Raleigh, North Carolina

E. NEIL PELLETIER
Office of Pesticide Programs, Criteria and Evaluation Division, Environmental Protection Agency, Washington, D.C.

R. RODRIGUEZ-KABANA
Department of Botany and Microbiology, Auburn University, Auburn, Alabama

H. JEAN SHADOMY
Departments of Medicine and Microbiology, Virginia Commonwealth University, Medical College of Virginia, Richmond, Virginia

SMITH SHADOMY
Departments of Medicine and Microbiology, Virginia Commonwealth University, Medical College of Virginia, Richmond, Virginia.

HUGH D. SISLER
Department of Botany, University of Maryland, College Park, Maryland

ZVI SOLEL
Division of Plant Pathology, Agricultural Research Organization, The Volcani Center, Bet Dagan, Israel

ELVINS Y. SPENCER
Research Institute, Agriculture Canada, London, Ontario

ELEANOR E. STORRS
Division of Life Sciences, Gulf South Research Institute, New Iberia, Louisiana

R. H. STOVER
Division of Tropical Research, United Fruit Co., La Lima, Honduras

WILLIAM C. VON MEYER
Marketing Department—Agrochemicals, Rohm and Haas Company, London, England

---

[*]Present address: Department of Foreign Regulatory Affairs, Rohm and Haas Company, Philadelphia.

GERALD E. WAGNER
Department of Clinical Pathology, Virginia Commonwealth University,
Medical College of Virginia, Richmond, Virginia

CHARLES C. YEAGER
Ventron Corporation, Beverly, Massachusetts

CONTENTS OF VOLUME 2

Chapter 1

HISTORY OF FUNGICIDES

Elvins Y. Spencer

Research Institute
Agriculture Canada
London, Ontario

1

I. INTRODUCTION

A history of fungicides would not be complete and would lack perspective without some reference to ancient and medieval plant pathology. This has recently been ably reviewed by Orlob [1]. Early plant pathology was not so much concerned with any particular disorder as it was with the overall success of the crop. Very often plant diseases were attributed to supernatural and magical causes, and it is not surprising that protective measures were based on the same conceptions resulting in magical practices and rituals. The most valuable contribution to early plant pathology came from astronomy and meteorology where certain disease developments were correlated with atmospheric conditions.

A survey of the developments in methods of plant protection indicates definite eras. If one ignores the magical practices of ancient times, one might cite sulfur as the fungicide marking the first era from unrecorded times to 1882. This was followed by the copper era from 1882 to 1934 which then led into the initial organic fungicide era. The latter era coincided with similar major expansion in organic insecticides, herbicides, and antibiotics. It is of interest to note that in this development systemic insecticides appeared very early without a parallel one in fungicides. Although much effort has been devoted to the search for and development of systemic fungicides, it was noted in 1967 that there is ". . . no outstanding commercial control of plant diseases by systemic fungicides, interest continues and the practical solution is only a matter of time" (Ref. 2, p. 25). Almost simultaneously with that statement, a publication appeared on the systemic fungicidal activity of 1,4-oxathiin derivatives [3] of which two have since been commercially developed. This marked the beginning of a new "sub era" of organic fungicides, namely systemic fungicides, several of which are already in industrial production with more at the developmental stage.

The presystemic fungicides are largely protectants, the initial ones being inorganic like sulfur and various forms of copper, with Bordeaux mixture being the most common form. Then in 1934, the modern era of organic fungicides emerged with the discovery of the fungicidal properties of the dithiocarbamates [4]. They do not penetrate the plant cuticle and therefore are usually not effective in eradicating established infection. Application is by foliar spray, so that complete coverage is required for protection and, as new growth is vulnerable to infection, spraying has to be repeated.

For protectant fungicides, selective accumulation by the spores plays a dominant role in their toxicity [5]. Thus very high concentrations of fungicide are required. For a majority of the protectants ED_{50} values lie within the range of 100-100,000 $\mu g/g$ spore, which is greatly in excess of toxic ratios normally considered for insecticides [6]. They show a wide range of activity against fungi as they react without specificity at cell-receptor sites. Thus the continuous application in the field of these relatively low-toxicity fungicides has rarely resulted in the appearance of resistant strains of plant pathogens.

Selective interference with vital processes is very rare with protectant fungicides. Usually the fungus is inhibited before it enters the plant; thus the mode of action is primarily in protecting the plant rather than in curing the disease. A fungicide that is systemic in its action is absorbed by the plant and then translocated within it, rendering protection from fungal attack or eradicating a fungus already present. The search for compounds with these properties to offset some of the disadvantages of protectant fungicides has been underway for some time. Limited laboratory success has been achieved but it is only within the past ten years that there have been commercial developments. The classical protectant fungicides usually interfere with energy-producing processes, inhibiting cell respiration; while most of the systemic fungicides, with the oxathiins as exceptions, interfere with biosynthetic processes. The systemics tend to exhibit a narrow and specific structure-activity relationship and are active at a much lower level than the older protectant fungicides. Although this might be advantageous, it is not surprising to find reports of resistance developing. Thus a parallel, but delayed, story develops similar to that for the narrow-spectrum specific-acting insecticides, raising new problems [100].

The high activity and specificity of these new fungicides can lead to ecological shifts. This possibility is greatly enhanced since soil application is common and the ecological change can result in the appearance of plant pathogens not occurring in the usual soil ecosystem. Alternatively, the suppression of some pathogens allows some tolerant beneficial organisms to increase [8]. There is also the possible adverse effect on other beneficial soil organisms, such as earthworms [7].

In spite of the rapid progress in the last 40 years in the development of organic protectant and, more recently, systemic fungicides, it is of significance to note that among 100,000 classified fungal species no more than approximately 200 are known to cause serious plant diseases [9]. Evidently, higher plants have a very effective mechanism for protection against a majority of fungi and/or very few fungi have the appropriate attack mechanisms. The demonstration over 30 years ago of the induction of resistance to a virulent race following inoculation with a fungus to which the plant was resistant, resulted in the proposal of the "phytoalexin theory" [10]. Many antifungal compounds have since been isolated from plants following inoculation and much has been written concerning the part these might play in disease resistance or plant protection. Recent increased activity in this area of investigation may result in synthetic analogs for plant protection or lead to improved disease resistance.

II. PLANT PROTECTANTS UP TO 1882

The varied methods of plant disease control, from incantations to weird brews through recorded time, have been extensively reviewed by Orlob [1] and more concisely by McCallan [2]. The use of sulfur as a pesticide dates back to the ancient Greeks, but its effective use was only discovered at the beginning of the 19th century when the preparation of lime sulfur was reported in 1802 [11] to control mildew on fruit trees. Later, in 1821, it was rediscovered as a formulation of sulfur and soft soap [12]. Various formulations of elemental sulfur or crude lime-sulfur preparations were developed in Europe and America during the early half of the 19th century to control powdery mildews on a variety of crops and leaf curl of peach.

Copper sulfate as a fungicide preceded elemental sulfur. In 1761 it was reported to control bunt, although a systematic laboratory study to determine optimum conditions for seed treatment was not carried out until a few years later [2].

III. PLANT PROTECTANTS, 1882-1934

Through poor quarantine control the fungus Plasmopara viticola, which causes downy mildew of grape vine, was introduced about 1870 from the United States to France on nursery stock. The chemical mixture of lime and copper sulfate, by pure serendipity, was discovered in 1882 as a fungicide for this organism [2]. It was due to the keen observation of the French botanist Millardet with the subsequent collaboration of a chemist, Gayon, that a satisfactory formulation was developed by 1885 and finally introduced a year later as Bordeaux mixture. Much work was subsequently done to improve this "universal" fungicide, as well as develop spray methods for its application, until it was challenged by the dithiocarbamates.

Lime sulfur had a revival and largely replaced Bordeaux mixture for some fruit crops [13]. This was largely due to the development of a dry-mix lime sulfur in which a wettable concentrate was made by the incorporation of calcium caseinate [14].

As a matter of historical interest, formaldehyde was introduced in 1897 and used for some time as a seed treatment for wheat smut. This was then partially displaced by copper carbonate dust, first used in 1902 and rediscovered in 1915 [2]. Then followed the organic mercurials originally introduced in 1913. They subsequently lost their dominant position to the dithiocarbamates and to some systemics such as the carboxins. The ultimate death knell to the organic mercurials has been the banning of their use in many countries due to toxic residues.

IV. PLANT PROTECTANTS, 1934-1964

These dates mark off the start of two new eras; first (1934) the introduction of synthetic organic protectant chemicals, and second (1964) the emergence of the first systemic fungicide to be developed commercially. This does not imply that there are not more protectant fungicides being synthesized.

A. Dithiocarbamates

Although 1934 marked the issuance of a patent for derivatives of dithiocarbamic acid, including the oxidation product tetramethylthiuram disulfide and various metal salts, it was over six years before any extensive developments were reported with promising results for its use as a seed dressing in preventing seedling blight in flax [4]. Meanwhile these products, normally used as accelerators in the curing of rubber, were being developed and subsequently formulated in both Great Britain and North America as effective protectant sprays. The black iron salt, ferbam, gave good control of orchard diseases, while the white zinc salt was more effective on vegetable crops [15-17].

B. Quinones

Following the dithiocarbamates in development was another rubber accelerator tetrachloro-1,4-benzoquinone (chloranil), which was very effective in seed treatment [18]. Its failure as a foliage fungicide prompted the development of ferbam [15]. A closely related and more active quinone, 2,3-dichloro-1,4-naphthaquinone, tended to be phytotoxic and dermatitic.

C. Ethylenebisdithiocarbamates

In 1943 a new series of dithiocarbamates was reported, with ethylenediamine replacing the monoamine to yield an ethylenebisdithiocarbamate [19, 20]. The initial sodium salt, although water soluble, converted to an active protectant insoluble fungicide after spraying. Results, however, were unreliable and the zinc salt, zineb, was found to be a stable and useful fungicide. The manganese salt, maneb, used on many vegetable crops was even more effective. A manganese-zinc complex salt, mancozeb, with improved stability and activity has been developed [21]. Much has been reported on the mode of action of the different salts and identification of metabolic products. The literature covering the first 10 years of intensive study of dithiocarbamates as fungicides was well documented in 1962 [22]. A revised structure for the active principle, ethylene thiuram monosulfide (ETM), arising from air oxidation of nabam, disodium ethylenebis(dithiocarbamate), in the presence of catalytic quantities of manganous ion, was reported somewhat later [23] and confirmed by others. Early in degradation studies of the ethylene-bisdithiocarbamates, ethylenethiourea was one of the products identified [24]. It has subsequently been shown to increase on storage [25] as well as be taken into the plant [26]. Since ethylenethiourea has been shown to have carcinogenic and tumorigenic properties [27] and cause thyroid cancer in albino rats [28], the possible significance of residues has caused concern. Thus a class of fungicides in use for some time is now under critical review. A symposium was recently devoted entirely to this subject [29]. Methods have been developed for analysis of ethylene thiourea [30, 31] as well as for differentiating dialkyl dithiocarbamates and thiuram sulfides from ethylenebisdithiocarbamates [32]. Although the actual residues of ethylenethiourea may be low on the raw plant [97], it has been shown that in the cooking of vegetables containing ethylenebisdithiocarbamate, residues of ethylenethiourea from degradation of the parent compounds increase with heat and time [33].

D. Trichloromethylthiodicarboximides

Another new class of compounds [34], containing the $-NSCCl_3$ group, was announced in 1952. The first, trichloromethylthiocarboximide, captan, was followed by a closely related one, folpet. A number of mono- and di-$SCCl_3$ derivatives have recently been evaluated for fungicidal activity [35]. It appears that the monosubstituted compounds are more active and that the whole compound rather than any of its degradation products is responsible for the fungitoxicity. These compounds are used widely for seed and soil applications and they provide excellent control for a large number of foliage diseases. Here again with a group of fungicidal agents in wide use for a number of years, some serious toxicological data have been published which require careful examination [36]. Captan labelled with ^{35}S was administered to mice, and the isolation of alkylated products suggested that captan may be a potent carcinogen.

E. Miscellaneous Protectant Fungicides

A number of other protectant fungicides were developed in this period but are in less common use. They include substituted guanidines like dodine, copper oxinates, substituted nitrophenyl crotonates, dinocap with related dinitrophenols, and chromate complexes [2].

There was considerable activity in the search for effective soil fungicides. More recent studies have emphasized the overall or net effect on the soil microflora. A delay in maturing of tobacco was shown to be due to the adverse effect of the fumigant on a nitrifying organism, thus resulting in a delayed supply of nitrogen [37].

F. Antibiotics

Since Fleming's discovery of penicillin in 1929 [38], there has been much activity in the search for antibiotics to control plant diseases. In contrast with the commercial success in the control of human and animal diseases, there are few antibiotics of practical importance in plant protection, namely, streptomycin (a bacteriostatic compound) and cycloheximide and more recently, two in Japan.

V. PLANT PROTECTANTS, 1964-1975

A. Systemics

The previous 30 years saw a major change in the development of organic plant protectants by contrast with the inorganics. For years attempts had been made to develop materials that could be applied to the roots or leaves and translocated throughout the plant, eradicating established pathogens and protecting it from further attack. The subject was reviewed [39-41] and summarized in 1967 with the statement, "there is yet no outstanding control of plant diseases by systemics" [2]. That statement was probably written simultaneously with the publication of a paper on the systemic fungicidal properties of some 1,4-oxathiins [3]. One of them, carboxin, has since been commercially developed with worldwide distribution. Since then, a surprising number of successful systemic fungicides have appeared on the market [42]. They belong to quite different chemical classes, are often very specific in their action and are toxic at low levels as compared with many of the insecticides. A monograph published in 1972 includes an extensive outline of the subject by specialists for different aspects [43].

1. Pyrimidines

A few comments on some of the members of this new group of fungicides will illustrate the advantages and disadvantages of their use. The pyrimidine derivatives dimethirimol and ethirimol are effective against some powdery mildews. They are so specific that following soil application dimethirimol controls cucumber powdery mildew for six weeks, while ethirimol protects barley for an entire season following seed treatment [44]. They are sufficiently stable in the soil to allow gradual release to the plant, yet are degraded rapidly enough in the plant so that significant residues do not accumulate. The level in the plant is sufficiently low that predatory mites are not affected.

2. Carboxanilides

Carboxin and a variety of carboxanilide compounds show a marked specificity for Basidiomycete fungi [45]. Recent studies on the mode of action of the oxathiins and related compounds have narrowed the structural requirements and pinpointed the site of action [46, 47]. The carboxanilide group is common to several systemic fungicides attached to oxathiin, thiazole, furyl [48], dihydropyran [49] and a number of other cyclic groups [50, 51]. The unique specificity involves reaction with the mitochondrial succinic dehydrogenase complex of the sensitive fungi, inhibiting succinate oxidation. From the evaluation of 93 carboxanilides, it is reported that the fungitoxic activity is influenced not only by substitution in the aniline ring but by the ring attached to the carbonyl [52]; three of the carboxanilides had a spectrum of fungitoxic activity that extended beyond the Basidiomycetes.

3. Benzimidazoles

Almost simultaneously with the announcement of the oxathiin systemic fungicides there appeared another group, derivatives of benzimidazole [53, 54]. The first three reported, thiabendazole, fuberidazole, and benomyl, are highly active against a wide variety of foliage diseases including powdery mildews and soilborne pathogens. Shortly thereafter two substituted thioureas, thiophanate and thiophanate methyl, were also introduced. They exhibited systemic and antifungal activity somewhat similar to the substituted benzimidazoles, probably because benomyl and thiophanate methyl are both hydrolyzed to methyl benzimidazole-2-carbamate (MBC) [55-57], considered the fungitoxic principle. The different functional groups play a part in effecting the differential toxicity between compounds. For example, thiophanate methyl is usually slower acting, due presumably to the greater time interval required for conversion to the toxic principle.

Other heterocyclic compounds which show systemic fungicidal activity such as derivatives of piperazine, azepine, and morpholine immediately followed [58]. None seem to have achieved the wide application of the thiabendazoles as yet.

4. Organophosphorus Fungicides

Organophosphorus compounds with systemic action and toxicity to insects were reported as early as 1945 with further experimental work published in 1949 [59]. Since then a great many more have been developed commercially, some toxic in themselves, others requiring activation by the plant or animal. By contrast, antifungal organophosphorus compounds with systemic activity have been slower to reach a level of relative significance. A substituted triazole tetramethylphosphorodiamide, triamiphos, was reported in 1960 as having slight systemic activity [60]. There has been much activity in the search for organophosphorus compounds with fungicidal and bactericidal properties as evidenced by 450 publications up to 1972 [61]. Recently the main developments have been in finding compounds that control diseases of rice [62], such as Kitazin and Kitazin P (diisopropyl S-benzyl phosphorothiolate) which had by 1970 reached an output of 1,600 tons [61], and Hinosan, 700 tons. It is anticipated that further developments will take place in the synthesis of new fungicidal preparations as biochemical research on the mechanism of action of phosphorus-containing fungicides on the fungal wall is extended.

5. Antibiotics

Although the systemic activity of a number of antibiotics, such as chloramphenicol, griseofulvin, cycloheximide, and streptomycin, has been known for 20 years, their use has been limited by cost, phytotoxicity, or limited activity. However, in Japan two antibiotics have been developed for the control of paddy blast in rice, namely, Kasugamycin and Blasticidin S [58].

6. Inhibitors of Species of Phycomycete

It is of significance that new systemic fungicides developed to date have very little or no activity against phycomycetes with the exception of chloroneb [63, 64]. It was, therefore, of considerable interest to note a sulfonamide with marked activity against this group of fungi [65]. Unfortunately, the high hopes which this compound generated have since been disappointed, due to its high phytotoxicity. However, since the withdrawal of this sulfonamide several experimental compounds have been announced that are effective against different species of phycomycetes; two of these show systemic activity, prothiocarb and pyroxychlor. Prothiocarb (SN 41 703), S-ethyl-N-(3-dimethylaminopropyl)thiocarbamate hydrochloride, controls Phytophthora fragariae in strawberries and Pythium ultimum in crocus and tulip bulbs [66]. Studies of its mode of action indicate that it acts selectively on fungi with cellulose walls and only at a relatively high pH so that in its nonionized state it can readily penetrate [67].

Pyroxychlor (Dowco 269), 2-chloro-6-methoxy-4-(trichloromethyl)pyridine, exhibits excellent control of Phytophthora parasitica on tobacco, when plants are transplanted in water or treated with a foliar spray [68]. This compound also shows promising control against root rots of ornamentals and soybeans [69]. A unique property of this chemical is downward translocation, following foliar application by contrast with the usual upward translocation of all other systemic fungicides after root application.[*]

7. Mode of Action

With such biological specificity there is much activity in a search for the enzyme systems being inhibited. Table 1 summarizes the possible general mode of action of a number of systemic fungicides.

It has been pointed out that the protectant fungicides usually interfere with the energy-producing processes and so are strong inhibitors of cell respiration. The systemic fungicides on the other hand interfere with biosynthetic processes, except for the oxathiins [45]. They are thus much more active against the fast-growing pathogens than against the much slower growing host plants. (See Volume 2 for details on the mode of action of protectant and systemic fungicides.)

Note added in proof:
[*]Subsequently withdrawn due to toxic side effects.

TABLE 1

Site of Action of Systemic Fungicides[a]

Compound	System affected
Benomyl (and MBC[b])	
Thiophanates	Biosynthesis of DNA[c] (mitosis)
Thiabendazole	
Fuberidazole	
Chloroneb	Biosynthesis of DNA (cytokenesis)
Carboxin	Mitochondrial respiration
Dimethirimol	Folic acid mediated C-1 transfer reactions
Ethirimol	
Triarimol	Lipid metabolism; sterol biosynthesis
Triforine	
Cycloheximide	
Blasticidin	Biosynthesis of proteins
Kasugamycin	
Kitazin	Biosynthesis of glycolipids in cell membranes
Polyoxin D	Biosynthesis of chitin
Triamiphos	Unknown
Tridemorph	

[a]Adapted from Ref. 9 and revised.
[b]Methyl benzimidazolecarbamate.
[c]Deoxyribonucleic acid.

B.   New Protectant Fungicides

There are four protectant organic fungicides in the developmental stages:
DPX 3217, 2-cyano-N-(ethylaminocarbonyl)-2-(methoxyimino)acetamide,
has been shown to be an eradicant against Phytophthora infestans on potatoes
and tomatoes and Peronspora viticola on grapes from applications of up to
two or three days after infection [70]. Another compound suggested as
suitable for the control of species of Pythium and soilborne Phytophthora is

metazoxolin (PP395), 4-(3-chlorophenylhydrazono)-3-methyl-5-isoxazolone
[71]. The third compound, chlorothalonil (tetrachloroisophthalonitrile), is
effective against a great number of pathogens of many vegetable, tree and
small fruit crops, including Phytophthora infestans [72]. A fourth, hymexazol (3-hydroxy-5-methylisoxazole), controls damping-off of rice and sugarbeet seedlings when used as a drench or incorporated in the soil [73]. It
appears, therefore, that progress is being made in developing eradicant
fungicides against species of Phytophthora as well as new protective ones.

VI. RESISTANCE

The high specificity of action of a toxicant is both a strength and weakness.
With the earlier development of selective organic insecticides, the occurrence of resistance to them in many organisms was soon observed. On the
other hand, the protectant fungicides, such as the dithiocarbamates and
inorganic compounds, are rather nonspecific toxicants and do not penetrate
into the plant tissue. Thus the development of resistance was unusual [74].
Only later with the introduction of selective systemic fungicides has resistance developed significantly. This may be brought about in several ways.
The fungal cell may become resistant so that the fungicide cannot reach the
site of action. Alternatively, the fungus may develop an increased capacity
to detoxify the fungicide or a decreased affinity for the fungicide [58, 74-76].

Dimethirimol was introduced in Holland in 1968 to control powdery mildew in cucumbers. During the following two years control tended to be
less satisfactory. Even after removal of the fungicide, tolerant fungi were
reported in 1971 [77]. Benomyl was very effective in the control of a heart
rot in cyclamen. However, within a relatively short period a strain of
Botrytis cinerea developed that could tolerate 1,000 ppm while the wild type
was controlled at 0.5 ppm [78]. As would be predicted from the common
active principle, these tolerant strains showed cross resistance to thiabendazole, fuberidazole, and thiophante methyl. During 1970 and 1971,
sugar beet was well protected from Cercospora beticola with low levels of
benomyl. However, in late July 1972, resistance was noted and subsequently a large acreage of beets was lost [79]. Thiophanate methyl as one
would predict, was of no avail. However, the protective fungicide fentin
acetate gave satisfactory control.

Apart from finding field resistance to some of the systemic fungicides,
resistant mutants are readily isolated following ultraviolet (UV) irradiation.
This technique has been used to isolate strains of Ustilago maydis with a
high resistance to oxathiin and thiazole fungicides. There is evidence of a
single-gene mutation and subsequent confirmation of a single site of inhibition in the wild strain in the succinic dehydrogenase system [80].

A number of suggestions have been made to reduce fungicide tolerance
[74, 81, 82]. They include avoiding the use of one fungicide on an exclusive

basis. Preferably systemics with different modes of action should be alter-
nated along with a protectant fungicide. As high a level of fungicide dose
as consistent with economic feasibility, avoidance of phytotoxicity, and
toxic residues should be employed. Host varieties should be selected which
have at least moderate disease resistance in order to exploit the interaction
of host resistance and fungicide control and ensure against failure of both
host resistance and the fungicide.

The prolonged use of a selective systemic fungicide may result in loss
of effectiveness, which may be incorrectly diagnosed as fungicide tolerance.
Natural antagonists of the pathogen may be more susceptible to the fungicide,
permitting the pathogen to develop more readily thus stimulating the development of tolerance [83]. (See Volume 2, Chapter 12.)

VII. SCREENING METHODS

The development of systemic fungicides has resulted in a revision of
screening methods with a shift from in vitro tests, e.g., spore germination
inhibition, to in vivo tests on leaves, whole plants, or roots. This has
entailed a great deal more space, time, and labor. Thus a critical decision
has to be made to develop an effective screening system yet keep it within
economic bounds. However, a case has been made that these elaborate
screening systems, besides being costly, laborious, and wasteful of chemicals, also bias results in favor of types of activity already encountered. It
is suggested that simple, nonsystemic tests followed by exact observation
and correct interpretation will result in a much more economical and efficient method in screening chemicals for systemic properties [84]. However, a procedure should be developed for those compounds that are nontoxic in vitro and would otherwise be missed. (See Chapter 3.)

To complicate further the evaluation of compounds as potential fungicides, results from in vivo and in vitro experiments often are not parallel
for a series [100]. For example, in the evaluation of some benzimidazole,
pyrimidine, and piperazine fungicides against powdery mildew of black currents and marrows, the order of in vitro activity was the same for both
pathogens. However, the activity rarely paralleled the in vivo assay [85].
The difference could be due to differential penetration or metabolism of the
toxicant or, alternatively, it might be due to an induced change in the host
resulting in the production of an antifungal agent or phytoalexin. It has
been noted that some fungicides are more effective under field conditions
than would have been expected from in vitro values. To examine the possibility that this might be due to added protection by phytoalexin production,
soybeans were treated with an assortment of fungicides and nonfungicidal
chemicals. Following incubation, nine fungicides, ten of their decomposition products, and eight other chemicals were shown to induce the production

of the phytoalexin, then regarded as hydroxyphaseollin [86] but later identified
as an isomer [98]. Another recent paper discusses systemic dichlorocyclo-
propane fungicides as activating the plant's resistance [99].

VIII. FUTURE DEVELOPMENTS

The increasing information concerning biochemical mode of action together
with improved methods of screening should lead to the development of more
systemic fungicides. The slight indication that some materials are translocated in the phloem may give leads to the production of more versatile
systemic fungicides translocated in both directions.

Apart from the direct control of a pathogen by a specific fungicide there
appears potential for an indirect method by the controlled induction of phytoalexins as previously mentioned [86]. The discovery that protective compounds can be induced in plants exposed to nonpathogenic fungi or chemicals
might provide a fruitful exploratory area for developing new methods of
crop protection. The subject of plant disease resistance, including extensive discussions of phytoalexin production, has been thoroughly reviewed
[87]. (See Volume 2, Chapter 13.) Apart from induction of protective
agents in situ, the compounds themselves might also be a source of new
fungicidal materials. For example, capsidiol, a phytoalexin induced in
peppers [88], was found to be toxic to Phytophthora infestans. Prior spraying of tomato plants with capsidiol before inoculation with zoospores of
P. infestans effectively controlled the late blight under growth room conditions for the eight days of the experiment [89].

IX. ENVIRONMENTAL EFFECT

Normal precautions are employed in the use of fungicides within the knowledge currently available. However, additional information sometimes comes
to light that may cast suspicion on the safety of the continued use of some
even well-established materials. This has already been mentioned in the
case of captan and the dithiocarbamates [28, 33, 36]. Benomyl and related
compounds prevent completion of mitosis [90-92]. Therefore it is not surprising to find that some of the benzimidazolecarbamates and related compounds are reported to have deleterious cytogenetic effects on a variety of
organisms [91, 93]. In addition, where the effect on beneficial organisms,
particularly in soil applications, is significant, changes may be required
[94]. With awareness of the importance of causing minimum adverse effects
on the environment, any doubts raised must be followed up and the information gained used to remove the doubts, even if it may be necessary to modify
programs in some cases.

X. ANALYTICAL METHODS AND RESIDUES

Since systemic fungicides are translocated within the plant, there is an even greater need than formerly for the development of methods to determine precisely not only the original fungicide but any metabolites. Most are non-volatile or thermally unstable so that the convenient gas-chromatographic electron-capture techniques are unsuitable. However, satisfactory individual methods have been developed and are reviewed [95]. (See Chapter 14.) An example of the application of a newly developed high-speed liquid-chromatographic technique for determination of residues and metabolites is reported for benomyl [96].

XI. CONCLUSION

During the period of eight years since the last major historical review of fungicides [2], there have been significant developments. The most note-worthy are the systemic fungicides, paralleled by biochemical studies of modes of action and the development of resistant strains of pathogens. New and specialized analytical methods have had to be developed. With some of these new compounds and after a more critical look at some of the older ones, toxicological questions have been raised which will have to be solved. Obviously the field of plant pathogen control is not static and there are sug-gestions of possible new areas such as tapping the release of the plant's own protection agents.

ACKNOWLEDGMENTS

The author is indebted to Drs. L. T. Richardson, A. Stoessl, G. D. Thorn, and E. W. B. Ward of the Research Institute, Agriculture Canada for con-tributions and discussions which are gratefully acknowledged.

REFERENCES

1. G. B. Orlob, Pflanzenschutz-Nachr. Bayer, 26, 65 (1973).
2. S. E. A. McCallan, in Fungicides (D. C. Torgeson, ed.), Vol. 1, Aca-demic Press, New York, 1967.
3. B. von Schmeling and M. Kulka, Science, 152, 659 (1966).
4. W. H. Tisdale and I. Williams, U.S. Pat. 1,972,961 (1934).
5. E. Somers, World Rev. Pest Contr., 8, 95 (1969).
6. E. Y. Spencer, in Perspectives of Biochemical Plant Pathology (S. Rich, ed.), Conn. Agr. Expt. Sta. New Haven, Bull., 663, 96 (1963).
7. A. D. Tomlin and F. L. Gore, Bull. Environ. Contam. Toxicol., 12, 487 (1974).

8. L. T. Richardson, 2nd Intern. Congr. Plant Pathol. Minneapolis, Abstr., 1136 (1973).

9. G. J. M. van der Kerk, OEPP/EPPO Bull., 10, 5 (1973).

10. K. O. Müller and H. Börger, Arb. Biol. Reichsanst. Land. Forstwirt. Berlin-Dahlem, 23, 189 (1940).

11. W. Forsyth, A Treatise on the Management of Fruit Trees, etc., Nichols, London, 1802.

12. J. Robertson, Trans. Hort. Soc. London, 5, 175 (1821).

13. S. E. A. McCallan, Cornell Univ. Agr. Expt. Sta. Mem., 128, 25 (1930).

14. A. J. Farley, New Jersey Agr. Expt. Sta. Bull., 379, 1 (1923).

15. J. M. Hamilton and D. H. Palmiter, Farm Res. (N.Y.), 9, 14 (1943).

16. J. W. Heuberger and D. O. Wolfenbarger, Phytopathology, 34, 1003 (1944).

17. J. D. Wilson, Phytopathology, 34, 1014 (1944).

18. E. G. Sharvelle, H. C. Young, Jr., and B. F. Shema, Phytopathology, 32, 944 (1942).

19. A. E. Dimond, J. W. Heuberger and J. G. Horsfall, Phytopathology, 33, 1095 (1943).

20. W. F. Hester, U.S. Pat. 2,317,765 (1943).

21. E. Y. Spencer, Guide to the Chemicals Used in Crop Protection, Publ. 1093, 6th ed., Res. Branch, Agr. Canada, Ottawa.

22. G. D. Thorn and R. A. Ludwig, in The Dithiocarbamates and Related Compounds, Elsevier, Amsterdam, 1962, p. 298.

23. C. W. Pluijgers, J. W. Vonk, and G. D. Thorn, Tetrahedron Letters, 1317 (1971).

24. R. A. Ludwig, G. D. Thorn, and C. H. Unwin, Can. J. Botany, 33, 42 (1955).

25. W. R. Bontoyan and J. B. Looker, J. Agr. Food Chem., 21, 338 (1973).

26. J. W. Vonk and A. Kaars Sijpesteijn, Ann. Appl. Biol., 65, 489 (1970).

27. J. R. M. Innes, B. M. Ulland, M. G. Valerio, L. Petrucelli, L. Fishbein, E. R. Hart, and A. J. Pallotta, J. Natl. Cancer Inst., 41, 1101 (1969).

28. B. M. Ulland, J. H. Weisburger, E. K. Weisberger, J. M. Rice, and R. Cyther, J. Natl. Cancer Inst., 49, 583 (1972).

29. B. G. Tweedy, J. Agr. Food Chem., 21, 323 (1973).

30. R. Engst and W. Schnaak, Z. Lebensm.-Unters.-Forsch., 144, 81 (1970).

31. W. H. Newsome, J. Agr. Food Chem., 23, 348 (1975).

32. W. H. Newsome, J. Agr. Food Chem., 22, 886 (1974).

33. W. H. Newsome and G. W. Laver, Bull. Environ. Contam. Toxicol., 10, 151 (1973).

34. A. R. Kittleson, Science, 115, 84 (1952).

35. E. J. Lien, Cheung-Tung Kong, and R. J. Lukens, Pestic. Biochem. Physiol., 4, 289 (1974).

36. M. D. Anderson and H. S. Rosenkranz, Henry Ford Hosp. Med. J.,
    22, 35 (1974).
37. J. M. Elliott, C. F. Marks, and C. M. Tu, Can. J. Plant Sci., 54,
    801 (1974).
38. A. Fleming, Brit. J. Exptl. Pathol. 10, 226 (1929).
39. A. E. Dimond, in Plant Pathology, Problems and Progress (C. S. Hol-
    ton et al., eds.), University of Wisconsin Press, Madison, 1959, pp.
    224-228.
40. F. L. Howard and J. G. Horsfall in Plant Pathology (J. G. Horsfall
    and A. E. Dimond, eds.), Vol. 1, Academic Press, New York, 1959,
    pp. 563-604.
41. A. H. M. Kirby, PANS, 18, 1 (1972).
42. D. Woodcock, Rept. Progr. Appl. Chem., 57, 401 (1972).
43. R. W. Marsh, ed., Systemic Fungicides, Longman Group Ltd., Lon-
    don, 1972.
44. K. J. Bent, Endeavour, 28, 129 (1969).
45. A. Kaars Sijpesteijn, World Rev. Pest Control, 9, 85 (1969).
46. G. A. White, Biochem. Biophys. Res. Commun., 44, 1212 (1971).
47. J. T. Ulrich and D. E. Mathre, J. Bacteriol., 110, 628 (1972).
48. J. R. Hardison, Plant Disease Reptr., 56, 252 (1972).
49. B. Jank and F. Grossmann, Pestic. Sci., 2, 43 (1971).
50. P. ten Haken and C. L. Dunn, Proc. 6th Brit. Insectic. Fungic. Conf.
    Brighton, 2, 453 (1971).
51. D. Woodcock, in Systemic Fungicides (R. W. Marsh, ed.), Longman
    Group Ltd., London, 1972, pp. 34-89.
52. G. D. Thorn and G. A. White, Pestic. Biochem. Physiol., 5, 380 (1975).
53. T. Staron and C. Allard, Phytiatr. Phytopharm., 13, 163 (1964).
54. C. J. Delp and H. L. Klopping, Plant Disease Reptr., 52, 95 (1968).
55. G. P. Clemons and H. D. Sisler, Phytopathology, 59, 705 (1969).
56. H. A. Selling, J. W. Vonk, and A. Kaars Sijpesteijn, Chem. Ind.
    (London), 1625 (1970).
57. J. W. Vonk and A. Kaars Sijpesteijn, Pestic. Sci., 2, 160 (1971).
58. D. Woodcock, Chem. Brit., 7, 415 (1971).
59. S. H. Bennett, Ann. Appl. Biol., 36, 160 (1949).
60. B. G. van den Bos, M. J. Koopmans, and H. O. Huisman, Rec. Trav.
    Chim., 79, 807 (1960).
61. A. F. Grapov and N. N. Melnikov, Russ. Chem. Rev. (English Trans.),
    42, 772 (1973).
62. E. Yoshinaga, Proc. 5th Brit. Insectic. Fungic. Conf., 593 (1969).
63. I. E. M. Darrag and J. B. Sinclair, Phytopathology, 59, 1102 (1969).
64. L. T. Richardson, Plant Disease Reptr., 57, 3 (1973).
65. H. Kaspers and F. Grave, 7th Intern. Congr. Plant Pathol., 203 (1970).
66. M. G. Bastiaansen, E. A. Pieroh, and E. Aelbers, Meded. Fac.
    Landbouwwet Rijkuniv. Gent, 39, 1019 (1974).
67. A. Kaars Sijpesteijn, A. Kerkenaar, and J. C. Overeem, Meded. Fac.
    Landbouwwet Rijkuniv. Gent, 39, 1027 (1974).

68. R. L. Noveroske, Phytopathology, 65, 22 (1975).
69. H. A. J. Hoitink and A. F. Schmitthenner, Phytopathology, 65, 69 (1975).
70. E. I. du Pont de Nemours, Experimental Fungicide DPX 3217, Wilmington, Del.
71. Plant Protection Ltd., Experimental Fungicide PP395.
72. Diamond Alkali Co., Experimental Fungicide Daconil 2787.
73. Sankyo Co., Ltd., Tachigaren (hymexazol).
74. J. Dekker, in Systemic Fungicides (R. W. Marsh, ed.), Longman Group Ltd., London, 1972, pp. 156-174.
75. D. C. Erwin, Ann. Rev. Phytopathol., 11, 389 (1973).
76. J. Dekker, OEPP/EPPO Bull., 10, 47 (1973).
77. K. J. Bent, A. M. Cole, J. A. W. Turner, and M. Woolner, Proc. 6th Brit. Insectic. Fungic. Conf. Brighton, 1, 274 (1971).
78. G. J. Bollen and G. Scholten, Neth. J. Plant Pathol., 77, 83 (1971).
79. S. G. Georgopoulos and C. Dovas, Plant Disease Reptr., 57, 321 (1973).
80. S. G. Georgopoulos, E. Alexandri, and M. Chrysayi, J. Bacteriol., 110, 809 (1972).
81. M. S. Wolfe, Proc. 6th Brit. Insectic. Fungic. Conf. Brighton, 3, 724 (1971).
82. J. M. Vargas, Jr., Phytopathology, 63, 1366 (1973).
83. G. J. Bollen, Neth. J. Plant Pathol., 77, 187 (1971).
84. W. Koch, Pestic. Sci., 2, 207 (1971).
85. D. R. Clifford and E. C. Hislop, Proc. 6th Brit. Insectic. Fungic. Conf. Brighton, 2, 438 (1971).
86. J. J. Reilly and W. L. Klarman, Phytopathology, 62, 1113 (1972).
87. J. L. Ingham, Botan. Rev., 38, 344 (1972).
88. A. Stoessl, C. H. Unwin, and E. W. B. Ward, Phytopathol. Z., 72, 141 (1972).
89. E. W. B. Ward, C. H. Unwin, and A. Stoessl, Phytopathology, 65, 168 (1975).
90. R. S. Hammerschlag and H. D. Sisler, Pestic. Biochem. Physiol., 3, 42 (1973).
91. J. A. Styles, Mutation Res., 21, 203 (1973).
92. J. P. Seiler, Mutation Res., 17, 21 (1973).
93. J. A. Styles and R. Garner, Mutation Res., 26, 177 (1974).
94. A. Stringer and M. A. Wright, Pestic. Sci., 4, 165 (1973).
95. P. B. Baker and R. A. Hoodless, Pestic. Sci., 5, 465 (1974).
96. J. J. Kirkland, J. Agr. Food Chem., 21, 171 (1973).
97. R. G. Nash, J. Agr. Food Chem., 24, 596 (1976).
98. R. S. Burden and J. A. Bailey, Phytochemistry, 14, 1389 (1975).
99. P. Langcake and S. G. A. Wickins, Physiol. Plant Path., 7, 113
100. R. J. W. Byrde and D. V. Richmond, Pestic. Sci., 7, 372 (1976).

Chapter 2

FUNGICIDES IN PERSPECTIVE

William E. Fry

Plant Pathology Department
Cornell University
Ithaca, New York

I. INTRODUCTION

As the world population expands rapidly, agricultural scientists must devise means to increase food production. Plant pathologists are concerned with at least two means of providing greater amounts of food: (1) reducing current yield losses due to diseases, and (2) preventing diseases from

19

devastating certain crops on recently cultivated land previously considered marginal for agriculture. The success of either of these efforts will depend in part on fungicides. At the same time, the environment must not be allowed to deteriorate due to avoidable pollution. Thus plant pathologists are faced with the challenge of using fungicides (when necessary) as effectively and as efficiently as possible so that none is introduced unnecessarily into the environment. This chapter reviews the epidemiological bases of contemporary disease control procedures and the role of fungicides in these procedures. Then a management approach to pest problems and how fungicides might be used in such an approach will be discussed. The treatment is necessarily selective, but a determined effort has been made to identify and use principles with broad application. Many diseases are used to illustrate points and the name of the pathogen (or disease) is indicated parenthetically each time a disease (or pathogen) is mentioned for the first time.

II. STRATEGIES OF PLANT DISEASE CONTROL

In agricultural practice there are few absolute rules because the underlying bases for many phenomena are imperfectly understood. Because plant diseases involve interactions among complex organisms and their environment, absolute rules (laws) describing disease are even more obscure. And finally, because control of disease is generally directed at populations of two different organisms in a changing environment, laws describing disease control are difficult to find. In this situation, it is useful to resort to models which describe disease development (even though imperfectly) and to try to synthesize generalities by using the models. Disease control strategies can be developed by analysis of the models.

Two models of disease development, identified by Van der Plank [1, 2], have been endorsed and expanded upon by others [3-5]. In the first model, disease increases with time from a given amount of inoculum at the beginning of the season. The increase is attributable to enlargement of old infection sites and occurrence of new ones, but the amount of inoculum does not increase. Thus the final amount of disease is a function of the amount of inoculum at the beginning of the season. Van der Plank [2] refers to such diseases as simple interest diseases. Pathogens which cause simple interest diseases frequently are soilborne; they may produce only one generation (life cycle) per season; and they may produce structures which enable them to survive for long periods. Consequently, they may have slow "death rates." Some examples of simple interest diseases are bunt of wheat (Tilletia foetida), Fusarium wilt of tomato (Fusarium oxysporum f. sp. lycopersici), root and butt rot of conifers (Fomes annosus), and golden nematode of potato (Heterodera rostochiensis). Because disease increase depends on the amount of initial inoculum, the first strategy of disease control is evident; namely, the reduction of the initial inoculum.

In the second model, disease also increases with time because established lesions enlarge, and new infection sites may result from the initial inoculum; but more importantly, propagules produced on established lesions initiate new ones. Thus the amount of inoculum is compounded (compound interest diseases). Pathogens which cause such diseases frequently infect foliage, are disseminated through the air, and reproduce rapidly. Thus they may have high "birth rates." Some examples of compound interest diseases are stem rust of wheat (Puccinia graminis var. tritici), late blight of potato (Phytophthora infestans), southern corn leaf blight (Helminthosporium maydis), and cucumber mosaic (cucumber mosaic virus). Because the final amount of disease depends on the rate of increase of inoculum, the second disease control strategy is identified; namely, the reduction of the rate of epidemic development.

Van der Plank [2] has proposed that the formula for continuous compound interest [Equation (1)] might reasonably describe the early phases of an epidemic of a compound interest disease:

$$x = x_0 e^{rt} \tag{1}$$

where x = the amount of disease (or inoculum) at any time, x_0 = the amount of disease (or inoculum) at the beginning of the season, r = the rate of increase in appropriate time units, and e = the base of natural logarithms, 2.71828 ... .

Depending upon the epidemiological characteristics of a disease, either reduction of the initial inoculum (x_0) or reduction of the rate of epidemic development (r) may be the most efficient strategy for control. For diseases (i.e., late blight of potato) in which x_0 is normally quite small and r is large, x at the end of the season may be 10^6-fold greater than x_0. Use of the model in Equation (1) indicates that reduction of r is much more efficient than is reduction of x_0 for control of such diseases. This strategy is confirmed by experience. Periodic application of protectant fungicides is the principal control procedure for potato late blight. The major effect of this practice is reduction of r [6]. For other diseases the final amount of disease (x) may be less than 10^2-fold greater than the initial amount of disease. If for such diseases, r is small, then reduction of x_0 is an efficient control procedure. Validity of this prediction is also confirmed by practice. For example, the most important control procedure for black leg of cabbage (Phoma lingam) is to selectively kill the pathogen in infected seed. If seed lots containing up to 2% infected seeds are not treated (with hot water or fungicide), P. lingam can cause nearly total loss of cabbage in a field. Phoma lingam does not reproduce as rapidly as Phytophthora infestans; thus this procedure, which reduces x_0, is effective. Finally, if x_0 is quite high, then even a low rate of disease increase might prove devastating. Therefore, it becomes necessary to reduce the amount of

22 FRY

initial inoculum. Obviously both of these strategies are necessary when x_0 is high and r is also high [7].

The time scale is important when using these two models of disease epidemics for understanding of specific diseases. The simple interest model may lose validity when the time reference is many growing seasons rather than a single one. For example, epidemics of cotton root rot (Phymatotrichum omnivorum) approximate the simple interest model when single seasons are considered, but compounding of inoculum occurs if the term of measurement is years. Jordan et al. [8] measured cotton root rot in the same field from 1937 through 1940 after the field had been planted to corn in 1935 and 1936. At the end of each season the number of plants killed by P. omnivorum was recorded. The severity of disease increased from year to year and a graph of percentage of plants killed versus time in years has partial sigmoid character, a feature indicative of a compound interest disease (see Fig. 1). The important message for the grower is that some measure designed to prevent the build-up of inoculum from year to year is necessary for disease control. Thus with years as the units of time, a procedure which reduces r is necessary.

Diseases of perennial plants must also be considered within the time scale of years if the pathogen survives with its host during the dormant season. Oak wilt (Ceratocystis fagacearum) and Dutch elm disease (Ceratocystis ulmi) have compound interest character when considered over

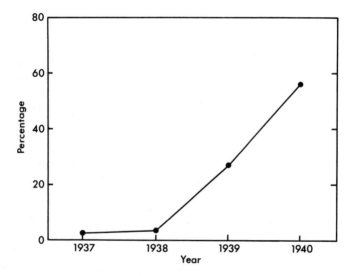

FIG. 1. Percentage of cotton plants killed by the end of the season (1937–1940), due to root rot caused by Phymatotrichum omnivorum. Data are from Ref. 8.

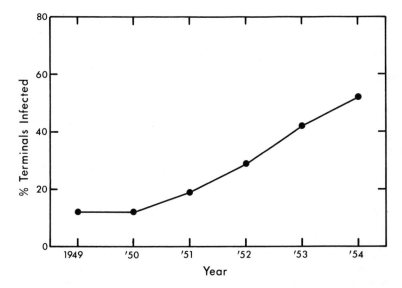

FIG. 2. Percentage of terminals infected by <u>Podosphaera leucotricha</u> in April from 1949 through 1954. Data are from Ref. 11.

years [9, 10] and procedures which reduce r over years are necessary to control the disease effectively. Apple powdery mildew (<u>Podosphaera leuco-tricha</u>) also illustrates the importance of preventing the increase of a pathogen of a perennial plant. Sprague [11] reported the percentages of terminals infected in unsprayed apple trees from 1949 through 1954. The percentage increased dramatically within each growing season and decreased between seasons. However, when the percentages at the beginning of each season (April) were compared (see Fig. 2), the initial inoculum increased each year. This disease must be treated to prevent year-to-year increase of the pathogen as well as to prevent an increase within a given year. Fortunately procedures which reduce the rate of epidemic development within a growing season also serve to reduce the rate of disease increase over the years.

To this point, we have examined two major strategies of disease control. The choice of strategy or combination of strategies depends on the epidemiological characteristics of each disease within and between growing seasons. For certain diseases reduction of the amount of initial inoculum is most efficient; for others reduction of the rate of epidemic development is most efficient; and finally for certain other diseases both strategies are needed. Now we shall consider some of the tactics within each strategy in relation to fungicide use.

A. Reduction of the Amount of Initial Inoculum

Tactics applicable in this strategy include: (1) quarantines, (2) pathogen-free propagating material, (3) cultural procedures, (4) biological control methods (including plant resistance), and (5) chemical and physical treatments of seed and/or soil. Fungicides are most commonly used to reduce pathogen populations around a germinating seed. This is accomplished by seed or furrow treatment, or by means of a general treatment such as fumigation, which reduces such populations in soil. Because tactics which apparently do not rely on chemicals, such as (1)-(4), may directly affect fungicide use, this relationship will be investigated in the following discussion.

The major relationship of quarantines to fungicide use is negative. Fungicides may be needed when a quarantine fails or is not implemented and a destructive pathogen reaches a previously isolated susceptible population. Phytophthora infestans, Peronospora viticola (downy mildew of grape) and Ceratocystis ulmi are such pathogens [12, 13]. The discovery and use of Bordeaux mixture was stimulated by the destruction due to potato late blight and downy mildew of grape. Even now control of these diseases depends on fungicides.

The development of pathogen-free propagating material for some crops has been influenced especially by systemic fungicides. For example, because treatment of barley and wheat seed with carboxin controls loose smut (Ustilago nuda, U. tritici) [14, 15], this chemical has aided in the production of certified seed. Black leg of cabbage (Phoma lingam) is controlled by treatment of seed with hot water or by soaking the seed for 24 hr in 0.2% thiram [16-18]. This selectively inactivates the pathogen. A 24-hr soak in 0.2% benomyl is even more effective [19]. But drying thousands of pounds of imbibed cabbage seed quickly and safely is a difficult task. Recent experiments indicate that application of benomyl to cabbage seed as a slurry can be as effective as a 24-hr soak in the fungicide [20] and, therefore, the problem of drying soaked seed is avoided.

For some other crops, fungicides have had a less prominent role in the production of pathogen-free propagating material. For example, use of cuttings from plants demonstrated to be free from pathogens has contributed markedly to the success of the florist industry. Institution of this technique saved the chrysanthemum industry in the United States after Verticillium wilt became a severe problem in the 1940s [21, 22]. To maintain the benefit of pathogen-free cuttings, commercial growers should use soil or potting mix which is also free of pathogens. Thus, rooting media which are reused should be treated chemically or physically. Finally, pathogen-free propagating materials are extremely important for control of diseases caused by viruses [22-25].

Frequently fungicides are used in conjunction with cultural procedures to achieve maximum disease control. In the northeastern United States Valsa canker of peach (Cytospora cincta and C. leocostoma) is controlled by pruning trees when they are growing vigorously instead of during the dormant season when the fungus, using pruning wounds as infection courts, is able to colonize host tissue [26]. Pruning wounds are treated with fungicide to assure maximum control.

For control of seed rots and damping off of various crops, other cultural procedures can be combined with fungicides for optimum control. Seeds should be planted when soil temperatures favor the host more than the pathogen [27] and such conditions need to be determined for each disease. For example, seedling blights of corn and wheat (Gibberella zeae) are affected differently by temperature. For corn, disease was most severe at 8-16°C; whereas for wheat, disease was most severe at 16-28°C [28]. Severity of these diseases can be lessened if wheat is planted in soils cooler than 16°C and if corn is planted in soils warmer than 16°C. This same rationale is appropriate for bunt of wheat [29] and for Rhizoctonia solani on cotton and spinach [30]. Application of a fungicide to the seed or furrow further favors the host. If, for market considerations, a grower plants when the soil temperature does not favor the host, the use of a fungicide becomes more important.

Biological control usually connotes any control procedure except the use of a chemical, but there are cases in which chemicals enhance biological control. In tests with carbon disulfide for control of root rot of citrus (Armillaria mellea), Bliss [31] found that the lowest dose resulted in the best control and that control improved with time after fumigation. Trichoderma viride, which suppressed growth of A. mellea, apparently rapidly recolonized treated soil and portions of the woody substrate inhabited by A. mellea. Both carbon disulfide and methyl bromide can be used for practical control of A. mellea [32]. Ohr et al. [32] demonstrated that A. mellea in roots treated with sublethal levels of methyl bromide was not recoverable when the roots were stored in field soil, but was recoverable from roots stored in sterile field soil. Decreased population of A. mellea was directly correlated with increased population of T. viride. And finally, T. viride was shown to be about one-half as sensitive to methyl bromide as was A. mellea [32]. Consequently, low levels of fumigant apparently stimulated biological control of A. mellea by T. viride. The same type of approach has been suggested to be feasible for biocontrol of many diseases caused by Sclerotium rolfsii [33].

For convenience, plant resistance will be considered here as part of biocontrol [34]. Oligogenic resistance, according to Day [35], or vertical resistance, according to Van der Plank [36], is effective in reducing the amount of initial inoculum because if races of the pathogen exist, some but not others are resisted. Because only a portion of the pathogen population is compatible with the host, the amount of effective initial inoculum is reduced. Diseases such as cabbage yellows (Fusarium oxysporum f. sp. conglutinans), Fusarium wilt of tomato, and stem rust of wheat are controlled by this type of resistance [35, 37, 38]. Unfortunately, oligogenic resistance has been unsuccessful if the pathogen population can adjust rapidly to changes in resistance in the host population. <u>Phytophthora infestans</u> rapidly overcame this type of resistance in potatoes and growers who planted cultivars with oligogenic resistance had to resort quickly to fungicides to achieve adequate control.

We have seen how the use of fungicides may be integrated with and influenced by other tactics of disease control which reduce the initial inoculum. Next to be examined will be interrelations between fungicides and other tactics of disease control which reduce the rate of epidemic development.

B. Reduction of the Rate of Epidemic Development

Using fungicides for this purpose is the disease-control tactic most familiar to growers as well as to the lay public. It is generally a highly visible activity. In addition to fungicides, however, various cultural procedures, plant resistance, and other biocontrols are important.

Because so much fungicide is used for control of compound interest diseases, it is worthwhile to demonstrate that periodic applications of a protectant fungicide indeed reduce the rate of epidemic development. Fungicide on a plant surface is effective against a proportion of the fungal propagules [2]. The goal is to apply enough fungicide to reduce the number of successful infections so that the epidemic does not progress at a rate which will result in a yield-reducing level of disease. Complete inactivation of all propagules is neither necessary nor desired and is probably impossible. The amount of fungicide required depends on the amount of inoculum, environmental conditions, and relative susceptibility of the host, that is, the fungicide square described by Van der Plank [2, 6]. Under environmental conditions conducive to disease, a fungicide must act on a higher proportion of fungal propagules than is necessary under conditions not conducive to epidemic development. As an example, the effects of a protectant fungicide (such as mancozeb, applied every 7 days) on development of late blight in the potato cultivar Russet Rural is illustrated in Figure 3. As the amount of applied fungicide increased, the rate of epidemic development decreased. The rate (r) of epidemic development was calculated [2] according to Equation (2)

$$r = \frac{1}{t_2 - t_1}\left(\log_e \frac{x_2}{1 - x_2} - \log_e \frac{1}{1 - x_1}\right) \tag{2}$$

where x_2 = the proportion of disease at time t_2, and x_1 = the proportion of disease at time t_1.

Systemic fungicides also reduce r and their availability has had and will continue to have a marked effect on disease control practices. Applications

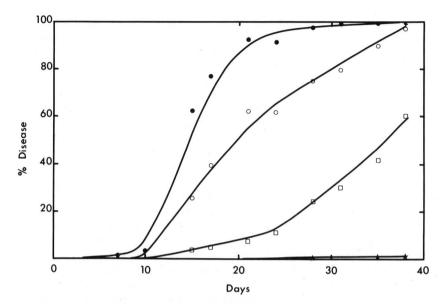

FIG. 3. Disease progress curves of late blight in the potato cultivar Russet Rural as affected by fungicide. Fungicide (mancozeb, a coordination product of zinc ion and maneb) was applied every seven days. Rates of epidemic development (r) were 0.753, 0.326, 0.245, and 0.068 per unit per day for plots treated with (●) 0.00; (○) 0.22; (□) 0.67; and (★) 1.79 kg fungicide/ha, respectively. Reprinted from Ref. 39.

of fungicides to cereals may become more common with systemic fungicides. Commercial applications of ethirimol (a systemic fungicide) to soil has led to long-term control of powdery mildew (Erysiphe graminis) on spring barley in Europe [15]. An experimental systemic fungicide, 4-n-butyl-1,2,4-triazole (RH-124), when applied early in the season, provided good protection against wheat leaf rust (Puccinia recondita) [40]. The soil probably served as reservoir from which the fungicide was continually taken up during the growing season [40].

Some cultural procedures may make the environment less conducive to disease development. If resistance of the host is constant, these procedures should reduce the need for fungicide. For example, alteration of the environment has adequately controlled rose powdery mildew (Sphaerotheca pannosa) [41]. Mildew is most severe on greenhouse roses in the northeastern United States during the spring and autumn when nights are cool and humid. When greenhouses were kept closed at night during these seasons the relative humidity rose and conditions conducive to S. pannosa were created. Growers were advised to open the vents in their greenhouses at dusk and turn on heaters. The result was that the moist air was replaced by

drier air and rose powdery mildew was controlled without the need for
fungicides [41]. The technique requires additional heating fuel and because
of the steep increase in energy costs in the past few years, the technique
is not now as widely used as before. Now growers rely on fungicides for
adequate control [42].

The relative contributions of fungicides and cultural procedures to reducing r have generally not been measured. However, there are several
examples of cultural procedures which reduce r. The combination of
these procedures with fungicide usage should achieve better control or need
for less fungicide. One disease for which cultural procedures are effective
is white mold of bean (Whetzelinia sclerotiorum). Alteration of the microenvironment by row spacing, by determinate plant growth, and by direction
of rows have all reduced the severity of white mold [43, 44]. But the effect
of these practices on the amount of fungicide needed for adequate control
has not been assessed. A second example concerns regulation of the timing
of sprinkler irrigation [45, 46]. If the microenvironment marginally favors
disease development, application of sprinkler irrigation may alter the
microenvironment to the extent that disease is favored [45].

The next corner in Van der Plank's fungicide square is resistance of the
host plant. The type of resistance discussed reduces the rate of epidemic
development. It has many names, including polygenic, horizontal, field,
partial, general, slow mildewing, or slow rusting [35, 36, 47-50]. Two
general characteristics of this type of resistance (hereafter referred to as
polygenic) are (1) it is not race-specific, i.e., is not effective against some
isolates while totally ineffective against other isolates; and (2) it rarely, if
ever, confers immunity upon a plant. Figure 4 illustrates the effect of this
type of resistance. The degree of late blight in plots of Russet Rural (a potato cultivar very susceptible to P. infestans) and Sebago (a potato cultivar
with slight polygenic resistance) was recorded during the growing season,
and r was calculated for each epidemic. In this case r = 0.753 for Russet
Rural and 0.254 for Sebago [39]. The relative contributions of fungicide and
polygenic resistance are beginning to be measured [39].

Both polygenic and oligogenic resistance have been utilized extensively
for the control of foliar pathogens. Cereal rusts have been controlled
mainly via oligogenic resistance [37] whereas foliar pathogens of corn have
been controlled via polygenic resistance [36, 51].

Because of the ability of pathogens to overcome oligogenic resistance
when these genes for resistance are all within the same individual plant,
interest has developed in using these genes in multiline cultivars [52, 53].
A multiline cultivar consists of individuals which are similar agronomically
but which differ from each other in specific oligogenic resistance genes. In
such cultivars, r is reduced when compared to pure line cultivars. It is
theoretically possible to combine the effects of oligogenic resistance in this
form with fungicide to achieve maximum control. Much work has centered
on multiline oats because of the relatively flexible market for agronomic

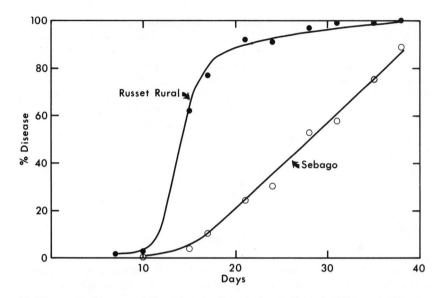

FIG. 4. Disease progress curves of potato late blight in plots of Russet Rural and Sebago potatoes. Rates of epidemic development (r) were 0.753 and 0.254 per unit per day for Russet Rural and Sebago, respectively. Data are from Ref. 39.

characteristics within this crop [54]. An additional use of the multiline is to prevent the predominance of any single pathogen race. This is accomplished by regulating the proportion of individuals with various genes [55]. Commercial multiline cultivars of wheat and oats have been released in Colombia [56] and in Iowa [57], respectively.

To this point, the emphasis of the chapter has been on analysis of disease control strategies and tactics and the relationship of fungicides to these tactics. The discussion illustrates that an integrated approach has been characteristic of disease control and that there has not been an exclusive reliance on fungicides. Nonetheless, fungicides are extremely important and the amounts used by farmers in the United States (for example) are increasing. In 1966, 33 million lb (15 million kg), active ingredient, were applied by farmers. This was 27% of total fungicides used. In 1971 farmers applied nearly 42 million lb (19 million kg) [58, 59]; much of the increase, however, is due to a different method of calculating the active ingredient in inorganic fungicides [59]. In 1971, nearly 16 million pounds of inorganic fungicide, nearly 26 million lb (11.7 million kg) of organic fungicide, and 112 million lb (54 million kg) of sulfur were applied [59]. The most commonly applied organic fungicides were captan (6.5 million lb; 2.9 million kg) and maneb (3.9 million lb; 1.75 million kg) [59].

Ordish and Mitchell [60] state that most fungicides used in the world
control diseases caused by only 12 fungi. Data for the United States emphasize this point. Disease control on apples and citrus accounted for 42% of
the total fungicide used in 1971 and areas in which these crops are concentrated (the Southeast and Northeast) accounted for 52% of the total fungicide
applied [59]. Nonetheless, the monetary value (ca. $33 million) of fungi-
cides applied in 1966 was relatively small compared to the total value of all
pesticides (ca. $500 million) applied by U.S. farmers in 1966 [61].

Fungicide use in the future will be strongly influenced by the philosophy
of pest management. Because of this influence, I shall analyze disease
control and fungicides in the framework of that philosophy in the remainder
of the chapter. Some ideas discussed are in use or are being made ready
for use. Other ideas are yet theoretical or experimental.

III. PEST MANAGEMENT, DISEASES, AND FUNGICIDES

Pest management is a relatively recent addition to the vocabulary of agricultural scientists. To some it is simply rational pest control with emphasis
on integrated control; it is a retrenchment from sole reliance on pesticides
for pest control [62]. The goal of this rational approach is to "optimize cost/
benefit ratios on a long-term basis for both the farmer and society" [63].
Thus pest management includes integrated control but also attempts to obtain optimum benefits (monetary, ecological, and societal) compared to
costs of pest control [64, 65]. While most research on pest management
techniques has occurred in countries with well-developed agricultural technology, the approach is also appropriate for less-advanced systems [66, 67].

The shift in emphasis from maximum control [68] of disease to pest
management [69] has been stimulated by three factors. The first and probably most significant factor is society's increased concern for the environment. Ecological awareness was stimulated in part by the publication of
Rachel Carson's Silent Spring [70] in 1962. The official position of many
countries now is that the environmental hazards of pesticides must be investigated thoroughly before their general use and that pesticides should be
used only for demonstrated needs by knowledgeable individuals. One result
has been a steady increase in the variety and rigor of tests designed to
predict the environmental impact of possible pesticides. Such tests have
contributed to the increased expense and time required for pesticide development. In addition, stringent new regulations concerning pesticide distribution and application have been promulgated [71].

The second factor favoring a pest management approach is reduced energy availability, although the total impact of this constraint on disease control procedures is not yet fully realized. Pimentel et al. [72] have concluded
that "the impressive agricultural production in the United States has been
gained through large inputs of fossil energy". Mechanized conventional

pesticide application is expensive in terms of energy requirements [72]. Clark [73] has warned that "the oil crisis may force a return to less ener- gy—intensive methods of pest control, even if environmental precautions don't." The absolute cost of production of food is about four times greater in the United States than in India, even though Americans spend a smaller proportion of their total income for food than do Indians [72].

A third factor stimulating a pest-management approach is the group of pest problems directly attributable to pesticide use, which, by analogy might be referred to as "iatrogenic" problems. These problems, which are well documented for insects and insecticide use, include (1) resistance to pesti- cides; (2) destruction of beneficial organisms so that there is a resurgence in the target pest population or so that a secondary pest is released from natural controls and becomes a primary pest; and (3) increased problems on nontreated crops or in nontreated areas [74].

Problems related to fungicide use have drawn less attention but such problems are increasingly significant. Resistance of fungi to fungicides was of little more than academic concern in 1967 [75] but by 1972 was eco- nomically significant [76]. (See also Volume 2, Chapter 12.) Increased disease of coffee trees due to application of fungicides has occurred in Ken- ya. Because of a yield increase attributable to a fungicide, trees were rou- tinely sprayed. Rust (Hemelia vestatrix) and coffee-berry disease (Colleto- trichum coffeanum) were aggravated by this practice [77, 78]. Applications of some fungicides have allowed other pathogens to cause greater amounts of disease. For example, applications of benomyl resulted in increased incidences of sharp eyespot disease of rye caused by Rhizoctonia solani [79], fruit rot of strawberry (species of Rhizopus) [80], and wet stem rot of cow- pea (Pythium aphanidermatum) [81]. Applications of nabam to tomato in- creased the intensity of gray mold (Botrytis cinerea) [82] and some fungi- cides used for control of peanut leafspot increased the amount of stem rot (Sclerotium rolfsii) on peanut [83]. Although most fungicides are relatively nontoxic to mammals, some such as the mercury-containing compounds are very toxic and human disasters occurred when mistakes in their use were made [84-86]. There is currently concern that carcinogenic compounds might be associated with ethylenebisdithiocarbamate fungicides [87, 88].

Thus environmental concern, energy considerations, and pesticide- stimulated problems have directed attention of agricultural scientists to a pest-management approach [89, 90]. Much of this effort has been directed by entomologists at insect pests. It is important to note here that although population dynamics of insect pests and fungal pathogens may have many similarities, the parameters controlling these populations may be quite different. For example, parasites and predators of foliar insect pests may be the most important factors affecting insect pest populations, but in most cases parasites and predators of fungi pathogenic to foliage are unlikely to be the most important factors affecting the pathogen population. Nonethe- less, the approach is applicable to disease management. Smith and

Huffaker [63] have identified several important steps in developing a work-able pest-management system:

A. Identification of critical parameters and interactions
B. Determination of need for a control action
C. Enhancement of natural mechanisms of pest suppression
D. Enhancement of pesticide selectivity
E. Development of new selective measures
F. Implementation of pest management systems

A. Identification of Critical Parameters and Interactions

The goal is to identify interactions among all significant organisms and the environment. Such information probably will be most useful as an aid to decision making if used in systems analyses. For certain diseases enough information is available for the construction of a working model of the system. Several simplified models (computer simulators of disease epidemics) have been constructed and can reproduce natural developments with reasonable accuracy [91-94]. Several benefits emanate from such simulators and investigators can (1) determine whether the understanding of effects of environment on disease epidemics is accurate; (2) determine which parameters need to be most accurately measured; and (3) predict the effect of control procedures, alteration of environment, etc. [95, 96].

For most diseases the effects of more subtle interactions, such as those involving the epiphytic microflora, are just beginning to be identified [97-99]. For example, a population of benomyl-resistant species of Penicillium prevented Botrytis cinerea (also resistant to benomyl) from causing severe disease on cyclamen [100]. The effects of fungicides on the epiphytic microflora are being investigated. Initial reports indicate that broad-spectrum protectant fungicides reduce the fungal flora [101-103] but that the bacterial flora is relatively unaltered [101].

Interrelationships among different types of major pests have only rarely been critically investigated. For example, on potatoes the interactions among the green peach aphid (Myzus persicae), various weed species, and P. infestans are not known. Consequently, the effect on late blight of a measure to reduce M. persicae populations is not known. Accurate description of these interactions are needed to construct more realistic and hence more useable simulators.

B. Determination of Need for a Control Action

Superficially it would seem that the decision as to whether or not a pest control procedure is necessary would be relatively straightforward. The reasoning is certainly logical. Ordish and Dufour [104] reduced this reasoning nearly to a syllogism:

1. Disease causes a loss of crop, the loss being valued at so much.
2. A remedy overcomes the disease and prevents all or some of the loss.
3. If the cost of 2 is less than 1, all is well and the procedure should be followed; if 2 is equal to, or more than 1, the work should not be undertaken.

The authors then immediately stated that the real world is not so simple as in the syllogism, and they discuss "complexities" which render the syllogism difficult to implement. Some costs and benefits affect mainly the grower (internal) and other costs and benefits affect mainly persons other than the grower (external).

1. Internal Costs and Benefits

Reliable disease assessments, prerequisite to determining losses and costs due to diseases, have been difficult to obtain, although certainly plant pathologists have considered the problem in detail [105–110]. Disease assessments and estimates of loss in the United States are probably as good as any in the world; yet for the country as a whole, even these may be questionable. For example, the average annual estimated losses (reduction from potential yield of corn and beans due to all pests from 1951 to 1960 added to about 50% for each crop [111], a percentage which some people find difficult to believe [112]. Unfortunately, disease-loss data for developing countries are even more difficult to estimate [113]. The expense of obtaining accurate and precise loss data with current technology [108] is the limiting factor. Systems analysis may be helpful in more accurately describing or predicting disease loss.

One of the very important concepts in pest management is that of the economic threshold. This concept has been defined (usually by entomologists) as the population of a pest which causes damage equal to the cost of control [114] or in a more precise way: "the population that produces incremental damage equal to the cost of preventing that damage" [115]. Thus the economic threshold determines whether an action should or should not occur.

The economic threshold is diagramatically illustrated in Figure 5. If disease pressure is very slight, the cost of control is greater than the benefit of that control and the ratio of return to cost is less than 1. A control procedure should not be used in this case. Point A (Fig. 5) is the economic threshold. Point B indicates the point of maximum efficiency of a control procedure. Inputs greater than B result in less efficient use of control procedures. Economic thresholds will vary depending upon such factors as climate, location, cultivar, purpose of crop, etc., and therefore will have to be determined for specific conditions [116]. Finally, the economic threshold should not be confused with the damage threshold (i.e., the pest population which reduces yield).

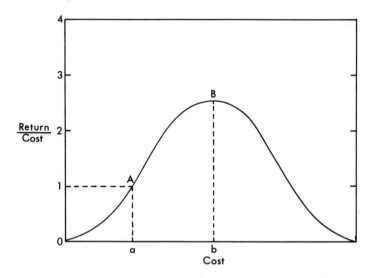

FIG. 5.   Idealized cost-effectiveness of a control procedure.

The next task is to determine how the concept of economic threshold relates to plant disease control. Several investigators have attempted to apply
the concept. Olofsson [117] related the earliness of potato late blight attack
to the need for prophylactic control. When late blight occurred very late in
the season, control sprays were not necessary. He projected that prophylactic control would have been effective in 71 of 75 cases in southern
Sweden but only in 13 of 38 cases in northern Sweden. This approach might
be expanded to incorporate the intensity of disease so that a grower in
northern Sweden could know the probabilities of different amounts of loss in
the absence of a prophylactic spray program. Reschke et al. [118] determined that for mid-season potatoes in Germany measures to prevent late
blight were economical only to about 74 days after emergence. In Montana,
the economic threshold of seed rot or seedling decay of wheat was not
reached, so that treatment of seed with a fungicide was not advised [119].
However, for control of covered smut of barley (Ustilago hordei) and common stinking bunt of wheat (Tilletia caries) the situation was quite different.
Smutty wheat commanded a lower price depending on the extent of contamination. This loss ranged from about $0.50 to $2.80 per acre ($0.22 to
$1.12 per hectare) which was well above the $0.057 per acre ($0.023 per
hectare) cost of control [120].

2.   External Costs and Benefits

The discussion to this point has dealt with specific grower problems and decisions, that is, the effect on a grower of a specific course of action. Even

more complex is the web of costs and benefits associated with pests and pest-management procedures which affect individuals other than those making the decisions, that is, external costs and benefits, or externalities [121, 122].

Because pests and control procedures affect persons in addition to those directly involved with managing the pests, society influences management activities and decisions. Restricted availability of certain pesticides and certification of pesticide applicators are two techniques by which U.S. authorities assure that especially hazardous pesticides are used responsibly. These regulations are costs associated with pesticide use. In addition to regulation by legislation, externalities can be influenced by "socialization" and by "modification of the incentive framework" [121]. Modification of the incentive framework would involve such approaches as taxes on externalities or subsidies to parties involved or damaged [121]. An example of a situation in which a tax could serve to internalize an externality concerns nitrate nitrogen in effluents from irrigated fields. The most effective means of reducing the nitrate effluent was to assess charges based on the effluent [123]. A similar approach might be used in pest management procedures.

C. Enhancement of Natural Mechanisms
of Pest Suppression

For entomologists this approach includes identifying and using natural ene-mies (introduced or indigenous), using resistant cultivars, and assorted cultural procedures [59]. For plant pathologists this approach includes many biological and cultural control procedures, some of which are cur-rently in practice [34, 124, 125]. Except for the use of plant resistance most such procedures suppress the activities or reduce populations of pathogens which cause simple interest diseases.

Fomes annosus, for example, causes a root and butt rot of conifers and it requires an external food source to be able to colonize living trees. Wounded roots or recently exposed stumps in a plantation serve as infection sites and these may then serve as centers from which the pathogen spreads to adjacent trees [27]. Procedures for preventing the establishment of the pathogen in plantations have included application of chemicals to freshly cut stumps to prevent colonization by F. annosus [126]. Another effective pro-cedure is to inoculate freshly cut stumps with Peniophora gigantea, a fun-gus which rapidly colonizes the stump and is an antagonist of Fomes annosus [127]. Successful inoculations have been achieved using suspensions of oidia [127]. By 1970, P. gigantea had been used on about 100,000 acres (40,468 ha) of pine plantation [127]. An interesting technique has been the inoculation of stumps by incorporating the oidia in the oil used to lubricate the chain of a power saw [128]. A similar method, but using Trichoderma sp., is also being investigated [129].

D. Enhancement of Pesticide Selectivity

Mechanisms of fungicidal selectivity have been discussed and can be grouped into several categories such as differential response to the environment, differential effect on the host, or differential effect on the pathogen [130]. In this discussion selectivity or specificity will refer to differential effects on fungi.

A selective pesticide should affect only target organisms. Thus problems such as decreasing natural biological control and hazard to wildlife, to farm workers, and consumers would be decreased. But, there are problems associated with the development of selective pesticides. Probably the major difficulty is that the market for a selective pesticide is smaller than that for a broad-spectrum pesticide [60]. Thus chemical industries have less incentive to develop selective pesticides than nonselective ones. A second problem is that for many diseases, especially those associated with the soil, a complex of pathogens may be involved. Thus for bean root rot, dexon would be effective against Pythium sp.; pentachloronitrobenzene would be effective against R. solani; and benomyl would be effective against Fusarium solani. However, use of two or three fungicides of diverse specificities approaches the effect of a broad-spectrum toxicant. Finally, extensive use of selective fungicides might create additional problems including the increased possibility of resistance [76].

At present and in the immediate future, selective use of conventional fungicides seems to be the principal manner of fungicide selectivity in practical plant disease control. In this context, fungicides can be used selectively with respect to the amount or frequency of application.

1. Selective Amounts of Fungicide

A suggestion that has appeared in the past and a procedure which has been practiced with the wisdom of a farmer's experience is that less fungicide is necessary to control disease on cultivars of greater polygenic resistance than on cultivars of lesser polygenic resistance. A factor now being defined is the amount of fungicide necessary to achieve the desired control on a given cultivar. Generally, recommendations are based on fungicide efficacy tests on very susceptible cultivars. In an attempt to devise guidelines, Fry [39] applied different weekly doses of mancozeb to each of two potato cultivars; Russet Rural, a cultivar very susceptible to late blight and Sebago, a cultivar of slight polygenic resistance. The rate of epidemic development (r) was calculated for each treatment. A regression of $\log_{10} r$ on fungicide dosage indicated that the difference between Sebago and Russet Rural was equivalent to 0.42-0.67 kg mancozeb applied weekly per hectare (see Fig. 6).

With such data, one could predict that if 1.79 kg/ha of mancozeb on a 7-day schedule adequately controlled late blight on Russet Rural, then 1.3-1.4 kg/ha mancozeb should adequately control late blight on Sebago. Probably

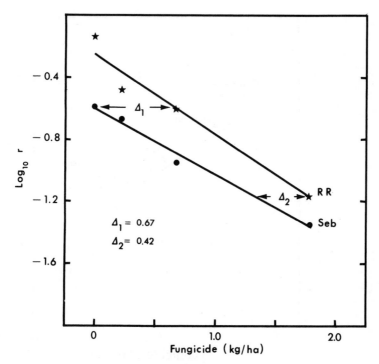

FIG. 6. Effect of a fungicide (mancozeb = coordination product of zinc ion and maneb) applied every 7 days on rate of epidemic development in plots of Russet Rural (RR) and Sebago (Seb) potatoes. Reprinted from Ref. 39, p. 910, by courtesy of The American Phytopathological Society, St. Paul, Minn.

a more important development would be to determine whether the same effect could be achieved by lengthening the interval between applications rather than using reduced individual doses.

   A major hindrance to combining the effects of polygenic resistance and protective fungicides has been that polygenic resistance has been difficult to measure. It may be possible to equate the effects of polygenic resistance and protectant fungicides by measuring the effect each has on the rate of epidemic development. For example, the rates of late blight development in seven commercial potato cultivars were plotted on a regression of $\log_{10}$ r for Russet Rural vs. fungicide concentration (see Fig. 7). Using this technique, the relative difference between Chippewa and Hudson was predicted to be equivalent to 0.2 kg mancozeb applied weekly per hectare and between Hudson and Sebago equivalent to 0.45 kg/ha. Thus, if 1.79 kg mancozeb/ha applied to Chippewa adequately controlled late blight, then 1.59 kg/ha should adequately control late blight on Hudson [39].

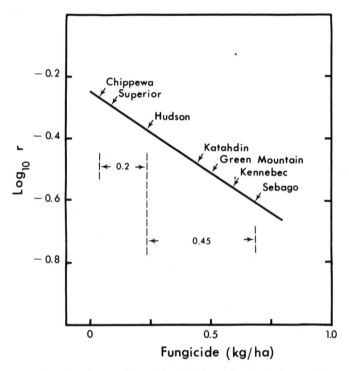

FIG. 7. Predicted relationship of fungicide and polygenic resistance among seven potato cultivars. Apparent infection rates were positioned on the regression of $\log_{10}$ r versus fungicide obtained from data for Russet Rural. Reprinted from Ref. 39, p. 911, by courtesy of The American Phytopathological Society, St. Paul, Minn.

This approach to integrating the effects of polygenic resistance and protective fungicides is presumably applicable to control of many diseases. The range of polygenic resistance to <u>Alternaria solani</u> among tomato cultivars or breeding lines is apparently as large as or larger than the range of resistance to late blight among commercial potato cultivars. In a test to evaluate the field resistance of 30 tomato cultivars or breeding lines, Barksdale [131] determined that the rate of epidemic development ranged from r = 0.169 for the most resistant (line 67B833-1) to 0.432 for the most susceptible (line 66B759-1-1). With the appropriate fungicide it should be possible to control early blight on line 67B833-1 with less fungicide than is required on line 66B759-1-1. Precise data are needed, however, to provide guidelines for the different dosages of fungicides required for adequate control on each cultivar or breeding line in various regions.

Another integrated method employed by Engelhard and Woltz [132] was to combine the effects of plant nutrition, soil pH, and fungicide for prevention of the symptoms of Fusarium wilt of chrysanthemum. The effects of nitrate nitrogen and lime in reducing symptom severity of Fusarium wilt of chrysanthemum were additive, although symptoms were not completely suppressed [133]. The combination of benomyl, nitrate nitrogen, and high lime completely prevented symptoms in the very susceptible cultivar Yellow Delaware. Presumably less fungicide would be needed for suppression of symptoms in less susceptible cultivars.

Adjuvants of various types might increase the efficacy of fungicides, thus reducing the amount of toxicant necessary for control. A current example is that of paraffinic oils which enhanced the uptake and translocation of benomyl and thiabendazole and increased the control of Verticillium wilt of cotton achieved by each compound in greenhouse tests [134]. The beneficial effect of oils on uptake of fungicides has been demonstrated for the control of other diseases as well [135, 136].

Finally, the method of fungicide application may affect the amount necessary for control. Several authors have recently considered the possibility of practical low-volume applications [137-139]. Grower practice and research indicate that with the appropriate fungicide, complete coverage is not always necessary because the fungicide will be redistributed in rain or irrigation water [140].

2. Selective Frequency of Fungicide Application

Logically, fungicides should be applied only when there is a need for them. Disease forecasts have aided growers by indicating whether or not control procedures were necessary [141]. Most forecasts have predicted the likelihood of disease occurrence and if disease was likely, a control procedure was recommended.

The approach to forecasting depends on which epidemiological model the specific plant disease most nearly resembles. For simple interest diseases, parameters which measure the amount or the efficacy of the initial inoculum are monitored. For example, the forecast for Stewart's wilt of corn estimates indirectly the numbers of overwintering flea beetles which transmit the pathogenic bacterium Erwinia stewartii [142, 143]. For compound interest diseases, parameters which determine the rate of epidemic development are measured. For example, potato late blight forecasters monitor temperature, relative humidity, and rainfall during the growing season [144].

Forecasting techniques for more and more diseases are being developed. For example, forecasting is now possible for northern corn leaf blight (Helminthosporium turcicum) on sweet corn [145], Cercospora leafspot of peanuts (C. arachidicola) [146], and leaf blight of onions (Botrytis squamosa) [147]. Progress is also being made on predicting wheat stem rust severities so that fungicide applications are becoming economically feasible [148,

149]. One of the disappointments to pathologists has been that few, if any, of the forecasting techniques have been utilized fully by growers. Of the many factors contributing to this situation probably the most significant is that application of fungicides is more easily managed if the chemical is applied on a regular schedule rather than on "demand." Environmental, economic, and energy considerations may cause growers to reevaluate this factor.

Finally, the recent utilization of computers in disease forecasting has generated interest among growers, researchers, and the lay public. In Georgia, a computer program uses meteorological data to develop a peanut leafspot spray advisory which is subsequently broadcast by television and radio [150]. In Pennsylvania, potato growers monitoring environmental conditions on their farms phone these data to "Blitecast" at the Pennsylvania State University. These data are telephoned to the computer terminal each week; a spray recommendation is developed and given to the grower by telephone [144].

E. Development of New Selective Measures

The possibility of using both chemical and biological procedures to achieve reliable, selective control is intriguing. A disease of apricot trees caused by Eutypa armeniacae appears to be a candidate for such an approach. Fusarium lateritium, a saprophyte on apricot trees, colonizes pruning wounds and apparently produces a substance which is toxic to E. armeniacae [151]. F. lateritium is also about one-tenth as sensitive to benzimidazole fungicides as is E. armeniacae. Therefore use of small doses of such fungicides should selectively favor growth of F. lateritium. Growth of E. armeniacae should be impaired both by the fungicide as well as by biological control [151].

One of the most vigorous research areas in physiology of plant disease has been centered around induced resistance and phytoalexins, but there is not yet an unanimous opinion about the role of phytoalexins in resistance to disease [152]. Nonetheless, these compounds are antifungal and they are induced in plants by a variety of physical conditions and chemicals including fungicides [153, 154]. One phytoalexin, capsidiol, has been used as a fungicide to control late blight of tomatoes [155]. The next step would be to determine whether induction of phytoalexins enhances the resistance of a plant. One can visualize that for control of certain diseases conventional fungicides might be replaced by or applied in conjunction with a treatment which could initiate or maintain phytoalexin biosynthesis.

Several reports indicate that virus infection of plants may affect resistance to fungi. In some cases systemic infection by a virus enhances resistance to a fungal pathogen. Potato plants infected with any of several viruses were more resistant to P. infestans and to Fusarium roseum

Avenaceum than were virus-free plants [156-158]. Virus infections of Italian prune trees and sugar beet plants have also enhanced the resistance of these plants to Cytospora cincta and Uromyces betae, respectively [159, 160]. Unfortunately, because virus infection does not always reduce fungal parasitism [161-164], each situation will have to be treated individually.

F. Implementation of Pest Management Programs

Assimilation of the various factors and components described previously into a workable, reliable pest management system is a task which is nearly overwhelming in magnitude. And only if implementation is successful will the effort expended on the various components of a pest-management system be fully worthwhile. Implementation is therefore the key step on which success of pest management relies. Unfortunately implementation of the system will be as complex as and probably more fraught with frustration than any other step considered in this discussion. However, some pilot projects have proved successful [165, 166]. The multitude of questions which a pest management scientist must address contributes to the frustration. For example, in an apple-pest-management system the following questions arose:

Should insecticide and fungicide be applied together?
Does the fungicide affect apple insect or mite pests?
Does the ground cover affect any of the pest species?
If the market for the crop is likely to be low, are the pesticide sprays economical?
What is the most efficient method for applying pesticides?
What is the interaction between pruning and pest populations?
Which cultivars are most appropriate based on market considerations?
Which ones are most appropriate based on climate and site?
Which spacing will be most appropriate?

These and many other questions will need to be addressed by the coordinators of the pest management system [167].

After the data concerning specific pests and their interactions have been incorporated, guidelines for decisions will have to be formulated and used to construct a decision-making process. Such guidelines are beginning to be developed [168, 169], although practical utilization in a total pest management system seems to be in the distant future.

Development and implementation of pest-management systems is expensive and one can reasonably ask, "who should pay?" Both the grower and society will benefit from operative pest management systems. Thus, each should probably accept a portion of the cost. Precedent has already been set in that pest-management consultants are being retained by growers in intensive agriculture and governmental agencies have established pilot projects in pest management [90]. An additional possibility might be to

develop crop insurance for participants in a pest-management program [90].
Thus the insurance of excessive pesticide could be abandoned.

Because pest management has become a relatively popular concept,
there is the hazard that societal expectations from it may be unrealistically
high [90]. Pest control is highly complex and established practices cannot
be drastically changed rapidly. Even so, adoption of reliable management
techniques is delayed too long after their development [90]. Insufficient
numbers of trained personnel who advise growers and who are well informed
of new developments probably account for the time lag. Among several
recommendations regarding implementation of pest management systems in
the United States was one that a position of "Pest Management Specialist"
be established in each state [90]. Hopefully, this individual would facilitate
the implementation of the philosophy and practice of pest management.

IV. SUMMARY

Plant pathologists are challenged to help increase food production by reducing losses due to diseases. The basic strategies of disease control are
to reduce the amount of initial inoculum and to reduce the rate of epidemic
development. The choice of strategy depends on the epidemiological characteristics of the disease. Use of fungicides is an important tactic within
each strategy, but often the most effective and efficient control procedure
involves more than one tactic, namely, an integrated approach.

Pest management has influenced the philosophy of disease control, as
illustrated by a shift in emphasis from maximum to optimum control. In
pest management, the need for control is based on economic, societal, and
environmental considerations. To attain more efficient use of fungicide,
there is likely to be a shift toward integrating several procedures such as
cultural, biocontrol, and forecasting methods. Implementation of pest
management systems should aid agricultural scientists to efficiently enhance
food production.

ACKNOWLEDGMENTS

I wish to thank P. A. Arneson, J. F. Rissler, W. A. Sinclair, and B. J.
Mosher for helpful suggestions, comments, and assistance in preparation
of this chapter.

REFERENCES

1. J. E. Van der Plank, in Plant Pathology: An Advanced Treatise (J. G.
   Horsfall and A. E. Dimond, eds.), Vol. III, Academic Press, New
   York, 1960, pp. 229-289.

2. J. E. Van der Plank, Plant Diseases: Epidemics and Control, Academic Press, New York, 1963.
3. A. E. Dimond and J. G. Horsfall, in Plant Pathology: An Advanced Treatise (J. G. Horsfall and A. E. Dimond, eds.), Vol. III, Academic Press, New York, 1960, pp. 1-22.
4. J. M. Thresh, Ann. Rev. Phytopathol., 12, 111 (1974).
5. D. Jowett, J. A. Browning, and B. Cournoyer Haning, in Epidemics of Plant Diseases: Mathematical Analysis and Modeling (J. Kranz, ed.), Ecological Studies, Vol. 13, Springer Verlag, New York, 1974, pp. 115-136.
6. J. E. Van der Plank, in Fungicides: An Advanced Treatise (D. C. Torgeson, ed.), Vol. I, Academic Press, New York, 1967, pp. 63-92.
7. J. E. Van der Plank, Pest Control Strategies for the Future, National Academy of Sciences, National Research Council, Washington, D.C., 1972, pp. 109-118.
8. H. V. Jordan, J. E. Adams, D. R. Hooton, D. D. Porter, L. M. Blank, E. W. Lyle, and C. H. Rogers, Tech. Bull. No. 948, U.S. Dept. of Agriculture, Washington, D.C., Feb. 1948.
9. W. Merrill, Phytopathology, 57, 1206 (1967).
10. W. A. Sinclair and R. J. Campana, Regional Research Publication in Project NE-25, N.Y. State Agr. Expt. Sta., Cornell University, Ithaca, N.Y. (in press).
11. R. Sprague, Washington Agr. Ext. Serv. Bull. 560, Washington State University, Pullman, April 1955.
12. A. E. Cox and E. C. Large, Handbook No. 174, U.S. Dept. of Agriculture, Agr. Res. Serv., Washington, D.C., 1960.
13. E. Gram, in Plant Pathology: An Advanced Treatise (J. G. Horsfall and A. E. Dimond, eds.), Vol. III, Academic Press, New York, 1960, pp. 313-356.
14. B. von Schmeling and M. Kulka, Science, 152, 659 (1966).
15. D. H. Brooks, in Systemic Fungicides (R. W. Marsh, ed.), Wiley, New York, 1972, pp. 187-205.
16. P. A. Minges, A. A. Muka, R. F. Sandsted, A. F. Sherf, and R. D. Sweet, 1975 Vegetable Production Recommendations, New York State College of Agriculture and Life Sciences, Cornell University, Ithaca, N.Y., 1975.
17. R. B. Maude, A. S. Vizor, and C. G. Shuring, Ann. Appl. Biol., 64, 245 (1969).
18. R. B. Maude, in Systemic Fungicides (R. W. Marsh, ed.), Wiley, New York, 1972, pp. 225-236.
19. B. J. Jacobsen and P. H. Williams, Plant Disease Reptr., 55, 934 (1971).
20. R. L. Gabrielson, M. W. Mulanax, K. Matsuoka, P. H. Williams, G. P. Whiteaker, and J. D. Maguire, Proc. Amer. Phytopathol. Soc., 2, 38 (abstr.) (1975).

21. A. W. Dimock, C. E. Williamson, and P. E. Nelson, in Chrysanthe-
    mums: A Manual of the Culture, Disease and Insects, and Economics
    of Chrysanthemums (R. W. Langhans, ed.), The New York State Exten-
    sion Service Chrysanthemum School with the cooperation of the New
    York State Flower Growers Association, Inc., 1964, pp. 116-160.
22. R. K. Horst and P. E. Nelson, Diseases of Chrysanthemum, Inform.
    Bull. 85, Plant Sciences, Plant Pathology 8, N.Y. State Agr. Expt.
    Sta., Cornell University, Ithaca, N.Y., 1975.
23. G. Nyland and A. C. Goheen, Ann. Rev. Phytopathol., 7, 331 (1969).
24. P. S. Deth, N.Y. Food Life Sci. Quart., 6, 5 (1973).
25. A. J. Braun, Research Circular No. 13, N.Y. State Agr. Expt. Sta.,
    Cornell University, Geneva, N.Y., 1968.
26. J. L. Brann, Jr., P. A. Arneson, and G. H. Oberly, Tree-Fruit
    Production Recommendations for Commercial Growers, New York
    State College of Agriculture and Life Sciences, Cornell University,
    Ithaca, N.Y., 1975.
27. S. D. Garrett, Pathogenic Root Infecting Fungi, Cambridge University
    Press, London, 1970.
28. J. G. Dickson, J. Agr. Res., 23, 837 (1923).
29. R. B. Stevens, in Plant Pathology: An Advanced Treatise (J. G. Hors-
    fall and A. E. Dimond, eds.), Vol. III, Academic Press, New York,
    1960, pp. 357-429.
30. L. D. Leach and R. H. Garber, in Rhizoctonia solani: Biology and
    Pathology (J. R. Parmeter, Jr., ed.), University of California Press,
    Berkeley, 1970, pp. 189-198.
31. D. E. Bliss, Phytopathology, 41, 665 (1951).
32. H. D. Ohr, D. E. Munnecke, and J. L. Bricker, Phytopathology, 63,
    965 (1973).
33. A. M. Smith, 2nd Intern. Congr. Plant Pathol., Minneapolis, Minne-
    sota, Abstr. No. 1137, 1973.
34. K. F. Baker and R. J. Cook, Biological Control of Plant Pathogens,
    W. H. Freeman & Co., San Francisco, 1974.
35. P. R. Day, Genetics of Host-Parasite Interaction, W. H. Freeman &
    Co., San Francisco, 1974.
36. J. E. Van der Plank, Disease Resistance in Plants, Academic Press,
    New York, 1968.
37. J. C. Walker, in Plant Pathology Problems and Progress 1908-1958
    (C. S. Holton, G. W. Fischer, R. W. Fulton, H. Hart, and S. E. A.
    McCallan, eds.), University of Wisconsin Press, Madison, 1959, pp.
    32-41.
38. P. Crill, J. P. Jones, D. S. Burgis, and S. S. Woltz, Plant Disease
    Reptr., 56, 695 (1972).
39. W. E. Fry, Phytopathology, 65, 908 (1975).
40. J. B. Rowell, Plant Disease Reptr., 57, 653 (1973).

41. A. W. Dimock and J. Tammen, in Roses (J. W. Mastalerz and R. W. Langhans, eds.), Pennsylvania Flower Growers, New York State Flower Growers Assoc., Inc., Roses, Inc., 1969, Chapt. 22.
42. R. K. Horst, private communication, 1975.
43. J. H. Haas and B. Bolwyn, Can. J. Plant Sci., 52, 525 (1972).
44. J. R. Steadman, D. P. Coyne, and G. E. Cook, Plant Disease Reptr., 57, 1070 (1973).
45. J. Rotem and J. Palti, Ann. Rev. Phytopathol., 7, 267 (1969).
46. J. Rotem, J. Palti, and J. Lomas, Phytopathology, 60, 839 (1970).
47. P. R. Day, in Pest Control Strategies for the Future, National Academy of Sciences, National Research Council, 1972, pp. 257-271.
48. G. Shaner, Phytopathology, 63, 1307 (1973).
49. R. D. Wilcoxson, A. H. Atif, and B. Skovmand, Plant Disease Reptr., 58, 1085 (1974).
50. H. D. Thurston, Phytopathology, 61, 620 (1971).
51. M. D. Simons, J. Environ. Qual., 1, 232 (1972).
52. J. A. Browning and K. J. Frey, Ann. Rev. Phytopathol., 7, 355 (1969).
53. N. F. Jensen and G. C. Kent, Farm Res., 29, 4 (1963).
54. K. J. Frey, J. A. Browning, and R. L. Grindeland, in Mutation Breeding for Disease Resistance, International Atomic Energy Agency, Vienna, 1971, pp. 159-169.
55. K. J. Leonard, Phytopathology, 59, 1851 (1969).
56. Anonymous, in Annual Report, Rockefeller Foundation 1964-1965, pp. 67-79.
57. K. J. Frey, J. A. Browning, and R. L. Grindeland, Iowa Farm Sci., 24, 3 (1970).
58. T. Eichers, P. Andrilenas, H. Blake, R. Jenkins, and A. Fox, Agr. Econ. Rept. No. 179, U.S. Dept. of Agriculture, Econ. Res. Serv., Washington, D.C., 1970.
59. P. A. Andrilenas, Agr. Econ. Rept. No. 252, U.S. Dept. of Agriculture, Econ. Res. Serv., Washington, D.C., 1974.
60. G. Ordish and J. E. Mitchell, in Fungicides: An Advanced Treatise (D. C. Torgeson, ed.), Vol. 1, Academic Press, New York, 1967, pp. 39-62.
61. H. T. Blake, P. A. Andrilenas, R. P. Jenkins, T. R. Eichers, and A. S. Fox, Agr. Econ. Rept. No. 192, U.S. Dept. of Agriculture, Econ. Res. Serv., Washington, D.C., 1970.
62. R. Van den Bosch, Organic Gardening and Farming, April 19, 1975, pp. 142-151.
63. R. F. Smith and C. B. Huffaker, OEPP/EPPO, Bull. 3, 31 (1973).
64. R. F. Smith, in Proceedings of Tall Timbers Conference on Ecological Animal Control by Habitat Management, No. 3, Tallahassee, Fla., 1972, pp. 53-83.

65. R. L. Giese, R. M. Peart, and R. T. Huber, Science, 187, 1045 (1975).
66. J. L. Apple, BioScience, 22, 461 (1972).
67. R. F. Smith, Bull. Entomol. Soc. Amer., 18, 7 (1972).
68. G. L. McNew, in Pest Control by Chemical, Biological, Genetic, and Physical Means, U.S. Dept. of Agriculture, Agr. Res. Serv., 33-110, Washington, D.C., 1966, pp. 41-52.
69. G. L. McNew, in Pest Control Strategies for the Future, National Academy of Sciences, National Research Council, Washington, D.C., 1972, pp. 119-133.
70. R. L. Carson, Silent Spring, Houghton Mifflin, Boston, 1962.
71. E. Deck, Ann. Rev. Entomol., 20, 119 (1975).
72. D. Pimentel, L. E. Hurd, A. C. Bellotti, M. J. Forster, I. N. Oka, O. D. Sholes, and R. J. Whitman, Science, 182, 443 (1973).
73. W. Clark, Smithsonian, 5, 59 (1975).
74. L. A. Falcon and R. F. Smith, Guidelines for Integrated Control of Cotton Insect Pests, Food and Agriculture Organization of the United Nations, Rome, 1973.
75. S. G. Georgopoulos and C. Zaracovitis, Ann. Rev. Phytopathol., 5, 109 (1967).
76. J. Dekker, in Systemic Fungicides (R. W. Marsh, ed.), Wiley, New York, 1972, pp. 156-174.
77. E. Griffiths, Trop. Sci., 14, 79-89 (1972).
78. J. N. Gibbs, Ann. Appl. Biol., 70, 35 (1972).
79. E. P. Van der Hoeven and G. J. Bollen, Acta Bot. Neerl., 21, 107 (1972).
80. V. W. L. Jordan, Plant Pathol., 22, 67 (1973).
81. R. J. Williams and A. Ayanaba, Phytopathology, 65, 217 (1975).
82. R. S. Cox, and N. C. Hayslip, Plant Disease Reptr., 40, 718 (1956).
83. P. A. Backman, R. Rodriguez-Kabana, and J. C. Williams, Phytopathology, 65, 773 (1975).
84. A. Curley, V. A. Sedlak, E. F. Girling, R. E. Hawk, W. F. Barthel, P. E. Pierce, and W. H. Likosky, Science, 172, 65 (1971).
85. F. Bakir, S. F. Damlugi, L. Amin-Zaki, M. Murtadha, A. Khalidi, N. Y. Al-Rawi, S. Tikriti, H. I. Dahir, T. W. Clarkson, J. C. Smith, and R. A. Doherty, Science, 181, 230 (1973).
86. N. A. Smart, Residues Rev., 23, 1 (1968).
87. J. P. Seiler, Mutation Res., 26, 189 (1974).
88. S. L. Graham, W. H. Hansen, K. J. Davis, and Carleene H. Perry, J. Agr. Food Chem., 21, 324 (1973).
89. Anonymous, U.S. Participation in the International Biological Program, Rept. No. 6 of the U.S. National Committee for the International Biological Program, National Academy of Sciences, Washington, D.C. (1974).

90. E. H. Glass (coordinator), Integrated Pest Management: Rationale, Potential, Needs and Implementation, Entomological Society of America, Special Publication 75-2, 1975.

91. G. E. Shaner, R. M. Peart, J. E. Newman, W. L. Stirm, and O. L. Loewer, Jr., Epimay: An Evaluation of a Plant Disease Display Model, Agr. Expt. Sta., Purdue University, RB-890, West Lafayette, Indiana, 1972.

92. P. E. Waggoner and J. G. Horsfall, Epidem: A Simulator of Plant Disease Written for a Computer, Bull. No. 698, Conn. Agr. Expt. Sta., New Haven, 1969.

93. P. E. Waggoner, J. G. Horsfall, and R. J. Lukens, Epimay: A Simulator of Southern Corn Leaf Blight, Bull. No. 729, Conn. Agr. Expt. Sta., New Haven, 1972.

94. R. E. McCoy, Ph.D. Thesis, Cornell University, Ithaca, N.Y., 1971.

95. P. E. Waggoner, in Epidemics of Plant Diseases: Mathematical Analysis and Modeling (J. Kranz, ed.), Ecological Studies, Vol. 13, Springer-Verlag, New York, 1974, pp. 137-160.

96. J. Kranz, in Epidemics of Plant Diseases: Mathematical Analysis and Modeling (J. Kranz, ed.), Ecological Studies, Vol. 13, Springer-Verlag, New York, 1974, pp. 7-54.

97. J. E. Crosse, in Ecology of Leaf Surface Micro-organisms (T. F. Preece and C. H. Dickinson, eds.), Academic Press, New York, 1971, pp. 283-290.

98. C. Leben, Phytopathology, 54, 405 (1964).

99. J. Van den Heuvel, Phytopathologisch Laboratorium Willie Commelin-Scholten, Mededeling 84, Baarn, Holland, 1970.

100. L. Van Dommelin and G. J. Bollen, Acta Bot. Neerl., 22, 169 (1973).

101. A. Bainbridge and C. H. Dickinson, Trans. Brit. Mycolog. Soc., 59, 31 (1972).

102. C. H. Dickinson, Pestic. Sci., 4, 563 (1973).

103. C. H. Dickinson, Trans. Brit. Mycol. Soc., 60, 423 (1973).

104. G. Ordish and D. Dufour, Ann. Rev. Phytopathol., 7, 31 (1969).

105. E. C. Large, Ann. Rev. Phytopathol., 4, 9 (1966).

106. Food and Agricultural Organization of the United Nations, FAO Manual on the Evaluation and Prevention of Losses by Pests, Diseases and Weeds, FAO, Rome, 1971.

107. K. S. Chester, Plant Disease Reptr. Suppl., 193, 190 (1950).

108. E. L. LeClerg, Phytopathology, 54, 1309 (1964).

109. W. C. James, C. S. Shih, W. A. Hodgson, and L. C. Callbeck, Phytopathology, 62, 92 (1972).

110. W. C. James, Ann. Rev. Phytopathol., 12, 27 (1974).

111. Anonymous, ARS Handbook No. 291, U.S. Dept. of Agriculture, Washington, D.C., 1965.

112. W. C. Paddock, Ann. Rev. Phytopathol., 5, 375 (1967).
113. J. Vallega and L. Chiarappa, Phytopathology, 54, 1305 (1964).
114. E. Sylven, Medd. Växtskyddsanst Stockholm, 14 (118), 69 (1963).
115. J. C. Headley, in Pest Control Strategies for the Future, National
     Academy of Sciences, National Research Council, Washington, D.C.,
     1972, pp. 100-108.
116. V. M. Stern, Ann. Rev. Entomol., 18, 259 (1973).
117. B. Olofsson, Medd. Växtskyddsanst Stockholm, 14 (119), 85 (1968).
118. M. Reschke, R. Heitefuss, and W. H. Fuchs, Z. Pflanzenkrankh.
     Pflanzenschutz, 81, 1 (1974).
119. D. E. Mathre, V. R. Stewart, R. H. Johnston, and D. E. Baldridge,
     J. Environ. Qual., 4, 117 (1975).
120. D. E. Mathre, W. G. Heid, Jr., and R. H. Johnston, Agr. Ext.
     Serv. Montana State University, and Farm Products Economy Div.,
     Bull. 669, U.S. Dept. of Agriculture, Econ. Res. Serv., Washington,
     D.C., 1973.
121. G. A. Carlson and E. N. Castle, in Pest Control Strategies for the
     Future, National Academy Sciences, National Research Council,
     Washington, D.C., 1972, pp. 79-99.
122. J. C. Headley, Ann. Rev. Entomol., 17, 273 (1972).
123. G. L. Horner, Amer. J. Agr. Econ., 57, 33 (1975).
124. K. F. Baker and W. C. Snyder, eds., Ecology of Soil-borne Plant
     Pathogens, Prelude to Biological Control, University of California
     Press, Berkeley, 1965.
125. G. W. Bruehl, ed., Biology and Control of Soil-borne Plant Pathogens,
     Amer. Phytopathological Society, St. Paul, Minn., 1975.
126. W. A. Sinclair, Memoir 391, Cornell University Agr. Expt. Sta.,
     Ithaca, N.Y., 1964.
127. J. Rishbeth, in Biological Control of Forest Diseases (V. J. Nordin,
     ed.), Congress of International Union of Forestry Research Organi-
     zations, Gainesville, Florida, 1971, pp. 29-35.
128. J. D. Artman, Plant Disease Reptr., 56, 958 (1972).
129. R. S. Hunt, J. R. Parmeter, Jr., and F. W. Cobb, Jr., Plant
     Disease Reptr., 55, 659 (1971).
130. J. G. Horsfall, in Pest Control Strategies for the Future, National
     Academy of Sciences, National Research Council, Washington, D.C.,
     1972, pp. 216-225.
131. T. H. Barksdale, Plant Disease Reptr., 55, 807 (1971).
132. A. W. Engelhard and S. S. Woltz, Phytopathology, 63, 1256 (1973).
133. S. S. Woltz and A. W. Engelhard, Phytopathology, 63, 155 (1973).
134. D. C. Erwin, R. A. Khan, and H. Buchenauer, Phytopathology, 64,
     485 (1974).
135. Z. Solel, J. Pinkas, and G. Loebenstein, Phytopathology, 62, 1007
     (1972).
136. D. H. Smith and F. L. Crosby, Phytopathology, 62, 1029 (1972).

137. D. L. Strider, G. C. Rock, and H. E. Blackwell, Tech. Bull. 222, N. C. Agr. Expt. Sta., Raleigh, N. D., 1973.
138. F. H. Lewis, D. Asquith, E. R. Krestensen, and K. D. Hickey, Progress Report 294, Agr. Expt. Sta., Penn. State University, College of Agriculture, University Park, Pa., 1969.
139. F. H. Lewis and K. D. Hickey, Ann. Rev. Phytopathol., 10, 399 (1972).
140. G. A. Brandes, Ann. Rev. Phytopathol., 9, 363 (1971).
141. P. R. Miller, in Plant Pathology Problems and Progress, 1908-1958 (C. S. Holton, G. W. Fischer, R. W. Fulton, H. Hart and S. E. A. McCallan, eds.), University of Wisconsin Press, Madison, 1959, pp. 557-565.
142. N. E. Stevens, Plant Disease Reptr., 18, 141 (1934).
143. P. R. Miller, Phytoprotection, 50, 81 (1969).
144. R. A. Krause, L. B. Massie, and R. A. Hyre, Plant Disease Reptr., 59, 95 (1975).
145. R. D. Berger, Phytopathology, 60, 1284 (abstr.) (1970).
146. D. H. Smith, F. L. Crosby, and W. J. Ethridge, Plant Disease Reptr., 58, 666 (1974).
147. P. B. Shoemaker and J. W. Lorbeer, Phytopathology, 59, 402 (abstr.) (1969).
148. M. G. Eversmeyer, J. R. Burleigh, and A. P. Roelfs, Phytopathology, 63, 348 (1973).
149. M. G. Eversmeyer, C. L. King, and W. G. Willis, Plant Disease Reptr., 59, 607 (1975).
150. D. W. Parvin, Jr., D. H. Smith, and F. L. Crosby, Phytopathology, 64, 385 (1974).
151. M. V. Carter and T. V. Price, Australian J. Agr. Res., 25, 105 (1974).
152. J. M. Daly, Phytopathology, 62, 392 (1972).
153. J. Kuć, Ann. Rev. Phytopathol., 10, 207 (1972).
154. J. J. Reilly and W. L. Klarman, Phytopathology, 62, 1113 (1972).
155. E. W. B. Ward, C. H. Unwin, and A. Stoessl, Phytopathology, 65, 168 (1975).
156. C. Fernandez de Cubillos and H. D. Thurston, Amer. Potato J., 52 221 (1975).
157. E. D. Jones and J. M. Mullen, Amer. Potato J., 51, 209 (1974).
158. J. Pietkiewicz, Phytopathologische Z., 82, 49 (1975).
159. A. W. Helton, Phytopathology, 64, 1410 (1974).
160. A. C. Magyarosy, B. B. Buchanan, and J. E. Duffus, Phytopathology, 65, 361 (1975).
161. S. P. S. Beniwal and R. Gudauskas, Phytopathology, 64, 1197 (1974).
162. A. A. Reyes and K. C. Chadha, Phytopathology, 62, 1424 (1972).
163. H. Koike and S. Yang, Phytopathology, 61, 1090 (1971).
164. J. C. Tu and R. E. Ford, Phytopathology, 61, 800 (1971).

165. J. E. Casey, R. D. Lacewell, and W. Sterling, <u>MP-1152</u>, Texas
     Agr. Expt. Sta., Texas A & M University System, College Station,
     Tex., June 1974.
166. P. L. Adkisson, in <u>Proceedings of Tall Timbers Conference on
     Ecological Animal Control by Habitat Management</u>, Tallahassee,
     Fla., 1973 (No. 4), pp. 175-188.
167. P. A. Arneson, private communication, 1975.
168. T. L. Vincent, <u>Biometrics</u>, <u>31</u>, 1 (1975).
169. G. A. Carlson, <u>Amer. J. Agr. Econ.</u>, <u>52</u>, 216 (1970).

Chapter 3

DETECTING POTENTIAL PROTECTIVE AND SYSTEMIC
ANTIFUNGAL COMPOUNDS

E. Neil Pelletier

Office of Pesticide Programs
Criteria and Evaluation Division
Environmental Protection Agency
Washington, D.C.

I. INTRODUCTION

Those who are interested in the historical aspects of phytopathology often
formulate cases for assigning paternal roles to those who have advanced
the science. Although cases have been made for establishing Bénédict
Prévost as the father of plant pathology, that eminence is now well fixed in
Anton De Bary. However, if spiritual roles were ever to be bestowed,
Prevost, at least among the practitioners of the art of evaluating antifungal
compounds, could properly be named the "godfather" or "patron saint" of
phytopathology. As set forth in his memoir of 1807 [1], Prevost was not
only the first to demonstrate by means of laboratory tests the efficacy of
chemicals against fungus spores, but he was able to correlate the laboratory
results with field studies. Furthermore, in his studies he was also concerned with the important factors of cost effectiveness, ease of handling
and application, and toxicology. Prevost's success in detecting an effective
compound among a handful was phenomenal, especially when compared to
present-day practitioners, who with the benefits of sophisticated chemistry
and an unending supply of compounds might expect to find one practical
antifungal compound in 5,000.

As in the days of Prevost, the discovery of fungicides must today still
depend on the empirical. The ideal of predicting activity on the basis of
structure or a quick chemical or physical measurement remains an elusive
goal. Neat structure-activity relationships are always found after the
activity was demonstrated.

Although antifungals are discovered by both individuals and institutions,
the empirical approach for detection, not to mention the subsequent development of compounds, requires the resources of large companies involved in
agrochemical research. The number of compounds screened by a large
company, whether they are obtained from a chemical synthesis program or
isolated from a biological process, may vary from less than 1,000 to
10,000 per year; the average company (if there is one) screens about 5,000
compounds per year. Despite considerable differences in approach, the
process of testing follows a similar pattern. Most companies test as potential fungicides any and all compounds that their chemists can synthesize or
isolate, regardless of the properties of the chemical or the rationale and
intent of the chemists. Despite gnawing uncertainties, a few companies
screen only compounds selected on a specific basis. Among the 5,000
compounds subjected to primary screening, perhaps 10% receive secondary
evaluation; among these 500, less than 5% will be candidates for preliminary
field testing. In this hypothetical scheme, seven compounds might be
chosen for secondary field testing and ultimately one compound would reach
advanced field testing and proven effectiveness. This one compound may
then fall to the wayside due to factors unrelated to activity.

II. CRITERIA FOR SELECTING
TEST METHODS

The proper conduct of any testing program demands the careful selection of methods, as well as an almost constant reexamination of the methods in use. The selection and reexamination of methods should be based on four criteria: suitability, reliability, efficiency, and economy.

Among these interrelated criteria, reliability is the most obvious. Although reliability is important at all phases of a testing program, it is most critical in the primary screening stage. At the primary test stage, it is necessary to have the greatest possible certitude that one has sieved through only the inactive compounds and not some potentially active compound that the design of the sieve mesh has failed to trap. Reliability as dependability and reproducibility, is best assured through good design and control. A test must be designed to be as simple as possible through the elimination of as many steps and variables as possible. In conducting the test, reliability is strived for through meticulous control of test conditions and materials. How many times have tests mysteriously failed in constancy, due to a cause which is laboriously traced to some factor such as an impurity in a new source of a growth-medium ingredient?

The measure of a test's reliability is the performance of standard compounds. If one or more standard compounds are not detectable, the reason for it should be determined. With an understanding of the reason for the failure of the test to detect exceptions, and barring the possibility of modifying the test, the options lie in utilizing an alternative more reliable test, an auxiliary method that detects types of compounds representing the exception, or in running the risk of missing compounds similar to the exceptions. Where there are no standards of comparison, confidence in a method depends entirely on design and precision.

The sensitivity of a test is also a factor in its reliability. A test should not only differentiate between active and inactive chemicals, but should differentiate degrees of activity among closely related chemicals and possibly qualitative and quantitative differences among samples of the same compound.

Unfortunately, methods of detecting antifungal activity can only approximate the system in which the antifungal will find its use. In selecting a test method, one has to consider if and to what extent the factors which are in operation in the ultimate use system are present in the test. Objectives and practicalities govern whether the factors can be contained in one test or whether the factors can be isolated into separate tests. In determining inherent fungitoxicity of a compound to a given fungus, the objective can be met with one test; the limitations are those imposed by the characteristics of the fungus and the conditions under which the test can be conducted. In determining fungicidal activity for the control of a plant pathogen, the objective can be met directly in one test by duplicating a disease system. However, it may be necessary or even desirable (due to limitations or complexity) to meet the objective through a number of tests in a step-by-step process measuring the separate responses that are active in the system.

The circumstances surrounding the detection of antifungals assume a program involving a massive number of compounds that individually are the product of test-tube-scale reactions or isolations. With a micro quantity of test chemical there must somehow be a sufficient amount to formulate it into a usable state, to apply it to a test medium, and to have a remainder available for further testing and possibly for chemical analysis. None of the test compound can be squandered on numerous tests or elaborate systems of application. Furthermore since all the steps leading to the commercialization of an antifungal compound require a sizeable investment on the part of the sponsoring institution or company (the screening step involves the smallest portion of the investment), some limit of resources has to be assumed. Presupposing a minute quantity of an immense number of compounds and finite resources, a program must be carried out with efficiency and economy.

Given a large number of compounds for screening, efficient testing requires that tests be simple and of a relatively short duration. Simplicity is achieved by minimizing manipulations, test concentrations, and the gathering of results. The time spent in interpreting test results should be devoted to those results which demonstrate the higher degrees of activity; little or no purpose is served in making tedious counts which represent inactivity or low degrees of activity.

III. IN VITRO TESTS

The outstanding features of in vitro tests are their efficiency and economy. Large numbers of compounds can be handled on a routine basis in one experiment and the results obtained within a relatively short period of time. The tests can be conducted with limited facilities and require only minute quantities of test compound. Under well-controlled conditions they are dependable and require little replication.

Their suitability is limited to the measurement of the response of a fungus to a chemical in an artificial system; however, this response can be measured under various conditions through manipulation and modification of the methods. Antifungal activity as produced by interactions beyond that of fungus and chemical must depend on in vivo tests.

A. Spore Germination Tests

The classic of the in vitro tests is the spore germination assay. This test goes back to Prevost and is probably the most extensively studied, highly standardized, and widely used of the fungicide test methods. The utility of the test was recognized in 1910 by Reddick and Wallace [2] and was developed and standardized by McCallan, Wilcoxon, Horsfall, and their coworkers [3-7]. Procedures for two variations of the test were published by the American Phytopathological Society [8-9].

Basically, the procedure makes use of glass slides on which drops of fungus spores are exposed to given concentrations of test chemical. After incubation in a moisture chamber for a time interval sufficient to allow for spore germination, counts of germinated or ungerminated spores are made with the aid of a microscope. The results, expressed as counts compared to untreated controls, are the measure of activity and can be expressed graphically as dosage-response curves [6, 10].

The flexibility of the test is such that its variations are nearly as numerous as the investigators who use it. As originally designed the test involved the deposit of a concentration of a toxicant on a glass slide and the addition of spore drops to the dried chemical. McCallan and Wilcoxon recommended a system whereby the chemicals and spores are mixed directly for deposit on the slide [10]. Investigators have advocated the use of a spray apparatus or a settling tower for deposit of the toxicant on glass slides [11, 12]. Suggestions for solving the difficulties associated with variations in drop size include the use of etched circles or raised cover slips on the slide, depression slides, or microbeakers [13-17]. Transparent materials such as plastic have been substituted for glass slides [18].

Since spores of many fungus species do not readily germinate in water drops, investigators have advocated several modifications, such as the use of spores seeded on disks of an agar-toxicant medium, seeded agar disks placed on deposits of chemical on slides, or spores on cellophane disks placed on filter paper disks of toxicant before transfer to slides [19-21]. A spore germination test utilizing dry spores on treated dried slides incubated in a moist atmosphere has been devised by Lukens and Horsfall [22].

Investigators have observed that a chemical may affect spore germination differently than it does mycelial growth or appressional formation and have designed tests to indicate spore and mycelial responses separately [22-25].

Modification of the spore-germination tests have been used to indicate the stability, volatility, persistence, and tenacity of antifungal compounds [26-29].

B. Agar Plate Tests

Agar-plate tests were born out of the need to detect antifungal activity against fungi whose characteristics of spore production, germination, or size do not lend themselves to spore germination assay. The tests measure the effect of a test chemical on fungus spores and/or mycelial growth on an agar medium. Basically, the procedure is to seed a given concentration of

spores or mycelial fragments on the surface of an agar plate. The test
chemical is contained in agar medium or is in intimate contact with the
agar. After incubation for a time sufficient to allow for measurable growth,
lack of growth or the extent of growth compared to untreated controls are
the measure of activity.

In 1927, Lee and Martin devised a method in which spores, after exposure to an aqueous concentration of test compound, are transferred to a
growth medium where activity is determined as colony counts compared to
an untreated control [30]. Palmiter and Keitt designed a test whereby fungus spores are seeded on the surface of an agar plate in which the test
chemical is incorporated [31]. Activity is determined by the number of
colonies produced. Other methods utilizing spores were published by
Mildner and co-workers, Thornberry, and Leben and Keitt [32-34]. These
methods involved placing concentrations of test chemical in holes in the
agar or on paper disks; the diameter of the zone of inhibition surrounding
the areas is the criterion of activity.

Procedures utilizing fungus mycelia have been devised. Carpenter
applied agar blocks of mycelial growth to the surface of agar in which test
chemical is incorporated [35]. Bomar, in a similar technique, introduced
hyphae growing on glass cover slips to treated agar [36].

Modified methods have been developed to measure the effect of chemicals in direct contact with the fungus and to overcome the restrictions
involved in testing materials which are insoluble in agar or which do not
diffuse through agar. Gattani recommended the use of pyridine purified
agar to minimize the chances of partial inactivation of chemicals which
react with agar [37]. Methods of treating cultures by spraying, dusting,
immersion, and exposure at a distance have been reported [38-41]. Forsberg introduced a method in which fungus-infested cotton thread is immersed in a toxicant and then transferred to the surface of an agar plate [42].
Sharvelle and Pelletier designed a method whereby paper disks are impregnated with toxicant, seeded with the test organism, and transferred to
the agar plate [43].

Further extensions of plate techniques have been developed to provide
greater sensitivity and detection of compounds applied to leaves and
seeds. Leben and Keitt suggested the use of chemically treated leaf disks
transferred to the surface of seeded agar [44]. This method was further
expanded by substituting treated seeds and leaf cuticle disks [45, 46].
Vaartaja developed a procedure to determine antifungal activity of compounds against seed-infecting pathogens; fungus cultures are overlaid with
an agar toxicant mixture, and the survival of seeds sown on the overlay is
the expression of activity [47]. In an effort to increase sensitivity of plate
tests, Edgington et al. designed a method in which bands of agar are used
to limit the agar diffusion of toxicant bidirectionally rather than through
360°. Treated disks of paper or leaf cuticle are placed on the surface of
the seeded agar bands, and the length of the zone of inhibition on the band
is the measure of activity [48].

C. Shake-Flask Tests

As an alternative to slide germination and agar plate assays, a shake-flask method has been developed [49]. Spores (fungal hyphae might also be used) are exposed to the test compound combined with a liquid nutrient contained in a shake flask. The resulting inhibition of germination or growth after a given time is the measure of activity.

IV. IN VIVO TESTS

The special virtue of in vivo testing is suitability. Because in vivo tests more nearly mirror the eventual use, results from in vivo tests can more nearly be expected to predict activity of the ultimate use. In vivo tests not only reflect the modes of action that come into play in the eventual use, but also reflect the factors which are destructive of activity. Furthermore, these tests can reveal positive and negative aspects of the practicality of use, e.g., a foliage disease assay measures fungicidal activity as interaction of the pathogen, host, and chemical, as well as measures possible phytotoxicity or favorable growth response.

In vivo tests are less economical and less efficient than in vitro tests. They require larger quantities of test compound, specialized application equipment, and preparation and holding facilities which are tied up for relatively long periods of time.

In vivo methods for detecting antifungals vary greatly, according to the type of end use. This review of in vivo methods will be limited to those for the evaluation of protectant, eradicant, and systemic agricultural chemicals; the methods for testing of compounds as preservatives for fabrics, leather, wood, adhesives, paints, etc., are excluded here. Those tests which are intermediate between the glass slide-plate techniques and the intact plant methods will be considered as in vivo tests since they involve functioning portions of plants.

A. Foliage Disease Assays for External
Protectants and Eradicants

Methods for protectant foliar fungicide evaluations make use of intact plants or detached plant parts which, after being sprayed or dusted with a test compound formulation, are inoculated with spores of a given pathogen and incubated under conditions conducive to disease development. An assessment of disease incidence is made and activity measured in terms of disease incidence compared to that in inoculated plants which have not been treated or treated with a standard fungicide. Assay for eradicant foliage disease activity is conducted in a similar way except that plants are inoculated at some time prior to application of test compound.

The introduction of synthesized organic fungicides gave impetus to the development of greenhouse foliar disease evaluations. Prior to that time, researchers such as Eyre and Salmon (procedure for evaluating eradicant activity against powdery mildew of hops) outlined methods as part of broader studies [50]. Hamilton in 1931 and Hamilton and Weaver in 1940 established methods for assay of compounds against apple scab and cedar-apple rust diseases [51, 52]. McCallan and Wellman in 1943 designed a method of evaluation of foliar fungicides by means of tomato foliage disease assays [53]. The methods of Hamilton, Weaver, McCallan, and Wellman served as models for numerous published and unpublished procedures.

When the objective of the in vivo portion of a screening program is to determine the spectrum of activity of test compounds, the most widely accepted approach is to select a limited number of test diseases which represent a major group of fungus diseases. When activity has been demonstrated in the group indicator, the spectrum is further investigated in members of the group. The choice of indicators depends on the limitations imposed by the causal organism and availability of host plants.

Among the more commonly used methods available for evaluation against leaf spot and blight diseases are: (1) McCallan and Wellman's [53] and Brook's [54] tomato foliage assays for early and late blight, Septoria and Botrytis; (2) the Atkins and Horn [55] cucumber anthracnose assay; (3) Crowdy and Wain's [56] broad bean Botrytis assay; (4) the methods of Hamilton et al. [57] for cherry leaf spot and peach brown rot; and (5) the rice-blast assay of Asakawa et al. [58].

As previously noted, Hamilton and co-workers devised tests for scab assays [51, 52]. Modification of these tests have been suggested [59].

Methods for assay against rust diseases utilize foliage of hosts such as apples, snapdragons, beans, wheat, oats, and rye [52, 60-64].

Assays for control of downy mildews using foliage of beans, grapes, and tobacco have been published [65-67].

Powdery mildew assays that make use of apples, beans, cucumbers, wheat, and barley have been recommended [57, 68-72].

In an effort to increase efficiency and economy of in vivo testing, researchers have designed tests which substitute detached leaves or portions of functioning plants for intact plants. In 1936, Marsh suggested a leaf method for assay of compounds for apple-scab disease [73]. In a comparison of the results with his method to those of a slide germination assay, he found that the three compounds tested demonstrated less fungicidal effect on the leaves. No difference in apple scab disease symptom was observed in comparing results from attached and detached leaves [74]. Among the diseases which have been utilized for detached-leaf techniques are tomato early and late blight, alfalfa black stem, and wheat rust [75-78]. Methods have been developed for the use of detached flowers, fruit, and etiolated hypocotyls [79-80].

Modifications of foliage disease assays have been employed to study the fate of fungicide deposits on leaf surfaces. At an interval between application of compound and inoculation, leaf surfaces are subjected to one or more weathering factors such as rain, temperature, wind, and light. The adaptations which have been published are designed to measure retention after exposure to rain. Foliage of intact plants or detached plants is subjected to water immersion or simulated rain; results such as disease incidence compared to foliage without water treatment is the estimate of tenacity [51, 81, 82].

B. Foliar Disease Assays for Systemic Activity

For testing antifungals as systemics against foliar diseases, the methods used for testing protectant and eradicant foliar fungicides are generally adaptable. After application of the compound to the foliage, a period of delay before inoculation with an appropriate pathogen allows for possible movement into and through tissues.

Methods have been designed which attempt to determine foliar disease control through root application. Crowdy and Wain devised a method in which the roots of broad bean plants are submerged in an aqueous formulation of compound for several days; activity is measured in terms of disease incidence after inoculation with Botrytis fabae [56]. In similar techniques, a test chemical is added to roots of plants established in various media. As indicators, these assays utilize early blight of tomato; powdery mildew, anthracnose, and scab of cucumbers; and rust of beans and wheat [83-86].

Methods are available which more specifically test for movement of compound from treated to nontreated parts of a leaf or foliage. El-Zayat et al. described a method whereby a compound is applied to one portion of foliage and disease severity in adjacent foliage is the measure of movement [87]. A method devised by Edgington and Schooley utilizes a compound applied as a band across a leaf surface; its movement is determined either by disease incidence in the untreated portion of the leaf or by bioassay of leaf tissue against an indicator organism [88].

C. Vascular Disease Assays for
Systemic Activity

Procedures for evaluation of systemic activity against vascular diseases make use of test chemical applied to root systems and subsequent inoculation. Diamond et al. designed a method utilizing soil drench of tomato roots followed by inoculation with the Fusarium wilt fungus [89]. Erwin et al. utilized cotton plants grown in soil amended with test compound followed by inoculation of stem tissues above the application level [90]. In a similar technique, leaf disks from plants placed in nutrient solution and test compound are bioassayed on agar plate seeded with an indicator fungus [91].

D. Translocation Assays

Laboratory tests have been devised to measure the movement of chemicals within a plant vascular system or through leaf surfaces and tissues. In 1954, Van Raalte designed a test to detect translocation through petiole sections [92]. Sections of potato petioles are placed upright on an agar surface seeded with an indicator fungus and a drop of test chemical is placed on the upper surface. A zone of inhibition of fungal growth on the agar is the criterion of translocation. In a similar technique devised by Reynolds the lower surface of carrot root sections is placed in a water formulation of test chemical and the upper surface is inoculated with a carrot root pathogen [93]. Modifications of these types of methods utilize woody stem sections and cotton bolls [94, 95].

Determination of transport through tissues after exposure of roots or cut surfaces to a test compound have also been made through chemical analysis and bioassay of extracts of plant parts distant from the application site [96, 97]. Use of radio-labeled compounds has also been suggested [98].

Solel and Edgington designed a method for detecting movement through isolated leaf cuticle [45]. Following placement of leaf cuticle on an agar plate seeded with an indicator organism, the test chemical is applied to the disk, the resulting zone of inhibition indicates transcuticular penetration.

V. SEED AND SOIL TREATMENT FUNGICIDE TESTS

Seed and soil treatments must perform under a wide and varied range of conditions that relate to the physical and chemical properties of soil, moisture, temperature, and interaction of microorganisms; consequently, procedures such as slide germination and agar plate tests are inadequate indicators of activity. They can serve only as tests preliminary to those which more nearly duplicate natural conditions. Also, because of the specificity of chemical compounds it is necessary to select methods which represent disease types.

Procedures which are specific for the detection of activity of seed and soil treatment compounds are a combination of in vitro and in vivo techniques. For the sake of organization they are not separated into these categories here.

A. Seed Treatment Tests

The general procedure for the evaluation of compounds, whether their activity is as external protectants, disinfectants, or as systemics, involves the uniform coverage of the surface of a variety of seed which is susceptible to one or more test pathogens. In the case of seed disinfectants, seeds which are naturally or artificially infested prior to chemical treatment are planted in soil or other media. In the case of seed protectants, healthy treated seeds are planted in soil or another medium which is naturally or artificially infested. The number of emerged healthy seedlings or plants, compared to nontreated and standard treatments, is the measure of activity.

The earliest methods for evaluation of seed treatments were concerned with disinfectant treatments for cereal smuts. In 1944, the American Phytopathological Society published greenhouse methods useful for evaluations against wheat bunt, loose and covered smut of oats, and loose smut of barley [99]. The disadvantage of such tests is the long period of time required for the development of discernible disease symptoms. A laboratory method that permits determinations within a short time was introduced by Muskett [100]. Treated oat seeds infected with Helminthosporium avenae, placed on moist paper, are irradiated to induce sporulation. The incidence of sporulation is the criterion of effectiveness. Methods have also been established for evaluations against such seedborne diseases as anthracnose of flax and cotton, and browning and stem break of flax (Polyspora lini) [101, 102].

Methods for the evaluation of seed protectants against damping-off diseases have been published as general descriptions [103, 104]. A standardized greenhouse method making use of pea seeds and naturally infested soil was developed by McCallan in 1948 [105]. This method is adaptable to the use of crops other than peas and soils artificially infested with specific pathogens.

In 1952, Hoppe first described the rolled-towel method of evaluating the efficacy of seed protectants under cool moist soil conditions [106]. This method, also known as the cold test, had been developed to determine the germination of corn seeds exposed to infested cold soils. In the rolled-towel method, treated seeds are placed on moist paper towels on which naturally infested soil is thinly spread. The towels are rolled and placed in incubators for a given time, at first at 10°C, and then at room temperature. Activity is in terms of germination of healthy seedlings compared to nontreated controls. Results from such tests were found to correlate with field tests [107]. Standardization of the cold-test method through the use of uniform inoculum has been suggested [108].

Among methods for evaluating specialized treatments is one suitable for testing seed treatments of water-seeded rice [109].

B. Soil Treatment Tests

The evaluations of chemical treatments for soilborne fungi are carried out through three approaches; agar plate procedures, soil-fungus systems, and soil-fungus-crop systems.

1. Agar Plate Procedures

Investigators generally agree that evaluations against soilborne pathogens
which employ spore germination or agar plate techniques correlate poorly
with greenhouse and field tests [110-112]. Evaluations against soilborne
pathogens of turf by means of plate tests are an exception to this poor correlation. They are an exception, in that, for effectiveness turf fungicides
need not enter and be subjected to the complexities of the soil system.
Klomparens and Vaughn reported good correlation between laboratory performance of fungicides tested on malt agar and field control of several turf
pathogens [113].

2. Soil-Fungus System Methods

Soil-fungus system methods measure the effect of the exposure of a test
fungus to a chemical in soil contained in a jar, cup, or a soil column
opened at one or both ends. After exposure, effect is determined by one
or more methods, such as counting the number of fungal colonies developed
from plating a sample of soil; determining the viability of the fungus;
measuring the zone of inhibition around soil samples placed on agar plates
seeded with an indicator fungus; and determining the viability of pathogens
in infested tissue.

In 1938, Ezekiel designed a soil-jar method whereby the test compound
is mixed or applied to the surface of soil infested with sclerotia [114]. The
viability of the sclerotia removed from the soil is the gauge of activity. In
1955, Zentmeyer provided a method in which inoculum, as a mycelial agar
disk, is placed in a vial of soil into which compound is added [111]. The
inoculum recovered after an incubation period is placed on an agar plate to
determine viability. Torgeson suggested a method for routine testing [115].
Artificially infested soil contained in a paper cup is drenched with test
compound; efficacy is estimated by incidence of mycelial growth on the soil
surface. Species of Fusarium, Pythium, Rhizoctonia, and Sclerotium are
suggested test fungi. Lingappa and Lockwood devised a method whereby a
spore suspension of test fungus is applied to the surface of treated com-
pacted soil. After incubation, a dye solution is placed on the soil surface
to kill and stain the spore, the spores are recovered on a film applied over
the soil surface, and the film examined for spore germination [116].
Thomas, in 1957, advocated the use of glass columns containing soil into
which inoculum disks are placed at various depths. The test compound is
added prior to filling the columns or as a drench after placement of the
inoculum. Toxicity is measured in terms of viability of the inoculum after
transfer to an agar medium [117].

Latham and Linn developed a method using segmented columns rather
than containers or continuous tubes [112]. After filling the joined segments
of the column with soil, the sections are separated and mycelial agar disks
(up to five pathogens can be tested together) are placed into the separations.

Chemicals are applied to the joined segments as drenches. The disks retrieved after incubation are transferred to an agar medium to determine viability. Since the disks are from various levels in the column, the test is also an indicator of chemical movement.

Several techniques have been suggested for the placement and maniupulation of inoculum; mycelium and/or spores can be added to soil as part of agar cylinders; as disks incorporated on paper, nylon, or membrane filters; enclosed in nylon envelopes; and as infested pieces of tissue [112, 117-120]. A technique has been described for infesting nonsterile soil with fungal cells without adding cultural substrate [121]. The technique was developed specifically for species of Fusarium.

Host plant tissues placed in infested treated soil have been used as fungus traps for determining a chemical's toxicity to soilborne Pythium, Rhizoctonia, and Sclerotium [122, 123].

Several tests, other than those already mentioned, can be utilized to measure a chemical's movement as well as its activity. Soil samples, retrieved by various means from given depths in a column, are assayed by placing samples of the soil on agar plates seeded with an indicator fungus [124, 125].

A number of procedures are used to determine the fungitoxicity of volatile compounds. These measure activity as the response of a fungus placed at a distance from treated soil [126-128].

3. Soil-Fungus-Crop Methods

Procedures for evaluation of soil fungicide treatments in a soil-fungus-crop system make use of a test chemical applied to infested soil planted to a susceptible host crop. The assessment of disease control is in terms of healthy plants compared to a standard fungicide treatment or to no treatment.

In 1953, Arndt utilized a mixture of test chemical and sand contained in aluminum dishes [129]. The sand is planted to cotton seed in a circle and the inoculum as Rhizoctonia solani is placed in the center. The number of healthy seedlings determines activity. A similar procedure utilizing treated media in dishes, flasks, and pots has been developed [130].

In 1960, Reinhart established a paper-cup method in which treated soil infested with Fusarium oxysporum is planted with cucumber seeds [131]. Control is expressed as a percentage calculated from the number of diseased plants in the treated and nontreated soil. Brinkerhoff and co-workers and Sinclair described methods utilizing flats of Rhizoctonia infested soil treated with a test chemical. Survival of healthy seedling from planted cotton seeds estimates activity [132, 133]. Ranney and Bird, in studies on the relationship of temperature to compound effectiveness, applied soil treatments directly into furrows planted to seeds [134].

A number of techniques for infestation of soil have been suggested. These include a method for infesting without disturbing the roots of established plants and a method for following the movement of pathogens from infested to noninfested soil [135, 136].

Greenhouse tests similar to those outlined have been developed for evaluation against the following soilborne diseases: Pythium damping-off of peas, Phytophthora root rot of avocado, Fusarium root rot of beans, and clubroot of cabbage [111, 137-139].

VI. FROM PROMISE TO PRACTICAL USE

Whether by means of a simple laboratory method or a complex greenhouse procedure, the promise of antifungal activity originates with the first positive test results. Prospects brighten with confirmatory assays, the establishment of a spectrum of activity and efficacy demonstrated by field trials. However, at some point, during or after the evaluation process, prospects may be dashed due to factors unrelated to, but as unpredictable as, activity itself. Despite proved effectiveness in field tests conducted under a wide and varied range of conditions, an insurmountable production problem or a toxicological hazard might destroy all hopes. For in the final analysis, it is only in the practical application of a compound that the promise of antifungal activity is realized.

REFERENCES

1. G. W. Keitt (transl.), Prévost, Memoir on the Immediate Cause of Bunt or Smut of Wheat, Phytopathological Classic No. 6, Amer. Phytopathological Soc., St. Paul, Minn., 1956.
2. D. Reddick and E. Wallace, Science, 31, 798 (1910).
3. S. E. A. McCallan, N.Y. State Agr. Expt. Sta. Mem., 128, 8 (1930).
4. S. E. A. McCallan, R. H. Wellman, and F. Wilcoxon, Contrib. Boyce Thompson Inst., 12, 49 (1941).
5. S. E. A. McCallan and F. Wilcoxon, Contrib. Boyce Thompson Inst., 11, 5 (1939).
6. F. Wilcoxon and S. E. A. McCallan, Contrib. Boyce Thompson Inst., 10, 329 (1939).
7. J. G. Horsfall, E. G. Sharvelle, and J. M. Hamilton, Phytopathology, 30, 545 (1940).
8. American Phytopathological Society, Committee on Standardization of Fungicidal Tests, Phytopathology, 33, 627 (1943).
9. American Phytopathological Society, Committee on Standardization of Fungicidal Tests, Phytopathology, 37, 354 (1947).
10. S. E. A. McCallan and F. Wilcoxon, Contrib. Boyce Thompson Inst., 9, 249 (1938).

11. S. E. A. McCallan and F. Wilcoxon, Contrib. Boyce Thompson Inst., 11, 309 (1940).
12. J. G. Horsfall, Chronica Botanica, 6, 292 (1941).
13. J. E. Young, Illinois Acad. Sci. Trans., 37, 59 (1944).
14. P. D. Peterson, Phytopathology, 31, 1108 (1941).
15. R. W. Barratt and J. G. Horsfall, Conn. Agr. Expt. Sta. Bull., 508 (1947).
16. W. B. Shafer, Phytopathology, 42, 519 (1952).
17. D. M. Spencer, Plant Pathology, 11, 41 (1962).
18. A. H. McIntosh, Ann. Appl. Biol., 49, 424 (1961).
19. A. J. Heyns, G. A. Carter, K. Rothwell, and R. L. Wain, Ann. Appl. Biol., 56, 399 (1965).
20. R. W. Pero and R. G. Owens, Phytopathology, 61, 132 (1971).
21. D. Neely and E. B. Himelick, Phytopathology, 56, 203 (1966).
22. R. J. Lukens and J. G. Horsfall, Phytopathology, 61, 130 (1971).
23. J. G. Horsfall and S. Rich, Phytopathology, 53, 476 (1963).
24. S. Rich, Phytopathology, 46, 24 (1956).
25. J. G. Horsfall, Principles of Fungicidal Action, Chronica Botanica, Waltham, Mass., 1956.
26. R. W. Barratt, Phytopathology, 36, 679 (1946).
27. G. Serra, Phytiatr.-Phytophar., 13, 107 (1964).
28. J. W. Heuberger, Phytopathology, 30, 840 (1940).
29. R. A. Chapman, J. G. Horsfall, and H. L. Keil, Phytopathology, 40, 4 (1950).
30. H. A. Lee and J. P. Martin, Phytopathology, 17, 315 (1927).
31. D. H. Palmiter and G. W. Keitt, J. Agr. Res., 55, 439 (1937).
32. P. Mildner, B. Mihanovic, M. Jusic, M. Hajsig, and V. Kuzmanovic, Antonie van Leeuwenhoek J. Microbiol. Serology, 29, 421 (1963).
33. H. H. Thornberry, Phytopathology, 40, 419 (1950).
34. C. Leben and G. W. Keitt, Phytopathology, 40, 950 (1950).
35. J. B. Carpenter, Phytopathology, 32, 845 (1942).
36. M. Bomar, Folia Microbiologia, 7, 185 (1962).
37. M. L. Gattani, Phytopathology, 44, 113 (1954).
38. B. W. Henry and E. C. Wagner, Phytopathology, 30, 1047 (1940).
39. C. Moreau and M. Moreau, Rev. Mycol., 24, 59 (1959).
40. A. J. Latham and M. B. Linn, Plant Disease Reptr., 49, 398 (1965).
41. P. H. Eastburg, B. L. McCaskey, and W. D. Thomas, Phytopathology, 47, 519 (1957).
42. J. L. Forsberg, Phytopathology, 39, 172 (1949).
43. E. G. Sharvelle and E. N. Pelletier, Phytopathology, 46, 36 (1956).
44. C. Leben and G. W. Keitt, Phytopathology, 39, 529 (1949).
45. Z. Solel and L. V. Edgington, Phytopathology, 62, 500 (1972).
46. J. L. Lockwood, C. Leben, and G. W. Keitt, Phytopathology, 42, 447 (1952).
47. O. Vaartaja, Phytopathology, 46, 387 (1956).

48. L. V. Edgington, H. Buchenauer, and F. Grossmann, Pestic. Sci., 4, 747 (1973).
49, R. T. Darby, Appl. Microbiol., 8, 146 (1960).
50. J. V. Eyre and E. S. Salmon, J. Agr. Sci., 7, 473 (1915).
51. J. M. Hamilton, Phytopathology, 21, 465 (1931).
52. J. M. Hamilton and L. O. Weaver, Phytopathology, 30, 7 (1940).
53. S. E. A. McCallan and R. H. Wellman, Contrib. Boyce Thompson Inst., 13, 93 (1943).
54. P. J. Brook, New Zealand J. Sci. Technol., A 38, 506 (1957).
55. J. G. Atkins, Jr. and N. L. Horn, Plant Disease Reptr., 37, 397 (1953).
56. S. H. Crowdy and R. L. Wain, Ann. Appl. Biol., 38, 318 (1951).
57. J. M. Hamilton, M. Szkolnik, and J. R. Nevill, Plant Disease Reptr., 48, 295 (1964).
58. M. Asakawa, T. Misato, and K. Fukunga, Noyaku Seisan Gijutsu, 6, 7 (1962).
59. A. E. Rich and M. C. Richards, Plant Disease Reptr., 43, 540 (1959).
60. S. E. A. McCallan, Contrib. Boyce Thompson Inst., 13, 367 (1944).
61. S. E. A. McCallan and R. C. Zingerman, Contrib. Boyce Thompson Inst., 21, 473 (1962).
62. J. E. Livingston, Phytopathology, 43, 496 (1953).
63. F. R. Forsyth and B. Peturson, Phytopathology, 49, 1 (1959).
64. H. L. Keil, H. P. Froshlich, and J. O. Van Hook, Phytopathology, 48, 652 (1958).
65. T. C. Pridham, L. A. Lindenfelser, O. L. Shotwell, F. N. Stodola, R. G. Benedict, O. Foley, R. W. Jackson, W. J. Zaumeyer, W. H. Preston, Jr., and J. W. Mitchell, Phytopathology, 46, 568 (1956).
66. J. Lafon, Rev. Viticult., 42, 174 (1946).
67. J. A. Pinckard, R. McLean, F. R. Darkis, P. M. Gross, and F. A. Wolf, Phytopathology, 30, 485 (1940).
68. A. H. M. Kirby and E. L. Frick, Ann. Appl. Biol., 51, 51 (1963).
69. D. C. Torgeson and C. G. Lindberg, Contrib. Boyce Thompson Inst., 21, 33 (1961).
70. C. E. Yarwood, Proc. 2nd Intern. Congr. Crop Protect., London, 1949, p. 500.
71. R. J. Smith and W. H. Read, Ann. Appl. Biol., 49, 233 (1961).
72. R. L. Wain, W. Sobotka, and D. M. Spencer, Ann. Appl. Biol., 51, 445 (1963).
73. R. W. Marsh, Trans. Biol. Mycol. Soc., 20, 304 (1936).
74. R. L. Nicholson, S. Van Scoyoc, J. Kuć, and E. B. Williams, Phytopathology, 63, 649 (1973).
75. P. H. Tsao, C. Leben, and G. W. Keitt, Phytopathology, 46, 26 (1956).
76. A. H. McIntosh and D. E. Eveling, Ann. Appl. Biol., 55, 397 (1965).
77. C. H. Ward, Phytopathology, 49, 229 (1949).

78. M. M. Payak, Experienta, 11, 9 (1955).
79. K. G. Rohrbach, J. E. Hunter, and R. K. Kunimoto, Plant Disease Reptr., 54, 694 (1970).
80. R. L. Nicholson, S. Van Scoyoc, E. B. Williams, and J. Kuć, Phytopathology, 63, 363 (1973).
81. P. H. Schuldt, Phytopathology, 44, 505 (1954).
82. H. P. Burchfield and A. Goenaga, Contrib. Boyce Thompson Inst., 19, 133 (1957).
83. J. Stubbs, Ann. Appl. Biol., 39, 439 (1952).
84. J. Dekker, Tijdschev. Plantenziekten, 67, 25 (1961).
85. O. M. van Andel, Nature, 194, 790 (1962).
86. T. C. Allen, Jr. and A. H. Freiberg, Phytopathology, 54, 580 (1964).
87. M. M. El-Zayat, R. J. Lukens, A. E. Dimond, and J. G. Horsfall, Phytopathology, 58, 434 (1968).
88. L. V. Edgington and J. Schooley, Phytopathology, 61, 1022 (1971).
89. A. E. Dimond, D. Davis, R. A. Chapman, and E. M. Stoddard, Conn. Agr. Expt. Sta. Bull., 557 (1962).
90. D. C. Erwin, J. J. Sims, and J. Partridge, Phytopathology, 58, 860 (1968).
91. R. G. Davis and J. A. Pinckard, Phytopathology, 59, 112 (1969).
92. M. Van Raalte, 3rd Intern. Congr. Phytopharmacie, 3, 76 (1954).
93. J. E. Reynolds, Plant Disease Reptr., 54, 223 (1970).
94. E. B. Himelick and D. Neely, Plant Disease Reptr., 49, 949 (1965).
95. H. S. Bagga, Phytopathology, 58, 1041 (1968).
96. L. V. Edgington and A. E. Dimond, Phytopathology, 54, 1193 (1964).
97. I. E. M. Darrag and J. B. Sinclair, Phytopathology, 59, 1102 (1969).
98. M. Snel and L. V. Edgington, Phytopathology, 58, 1068 (1968).
99. American Phytopathological Society, Committee on Standardization of Fungicidal Tests, Phytopathology, 34, 401 (1944).
100. A. E. Muskett, Ann. Botany (London), 2, 699 (1938).
101. A. E. Muskett and J. Colhoun, Ann. Botany (London), 6, 219 (1942).
102. C. H. Arndt, Phytopathology, 38, 978 (1948).
103. K. J. Kadow and H. W. Anderson, Illinois Univ. Agr. Expt. Sta. Bull., 439, 289 (1937).
104. C. F. Taylor and J. A. Rupert, Phytopathology, 36, 726 (1946).
105. S. E. A. McCallan, Contrib. Boyce Thompson Inst., 15, 91 (1948).
106. P. E. Hoppe, Univ. Wisc. Agr. Expt. Sta. Bull., 507 (1955).
107. P. E. Hoppe, Plant Disease Reptr., 42, 367 (1958).
108. M. V. Desai and C. S. Reddy, Phytopathology, 48, 386 (1958).
109. H. A. Lamey, Phytopathology, 55, 500 (1965).
110. J. B. Kendrick, Jr. and J. T. Middleton, Plant Disease Reptr., 38, 350 (1954).
111. G. A. Zentmyer, Phytopathology, 45, 398 (1955).
112. A. J. Latham and M. B. Linn, Phytopathology, 58, 460 (1968).

113. W. Klomparens and J. R. Vaughn, Mich. Agr. Expt. Sta. Quart. Bull., 34, 425 (1952).
114. W. N. Ezekiel, J. Agr. Res., 56, 553 (1938).
115. D. C. Torgeson, Proc. 6th Pacific Coast Res. Conf. on Soil Fungi, 1950 (C. E. Horner, ed.), p. 24.
116. B. T. Lingappa and J. L. Lockwood, Phytopathology, 53, 529 (1963).
117. W. D. Thomas, Jr., Phytopathology, 47, 535 (1957).
118. O. N. Nesheim and M. B. Linn, Phytopathology, 60, 395 (1970).
119. P. B. Adams, Phytopathology, 57, 602 (1967).
120. S. H. F. Chinn and R. J. Ledingham, Phytopathology, 52, 1041 (1962).
121. M. E. Corden and R. A. Young, Phytopathology, 52, 503 (1962).
122. P. L. Thayer and C. Weklberg, Phytopathology, 53, 391 (1963).
123. A. Zehara and P. Shacked, Phytopathology, 58, 410 (1968).
124. D. E. Munnecke, Phytopathology, 44, 499 (1954).
125. C. L. Maurer, R. Baker, D. J. Phillips, and L. Danielson, Phytopathology, 52, 957 (1962).
126. J. Oserkowsky, Phytopathology, 24, 815 (1934).
127. P. M. Miller and E. M. Stoddard, Phytopathology, 47, 25 (1957).
128. L. T. Richardson and D. E. Munnecke, Phytopathology, 54, 836 (1964).
129. C. H. Arndt, Plant Disease Reptr., 37, 397 (1953).
130. I. E. A. Darray and J. B. Sinclair, Plant Disease Reptr., 52, 399 (1968).
131. J. H. Reinhart, Plant Disease Reptr., 44, 648 (1960).
132. L. A. Brinkerhoff, B. B. Brodie, and R. A. Kortsen, Plant Disease Reptr., 38, 476 (1954).
133. J. B. Sinclair, Plant Disease Reptr., 41, 1045 (1957).
134. C. D. Ranney and L. S. Bird, Plant Disease Reptr., 40, 1032 (1956).
135. J. E. Halpin and E. W. Hanson, Phytopathology, 48, 481 (1958).
136. H. M. Elsaid and J. B. Sinclair, Plant Disease Reptr., 46, 852 (1962).
137. D. C. Torgeson, W. H. Hensley, and J. A. Lambreck, Contrib. Boyce Thompson Inst., 22, 67 (1963).
138. A. D. Davison and J. R. Vaughn, Plant Disease Reptr., 41, 432 (1957).
139. J. Colhoun, Ann. Appl. Biol., 41, 290 (1954).

Chapter 4

DEVELOPMENT OF CHEMICALS
FOR PLANT DISEASE CONTROL

Charles J. Delp

Biochemicals Department
Research Division
E. I. du Pont de Nemours & Co., Inc.
Wilmington, Delaware

I. DEVELOPMENT TODAY: A CURRENT EXAMPLE

The long, expensive route from discovery to commercial use of a new disease-control agent is filled with risks, detours, and occasional rewards; and although the way appears well charted, it is never the same for a new chemical. A typical development in the spring of 1977 could happen like this:

The institution sponsoring this development is an international chemical company with diversified activities. The company has a commitment to research and a dependence on new products and innovations to keep older products up-to-date and useful. They maintain a staff of technically trained field representatives familiar with orchards, paddies, fields, and research institutions around the world, and who can discuss problems of crop production with growers and agricultural scientists with equal facility. Theirs is a cooperative effort of many technical disciplines. Team members are dependent on the expertise, jargon, and language of other members, and all are joined by a common goal, namely, their dedication to the development of agricultural chemicals.

Organic chemists start the process by making thousands of new compounds for testing. Most are inactive, but a few show activity in controlling disease. Additional tests reveal unique properties of a specific structure, and interest grows. This chemical, coded CA-X (Control Agent-X), is a highly active systemic compound which alters the host-pathogen relationship enough to reduce disease damage. Most of the related compounds have some disadvantage, but CA-X and a few analogs are reasonably easy to make and preliminary toxicity observations suggest no hazard. The outstanding disease control demonstrated at company field stations is enough for a decision to initiate a complex development schedule where chemical, biological, and marketing studies are coordinated.

Analytical methods for this new structure are worked out so that fractional ppm quantities of residue can be detected on treated crops and soil and in animals. Breakdown or metabolic products are identified by ^{14}C-tracer techniques and synthesized for testing while extensive toxicological studies are started on CA-X and possibly some of its metabolites. The synthetic process and formulations are improved and evaluated. Economics are reviewed again. Then the management decides that CA-X can be sold if biological results continue to be favorable through the next few seasons of testing by company and government investigators.

Environmental studies go in many directions but will at least include the fate and bioinfluence in soil, water, and subsequent crops. For the next few years, results of CA-X treatments and samples for residue analysis will pour in from around the world. The decision whether to continue or discontinue development is based on many factors including:

1. Changing market potentials, including competitive products.
2. Rising cost of plant construction, raw materials, and energy.
3. Complexities of waste disposal.
4. Results from toxicology or environmental studies.
5. Changing government requirements for registration.

The results from disease-control, toxicity, residue, and environmental studies are submitted to government agencies corresponding to the U.S. Environmental Protection Agency (EPA) in countries where CA-X will be marketed. In many cases, some of the data must be generated in the country where CA-X will be registered for sale. If these data are favorable, researchers will ask for an experimental permit to allow expanded field testing and establish interim manufacturing facilities to produce CA-X for this program. New formulations will be evaluated for performance. Meanwhile, regulations of transportation agencies may restrict the shipping of samples to investigators.

Government officials review the data. The EPA usually poses questions which can mostly be answered only by more research and testing. This takes additional months, and sometimes the agency raises more questions which must be answered to comply with registration requirements. No product can be sold until there is official action to authorize registration and approve the product label.

The market potential must be substantial because this is an expensive task and the return of a profit is risky at best. The big risk, of course, is that some problem may develop which will preclude sale of CA-X. As much as $5-15 million and 5-10 years of company time, and thousands of research hours by agricultural investigators are at stake.

Marketing arrangements have been made with agricultural distributors, and their field personnel have been trained in the proper use of CA-X. As registration is authorized in various countries, the product is packaged in containers labeled with use directions and precautions in the native language. Growers expect to protect at least a test portion of the season's crop with the new systemic agent, and its sale must be accompanied by technical service to assure proper use.

This process is continually changing and differs from those presented by Wellman in 1959 [1] and 1967 [2]. More recent publications include a U.N. booklet [3], articles by Johnson (Dow Chemical Co.) [4] and Lever (ICI) [5], a publication by Eli Lilly [6], the 1973 Ernst and Ernst profile of 36 companies [7], recommendations in 1974 by Djerassi [8], and a review by Sbraga in 1975 [9]. The Agricultural Chemical Development Chart shown in Table 1 is a summary of the process today.

TABLE 1

Agricultural Chemical Development Chart

Years ———— 1st ————————————— 2nd or 3rd ——————

C H E M I C A L	Synthesis				
	~10,000 new compounds	Relatives of active structures	Field prep. 100–1000 g stability tests	Formulation and process studies	Residue analysis methods
M A R K E T I N G	General market analysis	Feedback to research	Patent applications in key nations	Patent support showings	Guidance for bio and residue data
B I O L O G I C A L	Primary screening (~500 active compounds)	Secondary evaluation (~5 effective compounds)	Mode of action studies	Toxicity LD_{50}, eye sensitization, subacute	Field evaluation (2 effective compounds)

Cost ($ million)   1/2

——— 4th ———————————————————			5th to 10th ———————		
Semiworks production 10–1000 kg	Formulation and process improvements	Residue analysis (metabolites)	Plant construction or remodel	Plant production	**P R O D U C T**
Government experimental approvals	Economic analysis decision to sell	Data for government regulatory agencies	Product label marketing plan	Education of distributors and users	**M A R K E T**
Toxicity metabolism, environmental fate studies	Field evaluation governments (1 compound)	Expanded field evaluation	Quality control tests	Research to support marketing	**G R O W E R U S E**
1	2	4	6 to 15 + cost of plant (5 to 50)		

II. CHEMICAL CONSIDERATIONS

A. Synthesis

New chemical disease control agents are discovered by design, inspiration, hard and persistent work and even by serendipity. Chemistry is a fascinating science of endless possibilities for discovering structures with new characteristics and potential use. The literature is full of structure-activity studies that give ideas to new synthesis; consequently some researchers have concluded that synthesis by design, from the facts generated in basic biochemical studies, is the best route to new discoveries. Perhaps some day that will be true; but today results are coming mostly from inspiration, persistence, and "good fortune."

Inspiration is still a nebulous factor associated as much with our intuitive nature as with our knowledge. Striking examples include the simultaneous, independent discoveries not explainable by access to the same information, and the unusual number of patent interferences and publication duplications. The empirical or "Edisonian" approach, modified to include up-to-date design information, is most effective in an analoging program where close relatives are synthesized to optimize desired qualities. In some cases, more than one analog may be needed to satisfy different biological needs. In most cases, the synthetic programs still contain surprises.

B. Formulation

A new chemical (active ingredient) is mixed with inert substances to improve stability, ease of handling, and performance because most compounds cannot be used effectively without formulation. The inherent activity can be influenced by the nature of the formulation, but each chemical presents a unique challenge and even close analogs offer individual responses to preparation techniques. Somers [10], in his presentation of the art and science or "cookery" of formulation, concluded that a clear definition of biological function is needed before more effective formulation is possible. However, when this information is not available, the development of formulations having an optimum balance of desirable characteristics is almost completely a matter of trial and error.

A detailed Pesticide Formulation text, edited by Van Valkenburg in 1973 [11], has a most appropriate preface, where he says:

> Once a candidate pesticide is mixed with anything (a slurry of the compound in water qualifies) the composition may be considered a pesticidal formulation. Hence, any research on a combination of an active compound and a second material is research on a pesticidal formulation. . . .

A few . . . principles include (1) the importance of solubility relationships . . . in selecting an emulsifier, (2) the effect of the phase inversion temperature on emulsion stability, (3) the dehydrating effect of fertilizer salts on emulsifiers and the resultant stability or instability of the emulsion, (4) the catalytic effect of clays and other carriers on the degradation of a pesticide, (5) the all-important flowability property of a mixture as it relates to the successful manufacture of a dry pesticidal formulation, and (6) the physical properties of formulations and how they affect the size of the particles in a spray and the movement of the particles in air.

The physical properties of a formulation and its resultant spray solution [or suspension] as indicated by particle size and surface properties (surface tension, contact angles, spreading coefficient) will dictate whether or not a spray droplet will adhere to a plant surface. Once a pesticide is on a target site it is faced with a lipid barrier which must be penetrated to affect biological activity. An understanding of the physical and chemical properties of this barrier can aid the formulator in designing his composition for optimum penetration and efficacy. And finally, when the formulation finds its way to the soil, the adsorption characteristics of the ingredients of the composition on soil colloids will dictate how far the pesticide will move with possible concomitant danger of contamination of soils and ground waters far removed from the site of application.

It is obvious that one should not routinely add inerts to pesticides just to get the right handling properties. One must consider the effect of the additives on the biological activity . . . on processing, and on the movement on or in the host and its environment. Through a better understanding of all the principles involved, one hopes that the formulator will be able to optimize his formulations in terms of efficacy and reduction of deleterious side effects.

Computerization is replacing some of the art of formulation [12] because the multitude of interdependent factors can be optimized more rapidly by computer than by trial and error. Cost, availability, and government clearance of the inerts—all limiting factors—are "computed" along with performance effects. Of course, cost, safety, and convenience for the user are considered in every formulation.

C. Analytical and Residue

The methods of analysis for each compound and its degradation products are developed so that fractional ppm quantities can be detected as residue in the crop and environment. This requires a variety of modern chemical techniques and, frequently, the development of entirely new methods.

More than one method may be needed to meet the requirements of analytical laboratories throughout the world. Actually, in many cases the methods are sensitive enough to detect small, insignificant amounts of chemical which create no hazard. Biological considerations are discussed later and residue analysis is reviewed in Chapter 14.

Residue information along with toxicity, determined by long-term feeding studies, is used to establish safe residue tolerances for use of the product. This information is also used for environmental studies. Samples of produce, plant parts, animal tissues, eggs, milk, feces, urine, soil, and water near treatment areas are collected and analyzed. Treatments at several times the use-rates are also made to obtain representative measurements of possible "misuse" situations.

D. Decomposition

It is not enough to know that the active ingredient is safe, present in low amounts, and disappears within a reasonable time. Decomposition products, including metabolites in living systems, must be identified and shown to be nonhazardous. Metabolites in plants may differ from those in animals; in some cases, they may be conjugated with or incorporated into natural components. The use of radiolabeled chemicals is invaluable in these studies. While some compounds decompose rapidly to natural usable components of the environment such as CO_2, others may form complex and hard-to-identify metabolites. Studies of these, sometimes elusive, metabolites may require the most sophisticated techniques of separation, isolation, and identification and several years of research effort.

E. Manufacture

It is a long, hard road from the initial laboratory synthesis to the development of a manufacturing process and construction of a plant for producing commercial quantities of a product. Alternate chemical routes for manufacture are explored, availability of raw materials determined, preliminary designs prepared, and economics of manufacture evaluated. Conditions of each operation are optimized for highest yield. These studies can take months, even years, of team effort before there is firm assurance that a new product can be economically manufactured. During this time, hundreds of pounds of the chemical must be made and formulated for broad field testing.

Management's decision regarding the size of the plant is critical because an oversized plant can make a product unprofitable, with excessive investment costs and overhead. Plants may have to be built in several countries for proximity to potential markets. Some countries require manufacture within their boundaries for sale authorization or maintenance of patent protection.

Little or no waste material can be discharged into natural waters from a plant. Typical waste-treatment processes include biodegradation, carbon absorption followed by thermal regeneration of the carbon, incineration, and chemical oxidation. Disposal and treatment of wastes can account for as much as a fourth of the design and plant-construction costs.

The safety of plant workers requires attention to mechanical, audio, thermal, and chemical hazards. The toxicology of intermediate chemicals is studied to help define the facilities required to ensure safe handling procedures. The analytical chemist must now come back into the picture with tools to monitor production for quality control.

Although production costs and schedules have always been critical, today's economy has added new dimensions where raw material costs and availability may change drastically with little notice. The cost-competitive system of a few years ago is modified with allocations of materials in short supply.

III. MARKETING CONSIDERATIONS

A. Market Analysis

Obviously, a market must be large enough to recover the cost of development, to ensure production and sale of a product, and to provide a reasonable return on plant investment. It is important, therefore, to study the economic significance of a disease and to guide screening and even synthetic programs. Diseases which threaten complete loss of high-value, large-acreage crops justify the development of superior products.

The estimated value of the loss per hectare due to a disease indicates the potential value-in-use of a disease-control treatment. The potential market is estimated by multiplying the value-in-use by the number of hectares with the disease problem, but is modified by competitive products, convenience of use, and grower practice. Some growers readily buy agents to insure maximum yields of highest quality; others limit their expenditures to a third or fourth of the likely loss; and some use chemicals only under the most adverse conditions. Of course, with broad-spectrum agents, several specific markets are combined for a larger potential.

Markets below $3 million annual sales at manufacturer's level seldom justify the search for a specific agent [13]. Some companies consider $10 or 20 million the minimum market [4]. In a description of the rapid growth of Japan's agricultural chemicals industry, Sugimoto [14] points out that it is now necessary to develop markets larger than that for rice-blast control because of the increasing cost of toxicity tests. Devastating diseases in local situations or on specialty crops can be of such a great concern that growers and pathologists try to find ways to subsidize registration costs of an agent known to provide economic protection. A U.S. government program (IR-4) is working toward this objective.

B. Patents

Industries seldom find it worth the effort to discover and develop a new compound without patent protection. A U.S. patent is a printed document in which a novel, useful, and unobvious invention is fully disclosed and its utility described. It grants the first inventor the right to 17 years of exclusive manufacture, use, and sale of new chemicals. New uses of old chemicals can be patented [15]. Congress was authorized in the U.S. Constitution (Article 1, Sec. 8) to establish a patent system ". . . to promote the progress of science and useful arts, by securing for limited times to authors and inventors the exclusive right to their . . . discoveries . . ." [16].

A new invention, disclosed in a patent application, is reviewed by examiners in the U.S. Patent and Trademark Office to be sure it is patentable. Some applications are rejected, others must be modified to remove claims which the examiner considers unpatentable. Fees to file and maintain a patent are modest in the United States, but in other countries can be as high as $14,000 (in East Germany) not including legal costs. Patent protection in 30 or more countries may cost $100,000 plus patent agent and translation fees. In some countries the patent is granted without examination and there is no assurance that it is valid until the published patent is contested or challenged and found valid in court. Because of the thorough examination procedure, U.S. patents are presumed valid. But if a case is contested in a federal court, the costs can exceed half a million dollars plus a vast amount of lost research and managerial time.

Patents granted to employees of government or public institutions may be licensed by others, but some agencies will not grant exclusive licenses. Nonexclusive licenses may not offer sufficient protection to justify development [17]. Since patents frequently issue before a product is registered, the additional time required for registration reduces time of patent protection. The temporary exclusion of competitors removes only a part of the risk of the development of an invention, but without this head start few new ventures could be profitable in a free society where a return on investment must be realized.

C. Decision to Sell

The decision to develop and, finally, to sell a new compound is seldom simple and obvious, but depends on the probability of profitable marketing. Market analysis is nebulous enough without the additional question of what portion of the market the new agent will "capture." This, of course, depends on price, competitive products, marketing strategy, and the effectiveness of the control agent.

The selling price of a product depends on the cost of research and development, manufacture, formulation, packaging, transportation, and on its value in the marketplace. Toxicology and hazard evaluations are of utmost importance in the decision to manufacture a product, as company managers are sensitive to their responsibility for the safety of their employees and the public. Implications of unacceptable toxic hazards are sufficient to stop development or sale.

A product may provoke unrealistic initial interest because it is new and different. But since grower use will be sustained only if there is some real advantage, accurate research data must determine the true efficacy of use. These data could result in a decision to drop a product. Any delay in that decision will be costly [4, 6, 18].

D. Registration

The effectiveness and possible hazards (benefit/risk) of a new agricultural chemical are reviewed by the regulatory agency of most countries—EPA in the United States—before it is registered for sale. When a company is satisfied with the performance and safety of its new product, information on biological efficacy, toxicity, residue levels, and possible impact on the environment is submitted to the government agency. Compliance with national regulations permits the registration of a product for sale and a specific use. Each new use for the product must be substantiated by effectiveness data and sufficient residue information to establish a residue tolerance and to define the methods of treatment. For example, registration of an agent for use on almonds does not authorize use on apples, or bananas, etc.

Registration authorizes use of an approved label affixed to every package of the product containing trade, common, and chemical names; percent composition of active and inert ingredients; and directions for use and cautions. Supplemental registrations and modifications to a label are applied for as new information becomes available. Most countries have similar requirements and labels are prepared in the languages of those countries. In the United States it is illegal to use a registered agricultural chemical in a manner inconsistent with the labeling instructions. The label is an important publication, for it not only directs the user but also serves as a legal document.

Registration is not a one-time effort. Old registrations are periodically reviewed and must be reissued in light of new data, changing use, and new toxicology regulations.

The EPA regulations and their administration have been widely discussed recently, because the agency has assumed tremendous responsibility and authority in making and enforcing regulations and then acted as judge on questions of fairness [19]. MacDougall [20] points to the problems of poor definition of requirement, serial-type requests for additional data, lack of consistency among reviewers, and excessive delays. The EPA is now in the process of reviewing over 30,000 pesticide products to determine which will be available for general use, which will be placed in a restricted-use category, and which will be taken off the market. This is a big task and the results will have a tremendous impact on U.S. agriculture.

E. Marketing

A potential customer must know about a product, be able to purchase it with ease, and have technical service available for effective use. The introduction of a new product proceeds best when each grower first uses it in a limited trial to prove its value in his situation. Growers are encouraged to convince themselves of the cost/benefit and they must be informed about the product and the correct method of use.

Advertisements in mass media, direct mail, and personal contacts are frequently used to inform prospective users about a new product, as well as detailed technical articles in farm publications, talks at growers' meetings, and local field demonstrations. Many growers learn about a new product from the tests and demonstrations by university, government, regulatory, and agricultural extension personnel. In the United States, the land-grant system facilitates an open line of communication between agricultural research and the grower.

Most disease control agents are purchased through local dealers or distributors where various amounts of storage, credit, and technical service are available. The manufacturer may supplement these efforts as needed. An increasing number of field representatives with formal training (many college graduates) in agriculture and pest control are on these marketing staffs to ensure sound use of the products. It takes specialized training and experience for effective service, considering the complex technology of balancing optimum control with minimum environmental effect. Consequently, the U.S. government is moving rapidly toward the certification of those who make recommendations or apply agricultural chemicals.

IV. BIOLOGICAL CONSIDERATIONS

A. Performance

Disease-control performance, the prime factor in a compound to be investigated, is evident only under the right test conditions. Among the millions of safe, inexpensive, patentable compounds that do not control disease, there

must be some excellent control agents gathering dust on storeroom shelves
for lack of a test to demonstrate effectiveness. The practical limitations of
time, space, availability of compound, etc., restrict the design of screening, so that each new compound cannot be evaluated under the infinite number of conditions required to demonstrate control of all the important plant
diseases. An attempt is made to test a maximum number of compounds by
methods applicable, within reasonable limits, to representative diseases.
Details of methods are discussed by Pelletier in Chapter 3 by Torgeson [21],
Horsfall [22], and others [23, 24].

In a screen, chemicals are prepared (formulated) for effective coverage, tenacity, penetration, and safety. Application is uniform, reproducible,
and efficient (a minimum of chemical and time). Readings are direct and
simple, but flexible enough to include the unusual or unexpected. Screens
can be adjusted to include a higher percentage of "active" compounds by
decreasing treatment rates, inoculum level, or criteria for selection.
Another factor, of course, is the source of compounds. A random selection of untested compounds will probably have fewer actives than the results
of a well-designed program for the synthesis of biologically active or disease control agents. Since the goal is continually higher and the competition more challenging, fewer "actives" qualify.

Laboratory and greenhouse tests are invaluable in the selection of compounds for field evaluation, but they can be misleading. Field performance
of compounds which appear equally promising in the greenhouse may vary
from poor to excellent. Field experiments are designed to include low
threshold rates, as well as excessive treatments, to see if higher quantities are required for control or if the host is injured. Variable environmental conditions make it necessary to reserve general judgment of performance until the chemical is tested for several seasons in several different locations.

Most larger companies are staffed and equipped to test a compound until
they are reasonably sure it is a commercial possibility. However, broad
evaluation under local conditions around the world is rarely possible without the participation of government and academic investigators. In some
cases, the company provides support for this help. In other cases, the
need for effective agents is great enough to justify local support. Smaller
companies or institutions with compounds to be tested may make contractual
arrangements with larger companies or with private research organizations.
Proof of effectiveness under broad field conditions is necessary before a
new material can be registered and sold. Although government agencies
and educational institutions may not participate in the discovery and early
development, they play a vital cooperative role in determining the value of
a promising candidate.

B. Mode of Action

The continued development of a compound may depend on an early understanding of its mode of action. The fact that a compound controls disease

is usually of more immediate practical concern than its mode of action, but some chemicals may not be used effectively until it is understood. For instance, a compound with direct-contact fungicidal action should be formulated and applied differently from one that controls disease by host modification. The rate of application, timing, and placement of treatments can be critical for optimum activity and should be experimentally varied to explore a compound's characteristics. For instance, standard foliar spray tests do not favor volatile action and a compound with strong curative effects may perform poorly if tested for preventive and residual properties. Seed and soil treatments can be the only practical methods of application for some chemicals.

Compounds with similar modes of activity can serve to complement one another if they also have some different characteristics. An example is a mixture of systemics which move and accumulate in different plant sites to give more complete protection. On the other hand, strains of a pathogen tolerant to one chemical may also be tolerant to another compound with a similar mechanism of action. An understanding of the biochemical, cellular-level mechanisms has been less important in the initial stages of development than the general mode of action described above. The basic mechanism of action studies can eventually help in the development of compounds or lead to improved new structures. The second volume of this text deals almost entirely with this subject.

C. Possible Hazards

Compounds active for disease control must be thoroughly tested for effects on plants, animals, and the environment to understand potential hazards. Most chemicals, even table salt, can have deleterious effects at some level. There is no harmless compound, only those that are more or less hazardous in relation to use and exposure.

Many chemicals control diseases only in association with some plant injury and a judgment must be made as between risk and benefit. Some obligate parasites, such as rusts, are dependent on a normal host physiology and when this is changed by even subtle physiological effects, the plant may no longer support the pathogen. In these cases there may be a fine line between acceptable plant injury and adequate disease control. A few of the disease-control agents now under development cause serious host injury at traditional use rates, but they can be used safely and effectively at lower rates because of their excellent activity. The method of application also influences the phytotoxic response. A chemical too injurious for use when sprayed on foliage may be safe as a soil treatment, or vice-versa. In this area of possible effects on crops, surprises are common under broad use conditions. A new chemical may appear to have little effect on a crop, but may severely damage a certain cultivar or cause excessive injury under unusual environmental conditions. Formulation, adjuvants, and endless combinations with other products can also change the effect on the host. It is impossible to test all these possibilities, but development programs are designed to discover as many potential hazards as is practical prior to commercialization.

During synthesis and initial screening of a new compound, the mammalian toxicity is not known and special care is taken to avoid exposure. Extremely toxic compounds are routinely handled safely under laboratory conditions because workers use precautions with all compounds. Acute toxicity is determined before larger quantities of a new compound are made or tested where some exposure is unavoidable. If the new compound causes irreversible damage to the eye, if it is a sensitizing irritant to skin, or if it is excessively toxic to test animals, it will be dropped from further consideration.

Disease control agents may need to be applied several times as foliage sprays in field experiments where repeated exposure of workers is difficult to avoid. It is, therefore, important to get information on possible cumulative toxicological effects from at least 10-day subacute tests before field evaluation. Fortunately, most antifungal compounds are relatively nonhazardous compared to insecticides, but each new compound is unique and must be carefully tested.

Up to this point in development the new compound has been handled by well-trained, careful investigators who know and respect the potential hazards. Beyond this point, a more complete understanding of possible hazards is necessary and must be determined by extensive testing because additional personnel will be involved.

Toxicity studies are conducted on a variety of animals, fish, and wildlife; they include chronic 2-year feeding, and 3-generation reproduction, teratogenic, mutagenic, carcinogenic, and inhalation tests. Additional environmental studies to determine movement (leaching), half-life, and possible effects on microorganisms in soil make frequent use of ^{14}C-labeled compounds. The results of some of the special laboratory studies are difficult to interpet in terms of hazard [25, 26]. This is a time-consuming, costly aspect of the development process and it may disclose a new problem which necessitates dropping the whole project [4]. It can cost over $1 million and three or more years of work to conduct and interpret the toxicological and environmental studies necessary for registration [25]. Details of safety, toxicology and environmental implications are presented in Chapter 15 of this volume and Chapters 14 and 15 of Volume 2. It takes extensive, scientific study to understand the potential hazard of a new compound and careful evaluation to decide what risk can be taken for a new benefit. The responsibility for this decision is shared by industry, government, the scientific community, and society in general.

D. Data Handling

The computer, which frees us of the labors of record manipulation, offers hope for the design of new effective structures. An almost insurmountable quantity of data is accumulated in a development program where thousands of compounds are made, identified and tested—some of them involving hundreds of experiments.

In 1965, Gorsline [27] summarized the introduction of the electronic digital computer for data manipulation and statistical reduction of agricultural experimental data. There are systems in use today where original biological data are recorded directly on punched or marked cards, immediately double checked, and then processed with chemical descriptions entirely by mechanical electronic hardware [28]. The accuracy, speed, and ability to store, correlate, and print reports are often indispensable. But programs can put limitations on future modifications of methods, codes, etc., unless given careful consideration. Although single-digit codes are more compact on cards, they lack the accuracy in the critical high- and low-disease control areas which is possible with the modified Barratt-Horsfall codes [29].

McDonald [12], in a paper titled "Using a Computer for Development of New Pesticides," describes the effective storage of data which can be retrieved, correlated, and used to design improved products. Saggers [28] described a computer system for the analysis of biological responses of a wide range of species of weeds. It often allows the prediction of activity in apparently unrelated areas, which may save time or indicate profitable lines of research. At the National Institute of Health, Chu and associates [30] are attempting to predict the pharmacological activity of drugs with an "empirical approach" which allows past biological data to guide current testing. Structure-activity predictions, even with computers, are still full of surprises.

E. Biosupport

Biological support is necessary for quality control in manufacturing and to solve marketing problems for the entire life of a product. It is misleading to imply that development of an agent stops when it is ready for sale. The dynamics of improvements can lead to changes in the product or its formulation which must be biotested along with regular physical and chemical monitoring for quality control. New supplies of starting ingredients or new manufacturing equipment necessitate checking the product for effectiveness and possible phytotoxicity. It is impossible to completely anticipate the problems which come from broad product use. Formulation changes and special combinations with other materials may require further biological evaluation. The monitoring of use areas to detect emergence of tolerant strains of a pathogen (common with miticides, insecticides, and drugs) must be considered. Most compounds continue to be evaluated and adapted to new uses in new areas throughout the life of the product.

V. DEVELOPMENT TOMORROW

A. Risk and Benefit

The problems of development are accelerating at an alarming rate, but so is the need for better disease-control agents. Some companies have left the pesticide business, others have stopped developing disease-control agents. This may be due to normal business adjustments, but it also points to the fact that there is no guarantee of permanence [18]. In a world where hunger and malnutrition are common and disease takes a substantial portion (12%) of crop yields [3], there is no question as to the need for more effective control agents.

The Ernst and Ernst Pesticide Industry Profile Study [7] shows the median cost of development of a typical pesticide increasing from $1.2 million in 1956 to $3.4 million in 1967 and $6.5 million in 1973. Fungicides commercialized in 1973 took about 7 years to develop from discovery. Johnson [4] points out that these costs do not reflect the cost of process development and waste control. He estimates one new pesticide for 10,000 compounds tested, at a cost of over $10 million in 8-10 years. It obviously takes more than 10,000 new compounds for one new antifungal agent, since disease control agents represent only a part of most agricultural chemical programs. Johnson emphasized the difference between a "winner" and the "losers" which can cost millions before they are abandoned. It takes a large, diverse chemical corporation to buffer the risks and maintain an effort of this scope.

Some reasons for increases in development time, cost, and risks are:

1. Fewer candidates because effective products have been developed for well-defined markets creating higher technical standards for more difficult disease-control situations.
2. Regulatory limitations and increased costs for experimental field testing.
3. Greater possibility of developing tolerant strains to the more efficient agents.
4. Added tests to demonstrate safety and increased control over potential environmental effects.
5. Higher probability of detecting questionable properties because of an expansion of long-term and highly specific toxicological tests.
6. More countries requiring data generated locally, thus more duplicate testing.

7. Increased costs of plant construction and manufacture due partly to
     reduction in environmental pollution.

These factors put agricultural chemical research and development under
increasing pressures. If programs are abandoned because of these pressures, a world full of hungry people may have to share more of the harvest
with blights, blasts, rusts, smuts, mildews, and rots. Surely we can put
the priorities of risk and benefit in proper perspective and make it possible
to continue the development of new products needed.

B. Fantasy of the Future

Some of us involved in the development of a new compound describe the
frustrations as a nightmare and occasionally indulge in a more pleasant
daydream of what it could be like. Let us try it. A futuristic fantasy
could be something like this:

The scene is filled with a heavy fog. Everything is dripping wet. As
the mist beings to clear, the assembled plant pathologists (who have
studied etiologies and made repeated isolations and inoculations) begin
to "feel" the presence of the pathogen. We are gathered in an endless
cereal field. The bread basket for a hungry world stretches in all
directions to the rolling horizon. Our observations and measurements
lead to the frightening conclusion: "The New Blight" is on the increase
and there is no resistant variety. The green, lush crop is doomed.
We must develop a control or we, too, are doomed.

Our prognosis is communicated to world leaders who grasp the
gravity of the situation. An emergency meeting with chemical industry representatives results in a plan of action. Computer programs are coordinated and fed the necessary data as synthesis chemists stand by with reagents poised to create a structure designed to
have the desired properties. The computer read-out is explicit,
simple, and now that we see it, obvious. There is no time to wait for
long-term or repeated tests results for millions of lives depend on
fast, effective development of this new agent. As studies are prepared
to confirm the expected activity, our engineers design and build manufacturing facilities. Next season's crop must not fail. Biologists
determine the best methods of treatment for disease control, the toxicological hazards, and the impact on the environment. Results from
these tests are just what we had anticipated.

Only minor modifications in existing low-energy manufacturing
facilities are necessary as the process requires only readily available, inexpensive raw materials and is self-contained with no waste
product for disposal. The selling price will guarantee a reasonable
profit to the manufacturer and is a bargain for the grower. Although

long-term testing will be necessary during the first years of use, the agent appears to present no hazard. Life forms cannot assimilate toxic quantities and the compound is found to decompose or metabolize rapidly to H_2O, CO_2, N_2 and a trace of O_2.

Application of 1 g of agent to 100 kg of seed is effective on all varieties including the new high-yield, complete-nutrition, short-season cultivar. As the seed germinates, a slight morphological change takes place in treated plants. The epidermal cell walls are thicker under moist conditions and the penetration peg of the infection hyphae is excluded. Infection by "The New Blight" pathogen can be thwarted.

Meanwhile, with the help of government agencies, the registration and label approvals are ahead of schedule and the distribution and education programs are running smoothly. Growers treat all of this season's seed. The crop is saved.

Fantasies may be unrealistic but some dreams do come true. Maybe in the future we will be able to assemble sufficient knowledge to solve our problems and maybe we will also be wise enough to do it effectively for the common good.

REFERENCES

1. R. H. Wellman, in Plant Pathology, Problems and Progress 1908-1958 (C. S. Holton, G. W. Fischer, R. W. Fulton, and S. E. A. McCallan, eds.), University of Wisconsin Press, Madison, 1959, Chap. 23.
2. R. H. Wellman, in Fungicides: An Advanced Treatise (D. C. Torgeson, ed.), Vol. 1, Academic Press, New York, 1967, Chap. 5.
3. Symposium, Pesticides in the Modern World, Cooperative Program of Agro-Allied Industries with FAO and other UN Organizations, Rome, Italy, 1972.
4. J. E. Johnson and E. H. Blair, Chem. Technol., Nov. 1972, p. 666.
5. B. G. Lever and W. M. Strong, OEPP/EPPO Bull., 3, 119 (1973).
6. Elanco, Agriculture's Partner Worldwide, Eli Lilly Co., Indianapolis, Ind.
7. Ernst & Ernst, Pesticide Industry Profile, National Agricultural Chemists Association, Washington, D.C., 1973.
8. C. Djerassi, C. Shik-Coleman, and J. Dickman, Science, 186, 596 (1974).
9. R. J. Sbraga, Ann. Rev. Phytopathol., 13, 257 (1975).
10. E. Somers, in Fungicides: An Advanced Treatise (D. C. Torgeson, ed.), Vol. 1, Academic Press, New York, 1967, Chap. 6.
11. W. Van Valkenburg, ed., Pesticide Formulations, Marcel Dekker, New York, 1973, Preface.
12. D. P. McDonald, Intern. Pest Control., 14, 11 (1972).

13. R. von Rumker, H. R. Guest, and W. M. Upholt, BioScience, 20, 1004 (1970).
14. Y. Sugimoto, Japan. Chem. Quart., 4, 69 (1968).
15. U.S. Dept of Commerce, Patent Office Pamphlet, Patents and Inventions: An Information Aid for Inventors, Washington, D.C., 1974.
16. C. H. Biesterfeld, Patent Law for Lawyers, Students, Chemists, and Engineers, Wiley, New York, 1949.
17. J. M. Utterback, Science, 183, 620 (1974).
18. M. B. Green, Weeds Today, 14 (Summer 1973).
19. Editorials, Amer. Fruit Grow., March and April 1975.
20. D. MacDougall, Chem. Technol., May 1971, p. 278.
21. D. C. Torgeson, in Fungicides: An Advanced Treatise (D. C. Torgeson, ed.), Vol. 1, Academic Press, New York, 1967, Chap. 4.
22. J. G. Horsfall, Principles of Fungicidal Action, Chronica Botanica, Waltham, Mass., 1956.
23. Fungicide and Nematicide Tests—Results of 1974, Vol. 30, The Amer. Phytopathological Society, St. Paul, Minn., 1974.
24. R. L. Wain and G. A. Carter, in Systemic Fungicides (R. W. Marsh, ed.), Wiley, New York, 1972, Chap. 2.
25. I. Schwartz, Chem. Week, Feb. 20, 1974.
26. M. C. Shurtleff, Flower and Garden, May 1974.
27. G. W. Gorsline, Agri. Sci. Rev., 2nd Quart., 1965, p. 1.
28. D. T. Saggers, Proc. 10th Brit. Weed Control Conf., Brighton, England, 1970.
29. I. F. Brown, Elanco Conversion Tables for Barratt-Horsfall Rating Numbers, available from Eli Lilly Co., Research Laboratories, Indianapolis, Ind.
30. K. C. Chu, R. J. Feldmann, and M. L. Spann, paper presented to the 169th Amer. Chem. Soc. Natl. Meet., Philadelphia, Pa., April 1975, Abstr. CHLT 34.

Chapter 5

THE THEORY AND PRACTICE
OF APPLYING FOLIAR FUNGICIDES

E. Evans[*]

Fisons Agrochemical Division
Chesterford Park Research Station
Saffron Walden, Essex
England

---

[*]Now deceased.

I. INTRODUCTION

The efficient use of any fungicide depends largely on dispensing the active
substance to the right place at the right time. The definition of efficient
placement and timing depends on the biological characteristics of both host
and parasite, the biochemical activity of the active ingredients, and the
physicochemical properties of the fungicide concerned.

Biochemically, fungicides are very active materials and only small
amounts of compound are required per unit area of crop. This is the overriding factor affecting application technology which presents considerable
problems in the attainment of a uniform distribution of fungicide on nonuniform targets such as the leaf surfaces of growing crops. Uniformity is
particularly important when dealing with nonsystemic fungicides but is
somewhat less critical with systemics.

As the physical characteristics of fungicidal compounds differ, so must
the methods of application. Thus solids, liquid, or gaseous compounds
require very different handling techniques, but as most present-day foliar
fungicides are highly insoluble, they are usually applied as aqueous suspensions. In this context, sprays may be defined as clouds of small droplets
suspended in air or any other gaseous medium. Fungicide sprays may be
simple solutions, emulsions, or aqueous suspensions of the active ingredient and in this state they are deposited in droplet form. Thus, the fungicide
encounters the pathogen either before or after infection has occurred and
thereby prevents or cures the disease depending on its mode of action. The
efficient application of any foliar fungicide depends on (1) the use of an appropriate formulation of the active ingredient; (2) selection of a suitable
machine to dispense the formulation; (3) ensuring maximum retention of the
spray cloud by the crop surface; and (4) the mobilization of the spray deposit
to improve contact between fungus and fungicide.

II. FORMULATION OF FOLIAR FUNGICIDES

The term "foliar fungicide" implies any fungicide, whether systemic or
nonsystemic, which is applied to the aerial parts of a crop for the control
of plant diseases. Most present-day fungicides are used in this fashion as
it is a rapid and convenient method of treating large areas of growing crops.
This is particularly true when small volumes of liquid are dispensed from
fast moving machines such as spray aircraft. As with all other methods of
application, the use of small spray volumes introduced specific problems
of formulation. It is therefore necessary to formulate any fungicidal compound in a fashion which is appropriate to the technique.

A. Basic Requirements of Formulation

As the physical characteristics of active fungicidal compounds are rarely
convenient for use per se and the amounts required per unit area of crop
are so small, it is essential that the compound is properly formulated.
Formulation problems may be concerned with (1) convenience, (2) stability,
(3) compatibility, and (4) biology, factors which should be considered whenever a new formulation is designed.

1. Convenience

Formulation should ensure that the product is conveniently packed and accurately dispensible under the crudest possible farm conditions. For example, when diluted to spray concentration a wettable powder (WP) should
be easily wettable and remain suspended throughout the spray period. It
would also be preferable if any sedimented deposits were freely resus-
pendable by simple agitation after standing for a few hours. Suspensibility
problems are most commonly tackled by particle-size reduction or microni-
zation of the active ingredient, frequently in the presence of an inert filler
such as "china clay." The inclusion of adequate amounts of appropriate
surfactants enables the WP to form a stable suspension for spraying. As
these surface-active substances may also create foaming problems in the
spray tank, the choice of type and amount of surfactant is an extremely
important formulation decision. In practical terms, suspensibility problems with wettable powders are considered to be so important as to warrant
the development of an international standard test procedure [1].

2. Storage Stability

The stability of a given molecule is obviously a function of molecular structure. This is of considerable concern to manufacturers quite early in the
development of any product. An inherently unstable molecule would be

likely to present storage stability problems which could be difficult or perhaps impossible to overcome by formulation technology. Thus, for example, a substance that was unstable in high pH conditions should not be mixed with a chalky or other alkaline filler of any kind. Storage stability problems may also be associated with recrystallization phenomena, and a substance which was quite active biologically when freshly prepared could deteriorate rapidly, as recrystallization or agglomeration of these particles occurred on storage. Rapid and frequent environmental changes are prone to exaggerate such effects and owing to the speed with which products may be transferred from one place to another this might easily occur. Accordingly, international test procedures for examining storage stability have been established [1].

3. Compatibility

This is of primary concern when two or more pesticides are mixed for application to a crop at the same time. Under these conditions, the spray tank may contain as many as three or four active materials plus a range of supplementary ingredients in an aqueous medium where both chemical and physical interactions may occur. The biological consequence may, therefore, consist of a loss of activity of one or other of the active ingredients or increased phytotoxicity. There may also be physical interaction due to the presence of different kinds of surfactants leading to coagulation of suspended particles and finally to their sedimentation.

Physical or chemical compatibility problems are usually very specific. For example, two products containing different active ingredients from the same manufacturer may be perfectly acceptable. However, similar products from different manufacturers might not be compatible because the ingredients used were incompatible, although the active ingredients were not. There is, however, a more obvious compatibility problem associated with the use of mixtures containing highly alkaline products such as Bordeaux mixture and lime sulfur. The chemical reaction within such mixtures is frequently such that both physical and biological incompatibility occurs, especially with many of the more sophisticated modern pesticide formulations.

4. Biological Considerations

Formulated products may include, in addition to the active ingredient, a range of substances variously designated as fillers, wetters, dispersants, stickers, solvents, and humectants.

Fillers are inert materials added for convenience to increase the bulk of the formulation or to assist in reducing the particle size of the active ingredient during manufacture. Wetters and dispersants are surface-active substances intended to ensure that the active ingredient is freely dispersed and remains suspended in the spray liquid. Stickers are adhesive materials said to increase resistance to weathering of surface deposits after spraying and solvents are, by definition, essential components of any emulsion or true liquid formulation.

As stated, fillers are generally added for nonbiological reasons but when present may modify the activity of the product as a whole. Various surfactants are good wetters and are themselves biologically active [2]; particularly toward the powdery mildew pathogens [3]. They also lead to a better distribution of the active compound and consequently increase the efficiency of nonsystemic fungicides in the laboratory [4].

Humectants such as glycerine are defined as substances which delay the drying out of surface spray deposits. Their use is said to increase the uptake of systemic substances such as streptomycin [5] but, as with all the other substances noted in this section, their use under field conditions seems of very limited biological significance.

B. Types of Formulation

1. The Significance of Solubility

In discussing the subject of formulating pesticides in general, Hartley [6] stressed the importance of realizing that certain parameters such as the melting points and vapor pressures of pure crystalline materials cannot be altered, although it is of course possible to vary the physical characteristics of products containing them. By contrast, the solubility characteristics of the active ingredients largely control the kinds of formulation that are possible. This concept is summarized in general terms in Table 1, which lists some of the common types of spray formulations according to the solubilities of their active ingredients in either oil or water.

Substances that are very soluble in water may obviously be stored as granular or powdered products or as concentrated solutions which can be diluted with water for application purposes. Active ingredients that are more soluble in oil than in water may conversely be formulated as emulsifiable concentrates or as stock emulsions (uncommon nowadays), both of which on dilution produce a dilute emulsion. These are liquid preparations in which droplets of one liquid are suspended in a second, the two being immiscible. Generally the oil phase is suspended in water and the active ingredient is present in the oil phase of the emulsion.

Most present-day fungicides are not particularly soluble in oil or water and are, therefore, formulated as wettable powders or suspension concentrates which are added to water immediately before applications. These are preparations containing surfactants in which the active ingredient is finely ground and frequently micronized to improve its suspensibility during the period of application, and which possibly increases biological activity [8].

2. Special Requirements for Low-Volume Spraying

Since the dosage rate of active ingredient per unit of crop required for controlling a given disease is usually the same whether it is applied in a large or small volume of water [9], sprays which are applied at low volumes need

TABLE 1

Some of the Commoner Types of Spray Formulations

Physicochemical characteristics of a.i.[a]	Possible formulations	Appearance at spray dilution
Substantially soluble and stable in water	Granular or powdered solids that dissolve in water	Dilute solution of a.i.
	Stock solution either in water or a water-soluble solvent	
Substantially soluble in oil and not in water	Emulsifiable concentrate[b]	Dilute emulsion of a.i.
	Stock emulsion[c]	
Not substantially soluble in oil or water	Wettable powder, freely dispersible in water	Dilute suspension of a.i.
	Suspension concentrate[d]	

[a] a.i. = active ingredient.
[b] Active ingredient dissolved in oil and emulsifier.
[c] Active ingredient dissolved in oil and emulsified in water.
[d] Finely ground active ingredient suspended in water.

to be highly concentrated. These suspensions of insoluble fungicides are prone to very rapid sedimentation and sprays to be used in this manner should be designed to avoid this problem whenever possible. Low-volume sprays are also characterized by the smallness of the droplets necessary in order to achieve adequate spray cover from a small volume of liquid. The production of a fine spray entails good atomization and the realization that the droplets produced are immediately prone to evaporation. At high temperatures small aqueous drops may evaporate to dryness and thus rapidly solidify unless special formulants such as trimethylamine stearate are added [10]. The same effect may be achieved by changing from an oil-based carrier [11] or by addition of other less volatile organic solvents that may also act as antievaporants.

III. SPRAY-CLOUD FORMATION

The spraying process is a simple method of applying small quantities of liquid to the large surface of a crop in a fairly uniform manner. Basically it consists of two phases, atomization of the spray liquid and the transfer of the spray cloud so formed to the target.

A. Atomization of Spray Liquids

This is the process whereby the continuous liquid phase is shattered into small droplets by the application of an external force in the form of hydraulic, mechanical, or even electrical energy. This results in the formation of a spray cloud consisting of a range of small droplets of liquid suspended in air. The devices used for this purpose vary according to the energy source applied, but the commonest in use is undoubtedly the nozzle. It basically consists of a small orifice through which the spray liquid is forced under pressure to emerge as a mass of drops of various sizes.

1. Hydraulic Nozzles

The nature of the spray produced by a nozzle varies according to the size and shape of the aperture and the pressure applied. The nozzles most commonly used are designed to give a fan-shaped or hollow-cone spray pattern, the main characteristics of which were listed by Evans [7]. Working with nozzles of different kinds, Frazer and Eisenklam [12] distinguished three methods of disintegration which they described as rim, perforated-sheet, and wavy-sheet mechanisms of atomization. These are illustrated in Figure 1.

2. Airblast Nozzles

These are structures in which high-velocity air currents are used to shatter the stream of liquid which may be delivered to the moving air by hydraulic pressure or simply by gravity. The airstream itself is usually provided from a high-speed fan. The higher the relative velocity of the air and liquid, the finer the spray cloud produced.

3. Rotary Nozzles

These are devices which consist of discs, cages, or cups that are spun at high speed by mechanical or electrical motors of various kinds. The spray liquid can be accurately dispensed to the rotating surfaces by gravity or other means, as the dispensing and disintegrating processes are executed separately. The rotating surfaces are initially covered with liquid films

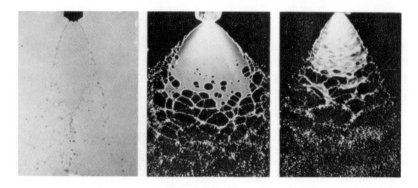

FIG. 1. The three basic methods of atomization developed by nozzles of different kinds: left, rim; center, perforated sheet; and right, wavy sheet. Reprinted from Ref. 12, p. 11, by courtesy of Blackwell Scientific Publications, Ltd., Oxford.

which are then disintegrated by centrifugal force. Such devices characteristically produce a spray with a narrower spectrum of drop sizes than is normally produced by hydraulic nozzles or airblast atomizers.

4. Special Purpose Atomizers

Although most commercial spraying machines are based on airblast, rotary, or hydraulic nozzles, certain other devices have been developed for specific purposes. Vibrating jets of spray liquid have been used in the field with the object of reducing spray drift, as a coarse spray cloud is produced with a minimum number of small satellite droplets. Similarly, impact nozzles have been used to give a very coarse spray. A jet of liquid is made to impinge on a flat surface producing a sheet of liquid of low kinetic energy that eventually shatters to produce droplets in a plane at right angles to the direction of flow of the spray liquid.

In general these devices are crude since they all produce a mixture of drop sizes and are therefore not easy to study. For laboratory purposes, vibrating needles have been used to produce droplets of uniform size. In principle the spray liquid is delivered from the end of a capillary glass tube with a vibrating needle point cutting in and out of the edge of this drop, discharging a stream of microdroplets of uniform size. With minor adjustments of flow rate and changes in the position of the vibrating needle, the size of these drops may be easily adjusted. Using such apparatus, Brunskill [13] was able to study the retention by pea leaves of droplets of any size.

B. Droplet Spectrum

All practical atomizing devices described in this chapter produce a cloud containing a variety of drop sizes, the range of which differs greatly from one to another. These differences can often be modified further by minor variations in the viscosity and surface tension of the spray liquid as well as the amount of energy applied. Thus the spectrum of droplet sizes emerging from a hydraulic spray nozzle can be varied simply by changing the operating pressure, while the size and speed of rotation of spinning disks are largely responsible for their droplet spectra. The range of droplet sizes delivered from a nozzle is generally much greater than from any rotary device.

On a quantitative basis, the mean droplet size in a given spray is usually expressed as the MMD (mass median diameter) or VMD (volume median diameter). If the spray liquid is water based and evaporation does not occur, then the arithmetic value of the MMD and VMD is the same, being calculated from sample measurements of droplet diameters. By definition, these values are calculated as the droplet diameter at which half the mass of the spray liquid is contained in droplets of smaller size and the other half in droplets of larger size than the MMD. Finding these values must, therefore, depend on recording cumulative totals as illustrated by Brown [14] and, more recently, by Amsden [15].

From the practical point of view, the droplet spectrum is extremely important since not only does it determine the final nature of the spray deposit produced but it also has a very big effect on crop penetration. A spray cloud containing a large proportion of very small droplets penetrates some crops more effectively than a coarse spray, as the smaller droplets have a low momentum (mass × velocity) and are thus less prone to impaction on the nearest part of the target; a feature obvious to anyone who has attempted to remain dry in misty conditions even under the densest forest canopy.

While crop penetration may be extremely desirable in certain situations, the use of fine sprays gives rise to spray-drift problems. The implications of this are generally less serious with fungicides and insecticides than with the hormone herbicides; even so the possibility of damage or undesirable residues on neighboring crops should always be borne in mind. These and similar practical considerations are discussed in the Pesticide Manual [16].

C. Transfer and Deposition of Spray Droplets

The mechanisms involved in the transfer of any spray cloud to the target depend largely on the method of atomization used and the growth habit of the

TABLE 2

Summary of the Most Commonly Employed
Commercial Techniques of Fungicide Application

Growth form of crop	Technique of interest	
	High volume	Low volume
Ground or field crop	Hydraulic sprayer with horizontal spray gear	Aircraft with conventional spray bar and nozzles or rotary atomizers
Bush or orchard plantation	Hydraulic sprayers with vertical spray bars or hand lances	Airblast machines and occasionally aircraft, e.g., for control of downy mildew of vines
Closed canopy plantations	Hydraulic sprayers with hand lances	Airblast machines from below and aircraft from above the canopy

crop concerned. Hydraulic spray systems of all kinds usually produce
droplets of very limited momentum with the result that the producing nozzles must be placed very close to the target. Owing to their small mass,
all spray droplets lose their kinetic energy very rapidly in air and their
final deposition is frequently a function of natural air and crop movements
or gravity. The arrangement of the spray nozzles will depend on the growth
form of the crop, a feature most clearly seen in relation to high-volume
spraying (see Table 2).

By definition, high-volume spraying involves the use of large volumes
of dilute spray solutions, which is a cumbersome process that is very demanding in terms of labor. For this reason, there is a marked tendency
either to reduce spray volumes applied with the traditional spray bar and
nozzles whenever possible or to use a fast moving airstream for atomizing
the spray liquid and carrying the droplets to the target. This is the principle of the airblast machine which contrasts with the low-volume mistblowers that are sprayers in which the atomizing and transfer mechanisms
are separated. The latter generally consist of a large fan that produces a
slow-moving airstream into which the spray is discharged for transport to
the target. An example of such a "low-volume" spraying machine is shown
in Figure 2.

Equally small volumes of liquid may be applied from spray aircraft of
various kinds using the traditional hydraulic spray bar and nozzles or rotary atomizers. In this case, the down-draft and the forward movement
of the aircraft are the forces ultimately responsible for the atomization and
transfer of the spray cloud, respectively (see Fig. 3).

FIG. 2. Airblast spraying machine showing the nozzle atomizers with large fan which aids in spray formation and in transferring the spray cloud to the target.

FIG. 3. Illustration of traditional spray bar and uniformly spaced nozzles mounted on a helicopter spraying potatoes at 2 gal/acre (approx. 25 liters/ha) for the control of potato blight (Phytophthora infestans).

Irrespective of how the droplets are produced, there are only two ways in which they are conveyed to the target, either by impaction on near-vertical surfaces close to the nozzle, due to the higher momentum of the fast moving drops, or by sedimentation of low-energy slow-moving on horizontal surfaces after penetration has occurred.

Interception and electrostatic attraction are sometimes mentioned as possible mechanisms of deposition, but in general these attractions are not thought to play a significant role in the formation of fungicide spray deposits.

D. Spray Systems Commonly Used

In its simplest form, a spraying machine consists of a reservoir of spray liquid, an atomizing device of some description, and one or more outlets for the spray cloud. The reservoir may be large or small, simple or complex, depending on its proposed method of operation. Similarly, any of the atomizing devices already noted may be used and this in turn affects the overall concept of the machine. Hydraulic systems of various kinds depend on pumping the spray liquid at pressure through nozzles. The pressure applied depends on the degree of atomization deemed necessary and the size and shape of the aperture involved. The latter will also affect the output of the system which in turn may determine the speed of operation of the sprayer in the field. Characteristically, the droplets produced by any hydraulic system have a limited energy potential and the outlets (in this case nozzles) are accordingly arranged to follow the contours of the crop.

By contrast, airblast machines tend to be more compact, having only a limited number of outlets that depend on blowing the droplets for some distance into the crop. For this reason, they tend to be a little more mobile than the hydraulic type with smaller reservoirs intended to carry more concentrated spray fluids. By definition, they must also provide some means of moving air at high speed in order to assist in the atomization and transfer of the spray. Used at a distance, they depend greatly on the degree of crop penetration achieved which, by implication, depends on having a fine spray.

Unless aided by fans or other similar mechanisms, all rotary devices produce droplets of comparatively low momentum; this is the reason such arrangements are frequently used on spray aircraft, the downwash of which is responsible for transferring the atomized liquid. The same applies when hydraulic nozzles are used on aircraft. Aircraft spraying generally is on the increase in the more sophisticated countries where large uninterrupted areas of one crop allow for speedy and effective operations to occur.

By varying the kinds of storage, atomizing, and discharging arrangements it is obvious that a very large range of machinery is available and the methods by which they may be classified are equally numerous, depending on the prime interests of those concerned. Thus, while the farmer may be concerned as to whether the implement is tractor drawn or mounted,

WHO [17] produced a review of the machinery available on the basis of the energy supplies used. From the biological point of view, however, it is more pertinent to follow the classification of spraying systems proposed by Courshee et al. [18]. They recognized three main techniques based on the spray patterns produced on the foliage.

1. High-Volume Technique. Any application that, when applied to the crop, induces "runoff." This involves large volumes of liquid.
2. Low-Volume Technique. Any technique that yields a discrete droplet pattern on the treated foliage. This usually involves small volumes of liquid.
3. Medium-Volume Technique. Any technique using intermediate spray volumes that yield a spray pattern where some of the droplets coalesce while the others remain discrete.

Recently, there has been a tendency to use the term "ultralow volume" to imply the use of undiluted products, but from the fungicide point of view this is best seen as the extreme limits of the low-volume techniques [19].

IV. RETENTION OF SPRAYS ON CROP PLANTS

A. Deposition Mechanisms

The average biological target is usually a very complicated physical structure and consequently the arrival of the spray cloud at the crop surface is subject to a multitude of factors that affect the retention of the droplets by the crop. Droplets of high momentum, due to their large size or to high velocity, impact on any small obstruction placed in their path, whereas those of low momentum, due to their small size or to low speed, are merely diverted around it. This is a fundamental statement of airflow dynamics which suggests that all solid objects are surrounded by a "layer of resistance" when placed in a mass of moving air. Droplets of high momentum are said to penetrate this layer, whereas those of low energy potential do not, being deflected to the sides of the obstruction. In this fashion, low-momentum droplets penetrate even the densest crop canopy but eventually, due to loss of velocity, they may settle out under gravity without impinging on the crop.

The effect of drop size and droplet velocity on dynamic catch is clearly seen in Figure 4. Here the dynamic catch is largely a measure of impaction on 1/8-inch cylinders, but under practical conditions the target itself is a much more complicated structure in terms of size and shape so that the filtration effect from even the simplest crop is extremely complex.

The phenomenon of sedimentation is comparatively simple, depending largely on the distance from the target and terminal velocity, which is the final velocity at which a droplet will settle in static air. It can be calculated

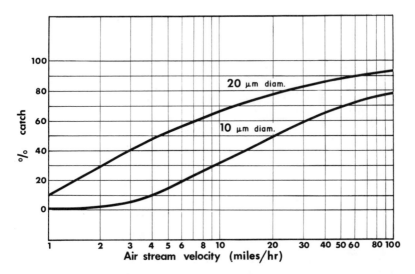

FIG. 4. The effect of droplet size and windspeed on droplet retention by a 1/8-in. diameter cylinder. Reprinted from Ref. 20, p. 22, by courtesy of the American Society of Agricultural Engineers.

empirically for spherical structures such as spray droplets, when the radius and density of the droplet is known and the viscosity of air is included in the calculation. In this fashion, Yeo [21] found that droplets above 100 μm in diameter settle very much more rapidly than those of 50 μm or less; thus, while a drop 10 μm in diameter has a terminal velocity of only 0.3 cm/sec, a 100 μm drop would descend at a rate of 27 cm/sec in static air conditions. This is a feature of considerable importance to those concerned with drift problems.

B. Retention of Spray Droplets

Whereas deposition is mainly a function of application technology, retention is primarily concerned with the nature of the surface being treated and the physical characteristics of the spray liquid used. Both Somers [22] and Evans et al. [23] found that different plant surfaces retained different quantities of spray liquid under identical conditions of application. In general it can be stated that simple factors such as the presence or absence of hair and perhaps more particularly the quantity, shape, and distribution of surface waxes greatly affect the amount of liquid retained per unit area of the surface.

Smooth waxy surfaces retain droplets quite freely while micro-rough surfaces, typified by peak and banana foliage, are highly reflective. Brunskill [13]

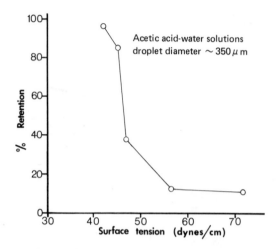

FIG. 5. The effect of surface tension on droplet retention of pea leaves. Reprinted from Ref. 13, p. 24, by courtesy of the British Crop Protection Council.

made a detailed study of this phenomenon on peas when he showed that drop size and surface tension had a big effect on droplet retention. Using the vibrating needle drop generator to produce droplets of uniform size he produced the data shown in Figure 5.

The dramatic change in droplet retention over such a small range in surface tension as 10 dyn/cm is of considerable practical importance, suggesting that any excessive dilution of a formulation containing the appropriate level of surfactant might lower the effectiveness of that treatment on certain crops by reducing the amount of active ingredient retained.

From studies of this kind it is now known that certain plants, such as temperate-zone cereals grown under glass, are particularly unretentive, an effect much less obvious when similar plants are grown out of doors.

Nevertheless, the significance of the problem is reflected in the fact that all fungicide recommendations for use on cereals stress the need for medium- rather than high-volume treatments. Low-volume sprays are usually very finely atomized so that the smaller droplets are well retained by the foliage. Cereals also present application problems in terms of target presentation. The long strap-shaped leaves are usually presented end on to the descending spray droplets, but this may be partially overcome by the forward movement of the spraying machine discharging a finely divided spray cloud containing an appropriately high concentration of surfactant.

Fortunately most agricultural crops do not present problems of this kind and the foliage of crops such as potatoes and tomatoes are easily wetted so that most droplets are freely retained. When sprayed at high volume,

however, even these surfaces become reflective as soon as some liquid
has been retained and newly arriving droplets can be seen to bounce freely
from the wet surface of the leaf [7]. Under these conditions the amount of
spray liquid retained by a leaf is related to the thickness of the surface
film [4]. Rich [24] had previously noted differences between the amounts of
zineb retained and the amounts of Bordeaux mixture held by foliage and
attributed such effects to different electrokinetic charges borne by the particles of zineb and Bordeaux mixtures, respectively. This has since been
questioned by Burchfield [25] and Evans et al. [23] who attributed these differences to the colloidal characteristics of Bordeaux mixture and the particulate nature of zineb.

C. Distribution of Foliar Fungicides

Irrespective of the complications involved in applying a foliar spray, its
efficiency must eventually depend on the effective distribution of the active
ingredient. "Distribution" is a term used to describe the overall placement
of a chemical on the crop, and if all susceptible parts receive a reasonable
uniform level of fungicide it is described as a good distribution. Conversely,
an erratic pattern is said to have a bad distribution. On a crop basis, distribution may be considered as having a vertical and horizontal component,
the significance of which will differ according to the biology of pathogen,
the structure of the crop concerned, and the method of application used.
Thus when dealing with a horizontal spray bar on a ground crop sprayer
fitted with a series of nozzles for the control of potato blight, it might not
be necessary to have a particularly uniform vertical distribution [9]. Any
missed strips resulting from a failure to ensure adequate overlap of swathes
would soon become biologically obvious due to lack of control.
    Such considerations apply equally well to plantation "closed canopy"
crops such as rubber, and simple tree or bush crops such as apples or
black currants. In the absence of evidence to the contrary, it is assumed
that the design and operation of all spray machinery should aim at achieving
the most uniform distribution possible, both vertically and horizontally.
This statement merely stresses the need for crop penetration which is in
turn a reflection of the droplet spectrum and a consequence of the spraying
system used.
    In classifying the various spraying systems, Courshee et al. [18] recog-
nized the importance of the spray pattern on foliage. The biological significance of this concept has since been stressed by Evans [19] who pointed
out that, although the mode of action of the discrete spray patterns of both
low- and ultralow-volume fungicide deposits generally are not well understood, any logical explanation of their good performance vis-à-vis equivalent
high-volume treatments must entail some effective mechanism of redistribution whether by the internal systemic movement of the active ingredient

TABLE 3

Effect of Distribution of the Efficiency
of Dinocap and Benomyl Deposits [19]

Concentration of applied droplets, %	Number of drops per leaf	Disease control, %	
		Dinocap	Benomyl
0.025	1	5	20
0.0125	2	10	40
0.0067	4	20	100
0.0033	8	55	100
0.0017	16	100	100

or its superficial movement in the liquid, solid, or vapor phase. The term "redistribution" is taken here to imply the movement of the active ingredient on the surface or inside the plant at any time after the spray deposit is formed by the drying out of the spray droplets.

The significance of a uniform deposit was demonstrated by Morgan [26] using a nonsystemic copper deposit for the control of Botrytis fabae on beans. A similar experiment was done by Evans [19], comparing the activity on nonsystemic dinocap and systemic benomyl against cucumber mildew Sphaerotheca fuliginea. The data in Table 3 were obtained by placing the same amount of active ingredient on each leaf in the form of 1, 2, 4, 8, or 16 droplets, keeping drop sizes as uniform as possibly by means of a micropipette. In the case of dinocap, the leaf area protected was roughly proportional to the number of droplets applied, whereas massive systemic redistribution had obviously occurred with benomyl.

V. THE FATE OF SPRAY LIQUID

Having been exposed to the atmosphere for a reasonable period of time, spray liquid evaporates on the leaves to yield initial spray deposits, the size and nature of which are a function of (1) the amount and type of formulation employed; (2) the method of application used; and (3) both the micro- and macrocharacteristics of the treated surface. Thereafter, the amount of active material present decreases due either to biochemical reactivity of the fungicide within or on the plant or its physical removal from the plant surface. On this basis the amount of fungicide present on a treated crop in the field at any point in time would be determined by the interaction of these two phenomena.

A. Initial Spray Deposits

Spraying to runoff, using normally retentive leaf surfaces such as bean and celery foliage, Rich [24] showed that the size of an initial deposit of zineb increased linearly with the concentration of the spray liquid. He found, however, that this did not apply to Bordeaux mixture, which reached a limiting value beyond which the increase in concentration did not increase the size of the initial deposit. This difference he attributed to differences in electrical charges between the organic zineb and the inorganic Bordeaux but both Burchfield [25] and Evans et al. [23] have questioned this interpretation, suggesting that the unexpected pattern for Bordeaux mixture is probably associated with the thixotropic nature of this product. By comparing the pattern of buildup of copper applied as colloidal Bordeaux mixture and particulate noncolloidal copper oxychloride, Evans et al. [23] showed that the amount of Bordeaux retained per application increased consecutively over a series of five treatments but that this did not occur with the oxychloride, a feature again attributed to the colloidal characteristics of Bordeaux. This is shown in the data given in Table 4.

B. Fungicide Residues

In discussing the subject of fungicide residues in crop plants it is essential to distinguish between systemic and nonsystemic materials which are basically exposed to very different environmental conditions from the moment they are applied. Thus while systemic materials are exposed to the

TABLE 4

Increments of Copper Retained from Consecutive Applications of
Bordeaux Mixture and Copper Oxychloride to Bean Leaves, Vicia faba

Spray application	μg Cu/cm^2 leaf 10:10:100 Bordeaux mixture	Copper oxychloride,[a] 0.25% Cu
1	8.6	7.5
2	7.7	7.8
3	9.0	6.9
4	10.6	5.7
5	12.7	7.7

[a]Sprayed at 100 gal/acre (935 liters/ha).

comparatively steady internal biochemical environment of the plant, non-systemics are subject to the extremely variable external environment of the plant itself.

1. Nonsystemic Fungicides

Surface deposits of all kinds are subject to a variety of weathering factors such as wind, rain, plant movements and growth, and to wide variations in temperature, each of which have been considered to be effective weathering factors at one time or another. Horsfall [27] measured such losses and noted a logarithmic pattern of loss for both Bordeaux mixture and cuprous oxide formulated as a wettable powder. This was confirmed by Evans et al. [23] who suggested that while Bordeaux deposits tended to flake away from the surface, wettable powders tend to disintegrate to their original particulate condition.

In order to minimize the variability due to diverse weather conditions, considerable effort was at one time devoted to the study of supplementary formulating agents that might improve the tenacity of such deposits. These included polymeric emulsions of polyvinylacrylate (PVA) and polyvinylchloride (PVC) [28] which Evans et al. [4] found to be of little biological consequence in the field, despite their apparent success in improving the persistence of copper deposits in the laboratory.

2. Systemic Fungicides

By definition, systemic fungicides are chemicals which are taken up by plants and redistributed within their tissues. The mechanisms of uptake are far from clearly understood since the rate and possibly manner of uptake differ from crop to crop and from one fungicide to another. Once inside the plant the fate of the fungicide varies according to its structure and consequently to its biological and chemical reactivity. The actual concentration within the tissues is decided by the balance between its rate of uptake and its rate of degradation. However, when further supplies are not forthcoming, the quantity present declines. Wain and Carter [29] suggested that although the decline of some organic fungicides might arise by simple chemical interactions within living tissues most degradations are likely to be enzymatic.

In reviewing the fate of some of the commercially used systemics, Kaars Sijpesteijn [30] noted that the benzimidazoles and also thiophanate-methyl are particularly stable in plant tissues and soil; both compounds appear to exist as high-molecular-weight complexes of some kind. By contrast, the pyrimidines ethirimol and dimethirimol are comparatively unstable, being eventually converted to water and soluble glucosides [31]. The oxathiin carboxin is only partially degraded to the nonfungitoxic sulfoxide and sulfone [32].

Since by implication all systemic substances are present within plant tissues, they all present potential residue problems at harvest and for this reason they are all subjected to intensive breakdown studies before they are ever marketed.

VI. SOME PRACTICAL ASPECTS OF APPLYING FOLIAR FUNGICIDES

Various organizations, representing both government and industry within the United Kingdom, have issued a series of proposals known as the Pesticides Code of Conduct. It deals with most aspects of using chemicals and particularly stresses the need to (1) correctly identify the problem of concern, and (2) choose the correct chemical and method of application for effective control.

A. Choice of Product and Method of Application

Although the average farmer may find it difficult to identify many plant diseases, he is usually only too well aware of the ravages of those that commonly occur on his own crops. Even so, it might be advisable to seek the assistance of a specialist if there is any doubt, since the identity of the pathogen concerned must affect both the choice of chemical and its method of application. There are various texts, such as the Pesticide Manual [16], that provide detailed information on the chemicals available for controlling diseases of many temperate crops but the decision on which to use may still depend on economic considerations. The choice of technique depends largely on the method of dispersal of the pathogen concerned, e.g., seedborne diseases are usually controlled with seed dressings, soilborne diseases by soil treatments, and airborne diseases by the application of foliar fungicides of various kinds. With the advent of systemic fungicides this distinction is not absolute, as seed dressings and soil treatments are now sometimes used to control certain airborne pathogens (with, e.g., ethirimol, benomyl). However, despite these possibilities, foliar pathogens are still mainly treated with fungicide sprays of the appropriate spectrum of activity.

In preparing to apply any spray it is essential to follow the instructions given on the label to the greatest possible detail as failure to do so could lead to an inaccurate and nonuniform distribution. Instructions should specify the concentration of product recommended and the volume per unit area of crop necessary for effective control. They should also specify spray timing and may even prescribe the precise order in which one or more products should be added to the spray tank. The most important item should be a list of precautionary measures to protect both user and consumer from any potential hazards.

B. Choice of Spraying Machine

The choice of sprayer is governed principally by the growth form of the crop to be treated (see Table 5) and the scale of operation proposed. Very large areas are often best treated by fast moving aircraft but this method might not be acceptable in uneven terrain or on fragmented plantations. If the topography is suitably uniform, a large area might still be covered fairly rapidly by various lightweight machinery. This implies the use of low-volume treatments with traditional hydraulic spray systems and airblast machines which carry minimum amounts of concentrated liquids rapidly from place to place. Even manually transported airblast knapsack sprayers are now commonly used for bush and small tree crops, whereas there is fairly extensive use of aircraft sprayers on field crops of various kinds.

Despite these generalizations, it is true to say that most foliar fungicides are still applied with traditional spray nozzles and it is therefore important to know the major characteristics of two basic kinds of nozzles described as fan jet and hollow cone, respectively.

C. Fan Jet and Hollow Cone Nozzles

Fan jet nozzles are based on the fishtail-burner principle. They consist basically of an elliptical shaped orifice from which the spray liquid emerges

TABLE 5

Main Characteristics of Fan Jet and Hollow Cone Nozzles

Fan jets	Hollow cone
1. Provide minimum hindrance to the passage of liquid	Considerable energy losses incurred during rotation of liquid within nozzle
2. Droplets have high kinetic energy and impact readily on nearest target	Droplets have comparatively low kinetic energy and are thus susceptible to drift
3. Output rather unpredictable and increases rapidly with use	Output is well controlled and nozzle is not readily worn
4. Spray liquid emerges as a triangular sheet of liquid	Spray liquid emerges as a hollow cone that shatters at the edges
5. Spray patterns are easily arranged to give uniform distribution across swath	Spray patterns are difficult to overlap

as a sheet that eventually shatters at its edges to produce droplets of various sizes (see Fig. 1).

By comparison, the liquid passing through a hollow cone nozzle is made to rotate within the nozzle so that it emerges from the resultant vortex as a hollow conical sheet of liquid that disintegrates at its margin, again producing a range of droplets of different sizes. The rotating motion is produced by passing the liquid through two or more holes in the "swirl plate." This is fitted a short distance from the aperture and the spray quality can be altered by moving the plate toward or away from the aperture. The main characteristics of both fan and hollow cone nozzles are given in Table 5.

The information given in Table 5 implies that volume for volume and pressure for pressure hollow cone nozzles are more suitable for the application of foliar fungicides than fan jets because they provide finer sprays of greater penetrative power.

D. Nozzle Output and Arrangement for Use

Having selected nozzles which have spray characteristics necessary for the purpose at hand it becomes necessary to: (1) inspect the spray pattern they produce; (2) ensure that the nozzle arrangement is satisfactory; and (3) check the output of the machine.

By a visual inspection of the machine in operation, a faulty nozzle is soon found to be giving an irregular or deformed spray sheet. Such effects can be simply confirmed by examination of spray patterns induced by spraying over a dry concrete surface.

More detailed study of spray patterns require the use of a patternator. This device consists of a series of troughs that collect the spray droplets at different points across the swath of the nozzle and delivers the liquid received to individual tubes for measuring. In this fashion nozzle quality, spacing, and height above the surface to be sprayed may be evaluated before attempting to spray the chemical concerned under field conditions.

The final feature to be checked is the output of the spraying system per se. In the case of hydraulic systems this implies the total amount of liquid delivered per minute from all the nozzles, being the cumulative total of each and every nozzle on the machine. Since each measurement would have its own margin of error, the output is generally assessed in terms of losses from the storage tank. For practical purposes it is best to take two measurements, one with the tank full and the other with the tank half full of liquid. This is the method normally applied, since it is not possible to catch all the liquid emerging from an airblast machine. All measurements should be made with the atomizing fan in operation. Either a set volume of liquid is delivered from the tank and the time measured, or the amount of liquid delivered from the tank in a given time is measured with a dipstick or similar device.

Of the features that contribute to accurate application the most difficult to assess correctly is that of swath width. When spraying a comparatively uniform surface with a horizontal spray bar this is reasonably simple, but it is more difficult when a tree, bush, or plantation crop is sprayed with an airblast or other low-volume spraying device as this is dependent on the operators assessment of the efficiency of the spray distribution. There is no general rule which applies to all crops.

E. Calibration of Spraying Machine

The spray concentration is fixed during the mixing process according to the manufacturer's instructions, and the decision on how to apply the specified volume of liquid to a given area of crop has then to be made. As the nozzle is the dispensing unit, the first consideration is generally given to the kind and size of nozzle most suitable to deliver the desired volume per unit of time. With the range of nozzles commercially available, it should be possible to select one with an output that covers to within ±10% of the desired volume and minor adjustments of the order of ±5% are usually possible with pressure changes. When the choice and number of nozzles (according to swath width) has been made, the machine still has to be run at the correct constant speed if the correct application rate is to be achieved.

The rate of application of any sprayer is dependent on four main factors:

1. The number of nozzles arranged on the spray bar, N
2. The output per nozzle, t, in liter/min
3. The width of swath being treated, W, in meter
4. The speed of the sprayer, s, in km/hr

These are connected by the simple equation,

$$\text{liter/ha} = \frac{600 \times (t) \text{ liter/min} \times N}{(s) \text{ km/hr} \times (W) \text{ meter}}$$

By knowing any four of these values it is obviously possible to calculate the fifth.

F. Operational Procedures

Before treating any crop, it is essential to check the precise application directions provided by the chemical manufacturer and the environmental suitability for spraying.

The influence of climatic conditions during the spraying process may be considerable. Wind velocities in excess of about 5 mph may result in an uneven distribution of fungicide, particularly when dealing with field crops. Wind velocities greatly in excess of this value may cause significant spray drift problems.

With certain chemicals there may be temperature limitations. For example, it would be unwise to apply dinocap to apples at high temperatures for fear of causing foliar damage, although this compound would be perfectly safe when applied earlier or later on the same day.

Considering spray retention, it is always necessary to refrain from spraying when the crop is wet with dew or rain since droplets bounce freely from any wet surface and the retention of more liquid may eventually lead to "runoff." As the objective of all fungicide treatment should be to achieve maximum retention of the active ingredient, this limitation is particularly important.

When all the necessary environmental criteria have been considered and the machinery has been checked and calibrated to deliver the volume recommended by the manufacturer, it remains to "set the tractor" to the calculated speed to deliver the spray to a given area of crop in unit time. At this point particular attention should be paid to the manufacturer's instructions as well as to the necessity for using clean water and agitation of the spray tanks at all times. Experienced operators have their own methods of tackling problems specifically associated with a given crop. In general, however, corners and turns provide the greatest concern where overdosing and uneven spray patterns are commonly found. How these difficulties are overcome will vary from crop to crop and from field to field, depending on the presence or absence of obstacles.

The operation having been completed with due care on the part of the spray operator, attention should be paid to the disposal of the empty spray containers. If a wettable powder formulation has been used the filters in various parts of the machine will also need cleaning and refitting as a routine operation.

VII. SUMMARY

The development of a practical recommendation for the efficient use of any fungicide depends on three factors:

1. Applying the correct amount and type of formulation designed to match the physical and biological requirements of the situation
2. Using the appropriate machinery in a fashion suitable to the crop
3. Correctly placing the chemical and timing the treatment to facilitate maximum contact between the fungus and the fungicide

ACKNOWLEDGMENTS

The author wishes to thank the Directors of Fisons Agrochemical Division for permission to publish the information contained in this chapter, and would like to thank his specialist colleagues for their kindness in checking the completed manuscript.

REFERENCES

1. R. de Ashworth, Soc. Chem. Ind. (London) Monograph, 21, 145 (1966).
2. E. Evans, J. Marshall, B. J. Couzens, and R. L. Runham, Ann. Appl. Biol., 65, 473 (1970).
3. E. L. Frick and R. T. Burchill, Plant Disease Reptr., 56, 9, 770 (1972).
4. E. Evans, J. R. Cox, J. W. H. Taylor, and R. L. Runham, Ann. Appl. Biol., 58, 131 (1966).
5. R. A. Gray, Phytopathology, 46, 105 (1956).
6. G. S. Hartley, Soc. Chem. Ind. (London) Monograph, 21, 122 (1966).
7. E. Evans, Plant Diseases and Their Chemical Control, Blackwell Scientific Publications, Oxford, 1968.
8. F. Wilcoxon and S. E. A. McCallan, Contrib. Boyce Thompson Inst., 3, 509 (1931).
9. E. Evans, J. W. H. Taylor, R. L. Runham, and B. J. Couzens, J. Roy. Aeron. Soc., 69, 660, 855 (1965).
10. G. S. Hartley and R. Howes, Proc. 1st Brit. Insectic. Fungic. Conf., 2, 533 (1961).
11. H. Guyot and J. Cuillé, Fruite d'outre-mer, 10, 101 (1955).
12. R. P. Frazer and P. Eisenklam, J. Imp. Coll. Chem. Eng. Soc., 7, 52 (1953).
13. R. T. Brunskill, Proc. 3rd Brit. Weed Control Conf., 2, 593 (1956).
14. A. W. A. Brown, Insect Control by Chemicals, Wiley, New York, 1951.
15. R. C. Amsden, Pans, 21 (1), 103 (1975).
16. H. Martin and C. R. Worthing (eds.), Pesticide Manual, 4th ed., British Crop Protection Council, 1974.
17. Annonymous, Equipment for Vector Control, WHO, Geneva, 1964.
18. R. J. Courshee, R. C. Amsden, and H. J. Morris, Report to the Government Entomologist, Sudan (unpublished), 1957.
19. E. Evans, in Pesticide Application by ULV Methods, Monograph No. 11, British Crop Protection Council, 1974.
20. F. A. Brooks, Agr. Eng., 5, 233 (1947).
21. D. Yeo, Report of 1st Intern. Agr. Aviation Conf., Cranfield, England, 1959.
22. E. Somers, J. Sci. Food Agr., 8, 520 (1957).
23. E. Evans, J. W. H. Taylor, R. L. Runham, and M. McLain, Trans. Brit. Mycol. Soc., 45, 81 (1962).
24. S. Rich, Phytopathology, 44, 203 (1954).
25. H. P. Burchfield, in Plant Pathology, Vol. III, Academic Press, New York, 1960, pp. 472–520.
26. N. G. Morgan, Ann. Rept. Long Ashton Res. Station, 1952, p. 99.
27. J. G. Horsfall, Fungicides and Their Action, Chronica Botanica, Waltham, Mass., 1945.
28. E. Somers, J. Sci. Food Agr., 7, 160 (1956).

29. R. L. Wain and G. A. Carter in Fungicides (D. C. Torgeson, ed.),
    Vol. 1, Academic Press, New York, 1967, pp. 561-611.
30. A. Kaars Sijpesteijn, in Systemic Fungicides (R. W. Marsh, ed.),
    Longmans Press, London, 1972, pp. 132-155.
31. P. Slade, B. D. Cavell, R. J. Hemingway, and M. J. Sampson,
    Proc. 2nd Intern. Congr. Pestic., Chem., Tel-Aviv, 1971, p. 5.
32. M. Snel and L. V. Edgington, Phytopathology, 60, 1708 (1970).

Chapter 6

CONTROL OF SEED AND SOILBORNE
PLANT DISEASES

R. Rodriguez-Kabana, Paul A. Backman, and Elroy A. Curl

Department of Botany and Microbiology
Auburn University
Auburn, Alabama

I. INTRODUCTION

The subjects of soil and seed treatments for control of diseases are the oldest in plant pathology and they still represent very active areas of research. Well known is the treatment of wheat seed with salt water for control of bunt, started by farmers in England in 1670 [1]. The subject of seed treatment spans, in technology and time, from such fortuitous developments to the advent of modern contact and systemic fungicides. The development of fungicides for seeds and methods for their application therefore represent a large subject. We have of necessity had to limit this chapter essentially to a practical introduction to the subject. Since a recent review adequately covers the subject of machinery for application of fungicides to seeds [2], we have sought in this chapter to complement that review with recent developments in fungicides and pathology, and to introduce some novel ideas for control of seedborne pathogens.

Treatment of soil for control of pests became a reality in 1869 [3] with the use of carbon disulfide for control of Phylloxera in vineyards in France. The introduction of this fumigant resulted in development of many practical rules for its application before the end of the 19th century. It is ironic that these rules had to be "rediscovered" later for the application of other fumigants. As in the case of seed treatments, the subject of soil treatment encompasses a long period fraught with many ideas and chemicals. In this chapter we are restricting our discussion to the more pertinent materials and methods of application used today. Many of the older fumigants and fungicides that are currently not in widespread use are omitted in our discussion and reference is made to reviews for details on their use. At the same time, when a group of materials belonging to the same type of chemicals is discussed, a representative of the group is selected for detailed discussion of the effects obtained from its use in soil. Since soil represents a dynamic system with many interacting organisms, present at all times, it is appropriate to initiate the discussion with a brief summary on what is known about the major soilborne plant pathogens. This should help the reader in understanding and developing better means of utilization of materials and methods available for their control. With this in mind, we have given particular attention to the spectrum of activity of major soil fungicides not only against target fungi but also against some nontarget organisms.

II. ECOLOGICAL CONSIDERATIONS

Certain knowledge of the ecology of soilborne plant pathogens is necessary for developing successful and reproducible control measures. As Kreutzer [4] pointed out, the morphology, physiology, and ecological growth habits of fungi influence susceptibility or resistance to toxicants. Fundamentals pertinent to fungicidal control of seedling diseases relate to pathogen populations in soil, growth behavior and survival mechanisms, host range, inoculum density-disease relationships, and rhizosphere ecology. Fortunately, excellent books, reviews, and published symposia on the biology of root-infecting fungi are available.

Some representative species of soilborne fungi, notable for inciting seedling diseases, are listed in Table 1. Soil fungi in general are widely distributed geographically because of their ability to colonize many substrates [5]. Perpetuation of many root-infecting fungi also is favored by wide host ranges. Thus the only major factors to limit geographical distribution are temperature and moisture. Distribution in the soil profile is largely limited by oxygen and availability of plant roots. Most plant pathogens are found in the upper 15 cm of the "A" horizon where organic matter and the majority of roots of many row crops are concentrated [6]. Approximately 60% of the soilborne diseases are initiated in the upper 7.5 cm of surface soil where fungi that cause basal stem rots are most prevalent, principally R. solani and species of Pythium, Phytophthora, and Sclerotium. Fungi that primarily attack roots may function at lower depths in the soil profile, e.g., Phytophthora cinnamoni, Verticillium albo-atrum, Phymatotrichum omnivorum, and vascular wilt fusaria. However, none of these pathogens is restricted to a particular zone. Sclerotia of P. omnivorum can be found at considerable depths, but upon germination the mycelium moves upward and girdles the root near the soil surface [7].

Garrett [8] has categorized seedling disease fungi as soil-inhabiting "unspecialized parasites." In this respect he refers primarily to Rhizoctonia solani and species of Pythium, which are highly competitive saprophytes growing freely through soil from substrate to substrate. Phytophthora also should be included in this group. Certain other root pathogens make limited growth or remain relatively stationary as dormant propagules until contacted by a host rhizosphere. Unspecialized parasites are most damaging to germinating seed and young seedlings but also attack feeder roots of older plants. Root-inhabiting "specialized parasites" are generally weak saprophytes that either depend largely on protection from microbial competition inside host tissue or persist as resistant propagules after death of the host. These may attack seedlings also but are not restricted by more resistant tissues of mature plants. Some pathogens considered to be weak saprophytic competitors are Helminthosporium sativum, Ophiobolus graminis, Phymatotrichum omnivorum, Verticillium albo-atrum, and the vascular wilt fusaria. However, the ecological classification of root-infecting fungi is incomplete due to insufficient information.

A summation of the "attributes of a successful parasite" is provided by Baker and Cook [9]. Some essentials of the parasite are adequate population (inoculum density), high survival capacity, rapid growth, escape from or tolerance to antagonists, and availability of susceptible host. Baker and Cook also discussed and classified types of biological interactions involved in disease. The biotic and abiotic environment imposes limitations on the

TABLE 1

Soilborne Pathogens and Seedling Diseases

Pathogen	Host range	Disease	Distribution	Mode of survival	Primary inoculum	Optimal soil conditions for disease
Rhizoctonia solani	Many plants	Seed decay; damping-off; root rots	Worldwide in temperate zones and tropics	Sclerotia; colonized organic matter	Sclerotia or mycelia	18–30°C; 30–80% moisture; high N
Pythium aphanidermatum	Many plants	Seed decay; damping-off	Worldwide	Oospores; sporangia	Zoospores	32–36°C; high moisture
Pythium ultimum	Many plants	Damping-off	Worldwide	Oospores; sporangia	Sporangia; zoospores	18–24°C; wet soil
Pythium myriotylum	Many plants	Damping-off; stem rot; wilt	Subtropical to tropical	Oospores	Zoospores	23–35°C; high soil moisture
Phytophthora cinnamoni	Over 300 species, conifers, orchards, ornamentals	Damping-off; feeder root disease	Worldwide	Chlamydospores	Chlamydospores; zoospores	Heavy soil; poor drainage; 19–25°C
Phytophthora parasitica	Over 42 families of flowering plants	Root rot; damping-off	World tropical and subtropical areas	Chlamydospores in crop residues	Chlamydospores; zoospores	High moisture; 25–36°C

120

Fusarium oxysporum f. sp. vasinfectum	Cotton	Vascular wilt	Worldwide	Chlamydospores; colonized organic matter	Chlamydospores or mycelia	27–32°C; 80–90% MHC[a]; sandy, acid soils; low P and high N
Fusarium solani f. sp. phaseoli	Bean	Root rot	Worldwide	Chlamydospores	Chlamydospores	High N; late planting, soil > 18°C; high inoculum density (1000–3000 units/g)
Verticillium dahliae; V. albo-atrum	Cotton and many other plants	Wilt	Worldwide	Microsclerotia	Microsclerotia or conidia	Wet, cool (26°C), alkaline soils; 30–75% MHC
Aphanomyces euteiches	Pea, sugar-beet, and many others	Root rot	Worldwide; cool areas of high rainfall	Oospores	Zoospores	20–30°C; heavy, wet soils; 30–75% MHC
Helminthosporium sativum	Barley, wheat, other cereals, and grasses	Crown and root rot	Worldwide	Dormant conidia	Conidia	28°C; relatively dry conditions
Thielaviopsis basicola	Tobacco, cotton, bean, many other hosts	Black root rot	Worldwide	Chlamydospores	Chlamydospores or endoconidia	Cold, wet, alkaline soils, decomposing crop residues

TABLE 1 (Cont.)

Pathogen	Host range	Disease	Distribution	Mode of survival	Primary inoculum	Optimal soil conditions for disease
Ophiobolus graminis	Cereals and grasses	Take all	Worldwide	Mycelium in crop residue	Mycelium in organic matter	Light alkaline soils; 16–24°C
Sclerotium rolfsii	Many plants; most in Leguminoseae and Compositae	Sclerotial blight or stem rot	Tropical and subtropical countries	Sclerotia and colonized organic matter	Sclerotia; mycelium in organic matter	Wet summers; 28–30°C; light aerated soils high in organic matter; low pH
Phymatotrichum omnivorum	Cotton and 2,000 other broadleaf plants	Root rot	Southwestern U.S.	Sclerotia; colonized roots	Sclerotia	Warm, heavy, alkaline soils poorly aerated; phosphate fertilizers
Sclerotinia sclerotiorum (Whetzelinia sclerotiorum)	Many plants	Stem rot; white mold; wilt	Worldwide	Sclerotia	Ascospores; mycelia from sclerotia; seedborne	Cool, wet periods, winter, spring, summer; high crop density

122

Cylindrocladium scoparium	Over 66 genera of 31 families	Root and stem rot (black rot)	Worldwide	Microsclerotia	Microsclerotia	High soil moisture followed by low moisture stress and high temp.
Cephalosporium gregatum	Soybean	Brown stem rot	Midwestern and southern U.S.; Mexico	Mycelium in crop debris	Conidia	18–24°C
Macrophomina phaseolina	Soybean, corn, cotton, and many others	Root and stem rot; charcoal rot	Worldwide	Sclerotia	Sclerotia	Low moisture; 25–35°C
Armillaria mellea	Forest, fruit trees, and others	Root rot; damping-off	Worldwide	Colonized roots	Old infected roots; rhizomorphs	Cool, moist soil

[a]MHC = moisture-holding capacity.

rate of increase of organisms, thus maintaining a so-called biological balance. When this balance is not in equilibrium, e.g., when conditions prevail that weaken host resistance and favor pathogen activity, plant disease is most likely to occur. One must be constantly aware that pesticides applied to soil may also alter the ecosystem and start chain reactions that greatly affect both phytopathogens and associated microorganisms [4].

A. Pathogen Population and Survival Mechanisms

For successful disease control the researcher must know the life habits of a specific pathogen in its natural environment. Pathogen population refers to mass or number of propagative units of a pathogen per unit of soil [10] and this is synonymous with "inoculum density." The true relation between pathogen population and amount of disease is often poorly understood. High inoculum density does not necessarily result in more disease [3] and the propagule that survives for long periods may not be the unit that infects. Nevertheless, fungicidal dosage is usually based on the extent of field infestation as determined by appropriate population assessment methods [11-14]. A series of review articles has dealt with the significance of populations of Pythium and Phytophthora [15], pathogenic fusaria [16], Verticillium [17], and Rhizoctonia solani [18]. Baker and Cook [9] have compiled a table of population densities for selected soilborne pathogens.

Inoculum for host infection may consist of conidia, ascospores, basidiospores, zoospores, oospores, sporangia, chlamydospores, sclerotia, rhizomorphs, or mycelia embedded in organic matter. Some pathogens, such as R. solani, can produce as many as four of these types [19], although sclerotia and colonized organic matter are the most important. Menzies [10] has provided a very pertinent review of factors affecting inoculum production, dispersion, and survival. Most inoculum is associated with crop residues left in the field and could be much reduced by removing or destroying the debris but this is rarely considered practical. Dispersion and survival of inoculum are also related to crop residues as these are disked in preparation for the next seedbed. Menzies reminds us that much information on inoculum density has come from controlled experiments where the pathogen was added to soil, and surprisingly little is known of the true form and nature of primary inoculum in the field. Original inoculum, in response to a treatment or environmental changes, may produce secondary spores, sclerotia, or other propagules different from the original. For example, chlamydospores of fusaria may germinate and produce conidia, or macroconidia may convert to chlamydospores. Conidia of Verticillium albo-atrum have been observed to transform into chlamydospores or microsclerotia [20].

The modes of pathogen survival in soil in absence of a crop host are especially important to disease control. A fungus may survive by continued activity and growth on secondary hosts, nonsuceptible carrier hosts such as weeds, or saprophytically in crop debris [12]. Strong saprophytic competitors in the genera Rhizoctonia, Pythium, and Phytophthora continue to grow on the dead host or other organic matter. However, these and most other soil pathogens enter into an inactive phase of the life cycle. This may be inactivity imposed by environment or dormancy in resting structures. The many interacting factors of environment that influence the survival of fungi in soil have been thoroughly reviewed [8, 12, 22–24].

The significance of weeds as symptomless carriers of potential pathogens was demonstrated by Wilhelm [25] using a sand culture isolation technique. Solanum sarachoides, a weed in strawberry fields, harbored a number of pathogens, including V. albo-atrum, Pythium ultimum, Macrophomina phaseoli, and Fusarium roseum. The rhizosphere effect may be involved in perpetuating fungi of intermediate competitive saprophytic ability, as found in F. oxysporum and F. solani f. sp. phaseoli. Many other examples of pathogen survival on roots of nonsusceptible plants have been cited [12, 21], thus pointing up the need for combined weed and pathogen control measures.

Most fungi are present in soil in the inactive or dormant state with relatively short periods of activity occurring when conditions are favorable for propagule germination or continued growth of quiescent mycelia. Two types of dormancy have been recognized [26]. Constitutive dormancy is a condition of inactivity or restraint upon development of a cell which is imposed by factors relating to cell structure or function. Oospores of phycomycetes and sclerotia with rind coverings, as in Sclerotium rolfsii [27], belong in this category. Exogenous dormancy is imposed by unfavorable environment and applies to most asexually produced spores and resting hyphae [24]. Survival of dormant propagules may cease upon exposure to extreme changes in the physical or chemical environment or upon contact with lethal biocides. Sussman [28] provides a table showing the longevity of fungal spores and the conditions under which they survive. It should be noted that most data on survival are obtained under artificial environments and we still know relatively little about survival of most pathogens in nature.

Exogenous dormancy is directly related to soil fungistasis, that widespread phenomenon of natural soils which prevents propagule germination under conditions that are otherwise favorable for germination. Factors contributing to fungistasis have been reviewed [29–31] and its significance to pathogen activity and survival has been discussed in numerous publications. There is little doubt that a combination of antibiosis and nutrient deficiency at the spore surface created by microbial competition is the major cause of fungistasis. This being the case, the prominence of plant rhizosphere influence on spore germination becomes clearly evident. For in this zone nutrient concentration is higher, due to root exudations, and microbial activity is intensified [32–34]. The qualitative nature of root exudates and microbial metabolites varies somewhat with the kind of plant and changes in soil environment; consequently, the response of fungal propagules will vary.

The principal genera of the Oomycetes that incite seedling diseases
have some survival mechanisms in common, but generalizations can be
hazardous. For species of Pythium the chief mechanisms of survival are
by zoospores and sporangia for very short and intermediate periods and
oospores for longer periods [35]. Although Pythium is generally considered
to belong to the soil-habiting classification according to Garrett, long-term
persistence of mycelium in field soils is not believed to occur [36]. Several species of Phytophthora exist in soil as chlamydospores and require
certain exogenous nutrients for germination [37]. Less is known about the
role of oospores in survival, but it has been suggested for P. cinnamoni
that these structures are important in sandy loam where moisture content
is in excess of 3% [38]. Our knowledge of the survival of Aphanomyces
euteiches and A. cochlioides in soil is incomplete [39]. Since mycelium
and zoospores apparently have little survival value, there is good circum-
stantial evidence that known survival of these fungi in fields for at least 10
years may be attributable to oospores. There is considerable evidence that
numerous alternate hosts serve to perpetuate these pathogens and increase
oospore concentration.

A great many soil fungi produce chlamydospores. Among potential
seedling pathogens, chlamydospore survival and function have been most
studied for formae of Fusarium oxysporum, F. solani, and F. roseum.
Such spores are usually produced in or on diseased host roots or in hypo-
cotyl tissue. Survival value apparently varies considerably among the
former. Chlamydospores of F. solani f. sp. cucurbitae (race 1) are short-
lived in soil as compared with the long survival of these structures in
F. solani f. sp. phaseoli [40]. Most propagules of F. oxysporum formae
disappear soon after a host crop is removed but some persist for long
periods. When macroconidia of F. solani f. sp. phaseoli are added to
moist, unsterilized field soil they germinate to form short germ tubes
which then produce chlamydospores, or the conidia may convert directly to
chlamydospores [42]. Both spore germination and production of new chlamydospores are closely related to the availability of exogenous carbon and
nitrogen [43]; glucose and ammonium nitrogen were particularly conducive
[44].

Thielaviopsis basicola, which attacks roots and hypocotyls of many
plants, also exists and survives in soil primarily as chlamydospores [45].
The endoconidia of this fungus generally are considered to be short-lived,
but studies have shown that they may live up to 8 months at low temperatures, suggesting overwintering capacity and a potential source of primary
inoculum in the spring [46].

Several root-infecting fungi produce sclerotia; prominent among these
are Rhizoctonia solani, Verticillium albo-atrum (or V. dahliae), Sclerotium
rolfsii, Sclerotinia sclerotiorum, Phymatotrichum omnivorum, Macrophomina phaseoli, and Cylindrocladium scoparium. Coley-Smith and
Cooke [23] reviewed and discussed the types, structure, survival, and

germination of fungal sclerotia and provided a table of survival periods under various field or experimental conditions. The influence of environmental factors on sclerotial survival seems to vary as much with researchers doing the study as with the specific fungi involved. It is seldom safe to say that soil temperature, moisture, aeration, organic matter, or pH alone limits survival of sclerotia, since all of these factors and the microbial population are interacting. Since most sclerotia can live under certain environmental conditions for several years, long rotations of 3-5 years free of susceptible crops are usually necessary for effective disease control [9]; in most cases this is not considered economically desirable. The difficulty of controlling sclerotial fungi with fungicides lies principally in failure to obtain adequate contact at all depths. While most sclerotia are in the upper 30 cm of soil, those of Phymatotrichum omnivorum have been found at depths to 2.5 m [47]. The more complex sclerotia of Sclerotinia sclerotiorum, Sclerotium cepivorum, and S. rolfsii have rind coverings which probably make them more resistant to pesticides. However, damage to the rind may induce the sclerotia to germinate or lead to colonization by other microorganisms [23]. Because of frequently observed action of other organisms on sclerotia, the idea of integrated control of pathogens has been receiving more attention. The selective action of sublethal concentrations of fungicides can induce biological suppression or destruction of propagules by antagonistic nontarget microorganisms [48, 49].

Fungi which produce microsclerotia (species of Verticillium) have the advantage of ready dissemination by runoff water and by windborne dust, and their large numbers provide many potential infection sites. Microsclerotia of V. albo-atrum possess exceptional longevity [50] and their survival ability apparently is highly stable with a low attrition rate of viable propagules in soil [51]. This pathogen, like several others, is also perpetuated on weed hosts and to some extent in colonized organic matter. Conidia produced in soil are usually short-lived.

Competitive colonization of organic matter is often a better survival mechanism than sclerotium production. All strains of R. solani do not readily produce sclerotia. There is sufficient evidence that this pathogen persists in soil primarily as a saprophyte in killed host tissues or in colonized crop debris [52, 53]. Perpetuation on weed hosts also is significant. One must keep in mind the great variability among strains [54], resulting in varied capacity for colonization, growth, survival, and infection. Ophiobolus graminis (Gaeumannomyces graminis) is not known to produce any kind of dormant propagule and persists almost entirely as mycelium in infected plant tissue [8, 12, 24]. The fungus suffers less loss of viability during low temperature or drought periods when other microbial activity in colonized tissue is low. Survival of O. graminis apparently is enhanced by added nitrogen, while survival of Cochliobolus sativus (Helminthosporium sativum) is reduced [8]; the latter, however, does have the advantage of dormant conidia.

Several other fungi should be mentioned as potential seedling pathogens. Plasmodiophora brassicae may enter long periods of complete dormancy of resting sporangia, which causes the release of many zoospores for infection of cruciferous host roots [9]. Cephalosporium gramineum becomes established in wheat straw after death of the host. This organism probably dies before complete disintegration of the straw [55] as stronger competitors such as species of Trichoderma take over. Another vascular pathogen, C. gregatum, causes a known stem rot of soybeans. Propagules of the onion pink root pathogen, Pyrenochaeta terrestris, were found most often associated with soil particles 0.5-1.0 mm in diameter and survived to a depth of 45 cm in soil [56]. Phoma betae attacks sugar beet seedlings; the fungus was recovered from North Dakota and Minnesota soils 26 months after planting but not in the third year [57]. Colletotrichum coccodes (C. atramentarium) attacks solanaceous plants and produces sclerotia on which spore-bearing acervuli may form. C. Gossypii has long been a serious pathogen of cotton seed and seedlings. Even some soil fungi generally known as saprophytes can be devastating to seedlings in certain climates. Preemergence and crown rot diseases of peanut inflicted by Aspergillus niger have caused great economic losses [58].

B. Inoculum Potential and Disease

Rather broad definitions have been applied to the term "inoculum potential." For our purpose, it means the capacity of a given inoculum density (mass of inoculum per unit of soil) to infect and cause disease. Relating most critically to this function are nutrient availability and genetic capacity to infect in a particular environment. As expressed by Wilhelm [3], propagule numbers (density) can be estimated independently, but factors of infection and disease are dependent upon external conditions. These are many and varied, and inocula of different pathogens respond differently to the prevailing environment. Aside from the effects of the common environmental components, temperature, moisture, aeration, carbon dioxide, and soil pH, perhaps the greatest influence on infection and disease severity is imposed by nutrient availability and microbial interactions at the root surface. Here is where dormant propagules must germinate or mycelia in organic matter become activated in response to root exudates which provide the energy for infection. The review by Baker and Martinson [19] on epidemiology of diseases deals with some of the factors affecting disease incidence and severity. The influence of plant root exudates on spore germination, microbial activity in the rhizosphere, host response, and pathogenesis has been well established [33, 59-61]. Development from spore germination to infection is particularly influenced by carbon and nitrogen sources available to the fungus [62]. The possible interactions between various pathogens and nitrogen amendments have been summarized in a table by Henis and Katan [63].

Some diseases increase in incidence while others decrease under the same conditions. Smiley [64] also summarized the influence of nitrogen form and soil pH on development of root diseases and concluded that ammonium nitrogen acts to reduce the rhizosphere pH while nitrate nitrogen increases it. Other mineral nutrient effects have been considered by Sadasivan [65] and by Curl and Rodriguez-Kabana [66]. The ratio of carbon to nitrogen seems to be very critical in determining whether certain organic supplements applied to soil suppress or stimulate pathogens [67]. High carbon-nitrogen ratios usually deter pathogen activity, whereas addition of readily utilizable nitrogen favors disease.

The dynamics of cotton seedling disease discussed by Bird [68] could well be essentially the pattern for many other seedling-pathogen relationships. Trouble begins when the seed is planted, as seed rot, preemergence damping-off, postemergence damping-off, and root damage occur in sequence involving any one or combinations of five to ten organisms, including R. solani, species of Pythium, Thielaviopsis basicola, species of Fusarium, Verticillium albo-atrum, bacteria, and nematodes. Frequencies of pathogens and associated microorganisms vary with seed quality, dates of planting, and other factors. A high frequency of species of Fusarium in the complex was associated with reduced disease, suggesting control by competition. A critical time for initiation of disease is at seed germination when higher amounts of carbohydrates and ninhydrin-positive materials are released, particularly at low soil temperatures [61].

C. Methodology

The extreme complex nature of the soil environment and microbial interactions virtually precludes standardization of techniques for root disease investigations. Methods usually must be deivsed or modified for specific pathogens and specific objectives. Individual published accounts of experimental procedures dealing with assessment of pathogen populations, growth and survival in soil, isolation and culture, and host inoculation and disease measurement are far too numerous to relate here. Many of the methods used most for investigating soil-fungus ecology and root diseases can be found in the compilations of Parkinson et al. [13] and Johnson and Curl [14]. Certain methods and media used in general plant pathological research [69] also are applicable. Sinclair [70] has provided a comprehensive review of special methods for the study of Rhizoctonia solani. Of these, the methods of assessing Rhizoctonia concentration in soils [52, 71] and methods of evaluating inoculum density [72] are especially useful. Selective media for isolation of pathogenic fungi are quite essential; these have been discussed by Tsao [73]. Techniques that have been suggested for herbicide-soil microorganism interaction studies [74] and for various aspects of biological control work [9] may be adaptable to fungicide investigations.

Ultimately, it is desirable to determine plant disease severity and losses following fungicide treatments. An old standard source for appraisal of disease losses [75] is still extremely useful. The procedure of Sherwood and Hagedorn [76] for determining root rot potential in pea fields is applicable for many situations; similar rating systems have been described for Aphanomyces root disease of peas and sugarbeets [39]. The subject of assessment of plant disease and losses has been recently reviewed [77].

III. SOIL TREATMENT

The addition of any compound to soil results in changes in its microbial life which depend not only on the nature of the compound but also on the form (formulation) and manner of application of the compound to the soil. While the spectrum of antifungal activity of fungicides depends primarily on the physical and chemical properties of the compound in relation to the target organism, it is not always possible to predict from results with in vitro tests its effectiveness against pathogens in the soil environment. The subject of the behavior of fungicides in soil has been reviewed by Kreutzer [4, 6] and more recently by Munnecke [78, 79]. Although it is not appropriate to cover the subject here, some general rules on the effectiveness of fungicides in soils against plant pathogens with respect to pathogens and also with reference to the behavior of fungicides in soil are as follows: (1) fungi with sclerotial structures are more resistant to toxicants than those without them [4]; however, there is no rule with respect to the relative effectiveness of a toxicant against conidia or spores of a target fungus in comparison to its hyphae; (2) the greater the degree of embedment of an organism in organic material, the higher the probability of escaping the effects of soil toxicants; and (3) the broader the spectrum of activity and the greater the effectiveness of a compound against fungi, the more perdurable its effects on the soil microflora.

The fate and effectiveness [4, 6] of a compound added to soil depends on the reaction of the soil with the chemical; the degree of sorption by the soil; and the rate of degradation of the compound in the soil. Degree of sorption in turn depends on the type of clay present and perhaps, more importantly [78], on the amount of organic matter present. Montmorillonitic expanding-lattice (hydrated) type clays offer less sorbing surface for nonpolar compounds than fixed clays such as kaolinite. Also as a rule sorption of a chemical in soil increases with its content in organic matter. A corollary of sorption is that the greater the degree of sorption by a soil for a compound, the lower the efficiency of the toxicant in that soil against pathogens.

A. Soil Biocides and Fumigants

Soil biocides are materials with broad-spectrum activity which, upon application to soil at sufficiently high dosage, eliminate or reduce microbial

populations, permitting subsequent development of seedlings in an environment free of pathogens. Typically, the reduction in microbial numbers is
short-lived and is followed by a burst of biological activity reflected in
increments in bacterial numbers, increased respiration, and colonization
of organic debris by fast-growing fungi such as those in the genera Trichoderma, Aspergillus, and a number of Mucorales. This state of biological
hyperactivity gradually disappears and microbial activity returns to pretreatment conditions.

The word fumigant is from the Latin "fumigare" and implies "to treat
with smoke." As such, a fumigant is defined as a material that when added
to soil would be a gas or will release a gas (vapor). Among the earliest
fumigant-type materials in agricultural use were carbon disulfide and
hydrocyanic acid [80]. A number of other materials quickly followed, such
as formaldehyde, chloropicrin, etc. Reviews on the early fumigants have
appeared and it is beyond the scope of this chapter to cover such early work
[81, 82]. Our discussion will instead be limited to representative fumigants
commonly used.

1. Halogenated Hydrocarbons

Among the commercially available materials several halogenated hydrocarbons in their various formulations occupy a key role in soil treatments
today.

a. Dichloropropenes and Mixtures. Dichloropropene was first introduced
by Carter [83] in Hawaii for control of soilborne diseases in pineapples.
Carter's material consisted of a mixture of 1,2-dichloropropene and 1,2-
dichloropropane. Later work demonstrated that the dichloropropene was
the active ingredient in the mixture [84]. In 1956 the Dow Chemical Company introduced a commercial formulation (Telone) which is for the most
part 1,2-dichloropropene. This compound is principally active against
nematodes at rates of 8-10 gal/acre (75-93 liters/ha). At higher rates,
fungicidal activity has been reported against Sclerotium rolfsii [85], species
of Fusarium [86], and species of Phytophthora [87]. Parris [88] found very
little effect of the dichloropropane-dichloropropene (DD) mixture on species
of Rhizoctonia and Fusarium, even at rates of 100 gal/acre (935 liters/ha).
Since these early investigations, it is generally recognized that dichloropropene is a weak fungicide and the only antifungal use recommended by the
company producing DD (Shell Chemical Co.) is for reduction of verticillium
wilt in mint.

In an effort to increase the antifungal properties of dichloropropenes,
formulations have been prepared consisting of mixtures of dichloropropene
with 15% chloropicrin (DD/Pic, Telone C) or with 20% methyl isothiocyanate
(Vorlex, Trapex), and even in a formulation containing 17% methyl isothiocyanate and 15% chloropicrin. These mixtures combine the well-known
antifungal properties of chloropicrin [81, 82] and methyl isothiocyanate [89].
These mixtures are very effective broad-spectrum fumigants and require,

as does dichloropropene alone, application to the soil one to several weeks ahead of planting to avoid phytotoxicity.

b. **Methyl Bromide.** The use of this sterilizing gas has been restricted to the treatment of nursery beds, vegetable and fruit production, and ornamentals. Its first use as an insecticide was reported by Le Goupil [90]. Its effectiveness as a soil fumigant was soon recognized, and by 1955 [81] it had an important place in the pharmacopoeia of agricultural chemicals. Unlike DD, methyl bromide has a low boiling point (4.5°C) and requires either a plastic cover for effective confinement of the gas, or application with deep injectors [91, 92]. The degree of confinement also depends not only on the type of plastic used but, for polyethylene tarps, on its thickness [3].

Although methyl bromide is a good nematicide, its performance as a soil fungicide is not considered as good for some applications [81]. Consequently, formulations containing methyl bromide and chloropicrin have been made available. Of these, a mixture of 67% CH_3Br and 33% chloropicrin is the most effective.

c. **Other Halogenated Hydrocarbons.** Other halogenated hydrocarbons commonly used for soil fumigation are almost exclusively restricted to nematicidal activities. Ethylene dibromide, like the dichloropropenes, is generally considered a mild fungicide although its value in controlling certain nematode-fungal complexes was recognized early [81]. Perhaps of more interest is the fungicidal value of the nematicide 1,2-dibromo-3-chloropropane (DBCP). This compound was introduced by McBeth and Bergeson in 1955 [93]. Since then, its antifungal effects have been reported for Rhizoctonia solani [94] in tomato seedlings and against Pythium ultimum [95]. In Alabama, its use is recommended for reduction of peanut pod rot primarily caused by R. solani [41].

2. Methyl Isothiocyanate and Precursors

The toxicity of alkyl isothiocyanates has been known for some time. However, it was only in the late 1950s that commercial use was made of methyl isothiocyanate (MII) alone or in combination with dichloropropene. Methyl isothiocyanate is a nematicide [96] as well as a good fungicide [97]. The activity of air passing through soil columns treated with MII against mycelium of Rhizoctonia and Pythium and spores of Myrothecium verrucarium was reported by Munnecke et al. [98]. These authors found that MII activity was greatest against Pythium and least for control of Rhizoctonia.

The fungicidal activity of MII, and of compounds which release it when added to soil, is interesting in that isothiocyanates are liberated from soil during crucifer residue decomposition [99] and are active against species of Aphanomyces [39].

Sodium N-methyldithiocarbamate (SMDC, metham sodium, Vapam) and 3,5-dimethyltetrahydro-1,3,5,2H-thiadiazine-2-thione (DMTT, dazomet, Mylone) are compounds that produce MII during their decomposition in soil [89, 100]. Although their behavior and mode of action is not identical with that of MII [79], they are considered here as having very similar antifungal properties and are sold today as general-purpose soil fumigants. For example, SMDC is formulated as a 31-33% (anhydrous basis) material and is used in broadcast treatments at rates of 40-100 gal/acre (374-934 liters/ha), depending on the method of application. Dazomet is formulated as 25-50% dusts, 50-98% wettable powders, and 19-24% liquid concentrates. This fungicide is used at rates between 75-1000 lb/acre (84-1,118 kg/ha) most commonly in the range of 250-300 lb/acre (280-360 kg/ha) 4 weeks or more before planting. It is not recommended for use above 32°C, and waiting time before planting should be increased when soil temperatures are below 15°C.

Potassium N-hydroxymethyl-N-methyldithiocarbamate (Bunema) is a compound like methan sodium which is currently being explored for possible agricultural use [101]. Very little is known regarding the mode of action of this compound, which presumably should parallel that of metham-sodium. There are apparently differences in activity, since this compound appears to be active at lower rates than metham-sodium.

3. Azides

These long-known compounds are the classical enzyme inhibitors of cytochrome oxidases, peroxidases and, in general, of metal-containing enzymes. Because of the basic nature of these enzymes to metabolic processes, azides can be expected to have a very broad-spectrum of activity. The agricultural uses of these compounds have been explored only relatively recently, and sodium and potassium azides are currently under review for use as soil treatments. Sodium or potassium azide (NaN_3 or KN_3) have been studied for activities against nematodes [102, 103], damping-off organisms (species of Cylindrocladium, Sclerotium rolfsii), and other soilborne plant pathogenic fungi [104-110]. The activity of NaN_3 or KN_3 depends on the liberation of hydrazoic acid (HN_3) when the salts come in contact with acidic soils [111, 112]. As such, these azides can be considered as fumigants. However, these materials do not require sealing with a plastic cover after application, and a water seal or thorough incorporation into the soil is sufficient for good results in preplant applications. Because of the extreme phytotoxicity of these compounds, they must be applied to the soil 2-3 weeks before planting or as a postemergence to the soil surface when plants are sufficiently large as to avoid serious injury to the root system [106, 107]. Sodium azide is formulated in a clay granule containing 2% Na_2CO_3, as a stabilizing buffer, to avoid rapid release of HN_3 on contact with acidic environments during transportation. Rates for use of NaN_3 are still under investigation, but generally a 30-60 lb/acre (33-67 kg/ha) for preplant applications and 8-12 lb/acre (9-13.5 kg/ha) for postemergence applications on peanuts and potatoes [106, 107] are adequate.

B. Methods of Application of Fumigants

Several recent publications have described the various methods of application of fumigants. Among them are those of Purdy [2] and, more recently, a discussion by Good [113] on fumigants with nematicidal activities. The following discussion will then be limited to brief descriptions of the various methods available. For more details, either Good's review or specific citations given in connection with the method should be consulted.

1. Small Areas

The methods available for use in gardens, experimental plots, or in limited areas (where large machinery cannot operate or it is uneconomical to operate) are for most fumigants limited to trench applications, watering in, or the use of injection guns.

a. Trench Application. This method consists of simply opening furrows in the previously loose and well-worked soil to a depth of 14-20 cm and trickling in the material at the recommended rate by means of a perforated can or similar type applicator. After application, the trench is quickly covered with soil to seal in the material. For most garden-type operations, furrows should be spaced every 25-30 cm. This type of procedure is well suited for application of relatively high-boiling fumigants such as dichloropropene and mixtures of this material that do not contain irritants, e.g., chloropicrin. Soil moisture should be about 50% of field capacity. For best results, after covering with soil and lightly packing the surface, a small amount of water should be added to cover the surface and form a water seal. In general, most fumigants should not be applied if the soil temperature exceeds 32°C. The wait period for planting after application varies with the material, the dosage, and the prevailing soil temperatures. In general, a period of 1-2 weeks is sufficient for most commercial dosages applied at temperatures in the range of 15-27°C; if the temperature is below 15°C, the length of time should be increased.

b. Watering In. These applications are limited to materials that are soluble in water (e.g., metham sodium or Bunema) or are available in emulsifiable formulations. The material can be metered in during irrigation by means of a hose equipped with a venturi-type valve. Soil should be lightly wetted first and the material should then be metered into the hose. For deep penetration of the material, the total dosage for the treated area should be divided into two or three portions and these applied at intervals followed by water alone between portions. This system permits penetration 40-50 cm deep when done correctly [113].

c. Injection. Injection of fumigants is most easily performed with the various types of commercially available guns. These are syringe-like instruments that permit, by different settings in the plunger, injections into the soil of accurately given doses of fumigants. For best results, a grid can be traced over the area to be treated so that injections are effected every 30-40 cm. The type of diffusion pattern obtained with this method of application is that described by Goring [114] for single-point injection. This results in a distorted sphere with its center at the point of delivery of the fumigant into the soil. The sphere is skewed downward from the point of injection. When a grid is used for multiple-point applications so that the injection points are sufficiently close to each other, the spheres of diffusion overlap and a broadcast treatment effect is obtained after adequate time is allowed for horizontal diffusion of the fumigant.

2. Field Applications

Large areas are generally treated with fumigants in broadcast or row patterns. This is accomplished with the use of chisels, blades, sweeps, spray and cover operations, application of granules, or through irrigation systems.

a. Knives and Chisels. For applications through chisels or knives, the material is metered from a reservoir by means of a gravity-flow valve when large (greater than 0.5 gal/acre; 4.7 liters/ha) volumes of fumigants are delivered per chisel. Gravity-flow valves are of relatively simple construction and consist of a tube with a small opening to the atmosphere to permit air flow and a disk with a hole the diameter of which varies according to the rate to be delivered. There are connector tubes on both ends. One tube goes to the reservoir, which is kept tightly closed except for a vlave that serves as a level indicator. The other tube is connected to the application point in the soil. The pressure head necessary for adequate delivery of recommended rates varies according to the density of the fumigant. Thus for dichloropropene (DD) the head is 30 cm higher than for DBCP since lower rates are needed for the latter material and its density is higher. A spring-loaded valve permits turning the system on or off for application of the fumigant. The fumigant is delivered to the soil through plastic tubes connected to a piece of metal tubing soldered to the back side of the chisel or knife. In row or bed treatments one or more chisels are set per row or bed, spaced to provide enough distance between lines of application. Under these conditions, diffusion and control patterns do not overlap. In broadcast treatments chisels are closely spaced so that the patterns of diffusion and control overlap. The row pattern of diffusion is typically a pear-shaped cylinder skewed downward from the line of injection [114]. In broadcast applications diffusion patterns overlap. In discussing the effectiveness of row and broadcast treatments, Goring [114] states that, as the lines of fumigant application to the soil are spaced closer together, pest control becomes uniform horizontally; variation in control occurs only in the vertical direction; and the zone of control below is usually larger than that above the injection lines. A change from a control pattern typical of row applications to that typical of broadcast fumigation can result from gradual increase in exposure time or rate of fumigant applied. Once the broadcast control pattern has been established, additional increments in fumigant increase both depth of control and control near the soil surface.

In using chisels the tractor should be moving when the chisels are pulled out of the soil at the end of the field. This reduces the chances of clogging the tip of the delivery tube behind the chisel. Considerable difficulties are sometimes encountered with the use of chisels in rocky or hard soil and in fields where large amounts of debris are present. In such cases, the use of a colter ahead of the chisel eliminates some of the difficulties by precutting the soil. The use of anhydrous ammonia applicators for delivery of fumigants in heavy or rocky soil has been suggested [113]; the springs in the tines of these applicators make such a system particularly well suited for rocky soils. Depth of placement of fumigants should be at least 20 cm and optimally 20-30 cm. The fumigant, after delivery to the soil, should be covered immediately by disk hillers, and a press-wheel cultipacker or soil drag should be used to seal the chisel channels effectively.

Fumigants can be added during subsoiling operations behind chisels or delivered behind moldboard plows. In using this type of equipment for broadcast treatments, care should be taken that the desired amount is delivered, since the area covered by the plow is larger than that covered by a chisel. The fumigant in these cases can be delivered with pressure equipment and a nozzle attached to the plow to obtain greater rates of delivery or several gravity-flow lines can be attached per plow.

b. Blade-type Fumigant Applicators. These applicators were proposed as early as 1954 by Jensen and Page [115] for broadcast applications. The device consists of a hollow wedge inside which are placed a number of nozzles. The blade (wedge) is pulled by the tractor into the soil to a depth of 15-20 cm; the chemical is sprayed and covered by the soil as the blade moves through the soil. Presumably this type of applicator results in a more homogenous distribution of the fumigant and better control than the chisel.

c. Sweeps. An approach to broadcasting fumigants, similar to the blade-type applicator, is that of adapting sweeps for spraying fumigants with nozzles [116, 117]. The sweeps used for this purpose are triangular boxes with slanting sides and an open base. The nozzle is placed at the angle tip and the spray is directed backward from the direction of movement.

Power requirements for the blade-type and special sweep applicators are greater than for chisels; consequently, their use could be limited in heavy soils or soils with a great deal of obstruction.

d. Spray and Cover. This system, which has been useful for fumigants such as metham-sodium, consists of spraying the material into the soil immediately ahead of bedding equipment. The use of rototillers with mounted nozzles, as for herbicide application, has not been generally successful for application of fumigants. Most rototillers do not penetrate deeper than 12-15 cm or they have very large power requirements for deeper penetration. Since fumigant application requires placement to depths of at least 20-25 cm, this type of equipment has not been found to be very effective. In general, when the spraying is not done ahead of a bedding operation, the use of disking equipment after spraying of the fumigant affords a more effective seal than rototilling [113].

e. Application of Solid Formulations. Some halogenated hydrocarbons have been prepared in granular form and the use of these formulations has been suggested where chisel, blade, or other standard forms of applications are not feasible [118-120]. In addition, materials such as dazomet (Mylone) are available only in dust or granular formulations which are generally delivered to the soil surface with a fertilizer spreader or with special granular applicators available commercially. These consist of V-shaped boxes with a rotor at the bottom which is self-powered or driven with sprocket wheels by the equipment on which it is mounted. The applicator box is perforated at the bottom with diamond-shaped holes. Aperture of the holes is changed with a sliding plate, also perforated with diamond-shaped holes. Generally the aperture is graduated with a gauge built on the box to permit calibration of the equipment. Granules delivered to the soil surface must be well worked in. This is usually achieved through the use of rototillers or by disking operations which can be performed simultaneously to the delivery of the granules. In general granular formulations of highly volatile fumigants are not as effective as other forms of application [119] since these materials require deep incorporation that cannot be achieved through standard rototilling or disking.

Granular applicators or spreaders permit broadcast or banded applications. With fertilizer-type spreaders, banding is most easily achieved by blocking off holes on the underside of the box with metal clamps or lengths of insulating tape.

f. Application Through Irrigation Systems. Emulsifiable concentrates of halogenated hydrocarbons can be applied through irrigation systems [121, 122]. In addition, application of metham sodium or emulsifiable concentrates of dazomet are recommended for use through irrigation systems [123]. In flood-type irrigation the concentrates are added to the water at points of maximum turbulence using simple, constant-head metering devices. Metering emulsifiable concentrates into overhead irrigation systems generally results in great loss of fumigants through volatilization when the water droplets are very small [113]. However, fumigants have been delivered using perforated irrigation pipes [124] which result in much larger drops, with reduced fumigant loss, and effective delivery of the material.

g. Use of Soil Covers. The use of a plastic seal to cover fumigated soil is necessary for low boiling materials such as methyl bromide. Several devices are commercially available which permit simultaneous application of methyl bromide to seed or nursery beds by means of chisels and coverage with a polyethylene sheet from a tractor mounted roll of plastic. For other fumigants, a plastic sheet permits more efficient use of the material by reducing the amount needed. Recent supply shortages and price increases in petroleum-based materials may not permit as free a use of plastic sealing in the future; other alternative methods of sealing will have to be found.

C. Soil Fungicides

The application of antifungal compounds to soil for control of plant pathogens is not as old as the use of fumigants. Apart from occasional and mostly unsatisfactory use of Bordeaux mixture and mercurials [125] for soil treatment, the use of fungicides in soil did not start until the development of pentachloronitrobenzene (PCNB) and the dithiocarbamates in the 1930s. Typically, the use of a compound in soil followed its development as a foliar fungicide. In this section we have therefore included not only compounds that are almost exclusively used as soil fungicides such as PCNB, but also foliar fungicides that have been used in soil treatment. Also included in this section are systemic fungicides that show promise for this use although today they are of limited use for soil treatment.

1. Dithiocarbamates and Captan

Among the dithiocarbamates, thiram has been used most frequently for soil treatment. This fungicide is fairly nonselective and has been found to be active against species of Pythium, Phytophthora, Aphanomyces, R. solani, Phoma, and others [4, 39]. It appears to be more effective against facultative saprophytes than against obligate parasites [4]. Its major use today is not as a soil fungicide but as a seed protectant. Thiram is available as dust, granular, wettable powder, or emulsifiable formulations. Formulations combining the fungicide with Actidione are available and presumably extend the range of activity of thiram in soil [123].

The activity of captan is similar to that of thiram in its broad-spectrum activity. This fungicide has been found to be active against species of Pythium and less so against R. solani [4]. It does not appear to be effective

against species of Aphanomyces [39], Sclerotium [125], or Phytophthora [4].
The major use of captan in soil is as a seed-protectant fungicide. It is
available as dust, wettable powders, or emulsifiable formulations and is
used in combination with PCNB for soil treatment, a formulation which has
a considerable spectrum of activity [123].

2. Halogenated Nitrohydrocarbons

Research on the antifungal activities of nitrohydrocarbons has been historically most successful with aromatic compounds, particularly with chloro-
nitrobenzenes. The use of chlorinated nitropropanes was explored relatively recently and led to development of the fungicide 1-chloro-2-nitro-
propane. Although this compound was found to be active against organisms
causing pre- and postemergence damping-off of seedlings (Rhizoctonia,
Pythium, Thielaviopsis, Aphanomyces, and Fusarium) [126, 127], it is
the only one of its class to reach commercial application and its use has
been very restricted. In contrast, the antifungal activities of chloronitro-
benzenes have been known since 1935 from the work of Brown [128] on their
effect on Plasmodiophora brassicae and R. solani. It was during this period
that I.G. Farbenindustrie introduced PCNB and since that time this compound has been found to be active against R. solani on a variety of row [129,
130] and vegetable crops [128, 131, 132], in trees and ornamentals [133],
and in the laboratory [134, 135]. The activity of PCNB against species of
Sclerotinia has been amply demonstrated [134, 136-138]. It is also active
against Sclerotium rolfsii [125] and S. cepivorum [139]. In addition, PCNB
is active against Streptomyces scabies [140] and other actinomycetes. Reports of activity of this compound against species of Rhizopus, Macrophomina, Penicillium, Trichoderma, and Olpidium have been listed by
Kreutzer [4]. However, PCNB has been found ineffective against species of
Pythium [131, 134, 136, 141] and Fusarium [134]; lack of activity against
these organisms as well as species of Phytophthora has led to the development of selective media for isolation of these fungi [142-144]. Similarly,
PCNB also is not active against Aphanomyces [39] and probably is ineffective against Colletotrichum gossypii, V. albo-atrum, Typhula itoana,
and Tilletia contraversa [4].

The tetrachloro analogue of PCNB, 2,3,4,5-tetrachloronitrobenzene
(TCNB), received some attention earlier. There appeared to be some basic
differences in activity from that of PCNB, particularly in the control of
species of Fusarium [145] for which TCNB is effective. Eckert [146] noted
that progressive chlorination of benzene enhanced fungistatic activity against
Rhizoctonia, but not against Pythium, and decreased phytotoxicity.

The effect of PCNB on plant parasitic nematodes has received some
attention in recent years. Boswell [147], working with peanuts, confirmed
earlier results by Miller on strawberries [148, 149] in which PCNB was
found to increase populations of the lesion nematode Pratylenchus
brachyurus.

At present, PCNB is marketed as a granular formulation in attapulgite-type clay, as a wettable powder, and a dust, and recently it has been made available as an emulsifiable concentrate. This array of formulations permits applications of granules or powders directly to the soil surface, as for the control of S. rolfsii on peanuts; application to the soil, followed by incorporation; in-furrow applications at planting time, in granule or liquid form; or even through irrigation systems, by injection into the water lines.

A common formulation of this fungicide with ethazol (ETMT) is much used in cotton as an in-furrow treatment for control of seedling diseases. The inclusion of ETMT broadens the spectrum of the formulation to include Fusarium and phythiaceous fungi.

Tolerance of susceptible pathogens to PCNB has been reported [150] and there are differences in susceptibility to it by various isolates of pathogens such as Rhizoctonia solani [151].

Recently, Richardson presented dosage-response curves to the vapor phase of 33 chlorine derivatives of benzene, nitrobenzene, and m-dinitrobenzenes against T. viride, R. solani, and P. ultimum [152]. Vapor action was greatest in the nitrobenzene series and least in the dinitrobenzene series. Within each series, derivatives with one or three chlorine substituents were generally more active than the parent chemical; with further chlorination, activity declined progressively as volatility decreased. There is sufficient vapor pressure for PCNB to support Goring's [114] earlier suggestion that this compound would be expected to diffuse appreciably through soil.

In soil, PCNB is converted to pentachloroaniline [153] which is somewhat less toxic than the parent compound. There is evidence [154] that PCNB may accumulate in bean plants to levels 0.1-2.0 times those in surrounding soil, and severity of disease in treated plants with PCNB is less than in plants not exposed to the fungicide.

3. Fenaminosulf (Dexon)

This narrow-spectrum soil fungicide was first described by Urbschat in 1960 [155]. It is particularly effective against the water molds, i.e., pathogens in the genera Aphanomyces, Phytophthora, or Pythium and relatively ineffective against R. solani [156-163]. The fungicide is decomposed by light [159]; however, this does not impair its effectiveness once incorporated in the soil. Thus, Khan and Baker [164] found that effectiveness of fenaminosulf against P. ultimum damping-off of cucumber and alfalfa did not decline 28 days following application. Effectiveness of the chemical, measured by colonization of potato discs by Pythium, declined almost linearly throughout the experimental period and was thought by the authors to represent more closely the residual fungicide in the soil. Alconero and Hagedorn [165] confirmed and explained the results obtained earlier by Mitchell and Hagedorn [161] in which the effect of fenaminosulf was found to be beneficial against pea root rot even after a series of plantings in infested soils treated only once with the compound. Chemical analysis of soils treated in the greenhouse showed substantial residues of fenaminosulf even after one year. The authors indicated that thorough mixing of the fungicide in the soil under field conditions was essential to maintain adequate protection of the roots.

Fenaminosulf is available as a dust or granular formulation and its application depends on thorough mixing with the soil. Although applications such as soil drenches [158] have been tried, these are generally considered to be too expensive as they require treatments at weekly intervals.

4. Ethazol-ETMT (Terrazole)

Introduced by Thomas [166] in 1962, ETMT has received relatively little attention from researchers. Its spectrum of activity is, in some respects, related to that of fenaminosulf and PCNB. Ethazol is active against various species of Phytophthora [167, 168] and Pythium [167, 169, 170]; at the same time, it has been reported to have systemic activity in tomatoes [171] and cotton, affording protection to seedlings against pre- and postemergence damping-off caused by R. solani [172]. Its major use has been in a mixture with PCNB intended to provide broad-spectrum activity against soilborne pathogens. This mixture is available in granular, dust, or emulsifiable formulations. It is principally used in the control of seedling diseases of cotton and for control of turf diseases.

5. Systemic Fungicides

Research on systemic fungicides received impetus in the past decade with the introduction of compounds like benomyl [173]. A large body of literature has accumulated on the subject, as indicated by the review of Erwin [174] and the coverage given to the subject in other chapters of this book. The application of systemic fungicides to soil and seeds has received considerable attention and the following discussion will illustrate the use of representative systemic fungicides from major chemical groups on the market.

a. Benomyl. This compound is a benzimidazole introduced by DuPont in 1968 [173]. It has the widest spectrum of activity of all the newer systemics and is the most effective of the benzimidazole fungicides [174]. It has been found to be active against several soilborne pathogenic fungi, notably R. solani [173, 175-177], V. albo-atrum [173, 178-180], Fusarium species [173, 181-183], T. basicola [184-186], S. Sclerotiorum [187, 188]. S. cepivorum [189], Cercosporella herpotrichoides [190], C. scoparium [191, 192], C. floridanum [192], and P. omnivorum [193]. Benomyl is not effective against pythiaceous fungi in the genera Pythium, Phytophthora, or Aphanomyces [173, 194], and not effective against S. rolfsii, Helminthosporium species and Alternaria species [173]. A study on the relation of taxonomic groups and toxicity of this compound has been published [195].

The lack of activity against pythiaceous fungi and S. rolfsii has resulted in
marked increments in disease, caused by these pathogens, in fields treated
with benomyl and related fungicides [196, 197]. Since benomyl is heat
stable, it has been utilized in selective media for isolation of pythiaceous
fungi and Armillaria mellea [14, 69].

Benomyl has been applied at planting time on seeds for control of soil-
borne pathogens [198], as in-furrow spray [194, 198], by injection into
soil [179, 180, 198], or to cotton in horizontal bands at planting time and
later in vertical bands as a side dress under irrigation furrows [179]. In
addition, postemergence applications have been tried as foliar sprays for
control of seedling diseases [198, 199]. Other applications to soil have
been in the form of sprays or as soil drenches [200, 201]. The use of surfactants for soil applications of benomyl generally increases its effectiveness [179, 180, 202]. Studies on the movement of benomyl through soil
[203] indicate that acidification and hence solubilization through conversion
to MBC, and the appropriate use of a surfactant, such as Tween 20, markedly enhance the movement of benomyl in soil. The possibility of applying
benomyl to the soil to control above-ground diseases has been explored for
the control of dutch elm disease and powdery mildews among others [201,
204]. However, this procedure while attractive is uneconomical because of
the large amounts of material required to attain acceptable effectiveness.

b. Carboxin and Oxycarboxin. These were probably the first of the modern systemic fungicides to succeed in large-scale use [174]. They were
first reported as fungicides by von Schmeling and Kulka [205]. These compounds were found to be active for seed treatment against loose smut of
barley (Ustilago nuda) and to be translocated from the roots of bean plants
to control Uromyces phaseoli. The activity of these compounds has been
reported to be restricted entirely to the Basidiomycetes [206, 207]. However, several isomers have been found to be active against a broader
spectrum of fungi including Phycomycetes and Fungi Imperfecti [206]. A
study by Snel et al. [208] on structure-activity relationships to fungitoxicity
by oxanthiin compounds indicated that substitutions in the carboxin molecule
did not increase the fungitoxic spectrum. Among the soilborne pathogens,
the oxanthiins are particularly active against S. rolfsii [209] and R. solani
[207, 210].

The use of oxanthiin fungicides has for the most part been reduced to
seed treatments for control of loose smut of cereals and the treatment of
cotton or sugarbeet seeds for control of R. solani [198] or S. rolfsii [209].

c. Chloroneb. Ryker [211] introduced this systemic fungicide for control
of seedling diseases of cotton. It has been found active against R. solani
[212], S. rolfsii [209], P. aphanidermatum [213], and Phytophthora cinnamoni [214]. The fungicide is unusual in that its activity does not follow
clear-cut taxonomic divisions such as is the case for ETMT or fenaminosulf.

Thus, while it is reportedly active against P. aphanidermatum when applied as a foliar spray in rye, it is not active against P. debaryanum or P. ultimum [174]. This fungicide has been used for the most part as a seed or in-furrow treatment for cotton. Over-the-top sprays for control of cotton seedling diseases have also been tried [215], as well as foliar sprays for control of P. aphanidermatum [213].

D. Methods of Application for Soil Fungicides

The appropriate method of applying a fungicide to soil depends on various factors: (1) chemical and physical properties of the compound which limit the type of formulations available and the type of applications that are possible; (2) relative activity of the compound against the target organism, e.g., where large dosages are required a concentrated formulation can be used; (3) the nature and ecology of the target pathogens; (4) tolerance of the host to the chemical; and (5) cost of the compound and formulation.

Some of the standard methods of applying fungicides to soil are discussed in the following section. Where appropriate, examples are given from recent literature to illustrate particular points.

1. Applications to the Soil Surface

Fungicides may be applied to the soil surface as dusts, using rotary-drum-type applicators; wettable powders or emulsions, using standard spraying equipment; granules with granular applicators; or with the use of fertilizer spreaders. Fungicides may be delivered through aerial applications [187] or mixed and applied with fertilizers. Perhaps the outstanding example of surface application of a fungicide is the use of PCNB for control of Southern blight (S. rolfsii) on peanuts. The pathogen is restricted to the surface or upper layers of soil and consequently the fungicide is applied without incorporation into the soil [125]. Since the disease appears in Alabama 40-60 days after the plants emerge, the fungicide is applied in a wide band over-the-top. Because of the insolubility of PCNB in water, little horizontal diffusion is obtained on the soil and it is essential to obtain as good coverage of the surface as possible. Considerable differences in performance have been noted between the various formulations of PCNB; at equivalent rates, those resulting in better distribution of the fungicide on the soil surface (such as sprays) generally result in much better control than granular formulations which leave areas of the soil surface untreated [41].

2. Surface Application Followed by Incorporation

When the pathogen is located throughout the soil profile and is not restricted to the soil surface or the nature of the fungicide does not permit its presence on the soil surface (fenaminosulf), or a systemic fungicide is used, the fungicide must be incorporated into the soil. The actual application to the soil can be performed with any of the equipment listed for surface applications and is dependent on the type of formulation used. Incorporation of the fungicide into the soil is effected with rotary-type tillers or by disking [216], an even better method.

3. In-furrow or in the Row Applications

These types of applications are specifically designed for control of seedling diseases. Three techniques can be considered: granular or dust applications; in-furrow spray applications; or planter-box dust applications. Because of the ubiquity of seedling diseases, the relatively small amounts of materials required, and the possibility of attaining accurate positioning of the material where it is most needed, these applications have received a great deal of attention.

a. <u>Granular or Dust Applicators.</u> Machinery of this type consists of a tractor-mounted hopper with a powered slow-moving agitator (for granules) or rotary drum (for dusts) with air hoses. This equipment can be calibrated for accurate delivery of desired amounts of fungicide to the seed furrow. These methods have been studied particularly for cotton-seedling disease control. Dust or granular in-furrow applications are considered as good or better than in-furrow spray, planter-box dust applications, or seed treatments [217-219]. Applications of dusts or small-mesh granules under windy conditions result in large amounts of material being lost, thereby reducing the amount of fungicide available for control in the seed furrow. This was especially true in the old-type dusters [220]. However, with modern dusters these losses have been reduced by incorporating into the design multiple outlets for dust dispersion in the row, a cover at the delivery end of the discharge tubes, high air velocity in the tubes to maintain the dust in suspension, and low air velocity at the discharge point to eliminate losses due to drift. The amount of material that can be delivered in dust or granule by furrow applications is limited by the equipment (15-20 kg/ha). When relatively large amounts of dusts (15-20 kg/ha) are delivered to the seed furrow, the drying effect of the dust can cause considerable damage to the seedlings [219].

Combinations of insecticides and fungicides in dusts or granules are possible and are regularly used in furrow treatments in crops such as cotton or peanuts. Sometimes these combinations cannot be mixed or sprayed in-furrow, whereas the use of granules or dusts permits such application and avoids the phytotoxic effect of sprayable formulations.

b. <u>In-furrow Spray Applications.</u> This is a relatively old method of applying fungicides to soil [221] used mostly for the control of seedling diseases in cotton and sugarbeets [217]. Equipment consists of a mixing tank where the fungicide emulsion or dispersion is maintained and agitated, a pump powered from the tractor's power take off or independently with a gasoline engine, a manometer, a control valve, a by-pass line to return to the tank, and a manifold with as many outlets as rows to be treated. The number of nozzles used to effect the spraying into the row, as well as the placement of these nozzles, depend on the crop. Single-nozzle systems have a nozzle with a cone-type spray pattern placed between the seed press wheel and the closing disks of the planters. The most common arrangement for cotton disease control consists of two nozzles [217, 219]. The front nozzle with a cone type of spray pattern is placed behind the seed tube and in front of the seed press wheel. The second nozzle with a fan type of pattern is placed behind the seed press wheel, between the closing disks and in front of the row press wheel. The purpose of the second nozzle is to obtain good treatment of the soil above the seed before it is placed on the seed when the row is closed. Other arrangements with three nozzles have been proposed for cotton but are not widely used [222].

c. Planter-Box Dust Applicators. The simplicity of this type of application resides in the fact that seed and fungicide can be delivered simultaneously without additional equipment. Dust or wettable powder formulations of fungicides are applied simultaneously with the seed and the two are mixed in the planter box or hopper. The seed is partly coated and falls together with some fungicide into the seed furrow in the planting operation and protection is thus easily obtained. Despite the obvious advantages of simplicity of operation and equipment, this method of application has some serious disadvantages that render it undesirable for most crops. Small seeds are not as well suited for this type of application as larger seeds. Large fuzzy seeds, such as nonacid delinted cotton that retain a good amount of fungicide, are ideally suited. However, with most seeds the fungicide dust tends to gradually settle at the bottom of the planter box [217-219]. This type of application is not conducive to uniform application of fungicide to the soil and when planting depth exceeds 3-4 cm, distribution becomes erratic. In summary, this method of application is poor in performance and simplicity is probably its sole advantage.

4. Dips and Drenches

Dipping seedlings in transplant water containing fungicide or applying drenches of fungicide water to soil in seedling beds are two methods of application which are generally restricted to ornamental, vegetable, or specialty crops such as tobacco. The limiting factor in these methods is the large volume of water required and the cost of transportation. With the development of systemic fungicides, a great deal of interest has centered around these methods of applications. Benomyl drenches have been used on tobacco [184, 188], musk melon [181], and azaleas [191], among others. Root dips in benomyl suspensions have been tried for control of white rot of onion [189] and for control of Cylindrocladium in azaleas [191].

IV. SEED TREATMENT

For many years seeds have been treated with fungicidal dusts and liquids for the control of seedborne pathogens and seedling diseases. During most of this period, the mercurials were the products of choice because they are highly fungicidal. Many of them possessed a volatile component that effectively sterilized the seed coat by diffusion [223] and untreated portions of the seed by its vapor action [224].

The value of seed treatments to increase seed emergence has not been contested. Virtually all the seed of many crops are treated with fungicides (see Table 2). The major exception to this rule is found in leguminous crops in which there is a widespread feeling by farmers that seed treatment may interfere with Rhizobium nodulation. In addition, these crops are only infrequently covered by proprietary (varietal) patents that induce the seed producer to pretreat his seed in order to sell the best available product. Too often the farmer must apply seed treatments in the seed hopper of his planter and it is often easier to overplant than treat.

The number of fungicides available to the farmer has grown appreciably since the 1960s. The loss of methyl and ethyl mercury formulations stimulated chemical companies to develop and market a number of new organic seed fungicides that often perform as well as their mercurial predecessors. Systemic fungicides (carboxin, benomyl) have allowed for removal of systemic and deep-seated infections. Contact fungicides (thiram, captan) which are usually effective against a broad-spectrum of fungi are still the products with widest acceptance. Mixtures (e.g., PCNB and ETMT) can increase the fungicidal spectrum of the more narrow-ranged fungicides. For example, mixtures have become very important for peanuts where Rhizopus and Aspergillus, the dominant pathogens, have been difficult organisms to control. Seeds are also commonly treated with insecticides (e.g., methoxychlor) and compatibility is important also here.

Equipment for application of seed treatment materials was covered very well by Purdy [2]. In this section equipment will be considered only with regard to effects on fungicide coverage and adhesion to seeds.

A. Seed Fungicides

The most extensively used seed treatment fungicides are captan, thiram, and PCNB-ETMT combinations (see Table 2, p. 148). All are contact fungicides without fumigant activity and have a broad-spectrum of fungicidal activity. The frequency of their use is reflected in their broad-spectrum, low cost, and phytotoxicity. The probability of resistance developing in target fungi is very unlikely due to multisite activities and to the prolonged period of use without the detection of resistant populations. Compatibility and performance, particularly when mixed with insecticides (e.g., methoxychlor), is also important; in several crops such as corn, rice, and vegetables, seeds are treated with insecticides. As yet, the recent development of systemic fungicides has yielded few commercial uses for these materials as seed treatments other than as chemicals with particular selectivity for hard-to-control organisms (e.g., carboxin for smuts). The one area where a significant market seems to exist is in treatment of ornamentals (bulbs, corms, and cuttings).

Recent work on soybean seed treatments indicated that benomyl was the only fungicide tested which could eradicate fungal infections of the embryo [225]. Benomyl, along with captan and thiram, was effective against infections of the testa by Diaporthe and Penicillium. However, benomyl proved to be somewhat phytotoxic to the extremely sensitive soybean.

The problem of resistance to systemics has been examined by several research groups [150, 226, 227] and, though it has been found to occur among foliar pathogens, resistance has yet to be proved for the seed and soilborne diseases. Ben-Yephet et al. [226] examined the problem of resistance of Ustilago hordei to benomyl and carboxin. Their findings indicated that resistance to benomyl was under polygenic control, but the fungus was unlikely to remain a dominant phenotype because of the disappearance of the resistant type when selective pressure was removed. U. hordei, on the other hand, was found to develop a stable polygenic resistance to carboxin, resulting in the continued presence of the resistant phenotype even if selective pressure were removed. The findings for carboxin may have future bearing on smut and bunt control in grains. If enough pressure is provided by frequent and extensive carboxin treatments, control of seed and seedling phases of these diseases with carboxin and its analog oxycarboxin could be lost.

Seed treatment through gaseous action reached a peak with the use of ethyl and methyl mercury. These fungicides possessed the capability of controlling pathogens in nontreated areas through their fungitoxic vapors which also diffused into shallow but internal infection sites. Removal of the mercuries from the list of EPA approved products left a need for fungicides with the combination of contact and gaseous action; to date, no likely candidates have been developed to fill the void.

B. Biological Seed Treatments

The use of microorganisms as seed treatments has been known for many years. Classically, the treatment has dealt with the addition of Rhizobium to legumes, but in recent years other species of bacteria have been added to a broad-spectrum of plant seeds resulting in positive growth responses [228]. Many of these responses are thought to be based on hormone production, nitrogen fixation, and phosphate release; but some are apparently based on disease control.

TABLE 2

The Use of Seed Treatment Fungicides by Crops in the United States, and the Pathogens to Which They are Directed

Crop	U.S. acres[a] × 10^6	lb[b] × 10^6 seed	Major pathogens		Fungicide treatment, %				Major fungicides		Rate		Treated crop, %
			Seed	Soil	Total	Pre-treated	Hopper	Insecti-cide, %					
Corn (field)	79 (32)	1,000 (450)	Diplodia Fusaria Giberella Penicillium Aspergillus Colletotrichum	Pythium Helminthosporium	100	100	v.f.[c]	80	Captan Thiram	1 2	oz/bu[d] oz/bu	(0.8) (1.6)	95 5
Sorghum	19 (7.7)	131 (59)	Fusarium Aspergillus Rhizopus	Rhizopus Fusarium Pythium Smuts	100	100	v.f.	25	Captan PCNB-ETMT	2-3 2	oz/bu oz/bu	(1.6-2.4) (1.6)	90 10
Soybeans	55 (22)	3,600 (1,640)	Diaporthe Cercospora Fusarium	Rhizoctonia Pythium Macrophomina	35	0	35	0	Captan Thiram PCNB-ETMT	2 2 2	oz/bu oz/bu oz/bu	(1.6) (1.6) (1.6)	55 15 25
Peanuts (groundnuts)	1.6 (0.65)	175 (79)	Rhizopus Aspergillus niger A. flavus Penicillium	A. niger Rhizoctonia Pythium Fusarium	100	100	0	0	Captafol-DCNA Captan-maneb Captan-DCNA Thiram	3-5 4-5 5 6	oz/cwt[e] oz/cwt oz/cwt oz/cwt	(1.8-3) (2.4-3) (3) (3.6)	20 50 10 5
Small grains wheat oats barley rye	70.7 (28.4) 18.9 (7.6) 9.5 (3.8) 3.6 (1.4)	77.8 (35.0) 48.6 (22.8) 15.4 (6.9) 5.2 (2.2)	Helminthosporium Rhyncosporium Ustilago Urocystis Sphacelotheca	Smuts Blights	50	45	5	15	Phenylmercury-L PCNB-L PCNB-ETMT-L Captan Carboxin-thiram BHC	0.75 2 2 2 2 2	oz/bu oz/bu oz/bu oz/bu oz/bu oz/bu	(0.46) (1.6) (1.6) (1.6) (1.6) (1.6)	45 15 10 15 5 10

Crop											
Rice	2.5 (1.0)	7.3 (3.3)	Helminthosporium Pyricularia Curvularia Fusarium	Achlya Pythium Rhizoctonia	95	v.f.	95	PCNB-ETMT-L Captan-L Thiram Captafol Mancozeb	2 oz/bu (1.6) 2 oz/bu (1.6) 2 oz/bu (1.6) 2 oz/bu (1.6) 2 oz/bu (1.6)	15 50 5 15 5
Vegetables	340 (153)									
peas	208 (93)		Fusarium	Pythium	60	60	50	Captan	2–4 oz/cwt (1.2–2.4)	70
beans	66 (29)		Ascochyta	Rhizoctonia	60		50	Thiram	2–3 oz/cwt (1.2–1.8)	30
corn	20 (9)		Bacteria	Fusarium	80		90			
others	44 (19)				100					
Leguminous forages	170 (77)			Pythium Rhizoctonia Phytophthora	v.f.	v.f.	v.f.			
Potatoes	1.3 (5.25)	2,200 (990)	Fusarium Streptomyces Bacteria Rhizoctonia	Rhizoctonia Pythium	80	80	0	Captan Maneb Polyram	16 oz/cwt (9.6) 16 oz/cwt (9.6) 16 oz/cwt (9.6)	50 25 25
Cotton	14.5 (5.95)	363 (166)	Anthracnose Fusaria Bacteria	Rhizoctonia Fusarium Pythium	100	85	25	Captan PCNB-ETMT-L Chloroneb-thiram Carboxin Fenaminosulf	2–4 oz/cwt (1.2–2.4) 12–16 fl oz/cwt[f] (7–10) 10–13 oz/cwt (6–7.8) 6 oz/cwt (3.6) 2–3 oz/cwt (1.2–1.8)	60 15 3 3 2

[a]Hectares in parentheses.
[b]Kilograms in parentheses.
[c]v. f. = very few seeds treated.
[d]bu = bushel; g/liter in parentheses.
[e]cwt = 100 lb; g/kg in parentheses.
[f]cm^3/kg in parentheses.

Seedling blight of maize was controlled in pots by coating kernels with a suspension of Bacillus subtilis. Performance under field conditions was similar to that found for seed treated with captan [229]. Chang and Kommedahl [230] reported similar results in corn, with the indication that antibiosis was the mechanism involved in increased emergence. Arthrobacter was reported to reduce tomato seedling losses from Pythium debaryanum by lysing the mycelium [231]. Rhizoctonia solani mycelium was lysed by the addition of B. subtilis [232]; this bacterium applied to wheat and carrot seed increased final yield by 40% in each case [233]. Both biological control and a hormonal response were believed responsible for the effect.

Recently, research has been conducted in the United States using Bacillus uniflagellatus for seed and seedling disease control in a variety of crops. Results indicate that several grain diseases can be suppressed [234, 235] and that the dried organism is compatible with captan dust as a combination seed treatment [41].

Use of antagonistic fungi for disease control has not reached the developmental stage comparable to seed treatments. However, recent reports indicate potential for use of Trichoderma as a seed treatment [236], Kommedahl and Mew [237] and Tveit and Moore [238] have found suppression of Helminthosporium sativum and Fusarium roseum in corn using an antagonistic isolate of Chaetomium globosum as a seed treatment.

C. Adhesion and Coverage

Mills [239] related adhesion of dry fungicides to three qualities: gross seed surface area per bushel; seed-coat surface characteristics; and fungicide qualities (mesh size, aggregation tendency, etc.). In general he found that small, rough seed treated with a fungicide of small particle size had the best adhesion. Lord et al. [240] comparing liquid and dry materials found that seed treated with liquids had less uniform distribution but a larger percentage of fungicide adhesion, than did seed treated with dry materials. These studies conducted with commercially treated seeds indicated that seed "loading" with fungicides was independent of the type of machinery used for treatment and was not consistently affected by formulation, weather, or variety of the seed.

When Griffiths et al. [241] compared adhesion of fungicides and insecticides on wheat seed, they found that commercially treated seed retained only 30-60% of the target dose of fungicides (mostly dust formulations) but more than 80% of the liquid insecticides. If insecticides were applied first, adhesion of fungicides was improved.

Clinton [242] coated peanut seed with a methyl cellulose "artificial testa" that could itself contain a fungicide. His results indicated some reduction of Rhizopus and Aspergillus flavus but showed no overall emergence benefits. Coatings have also been developed to aid in Rhizobium establishment and improve the ballistic properties of aerially sown seed [243].

D. Compatibility of Seed Fungicides with Rhizobium

Many studies conducted over the years have indicated that seed fungicides greatly increase the emergence of legumes; yet, this group of crops is most commonly not treated with these fungicides (see Table 2). The reason for the poor acceptance of seed treatment is that many seed fungicides are toxic to Rhizobium as shown in culture [244, 245] and in greenhouse studies [246, 247].

Curley and Burton [246] determined that thiram was least toxic to Rhizobium after a 24-hr preplant period on the seed, followed by treatment with carboxin and captan; PCNB was the most toxic of the four fungicides tested against the moist, peat-base Rhizobium inoculum used in this study. Freeze-dried and oil-dried Rhizobium formulations are also available, the latter in a prepackaged mix with thiram. No studies are available at present on differential responses of the various types of Rhizobium formulations to fungicides.

Curley and Burton [246] further demonstrated that applying fungicide before, after, or as a mix with Rhizobium made no difference on nodulation of test soybeans. Their methods indicated that nodulation efficiency of a fungicide-Rhizobium mixture on soybeans could best be determined by counting the number of nodules on the tap root 2 weeks after planting.

E. Methods of Application

The first decision of every seed producer is to choose seed lots that are acceptable for commercial sale. Often this is partially done by selecting lots that have been produced on certified and inspected farms. The usual procedure is then to determine what seeds have highest germinability. Tests are usually conducted in a high-humidity seed germinator operating at near optimal temperature for the crop species. Tetrazolium tests (dehydrogenase activity) are often conducted to determine if seeds have been damaged (mechanically or by insects and fungi) and their vigor reduced. For crops such as peanuts, where the predominant seedborne pathogens may vary from year to year, different fungicide combinations should be tested to determine which gives optimal performance.

The selection of seed lots by a "vigor-rating" system is finding acceptance in the seed industry. This system relies on a quantitative measure of growth, as well as an evaluation of germination [248]. The vigor index developed allows selection of seed lots and fungicide treatments that give best performance in both variables.

Mechanical injury of seed during processing is a major problem in some crops. The work of Bell [249] has demonstrated that abrasion and cut-type injuries, like those that occur in peanut shelling, are important sites of fungal invasion and result in lower germination. In addition, peanuts destined for seed channels usually have to wait 3-8 weeks after shelling before treatment with fungicides. This prolonged period between shelling and treatment was found to significantly decrease seed germination [250]. Immediate treatment after shelling maintained acceptable germination levels. Decontamination of the sheller by fungicides before shelling did not change this trend.

Acceptance of this obvious benefit is impaired by the seedsmen's desire to run a germination check after shelling but before treatment to check for germination reductions during shelling.

Seed treatment fungicides when applied before planting are usually applied to the seed as liquids or liquid slurries. The major exception to this rule is in peanuts where the testa is loosened and often lost when treated with liquids, resulting in lower germination.

The addition of fungicides to seed in the planter box (hopper box) is limited primarily to soybeans and cotton. This method is fraught with several basic flaws. The fungicide is placed on top of a planter box full of seed and must settle by gravity to treat the seed at the bottom. The resulting treatment patterns are erratic, with some seed receiving excessive fungicide while others receive too little. For soybeans, where virtually no seeds are pretreated, it is the only method available to the farmer. In the future this may be rectified by the development of protected varieties which will induce seed companies to pretreat in order to out-perform their competitors.

REFERENCES

1. E. C. Large, The Advance of the Fungi, Dover Publications, New York, 1962, p. 488.
2. L. H. Purdy, in Fungicides: An Advanced Treatise (D. C. Torgeson, ed.), Vol. 1, Academic Press, New York, 1967, pp. 195-237.
3. S. Wilhelm, Ann. Rev. Phytopathol., 4, 53 (1966).
4. W. A. Kreutzer, Ann. Rev. Phytopathol., 1, 101 (1963).
5. K. A. Pyrozynski, in The Fungi (G. C. Ainsworth and A. S. Sussman, eds.), Vol. 3, Academic Press, New York, 1968, pp. 487-504.
6. W. A. Kreutzer, in Plant Pathology: An Advanced Treatise (J. G. Horsfall and A. E. Dimond, eds.), Vol. III, Academic Press, New York, 1960, pp. 431-470.
7. S. D. Lyda, in The Relation of Soil Microorganisms to Soilborne Plant Pathogens (G. C. Papavizas, ed.), South. Coop. Series Bull. 183, Virginia Polytechnic Institute, Blacksburg, Ba., 1974, pp. 64-73.

8. S. D. Garrett, Pathogenic Root-Infecting Fungi, Cambridge University Press, London, 1970.
9. K. F. Baker and R. J. Cook, Biological Control of Plant Pathogens, W. H. Freeman & Co., San Francisco, 1974.
10. J. D. Menzies, in Root Diseases and Soil-Borne Pathogens (T. A. Toussoun, R. V. Bega, and P. E. Nelson, eds.), University of California Press, Berkeley, 1970, pp. 16-21.
11. J. D. Menzies, Ann. Rev. Phytopathol., 1, 127 (1963).
12. J. D. Menzies, Bot. Rev., 29, 79 (1963).
13. D. Parkinson, T. R. G. Gray, and S. T. Williams, Methods for Studying the Ecology of Soil Microorganisms, 1BP Handbook No. 19, Blackwell Scientific Publications, Oxford, 1971.
14. L. F. Johnson and E. A. Curl, Methods for Research on the Ecology of Soil-Borne Plant Pathogens, Burgess Publishing Co., Minneapolis, Minn., 1972.
15. A. F. Schmitthenner, in Root Diseases and Soil-Borne Pathogens (T. A. Toussoun, R. V. Bega, and P. E. Nelson, eds.), University of California Press, Berkeley, 1970, pp. 25-27.
16. S. N. Smith, in Root Diseases and Soil-Borne Pathogens (T. A. Toussoun, R. V. Bega, and P. E. Nelson, eds.), University of California Press, Berkeley, 1970, pp. 28-30.
17. R. L. Powelson, in Root Diseases and Soil-Borne Pathogens (T. A. Toussoun, R. V. Bega, and P. E. Nelson, eds.), University of California Press, Berkeley, 1970, pp. 31-33.
18. Y. Henis, in Root Diseases and Soil-Borne Pathogens (T. A. Tousson, R. V. Bega, and P. E. Nelson, eds.), University of California Press, Berkeley, 1970, pp. 34-41.
19. R. Baker and C. A. Martinson, in Rhizoctonia solani, Biology and Pathology (J. R. Parmeter, Jr., ed.), University of California Press, Berkeley, 1970, pp. 172-188.
20. S. Wilhelm, Phytopathology, 44, 609 (1954).
21. D. Park, in Ecology of Soil-Borne Plant Pathogens (K. F. Baker and W. C. Snyder, eds.), University of California Press, Berkeley, 1965, pp. 82-98.
22. A. S. Sussman, in Ecology of Soil-Borne Plant Pathogens (K. F. Baker and W. C. Snyder, eds.), University of California Press, Berkeley, 1965, pp. 99-100.
23. J. R. Coley-Smith and R. C. Cooke, Ann. Rev. Phytopathol., 9, 65 (1971).
24. D. M. Griffin, Ecology of Soil Fungi, Chapman and Hall, London, 1972.
25. S. Wilhelm, Phytopathology, 46, 293 (1956).
26. A. S. Sussman and H. O. Halvorson, Spores, Their Dormancy and Germination, Harper and Row, New York, 1966.
27. I. Chet, Can. J. Bot., 47, 593 (1969).

28. A. S. Sussman, in The Fungi (G. C. Ainsworth and A. S. Sussman, eds.), Vol. 3, Academic Press, New York, 1968, pp. 447-486.
29. J. L. Lockwood, Ann. Rev. Phytopathol., 2, 341 (1964).
30. R. M. Jackson, in Ecology of Soil-Borne Plant Pathogens (R. F. Baker and W. C. Snyder, eds.), University of California Press, Berkeley, 1965, pp. 363-373.
31. A. G. Watson and E. J. Ford, Ann. Rev. Phytopathol., 10, 327 (1972).
32. A. G. Lochhead, in Plant Pathology, Problems and Progress (C. S. Holton, G. W. Fischer, R. W. Fulton, H. Hart, and S. E. A. McCallan, eds.), The University of Wisconsin Press, Madison, 1959, pp. 327-338.
33. A. D. Rovira, in Ecology of Soil-Borne Plant Pathogens (K. F. Baker and W. C. Snyder, eds.), University of California Press, Berkeley, 1965, pp. 170-186.
34. A. D. Rovira and C. B. Davey, in The Plant Root and its Environment (E. W. Carson, ed.), University Press of Virginia, Charlottesville, 1974, pp. 155-204.
35. F. F. Hendrix, Jr. and W. A. Campbell, Ann. Rev. Phytopathol., 11, 77 (1973).
36. M. E. Stanghellini, Proc. Amer. Phytopathol. Soc., 1, 211 (1974).
37. S. M. Mercetich and G. A. Zentmyer, in Root Diseases and Soil-Borne Pathogens (T. A. Toussoun, R. V. Bega, and P. E. Nelson, eds.), University of California Press, Berkeley, 1970, pp. 112-115.
38. S. M. Mercetich and G. A. Zentmyer, Phytopathology, 56, 1076 (1966).
39. G. C. Papavizas and W. A. Ayers, USDA Tech. Bull., 1485 (1974).
40. S. M. Nash and J. V. Alexander, Phytopathology, 55, 963 (1965).
41. R. Rodriguez-Kabana and P. A. Backman, unpublished work, 1975.
42. S. M. Nash, T. Christou, and W. C. Snyder, Phytopathology, 51, 308 (1961).
43. R. J. Cook and M. N. Schroth, Phytopathology, 53, 254 (1965).
44. G. C. Papavizas, P. B. Adams, and J. A. Lewis, Phytopathology, 58, 414 (1968).
45. P. H. Tsao and J. L. Bricker, Phytopathology, 56, 1012 (1966).
46. P. B. Adams, in The Relation of Soil Microorganisms to Soil-Borne Plant Pathogens (G. C. Papavizas, ed.), South. Coop. Series Bull. 183, Virginia Polytechnic Institute, Blacksburg, Va., 1974, pp. 7-12.
47. C. H. Rogers, Texas Agr. Expt. Sta. Bull., 614 (1942).
48. R. A. Ludwig, in Ecology of Soil-Borne Plant Pathogens (K. F. Baker and W. C. Snyder, eds.), University of California Press, Berkeley, 1965, pp. 471-478.
49. C. D. McKeen, in Biology and Control of Soil-Borne Plant Pathogens (G. W. Bruehl, ed.), The Amer. Phytopathological Soc., St. Paul, Minn., 1975, pp. 203-204.
50. S. Wilhelm, Phytopathology, 45, 180 (1955).
51. R. J. Green, Jr., Phytopathology, 59, 874 (1969).

52. M. G. Boosalis and A. L. Scharen, Phytopathology, 49, 192 (1959).
53. G. C. Papavizas, in Rhizoctonia solani, Biology and Pathology (J. R. Parmeter, Jr., ed.), University of California Press, Berkeley, 1970, pp. 108-122.
54. N. T. Flentje, Helena M. Stretton, and A. R. McKenzie, in Rhizoctonia solani, Biology and Pathology (J. R. Parmeter, Jr., ed.), University of California Press, Berkeley, 1970, pp. 52-65.
55. P. Lai and G. W. Bruehl, Phytopathology, 56, 213 (1966).
56. S. R. Siemer and E. K. Vaughan, Phytopathology, 61, 146 (1971).
57. W. M. Bugbee and O. C. Soine, Phytopathology, 65, 1258 (1974).
58. R. B. Morwood, Quart. J. Agr. Sci., 10, 222 (1953).
59. Yung Chang-Ho and C. J. Hickman, in Root Diseases and Soil-Borne Pathogens (T. A. Toussoun, R. V. Bega, and P. E. Nelson, eds.), University of California Press, Berkeley, 1970, pp. 103-108.
60. J. R. Coley-Smith and J. E. King, in Root Diseases and Soil-Borne Pathogens (T. A. Toussoun, R. V. Bega, and P. E. Nelson, eds.), University of California Press, Berkeley, 1970, pp. 130-133.
61. D. S. Hayman, in Root Diseases and Soil-Borne Pathogens (T. A. Toussoun, R. V. Bega, and P. E. Nelson, eds.), University of California Press, Berkeley, 1970, pp. 99-102.
62. T. A. Toussoun, in Root Diseases and Soil-Borne Pathogens (T. A. Toussoun, R. V. Bega, and P. E. Nelson, eds.), University of California Press, Berkeley, 1970, pp. 95-98.
63. Y. Henis and J. Katan, in Biology and Control of Soil-Borne Plant Pathogens (G. W. Bruehl, ed.), The Amer. Phytopathological Soc., St. Paul, Minn., 1975, pp. 100-106.
64. R. W. Smiley, in Biology and Control of Soil-Borne Plant Pathogens (G. W. Bruehl, ed.), The Amer. Phytopathological Soc., St. Paul, Minn., 1975, pp. 55-62.
65. T. S. Sadasivan, in Ecology of Soil-Borne Plant Pathogens (K. F. Baker and W. C. Snyder, eds.), University of California Press, Berkeley, 1965, pp. 460-470.
66. E. A. Curl and R. Rodriguez-Kabana, in The Relation of Soil Microorganisms to Soil-Borne Plant Pathogens (G. C. Papavizas, ed.), South. Coop. Series Bull. 183, Virginia Polytechnic Institute, Blacksburg, Va., 1974, pp. 47-50.
67. G. C. Papavizas, in The Relation of Soil Microorganisms to Soil-Borne Plant Pathogens (G. C. Papavizas, ed.), South. Coop. Series Bull. 183, Virginia Polytechnic Institute, Blacksburg, Va., 1974, pp. 25-29.
68. L. S. Bird, in The Relation of Soil Microorganisms to Soil-Borne Plant Pathogens (G. C. Papavizas, ed.), South. Coop. Series Bull. 183, Virginia Polytechnic Institute, Blacksburg, Va., 1974, pp. 75-79.
69. J. Tuite, Plant Pathological Methods, Burgess Publishing Co., Minneapolis, Minn., 1969.

70. J. B. Sinclair, in Rhizoctonia solani, Biology and Pathology (J. R. Parmeter, Jr., ed.), University of California Press, Berkeley, 1970, pp. 199-242.

71. G. C. Papavizas and C. B. Davey, Phytopathology, 52, 834 (1962).

72. B. Sneh, J. Katan, Y. Henis, and I. Wahl, Phytopathology, 56, 74 (1966).

73. P. H. Tsao, 1970, Ann. Rev. Phytopathol., 8, 157-186.

74. E. A. Curl and R. Rodriguez-Kabana, in Research Methods in Weed Science (R. E. Wilkinson, ed.), POP Enterprises Inc., Atlanta, Ga., 1971, pp. 161-194.

75. K. S. Chester, Plant Disease Reptr. Suppl., 193, 191 (1950).

76. R. T. Sherwood and D. J. Hagedorn, Wisc. Agr. Expt. Sta. Bull. 531 (1958).

77. W. C. James, Ann. Rev. Phytopathol., 12, 27 (1974).

78. D. E. Munnecke, Ann. Rev. Phytopathol., 10, 375 (1972).

79. D. E. Munnecke, in Fungicides: An Advanced Treatise (D. C. Torgeson, ed.), Vol. 1, Academic Press, New York, 1967, pp. 509-559.

80. H. Martin, The Scientific Principles of Crop Protection, Edward Arnold, Ltd., London, 1959, p. 359.

81. A. G. Newhall, Bot. Rev., 21, 189 (1955).

82. O. Vaartaja, Bot. Rev., 30, 1 (1964).

83. W. A. Carter, Science, 97, 383 (1943).

84. W. A. Carter, J. Econ. Entomology, 38, 35 (1945).

85. E. E. Clayton, T. W. Gaines, and F. A. Todd, Phytopathology, 39, 4 (1949).

86. F. L. Stark and B. Lear, Phytopathology, 37, 688 (1947).

87. G. A. Zentmyer and J. B. Kendrick, Phytopathology, 39, 864 (1949).

88. G. K. Parris, Phytopathology, 35, 771 (1945).

89. D. E. Munnecke and J. P. Martin, Phytopathology, 54, 941 (1964).

90. Le Goupil, Rev. Pathol. Vegetale, 19, 169 (1932).

91. M. J. Kolbezen, D. E. Munnecke, W. D. Wilbur, L. H. Stolzy, F. J. Abu-El-Haj, and T. E. Szuszkiewicz, Hilgardia, 42, 465 (1974).

92. Cornell Univ. Agr. Expt. Sta. Bull., No. 850, 1948.

93. C. W. McBeth and G. B. Bergeson, Plant Disease Reptr., 39, 223 (1955).

94. L. J. Ashworth, B. C. Langley, and W. H. Thames, Jr., Phytopathology, 54, 187 (1964).

95. B. Brodie, Phytopathology, 51, 798 (1961).

96. E. A. Pieroh, H. Werres, and K. Rashke, Anz. Schaedlingskunde, 32, 183 (1959).

97. C. Kotter, Z. Pflanzenkrankh. Pflanzenschutz, 70, 342 (1963).

98. D. E. Munnecke, K. H. Domsch, and J. W. Eckert, Phytopathology, 52, 1298 (1962).

99. J. A. Lewis and G. C. Papavizas, Soil Biol. Biochem., 2, 239 (1970).

100. G. A. Lloyd, J. Sci. Food Agr., 13, 309 (1962).

101. Anonymous, Bunema, a Bactericide, Fungicide and Nematicide Concentrate, Agriculture Release No. 473, Buckman Lab., Inc., 1973.
102. I. R. Bradbury, A. Campbell, C. W. Suckling, H. R. Jameson, and F. C. Peacock, Ann. Appl. Biol., 45, 241 (1957).
103. W. J. Birchfield, F. Parr, and S. Smith, Plant Disease Reptr., 53, 923 (1969).
104. J. R. Davis, R. H. Callihan, and M. D. Groskopp, Plant Disease Reptr., 56, 400 (1972).
105. J. D. Gay, Plant Disease Reptr., 54, 604 (1970).
106. R. Rodriguez-Kabana, P. A. Backman, H. Ivey, and L. L. Farrar, Plant Disease Reptr., 56, 362 (1972).
107. R. Rodriguez-Kabana, P. A. Backman, and P. S. King, Plant Disease Reptr., 59, 528 (1975).
108. R. C. Rowe, M. K. Beute, J. C. Wells, and J. C. Wynne, Plant Disease Reptr., 58, 348 (1974).
109. D. J. Weaver, Plant Disease Reptr., 55, 1094 (1971).
110. M. C. Rush, C. W. Averre, S. F. Jenkins, and A. Khan, Phytopathology, 60, 1311 (1970).
111. P. B. Adams, H. L. Warren, and C. J. Tate, Plant Disease Reptr., 55, 1077 (1971).
112. J. V. Parochetti and C. F. Warren, Weed Sci., 18, 555 (1970).
113. J. M. Good, in Nematodes of Tropical Crops (J. E. Peachey, ed.), Tech. Commun. No. 40, Commonwealth Agr. Bureaux, England, 1969, pp. 297-313.
114. C. A. I. Goring, Ann. Rev. Phytopathol., 5, 285 (1967).
115. H. Jensen and G. E. Page, Plant Disease Reptr., 38, 401 (1954).
116. G. E. Page, Misc. Papers, Ore. Agr. Expt. Sta., No. 111, 1961, p. 5.
117. G. E. Page, Ext. Bull. Ore. Stat. Univ., No. 813, 1963, p. 14.
118. J. N. Sasser and C. J. Nusbaum, Plant Disease Reptr., 38, 65 (1954).
119. D. J. Morton, Plant Disease Reptr., 43, 248 (1959).
120. A. L. Taylor and A. M. Golden, Plant Disease Reptr., 38, 63 (1954).
121. J. H. O'Bannon, Plant Disease Reptr., 42, 857 (1958).
122. P. B. Brudenell, in Nematodes of Tropical Crops (J. E. Peachey, ed.), Tech. Commun. No. 40, Commonwealth Agr. Bureaux, England, 1969, pp. 324-332.
123. W. T. Thomson, Agricultural Chemicals. IV. Fungicides, Thomson Publications, Indianapolis, 1973, p. 207.
124. A. L. Taylor and J. H. O'Bannon, Plant Disease Reptr., 51, 218 (1968).
125. R. Aycock, N. Carolina State Coll. Agr. Expt. Sta. Tech. Bull. No. 174, 1966, p. 202.
126. J. W. Lorbeer, Fungicide-Nematicide Tests, 20, 66 (1964).

127. E. Bushong, Seed and Soil Treatment Newsletter, Kansas State University, 4, 39 (1962).
128. W. Brown, J. Pomol. Hort. Sci., 13, 247 (1935).
129. B. W. Kennedy and L. A. Brinkerhoff, Plant Disease Reptr., 43, 90 (1959).
130. C. H. Arndt, Plant Disease Reptr., 37, 397 (1953).
131. M. J. Smieton and W. Brown, Ann. Appl. Biol., 27, 489 (1940).
132. M. M. DeLint, Verslag. Plantenziekten Dienst. Wageningen, 130, 134 (1956).
133. I. A. S. Gibson, M. Ledger, and E. Boehm, Phytopathology, 51, 531 (1961).
134. J. B. Kendrick and J. T. Middleton, Plant Disease Reptr., 38, 350 (1954).
135. M. J. Reavill, Ann. Appl. Biol., 41, 448 (1954).
136. W. Buddin, J. Min. Agr. Eng., 44, 1158–1159.
137. H. Wasewitz, Kranke Pflanze, 14, 71 (1937).
138. A. Ylimäki, Acta Agral. Fennica, 83, 147 (1955).
139. L. D. Leach and W. S. Seyman, Phytopathology, 47, 527 (1957).
140. W. J. Hooker, Plant Disease Reptr., 38, 187 (1954).
141. R. P. Scheffer and W. J. Haney, Plant Disease Reptr., 40, 570 (1956).
142. R. F. Hendrix and E. G. Kuhlman, Phytopathology, 55, 1183 (1965).
143. R. A. Flowers and J. W. Hendrix, Phytopathology, 59, 725 (1969).
144. G. C. Papavizas, Phytopathology, 57, 848 (1967).
145. M. Brook and C. G. C. Chesters, Ann. Appl. Biol., 45, 498 (1957).
146. J. W. Eckert, Phytopathology, 52, 642 (1962).
147. T. E. Boswell, Ph.D. Thesis, Texas A & M University, College Sta., Texas, 1968.
148. P. M. Miller and P. E. Waggoner, Plant Soil, 18, 45 (1963).
149. S. Rich and P. M. Miller, Plant Disease Reptr., 48, 246 (1964).
150. S. G. Georgopoulos and C. Zaracovitis, Ann. Rev. Phytopathol., 5, 109 (1967).
151. M. N. Shatla and J. B. Sinclair, Phytopathology, 53, 1407 (1967).
152. L. T. Richardson, Phytopathology, 58, 316 (1968).
153. W. H. Ko and J. D. Harley, Phytopathology, 59, 64 (1969).
154. P. R. Bristow, J. Katan, and J. L. Lockwood, Phytopathology, 63, 809 (1973).
155. E. Urbschat, Angew. Chem., 72, 981 (1960).
156. L. D. Leach, R. H. Garber, and W. J. Tolmsoff, Phytopathology, 50, 643 (1960).
157. L. D. Leach, R. H. Garber, and W. H. Lange, Plant Disease Reptr. Suppl., 259, 213 (1959).
158. G. A. Zentmyer and J. D. Gilpatrick, Phytopathology, 50, 660 (1960).
159. F. J. Hills and L. D. Leach, Phytopathology, 52, 51 (1962).

160. J. Tammen, D. P. Muse, and J. H. Haas, Plant Disease Reptr.,
     46, 646 (1961).
161. J. E. Mitchell and D. J. Hagedorn, Plant Disease Reptr., 50, 91
     (1966).
162. D. J. Phillips and R. Baker, Plant Disease Reptr., 46, 606 (1962).
163. G. C. Papavizas, Plant Disease Reptr., 51, 125 (1967).
164. S. Khan and R. Baker, Phytopathology, 58, 34 (1968).
165. R. Alconero and D. J. Hagedorn, Phytopathology, 58, 24 (1968).
166. W. D. Thomas, Jr., P. H. Eastburg, and M. D. Bankuti, Phyto-
     pathology, 52, 754 (1962).
167. J. E. Wheeler, R. B. Hine, and A. M. Boyle, Phytopathology, 60,
     561 (1970).
168. A. L. Bertus, Plant Disease Reptr., 58, 437 (1974).
169. G. I. Robertson, Seed Nursery Trader, 71, 9 (1973).
170. J. F. Knauss, Plant Disease Reptr., 56, 1074 (1972).
171. G. J. Muller, M. B. Linn, and J. B. Sinclair, Plant Disease Reptr.,
     56, 1054 (1972).
172. A. S. Al-Beldawi and J. B. Sinclair, Phytopathology, 59, 68 (1969).
173. C. J. Delp and H. L. Klopping, Plant Disease Reptr., 52, 95 (1968).
174. D. C. Erwin, Ann. Rev. Phytopathol., 11, 389 (1973).
175. E. Shlevin and J. Katan, Plant Disease Reptr., 59, 29 (1975).
176. A. S. Al-Beldawi and J. A. Pinckard, Plant Disease Reptr., 52, 781
     (1968).
177. A. S. Al-Beldawi and J. A. Pinckard, Plant Disease Reptr., 54, 76
     (1970).
178. D. C. Erwin, H. Mee, and J. J. Sims, Phytopathology, 58, 528
     (1968).
179. J. A. Booth, T. E. Rawlins, and C. F. Chew, Plant Disease Reptr.,
     55, 569 (1971).
180. W. L. Biehn, Plant Disease Reptr., 54, 171 (1970).
181. R. N. Wensley and C. M. Huang, Can. J. Microbiol., 16, 615 (1970).
182. J. M. Vargas and C. W. Laughlin, Plant Disease Reptr., 55, 167
     (1971).
183. W. L. Biehn and A. E. Dimond, Plant Disease Reptr., 54, 136 (1970).
184. H. L. Szenger, Plant Disease Reptr., 54, 136 (1970).
185. G. C. Papavizas, J. A. Lewis, and P. B. Adams, Plant Disease
     Reptr., 54, 114 (1970).
186. S. K. Gayed, Can. Plant Disease Surv., 49, 70 (1969).
187. R. L. Gabrielson, W. C. Anderson, and R. F. Nyvall, Plant Disease
     Reptr., 57, 164 (1973).
188. W. F. T. Hartill and J. M. Campbell, Plant Disease Reptr., 57,
     932 (1973).
189. O. C. Maloy and R. Machtmes, Plant Disease Reptr., 58, 6 (1974).
190. P. Chidambaram and G. W. Bruehl, Plant Disease Reptr., 57, 935
     (1973).

191. A. W. Engelhard, Plant Disease Reptr., 55, 679 (1971).
192. R. K. Horst and H. A. J. Hoitink, Plant Disease Reptr., 52, 615 (1968).
193. R. B. Hine, D. L. Johnson, and C. J. Wenger, Phytopathology, 59, 798 (1969).
194. F. R. Harper, Plant Disease Reptr., 52, 565 (1968).
195. L. V. Edgington, K. L. Khew, and G. L. Barron, Phytopathology, 61, 42 (1971).
196. R. J. Williams and A. Ayanaba, Phytopathology, 65, 217 (1975).
197. P. A. Backman, R. Rodriguez-Kabana, and J. C. Williams, Phytopathology, 65, 773 (1975).
198. C. D. Ranney, Crop Sci., 12, 346 (1972).
199. J. J. Natti, Phytopathology, 61, 669 (1971).
200. J. S. Jhooty and D. S. Behar, Plant Disease Reptr., 54, 1049 (1970).
201. J. D. Gilpatrick, Plant Disease Reptr., 53, 721 (1969).
202. J. A. Booth and T. E. Rawlins, Plant Disease Reptr., 54, 741 (1970).
203. R. E. Pitblado and L. V. Edgington, Phytopathology, 62, 513 (1972).
204. D. Neely, Plant Disease Reptr., 58, 261 (1974).
205. B. von Schmeling and M. Kulka, Science, 152, 659 (1966).
206. L. V. Edgington and G. L. Barron, Phytopathology, 57, 1256 (1967).
207. L. V. Edgington, G. S. Walton, and P. M. Miller, Science, 153, 307 (1966).
208. M. Snel, B. von Schmeling, and L. V. Edgington, Phytopathology, 60, 1164 (1970).
209. A. N. Mukhopadhyay and R. P. Thakur, Plant Disease Reptr., 55, 630 (1971).
210. A. S. Al-Beldawi and J. A. Pinckard, Plant Disease Reptr., 54, 524 (1970).
211. T. C. Ryker, Proc. Cotton Disease Council, 25, 137 (1965).
212. I. E. M. Darrag and J. B. Sinclair, Phytopathology, 59, 1102 (1969).
213. R. H. Littrell, J. D. Gay, and H. D. Wells, Plant Disease Reptr., 53, 913 (1969).
214. W. K. Hock and H. D. Sisler, Phytopathology, 59, 627 (1969).
215. N. D. Fulton, Plant Disease Reptr., 55, 307 (1971).
216. J. D. Menzies, Amer. Potato J., 34, 219 (1957).
217. C. R. Maier and R. G. Bullard, Plant Disease Reptr., 51, 487 (1967).
218. C. R. Maier, Plant Disease Reptr., 45, 276 (1961).
219. C. R. Maier and E. E. Staffeldt, New Mexico Agr. Expt. Sta. Bull. No. 474, 1963, p. 41.
220. C. S. Holton and T. L. Jackson, Plant Disease Reptr., 36, 423 (1952).
221. L. D. Leach and W. C. Snyder, Phytopathology, 37, 363 (1947).
222. W. H. Aldred, L. H. Wilkes, and L. S. Bird, Texas Agr. Expt. Sta. Progr. Rept. No. 2181, 1961, p. 3.

223. D. Lindstrom, Agr. Food Chem., 7, 562 (1959).

224. S. G. Lehman, Phytopathology, 33, 431 (1943).

225. M. A. Ellis, M. B. Ilyas, and J. B. Sinclair, Phytopathology, 65, 553 (1975).

226. Y. Ben-Yephet, Y. Henis, and A. Dinoor, Phytopathology, 65, 563 (1975).

227. D. R. Mackenzie, R. R. Nelson, and H. Cole, Phytopathology, 61, 458 (1971).

228. M. E. Brown, Ann. Rev. Phytopathol., 12, 181 (1974).

229. P. Singh, R. S. Vasodeva, and B. S. Bajaj, Ann. Appl. Biol., 55, 89 (1965).

230. I. Chang and T. Kommedahl, Phytopathology, 58, 1395 (1968).

231. R. Mitchell and E. Hurwitz, Phytopathology, 55, 156 (1965).

232. C. M. Olsen and K. F. Baker, Phytopathology, 58, 79 (1968).

233. R. D. Price, P. R. Merriman, and J. K. Kellmorgan, 2nd Intern. Congr. Plant Pathol., Minneapolis, Contrib. No. 0666, 1973.

234. H. Cole, P. Sanders, C. W. Goldberg, and P. E. Nelson, Fungicide Nematicide Tests—Results of 1972, Vol. 28, The Amer. Phytopathological Soc., St. Paul, Minn., 1972, p. 290.

235. P. R. Merriman, R. D. Price, K. F. Baker, J. F. Kellmorgan, T. Piggot, and E. H. Ridge in Biology and Control of Soil-Borne Plant Pathogens (G. W. Bruehl, ed.), The Amer. Phytopathological Soc., St. Paul, Minn., 1975, pp. 130-133.

236. P. A. Backman and R. Rodriguez-Kabana, Phytopathology, 65, 819 (1975).

237. T. Kommedahl and I. C. Mew, Phytopathology, 65, 296 (1975).

238. M. Tveit and M. B. Moore, Phytopathology, 44, 686 (1954).

239. J. T. Mills, Can. J. Plant Sci., 52, 449 (1972).

240. K. A. Lord, K. A. Jeffs, and R. J. Tuppen, Pestic. Sci., 2, 49 (1971).

241. D. C. Griffiths, K. A. Jeffs, G. C. Scott, R. Gair, E. Lester, F. E. Maskell, and J. H. Williams, Pestic. Sci., 3, 609 (1972).

242. P. K. S. Clinton, Empire J. Expt. Agr., 28, 211 (1960).

243. D. Beller, Agrichemical Age, 17 (12), 7 (1974).

244. B. Jakubusiak and J. Golebiowska, Acta Microbiol. Polon., 12, 196 (1963).

245. J. Golebiowska, H. Kasubiak, and M. Pajewska, Acta Microbiol. Polon., 16, 153 (1967).

246. R. L. Curley and J. C. Burton, Agronomy J., 67, 807 (1975).

247. N. N. Afifi, A. A. Moharram, and Y. A. Hamdi, Arch. Microbiol., 66, 121 (1969).

248. J. A. Pinckard and D. R. Melville, Plant Disease Reptr., 59, 262 (1975).

249. D. K. Bell, Phytopathology, 64, 241 (1974).

250. P. A. Backman, J. M. Hammond, and R. Rodriguez-Kabana, Plant Disease Reptr., 60, 1 (1967).

Chapter 7

CONTROL OF VASCULAR PATHOGENS

Donald C. Erwin

Department of Plant Pathology
University of California
Riverside, California

I. INTRODUCTION

The control of vascular diseases by the use of chemicals differs from the control of foliar diseases because the vascular diseases are caused by microorganisms that have the particular ability to live and migrate in the xylem tissue of the plant. Thus they are protected from an externally applied toxic chemical by several layers of plant tissue. Furthermore, vascular diseases are, for the most part, single-cycle diseases that depend on inoculum deposited in soil or in a host during a previous season or seasons.

Because of these facts the strategy of chemical control of vascular diseases has been limited to the use of preventive biocides with eradicative properties which, when applied to the soil, reduce, suppress, or eradicate the primary inoculum. The only chemicals that have been proved to be capable of this task are the biocidal fumigants which are placed in the soil at various depths. This type of treatment is expensive, and control by these means has been limited to crops which are intensively produced, such as the strawberry.

There has been interest in therapy by means of systemic fungicides on plants affected by the vascular wilt diseases for many years, but the deep seated systemic nature of the pathogens and the lack of fungitoxic chemicals with the capability to translocate through plant tissue without harming the plant made this goal difficult to reach. Interest in this approach was revived when the systemic benzimidazole fungicides, thiabendazole and benomyl, each of which was relatively nonphytotoxic, were introduced. These fungicides and related derivatives are now being actively studied on a field scale principally for the control of Dutch elm disease caused by Cerato-cystis ulmi. Despite the progress in this field and the apparent possibilities for practical control, there are still many difficulties to overcome and there is concern that even successful treatment may be only a stop-gap measure that buys time before the diseased elm trees are eventually lost. While the systemic chemicals such as benomyl have eradicative properties, there has been little or no research to indicate that they can effectively control the primary inoculum in the field.

For the most part, the most successful practical measures employed for the control of vascular wilt diseases, at least in annual crops, has come about through the use of resistant varieties that have been selected either from natural populations or from progeny of crosses in a breeding program. This approach will undoubtedly remain the most important one. The success of this method of control is thoroughly documented by Walker [1] for the control of Fusarium wilt of tomato.

However, the gene pool of some crops is often limited, and sources of high level resistance may be nonexistent in the foreseeable future. Also, in the tree and forest crops the existing trees cannot readily be replaced by resistant seedlings; if they are, there is a loss of time and funds that have been invested in their growth. In the long-term development of resistance to certain crop pathogens, there is always the possibility that the fungal pathogen might overcome the genetically controlled resistance by selection pressure. Wilhelm [2] indicated that the practical success of preplant fumigation of strawberry land to control the vascular wilt caused by Verticillium, as well as several other root diseases of strawberry, allowed the plant breeder to concentrate more fully on selection and breeding for other important horticultural characters that lead to improvement of the strawberry crop. Although breeding for resistance is a fruitful and ecologically desirable approach, there is a limit to the number of progeny tests that can be conducted against several diseases of a single crop, and the chance of incorporating resistance to several diseases diminishes rapidly with the increase in numbers of disease problems. For these reasons, chemical control and mainly soil fumigation, where economically feasible, have attained a high level of practicality. While the technology is improving, this practice appears to be here to stay and will continue to have an important place in agriculture for the control of some vascular wilt diseases. The expense of fumigation is a factor that limits its use to disease problems of high-value crops. At present, fumigation of land with a mixture of chloropicrin and methyl bromide (1:1), including plastic tarps, costs about $1.50/lb (about $3.00/kg) [$1,505/ha].

The investigation of control of vascular wilts with systemic chemicals is still in the experimental stage, but for some diseases it seems to be on more solid footing now than at any time past. Agricultural chemical companies have much more interest in the development of systemic chemicals now than ever before despite the high cost. This attitude has been stimulated partly by several breakthroughs in the development of fungicides. First the carboxanilide compounds, e.g. carboxin (5,6-dihydro-2-methyl-1,4-oxathiin-3-carboxanilide), have the unique capacity to eradicate the systemic, though not vascular, loose smut fungus (Ustilago nuda) from barley seed [3]. Unfortunately, this family of compounds has not shown much promise for control of the vascular wilt fungi except for 3-dihydro-5-o-phenyl-carboxanilido-6-methyl-1,4-oxathiin against Verticillium albo-atrum in vitro (DM = dark mycelial form) [4].

The important development of the carboxanilide compounds was followed by the discovery that the anthelmintic thiabendazole, 2-(4-thiazolyl) benimidazole (TBZ), was both fungitoxic and systemic [5]. We found that TBZ had the unique capacity to translocate from the soil upward in cotton to control Verticillium inoculated into stems [6]. Next came benomyl, methyl-1-(butylcarbamoyl)-2-benzimidazole carbamate [7], which was more effective as a systemic fungicide than thiabendazole [6, 8]. The degradation product of benomyl in plant tissue proved to be methyl-2-benzimidazole carbamate (MBC) [9, 10]. Thiophanate methyl, 1,2-bis(3-methoxycarbonyl-2-thioureido)-benzene, a compound quite unlike benomyl originally, was found to change in soil and plant tissue to MBC [11-14]. There are several other compounds that have properties similar to benomyl, but as yet none has been proved to be superior in antifungal and systemic activity. For control of Verticillium wilt disease [215], MBC is equivalent to benomyl, and in several acidic forms it translocates more efficiently in plants than benomyl [15-23].

Although much was expected of these compounds when they were introduced, there were many problems in their use for control of vascular diseases. However, from the extensive amount of research that has been done, many principles have evolved. For the most part, translocation is essentially upward in the xylem stream into leaves as a consequence of diffusion pressure deficit resulting from transpiration [24]. Thus a systemic fungicide applied to leaves is not likely to move down the petiole and up the stem. Crowdy [25] reviewed the physiological concepts of translocation of solutes in plants and also showed why most of the translocated chemicals in the apoplast move upward. Lateral movement from one leaf to another could occur if transpiration were decreased drastically in the donor leaf or were increased in the receptor leaf. Movement of chemicals active against vascular wilt diseases in the symplast has not been reported.

In this chapter the control of vascular wilt fungi, by the currently acceptable but expensive practice of soil fumigation and by the experimental but promising use of systemic chemicals, will be discussed. Where possible, these methods will be placed in perspective with other methods of control. Another area, the experimental use of growth regulators which change the metabolic processes in the plants and appear to induce a type or types of resistance, will also be described. This aspect of chemotherapy was discussed by Horsfall and Dimond [26] and Horsfall [27, 28]. Evidence for this phenomenon was given by Davis and Dimond [29], but in 1962 Dimond commented in retrospect [30] that increasing host resistance to chemicals may not prove to be useful since the effect of growth regulators was too often short-lived and the side effects undesirable. He offered the speculation that a search for compounds that increase the content of polyphenols or other compounds known to confer resistance to disease might be worthy of investigation. Since then there have been several papers [31-35] that suggest that some growth regulators may increase the resistance of plants to some vascular diseases without harmful side effects. These and other studies of this nature will be described in this chapter.

II. STRATEGY OF CONTROL

A. Epidemiology as Related to Control

Van der Plank [36] makes the analogy that the rate of increase of the single-cycle disease such as Fusarium wilt of cotton caused by Fusarium oxysporum f. sp. vasinfectum resembles the increase of money at simple interest. He shows that sanitation, which is any measure including use of fungicides that suppresses the initial inoculum and rotation, where the initial inoculum cannot survive without the host, has a profound effect on the rate of disease development. He presents calculations which indicate that the amount of disease each year is directly proportional to the initial inoculum.

This analogy differs from the analogy of the development of the multiple-cycle disease (e.g., late blight of potato caused by Phytophthora infestans) with the growth of money at continuous compound interest. In multiple-cycle diseases, lesions on leaves "beget" inoculum in the form of sporangia (secondary inoculum) which are carried to noninfected leaf tissue. With this type of disease, suppression of the initial inoculum delays the epidemic, but under favorable environmental conditions the epidemic eventually reaches a peak as high as when the initial inoculum was not suppressed (no sanitation). This is due to the secondary spread of sporangia from a primary lesion to susceptible leaves.

Although it is beyond the scope of this chapter to review all of Van der Plank's [36] thought-provoking concepts of epidemiology, the analogy of the single-cycle vascular wilt disease with the growth of money at simple interest is pertinent here. As simple interest does not "beget" interest until the second year, so the vascular wilt disease seldom "begets" inoculum until the following year. The epidemic actually spans several successive years during which primary inoculum increases or decreases, depending on various sanitation effects.

Environment can affect the severity of a vascular wilt disease, and this can be confused with the effect of inoculum density in soil. It is well known that Fusarium wilt of tomato is more severe at relatively high temperature [1] and the Verticillium wilt of cotton is more severe at lower temperature [37]. It is also likely that temperature and inoculum density interact.

Van der Plank's concept [36], though perhaps simplistic to apply directly to the field, is a valuable model on which a discussion of the role of control measures can be based. If the increase in disease incidence in one year depends on the inoculum from the previous season or seasons, then it is only logical that effective control methods should be aimed at the initial inoculum in the soil, or, as in the case of the Dutch elm disease, at the initial inoculum in the diseased or dead trees. In the latter case early approaches to control include burning of diseased trees (eradication). Although the concept of the role of initial inoculum suggests that sanitation is much more important for the control of vascular wilt diseases than foliage diseases, there is still too little information on the level of initial inoculum in soil and the relationship of this level to vascular disease occurrence or to severity to make practical recommendations.

B. Inoculum Density as Related to Control

If we make the hypothesis that inoculum level in the soil is directly related to vascular disease incidence, as Powelson [38] has indicated for Verticillium wilt, then it is evident that more information on inoculum density would make evaluation of the effect of treatments of soil more meaningful. For instance, if the threshold inoculum level to attain a 50% incidence of disease (under optimum conditions for that disease) is a hypothetical 10 X (where X is a measurable unit of inoculum), it follows that the efficiency of a control measure in a soil that contains 100X inoculum cannot be measured by the amount or severity of disease, if the control measure reduces the inoculum level to only 50X.

Following the model of Van der Plank [36], the first principle of control of the vascular wilt disease is to reduce or suppress the initial inoculum in soil, or in the tree as in the case of Dutch elm disease. This is accomplished by various sanitation measures such as removal of diseased trees, incorporation of residues in soil, flood fallowing [39, 40], flaming [41-43], and by fumigation with a general biocide [44].

To follow this model further, the example of Verticillium wilt of cotton is pertinent. Verticillium wilt of cotton is a vascular disease that causes extensive yield loss in the San Joaquin valley of California and in other areas of Southwestern United States; exceptions are the Imperial, Coachella, Palo Verde, and Yuma valleys where summer temperatures are too high for the disease to cause more than minimal yield losses. This disease is caused by the microsclerotial form of Verticillium called Verticillium dahliae by some [45, 46] and V. albo-atrum by others [47]. To avoid confusion caused by this controversy in nomenclature when reading the literature, the simplest approach is to note whether the author is writing about the microsclerotial form (MS) or the dark mycelial form (DM). These forms are distinct enough to be differentiated. The MS form causes vascular wilt diseases and has a long residual life (up to 17 years) in soil and in plant debris and consequently is not easily controlled by crop rotation. The DM form also causes vascular wilt but has a shorter life in soil, and short rotations consequently control the disease [38]. The optimum and maximum temperatures for growth of the DM form are lower than for the MS form [46]. Schnathorst's review [46] presents a convincing argument in favor of using the name V. dahliae for the MS form. Since many researchers in the United States appear to be moving toward accepting this name, and because V. dahliae has long been the accepted name in Europe, for discussion the MS isolate will be referred to as V. dahliae and the DM as V. albo-atrum. However, when citing individual reports, in which these names have been used differently, MS and DM will be appended as appropriate to confirm the type being discussed.

In the San Joaquin Valley, the control of Verticillium wilt of cotton by V. dahliae (MS), which includes at least two pathotypes (a nondefoliating and a defoliating type [48]), has received the attention of numerous researchers. They have attempted to control the disease by using a number of cultural and chemical practices which include rotation and fumigation at both biocidal [49] and low levels [50] and with systemic fungicides [51-53]. In many of these studies the methods of measurement of disease included the incidence of disease as measured by vascular discoloration in the fall, the severity of disease as measured by the degree of chlorosis on leaves, and the percentage of petioles that yielded Verticillium when cultured [54].

It became evident that failure of many treatments to reduce the incidence of disease might have been due to the presence in soils of quantities of initial inoculum which exceeded the threshold level necessary to cause disease. If this were true, then incidence of disease would quantitatively underestimate the suppressive effect of a treatment on the initial inoculum per se, although it would be a quantitative measure of the effect on the total disease. Although few cultural treatments could be expected to have more than a partial effect on suppressing initial inoculum, nevertheless it is important to be able to quantitatively measure the degree of their effects.

Ashworth and co-workers reported that the wet sieve method of assaying soil provided population (propagules per gram of soil) data that were correlated with disease incidence (based on presence or absence of vascular discoloration) in the field [55]. The soil was first washed through a coarse sieve and then through a 37-μm sieve on which the microsclerotia were retained (wet sieve method). These were plated on a sugarless Czapek's agar covered with cellophane and the colonies counted after a suitable incubation period [56]. The inoculum density (microsclerotia per gram of soil) from 24 fields was directly correlated ($r = 0.9$) with the incidence of Verticillium wilt, as measured by vascular discoloration in the fall [55]. In the 24 fields they assayed, 100% incidence of disease occurred in soils containing 3.5 or more microsclerotia per gram and a 20-50% incidence of disease in soils containing 0.3 to 1.0 microsclerotia per gram (Fig. 1). The quantitative threshold detection level (15 g soil sample) of detectable microsclerotia ranged from 0.13 ± 0.02 to 46.97 ± 2.11 with a mean coefficient of variation of about 15.4% [56]. In a subsequent paper, the technique for plating the microsclerotia was modified to include use of sodium polypectate as the sole carbon source rather than cellophane [57].

When DeVay et al. [58] assayed 13 fields, using the Anderson air sampler method, in contrast to previous work [55], the numbers of propagules per gram of soil in infested fields were not correlated with the incidence of disease which they based on foliar symptoms. While their conclusions were valid, the assessment of disease incidence based on foliar symptoms quantitatively gives a more variable and a lower estimate of disease incidence than the vascular discoloration symptom used by Ashworth et al. The actual infection or noninfection of a plant, as indicated by the internal vascular discoloration symptom, may be directly related to inoculum density, but the degree of foliar symptoms expressed may be related to both soil and air temperatures and other factors as well as to inoculum density [58].

Ashworth et al. [59] also reported that harsh milling of soil fractured microsclerotia and thus increased the numbers. However Butterfield and DeVay showed that the ball mill they used for pulverizing dry soil did not fracture the microsclerotia in their experiments. The Anderson air sampler method, when compared to the wet sieve method, consistently increased the counts of microsclerotia per gram of soil by an average factor of 2.8 [60, 225]. Also there was a positive linear correlation between the number of microsclerotia per gram of soil (0-10) at planting time and

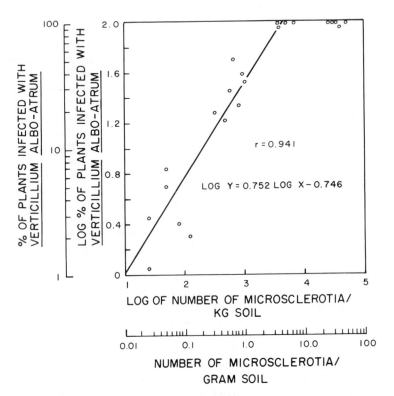

FIG. 1. The logarithmic relationship between the inoculum density of Verticillium albo-atrum (MS) in field soils and infection of cotton plants at the end of the growing season. Reprinted from Ref. 55, p. 902, by courtesy of the American Phytopathological Society.

the incidence of vascular discoloration at harvest [60, 225]. Although the incidence of disease was related to inoculum density, there were other factors as well that accounted for foliar symptoms [61, 226].

A satisfactory soil assay method should have value for predicting the success or failure of a susceptible high value crop; but there are complications. Pathogens such as Verticillium may be counted as discrete units in soil but actually exist as propagules clumped together in plant material. Schnathorst's data indicated that only the hyaline cells attached to plant debris and not microsclerotia were persistent units of inoculum [62, 63], but in soil plates most of the colonies of Verticillium arose from identifiable microsclerotia [225]. Another complication is that some isolates from soil are nonpathogenic, some are only moderately virulent and some are severe defoliators [58].

Despite these differences, the principle remains that the inoculum from the decaying plant exists for long periods of time in soil. Because of this, there will be a need in the future to predict the probability of amount or severity of disease before a crop is planted. It appears that the work reported here supports in some measure the contention that suppression of the initial inoculum in soil is of primary importance for control of Verticillium wilts.

The disadvantage of using any soil assay technique is the time-consuming procedure to obtain the data. Nevertheless, these reports present a practical way to measure the effects of sanitizing treatments much more accurately than measurement limited to disease incidence. When one contemplates the time and money invested in less quantitative studies that yield no data on the suppression of the initial inoculum, the effort required for such an assay might be justified.

There is reason to expect that methods for prediction of the density of propagules will eventually develop to the point where they are practical in modern agriculture for both research and crop production. Since it is beyond the scope of this chapter to review all of the methodology for determination of inoculum levels in soil, the reader is referred to the review by Tsao [64] and the chapter by Powelson [38].

III. CHEMICAL CONTROL

Chemical control of vascular diseases can be divided into preventive and therapeutic measures. Historically, prevention of plant disease has been the most successful. Although plant pathology deals more with populations than with individuals, the advent of the newer systemic antibiotics and fungicides has generated greater interest in therapy than ever before. For instance, there is now hope of avoiding the esthetic and economic loss entailed in the destruction of a favorite tree by Dutch elm disease.

Since the vascular wilt pathogens are systemic, the initial inoculum for the most part comes from the diseased plants that are in the process of decomposition in the soil and, in the case of Verticillium (MS) and Fusarium wilt, from residual inoculum that remains viable in the soil for long periods of time. Therefore, the most logical initial approach to control is to reduce the inoculum to the minimum [36]. Reduction of the initial inoculum in soil usually requires approaches more drastic than rotation to nonsusceptible crops, especially with Fusarium and Verticillium wilt (caused by the MS form), because of the longevity of the infective propagules in soil. Thus, to restore land that has been infested, either sterilization by steam, pasteurization by steam-air mixtures [65], or use of biocides must be used. The major problem with this strategy, however, is the expense involved to reduce the inoculum quickly by chemical means. The review of the factors affecting the efficacy of fungicides in the soil by Munnecke is pertinent [66].

A. Prevention

Prevention of vascular diseases involves several practices, of which the
fungicide or biocide is usually the last resort, except in high income crops
like strawberry. Obviously, vascular wilt diseases will not occur on noninfested new land, and as long as new land exists, simple rotation suffices.
Such was the practice in Central America for control of Fusarium wilt of
banana (Fusarium oxysporum f. sp. cubense). However, new land is becoming scarcer, not only in Central America, but all over the world.

Flood fallow and replanting every four to eight years became the only
economical method to control Fusarium wilt of banana in Central America
[39]. Fusarium oxysporum f. cubense, an aerobic fungus [40], was suppressed in saturated or flooded soils. Survival was greatest on the surface
of the water where aeration was the greatest.

In the case of Verticillium wilt caused by V. dahliae (MS), rotation to
nonsusceptible crops such as barley and sorghum has been successful in
California only in areas where long rotations of five years or more can be
practiced.

There is also the problem of weed hosts on which inoculum may increase.
Many weeds are symptomless, such as the nightshade plant Solanum sarachoides, but may nevertheless become infected and contribute to the primary inoculum in the soil [67].

Inoculum can be carried on seed. For example, Fusarium oxysporum
can be carried on bean or cowpea as chlamydospores, as colonized pieces
of plant residue from the threshing operation, or on vegetatively propagated
plant material such as Verticillium dahliae (MS) on potato "seed" tubers.
Control in these cases can be effected by a seed treatment fungicide.
Mercury-containing fungicides were found to be effective and were used in
the past. However, these fungicides have a very limited use and are no
longer registered because of their extremely high mammalian toxicity.

1. Soil Fumigation to Control Verticillium Wilt

One of the best examples of economically successful prevention of a vascular wilt disease by use of soil fumigation is the control of Verticillium wilt
of strawberry in California [2], one of the most intensively grown and potentially one of the most valuable crops in agriculture. Fumigation has markedly increased the yield, and fumigation with methyl bromide and chloropicrin mixtures has become profitable. In most strawberry cropland,
fumigation is repeated every year before planting pathogen-free nursery
stock.

There are several steps in the control of strawberry diseases, all of
which are based largely on preventive measures. Plants free of virus and
fungus are produced from asexually rooted cuttings in pathogen-free greenhouses and are transplanted to a field fumigated with a mixture of methyl
bromide and chloropicrin. Wilhelm et al. stated that one-third of the total

strawberry crop in the United States is grown on 12,000 to 15,000 acres
(4,860 to 6,070 ha) in California [2]. Diseases of strawberry and the methods used for field fumigation of soil are described by Wilhelm [67].

Verticillium wilt caused by Verticillium dahliae (MS form) is of major
importance on strawberry; many other root-rotting fungi include Cerato-
basidium sp. (Rhizoctonia fragariae), Pyrenochaeta sp., the gray sterile
fungus, Pythium ultimum and Phytophthora fragariae [2]. According to
Wilhelm et al. [2], before soil fumigation was developed strawberries
typically grew poorly or even failed after 1 year on the same land. Likewise, following certain crops which harbored Verticillium, such as tomatoes,
the strawberry crop often failed. On fairly normal fields, yields of 6,000-
10,000 kg/ha have now been increased to 40,000-60,000 kg/ha. This is
largely due to the control of the vascular wilt fungus Verticillium and other
soilborne fungi by soil fumigation with chloropicrin and methyl bromide.

The development of chloropicrin (trichloronitromethane), used as tear
gas in World War I, as a soil fumigant for control of Verticillium in the
1950s, is documented by Wilhelm et al. [2, 68]. In early work (1950-1954),
he and his colleagues determined that 480 kg/ha of chloropicrin, applied to
land by hand applicators and by a soil chisel, controlled Verticillium wilt.
Later the technology of using polyethylene sheeting to cover the land following application of the fumigant to prevent evaporation losses allowed a reduction in the effective dosage to 380 kg/ha.

Another development was the finding that a mixture of chloropicrin and
methyl bromide was superior to either chemical alone. Munnecke and
Lindgren [69] reported that methyl bromide was ineffective for control of
Verticillium dahliae (MS) in nursery soils, although it was extremely effective for Fusarium, Pythium, Sclerotinia, Sclerotium, and Rhizoctonia.
Methyl bromide applied to soil up to 201 kg/ha was also ineffective against
Verticillium in the field and in some cases Wilhelm et al. [70] found that
rates up to 180 kg/ha actually increased the severity of Verticillium wilt of
strawberry. However, when a mixture of 192 kg/ha of chloropicrin and
106 kg/ha of methyl bromide were injected together in soil and covered with
polyethylene sheeting after application, the control of Verticillium wilt was
equal to the control obtained by use of 480 lb/acre (537 kg/ha) of chloropicrin. They were not able to conclude whether the increase in efficacy of
the mixture was due to synergism or to the formation of new compounds [70].
Since the mixture of chloropicrin and methyl bromide was also more successful in controlling weeds than either chemical alone, in a broad sense
(its total beneficial effect on both weeds and pathogens) Wilhelm et al. [2]
considered the mixture to be synergistic. However, in a strict sense there
is not enough physical and biological data to conclude that they are synergistic [66]. The predominant practice in California is to use a 1:1 mixture
of methyl bromide-chloropicrin at rates of 325-450 lb/acre (364-504 kg/ha)
applied with soil chisels at a depth of 6-8 inches (15-20 cm) to land that is
subsequently covered for 48 hr with a polyethylene tarp. A 2:1 mixture at

275-375 lb/acre (308-420 kg/ha) suffices to maintain the land free of pathogens. In 1972, 9,000 ha of strawberry land was fumigated with a methyl bromide-chloropicrin mixture [2]. Costs of fumigation of soil with this 1:1 mixture at 400 lb/acre (448 kg/ha) including the polyethylene tarp is about $650/acre ($1,505/ha) (1977 data).

While the technique of soil fumigation with methyl bromide-chloropicrin has been successful for many years in the strawberry industry [2], fumigation to control Verticillium wilt of cotton has been fraught with many problems. Wilhelm [49] applied methyl bromide-chloropicrin (45:55) to highly infested land in Tulare County in California. In the first year Verticillium wilt was practically eliminated and in the following year the incidence was less than 10%, with one exception where it reached about 30%. Despite the success in the control of Verticillium wilt, the yield of cotton lint was reduced from 672 lb/acre (753 kg/ha) to 440 lb/acre (493 kg/ha) by the fumigation treatment. This adverse phenomenon was due to the initial stunting of cotton in the fumigated plots. Later in the season, cotton became more vegetative than normal and the boll set was delayed. The early stunting was considered to be associated with Zn^{2+} deficiency, since the condition could be corrected to some extent by foliar application of $ZnSO_4$ [216].

Although the percentage of plants with Verticillium wilt was reduced in chloropicrin-methyl bromide fumigated fields from nearly 100% to an average of less than 10%, the increased yields in the second year from 684 lb/acre (766 kg/ha) to 814 lb/acre (912 kg/ha) of cotton lint after fumigation did not offset the cost [49].

De Vay et al. [50] applied Telone (1,3-dichloropropene) and Telone C (85% 1,3-dichloropropene and 15% chloropicrin) by soil chisels before planting cotton in California. They spaced two shanks, at 20-22 inches (51-56 cm) depth, 36 inches (91 cm) apart, and three shanks at 8-10 inches (20-25 cm) in depth were offset so that each of the centers of application were 18 inches (46 cm) apart. The incidence of Verticillium wilt of cotton was reduced in one field experiment from 27 to 7% by chloropicrin (45 gal/acre; 426 liters/ha) and to 9% by Telone (61 gal/acre; 577 liters/ha). Both treatments increased the yield of cotton lint 61%. In another experiment, chloropicrin (45 gal/acre; 426 liters/ha) reduced the incidence of Verticillium wilt from 66 to 11%. However, in this test, early stunting and subsequent increased growth in fumigated plots occurred with a consequent reduction in yield of lint.

Hafez et al. [71] reported that fumigation with chloropicrin (420 liters/ha; 714 kg/ha) in a field in California reduced the incidence of Verticillium wilt from 27% to 4%, but the corresponding increase in yield of cotton lint of 15% was not statistically significant. However, addition of potassium (270 kg/ha as KCl) not only reduced the incidence of Verticillium wilt from 27% to 7% but also increased the yield of cotton lint by 62%. They suggested that the plants in potassium-treated soils were more resistant to Verticillium.

Later the problem of initial stunting of cotton and coincident Zn^{2+} deficiency was found by Hurlimann et al. [72, 73] to be due to the destruction of spores of <u>Endogone</u> sp., a Phycomycetous mycorrhizal fungus, by the chloropicrin-methyl bromide fumigant. They found roots of occasional plants in the fumigated field that had escaped the early stunting effect with large populations of <u>Endogone</u> spores, while the roots of stunted plants were devoid of them. In pot tests where <u>Endogone</u> spores were added to fumigated soil, subsequent growth of cotton was normal, but where no <u>Endogone</u> spores were added to fumigated soil, plants remained stunted.

The report of Hurlimann et al. [72, 73] is one of many examples of the important role of mycorrhizae on crops. Gerdemann [74] found a striking increase in growth of corn and tulip trees following reinfestation of sterilized soil with spores of <u>Endogone fasiculata</u>. A similar response by citrus seedlings was also shown by reinfestation of fumigated soil with mycorrhizae [75].

In contrast to the situation with cotton, fumigation has not caused stunting of strawberry plants. However, Wilhelm et al. [2] report that <u>Endogone</u> sp. is carried in all asexually propagated strawberry planting stock and is favored by the stronger strawberry root system which develops in fumigated soil. Perhaps this is the difference, that is, cotton and citrus are initially planted from seed on which one would not expect to have a native flora containing <u>Endogone</u>. If the technology of development of the beneficial endotrophic Phycomycetous mycorrhizal fungus for cotton were improved to the place where inoculum of efficient strains could be applied to cotton seed, perhaps fumigation with mixtures of methyl bromide and chloropicrin could be used to reduce the initial inoculum to a low enough level to control Verticillium wilt of cotton for several years.

Another problem has been the tendency for recurrence of Verticillium wilt in the second and third year after fumigation. This problem may be due to lack of understanding of the precise relationship of inoculum density to the dosage of a fumigant required for control. Wilhelm discussed inoculum density and fumigation and observed that heavily infested soils require more fungicide to achieve control than lightly infested soils [76]. Bald and Jefferson [77] stated that "moderately efficient soil treatments are often useless when applied to heavily infested soils." They presented a crude mathematical model to show the effects of dosage on control. From their theoretical data, it is extremely unlikely that all of the inoculum will be eradicated even by high dosages. Richardson and Munnecke [78] said that the "logarithm of the fungicide dosage at the 50% control level increases proportionately with the logarithm of the inoculum density." With cotton, a woody perennial-type plant that is grown as an annual, the microsclerotia form in the lignified xylem tissues as well as in parenchymatous tissues after the colonized plant dies. The microsclerotia which form in the plant during the first season are protected by several layers of tissue in a lignified stem, much of which does not decay until one to three years have passed. It is not unusual to find only partially decomposed cotton stalks containing microsclerotia as much as 1 year after they were disced or plowed into the soil.

To Verticillium (MS) infested soil used for potatoes, Easton et al. [79] applied (a) a mixture of Telone (1,3-dichloropropene and related hydrocarbons) and chloropicrin, and (b) DD (1,3,-dichloropropene, 1,2-dichloropropane, 3,3-dichloropropene, 2,3-dichloropropene, and related C_3 chlorinated hydrocarbons) plus chloropicrin. Treatment (a) was applied at 187 liters/ha and treatment (b) at 46.7 liters/ha before planting potatoes during each of several years. Fumigation with these relatively low dosages of fumigant reduced the numbers of Verticillium propagules in soil after each fumigation; however, by fall the numbers of propagules had increased. This unusual phenomenon was attributed to contamination of fumigated plots by irrigation water flowing from nonfumigated plots. In most years fumigation with both chemical treatments increased the yields of potato tubers significantly.

The nature of the propagules that were detected by his soil dilution assay method [80] was not determined; however, it is significant that the percentage of potato plants with Verticillium wilt was relatively low (34% in 1968) considering the high inoculum density in soil (as high as 1,066 propagules per g on April 10). Since Easton did not indicate that soil was dried before the dilution assay (drying is considered to kill conidia but not microsclerotia), perhaps some of the inoculum assayed was conidial. He [81] reported also that fumigation of soil with Terrocide 30 (ethylene dibromide plus trichloromethane), Terrocide 30D (DD plus trichloromethane), and Vorlex (chlorinated C_3 hydrocarbons plus methylisothiocyanate) controlled Verticillium in infested potato soils when used at dosages of 12.5 to 25 gal/ acre (118.2-236.5 liters/ha).

According to Easton [217], for control of Verticillium wilt of potato chloropicrin alone is insufficient and must be mixed with DD, EDB, or Telone.

Young [82] applied Vapam (sodium N-methyl dithiocarbamate dihydrate) at a 6-inch depth in soil at 190 lb/acre (225 kg/ha) 10 days before planting potatoes and obtained striking control of Verticillium wilt (early dying disease) caused by V. albo-atrum (MS), as well as weed control and a doubling in yield of potato tubers. At that time (1956), however, the increase was not considered to be economical.

In summary, fumigation with biocides such as methyl bromide and chloropicrin is an effective and practical control measure for Verticillium wilt of strawberry but not of cotton. In fumigation of nursery soils and field soils, the elimination of mycorrhizal fungi has caused problems which will have to be surmounted by proper reinfestation procedures.

The fumigation of land to be planted with potatoes by relatively low dosages of chloropicrin to attain control of the early dying—Verticillium wilt caused by both V. albo-atrum (DM) and V. dahliae (MS)—appears to be a practical method of control. It would appear that perhaps potato responds to a smaller reduction in the population of Verticillium than strawberry or cotton.

2. Control of Verticillium Wilt of Peppermint by Flaming

Although gas flaming is a physical method of control, this method indirectly arises from the chemical use of the hydrocarbon gas, propane. This method was reported as a way of reducing microsclerotial inoculum of Verticillium dahliae (MS) in stems of potato plants [43]. Mint is grown as a perennial in Oregon for about six years in one rotation. Since the plants reproduce each succeeding year from rhizomes, the plants infected in one year will also be infected the next. Although the spread of Verticillium dahliae (MS) in soil is extremely slow, the fungus is readily disseminated as microsclerotia after forming in diseased stem and leaf tissue. When this residue is left in the field, a single area of infested plants (6 m = 20 ft in diameter) may spread to as much as 247 ha (100 acres) [41]. Flaming with propane gas reduced the viable propagules 80-90% [41]. In another study, McIntyre and Horner reported that in fields with a minimal number of infected plants, propane flaming at speeds of 2.0-2.7 mph (3.2-4.3 km/hr) increased internal temperatures of stems to 60°C and incinerated the leaves [42]. Tissue-plating and dilution assay of stems indicated that the percentage of kill of the fungus was over 99%. This is a remarkable ecological accomplishment in control of a vascular disease on a perennial crop.

According to Horner [218], flame control has become practical in mint culture. He estimated that the loss due to Verticillium wilt in Oregon has been reduced from 15 to 3% per year allowing a gain of $3,600,000 at a cost of $400,000 ($20.00/acre; $49.40/ha).

Powelson and Gross [43] reported that propane flaming killed Verticillium in potato vines. They employed a novel technique of inserting "temp-pils," which are short tubes that melt at different temperatures, into the pith of potato stems. When the propane burner was passed over the field at a speed of 1.5 mph (2.9 km/hr), the temperatures attained were 52°C in 100% of the plants, 65.5°C in 90% of the plants, and 93°C in 53% of the plants. Overall, Verticillium was killed in 95% of the stems assayed.

These reports indicate that heat is capable of eradicating Verticillium in plants. Since the persistent microsclerotia do not form until after the plant has become moribund, application of heat or any eradicative factor would be most likely to be successful at this stage.

3. Soil Fumigation to Control Fusarium Wilt

From the voluminous literature on the interaction between Fusarium wilt and root-knot nematode, one might expect that Fusarium wilt does not occur without root-knot nematode; however, each disease can be damaging without the other and they do not always occur in the same fields.

In the Florida tomato growing industry, there appears to be a situation similar to that of the banana industry in Central America before the 1950s, in which the grower moved to new land to escape Fusarium wilt, Verticillium wilt, or the root-knot nematode disease [83]. "New land" is still available in many areas but is becoming scarce, and as the population of the world increases will become even more scarce.

It is well known that the persistence of the causal agent Fusarium oxysporum is due to the production of chlamydospores [84]. In California there are only about five properties in Kern County on which the Fusarium wilt disease of cotton has been found. Either the causal fungus has not been generally introduced or soil conditions do not favor it. In a recent paper, Smith and Snyder [84] reported quantitative data on the variation in the number of propagules of Fusarium oxysporum f. sp. vasinfectum in the soil in one of these fields as affected by different crop rotations. They showed that the actual numbers of pathogenic propagules did not decrease over a period of years even when a nonsusceptible crop such as barley was grown. They also found that the fungus reproduced in the crowns and roots of the weeds, yellow nut sedge, and several species of Malva. Although typical wilt diseases did not occur in these plants, invasion of the cortex of these plants was sufficient to allow an increase in the number of propagules. Numerous other similar examples are given in this literature review [84]. This is undoubtedly the reason why growers with severe Fusarium wilt problems must seek new land unless they can rejuvenate the old land by an economical and ecologically effective method.

Resistance to race 1 of F. oxysporum f. sp. lycopersici has been successfully incorporated into most commercial tomato varieties. However, in Florida, race 2 has become more important and, although the variety Walter is resistant to race 2 [1], fumigation might be considered as a choice for control. In cases where resistance is available, soil fumigation serves to control weeds, Verticillium wilt, nematodes, southern blight, and damping off [219].

Jones et al. [83] recently reported the efficacy of several fumigants to control Fusarium oxysporum f. sp. lycopersici in soil. Marketable yields of tomato were increased using allyl alcohol (2-propen-1-ol), 25 gal/acre (236 liters/ha), plus EDB, 6 gal/acre (57 liters/ha) or EP201, chloropicrin (15%) plus methylisothiocyanate (17%) plus DD (68%). The percentage of incidence of Fusarium wilt was decreased by the fumigants.

Several reports indicate that Fusarium wilt is more severe in acidic soils than in alkaline [1]. Jones and Woltz [85] found that liming the soil to pH 8.0 decreased Fusarium wilt of tomato. In later field experiments Jones and Overman [86] treated fine sand, artificially infested with the Fusarium race 2 the year before, with $Ca(OH)_2$ to obtain a pH of 7.0 and 7.5 and with sulfur to obtain a pH of 6.0. The calcium content was equilibrated by addition of $CaSO_4$ to plots adjusted to pH 6.0 and 7.0. Subplots were fumigated with Vorlex (methylisothiocyanate plus DD) and with a mixture of chloropicrin and DD two weeks before planting tomato seedlings. By the end of the season there was an 82% incidence of Fusarium wilt at pH 6.0, 60% at pH 7.0, and 41% at pH 7.5. The fumigants further reduced the incidence of Fusarium wilt to a range of 2.2 to 15.0% with no significant difference being due to the effects of pH. Yield of tomato fruit was increased in the soil fumigated at pH 6.0 from 793 bu/acre (1 bu/acre = 67.2 kg/ha) to 1,013 in the chloropicrin and DD treated soil and to 1,091 in the Vorlex and DD treated soil. Yields increased at pH 7.5 from 1,168 to 1,368 and 1,323 bu/acre respectively, for the two fumigant treatments. This finding appeared to be of great practical importance since the application of $Ca(OH)_2$ alone could be done for $30-50/acre ($74-124/ha) whereas the fumigation costs were about $250-300/acre ($620-740/ha). However, they estimated that the increased yield due to fumigation more than returned the cost. The probability of Verticillium wilt becoming active at the higher pH posed an alternative problem in their view.

4. Soil Fumigation to Control Vascular Wilt-Nematode Complexes

Root-knot nematodes, species of Meloidogyne, which are sedentary endoparasites, have been found to increase the susceptibility of several plants to Fusarium wilt caused by Fusarium oxysporum. Some species of Pratylenchus, which are ectoparasitic, also seem to interact but more strongly with Verticillium wilt than Fusarium wilt.

The experimental evidence that many pathogenic nematodes increase the susceptibility of plants to Fusarium wilt of cotton, tomato, and other crops is well summarized by Powell [87, 88] and for tomato by Walker [1]. Powell [87, 88] notes that the association of the root-knot nematode disease and Fusarium wilt of cotton was observed as early as 1892 by Atkinson.

There is no question that each pathogen has the ability to cause disease by itself and each can exert severe effects on a susceptible crop. However, the principal effect of the root-knot nematode has been the increase in susceptibility of varieties resistant to Fusarium wilt. Another phenomenon, the indirect control or suppression of Fusarium wilt by fumigation of soils with nematicides, has also called attention to the interaction since the nematicides in common use are not fungicidal to Fusarium [89].

The literature which contributes to proof that there is a valid fungus-nematode complex interaction has been reviewed by Powell [87, 88], and therefore these examples will not be covered here in great detail. Rather, the literature showing that nematicides actually suppress the severity of the vascular wilt disease will be emphasized.

In this sense, there is another strategy involved other than that of the suppressive effect of fumigants on the primary inoculum which was discussed earlier in this chapter. Certainly the nematicidal fumigants, ethylene dibromide (EDB), 1,2-dibromo-3-chloropropane (DBCP) and a mixture of 1,3-dichloropropene and 1,2-dichloropropane (DD) act on the primary nematode inoculum in the soil, but these fumigants appear to have little effect on the primary inoculum of Fusarium (chlamydospores) which persist in the soil [89]. Thus, the nematicide acts indirectly on Fusarium wilt by removing or reducing a predisposing factor. In this sense, the nematicide probably acts therapeutically in the control of Fusarium wilt on root-knot affected plants. Stark and Lear [89] compared the effects of several fumigants in soil infested with a Fusarium sp. capable of causing damping off of peas. They concluded from these studies (based on the percentage of emergence of peas in treated and nontreated soil) that chloropicrin was the most fungicidal of all fumigants tested. 1,2-Dichloropropane had some fungitoxicity, but only at dosages above those required to be nematicidal, while EDB had little or no fungitoxicity to the Fusarium sp. tested. If the extrapolation can be made that F. oxysporum inoculum is as insensitive to EDB as other species of Fusarium, then the many experiments in which EDB has been used to separate nematode effects from fungus effects are very meaningful.

There are many examples of the effects of nematicides on Fusarium wilt. Thomason and Erwin [90] reported results of a field experiment with cowpea (Vigna sinensis) in sandy soil with a history of the root-knot nematode disease, caused by Meloidogyne javanica and Fusarium oxysporum f. sp. tracheiphilum. In this study, the wilt-resistant variety Grant and the wilt-susceptible variety Chino 3 were planted in a 5 × 5 Latin square design after applying preplant applications of EDB at 7-inch (18 cm) depth and DBCP at 0.75 gal/acre (7.1 liters/ha). Another treatment with DBCP at 0.5 gal/acre (4.7 liters/ha) was made by side dressing 3 weeks after planting. When plants were rated for vascular discoloration caused by Fusarium wilt, the index (0-5) for Grant was reduced from 2.7 with no treatment to 0.8 in plots treated preplant with both nematicides. The postplanting treatment with DBCP reduced the Fusarium wilt index to 1.3. The vascular discoloration index for the susceptible Chino 3 was reduced from 4.4 to 2.4-2.8 by all the nematicidal treatments. Control of the nematodes reduced the severity of Fusarium wilt on both the Fusarium-resistant Grant and the Fusarium-susceptible Chino 3 varieties. In a greenhouse experiment when plants were grown in soil infested with both organisms, the resistance of the variety Grant was reduced to susceptibility by Meloidogyne javanica. When plants of the Grant variety were bare rooted and dip inoculated in a Fusarium spore suspension they remained resistant to Fusarium wilt. In the treatment in which the plants were grown in soil infested with both Fusarium and Meloidogyne, the variety Grant became susceptible. For this reason, it was concluded that the effect of Meloidogyne was due to a change in the metabolism of the host and probably was not due to wounding by the nematodes [90].

In 1948 Smith [91] conducted field experiments on light sandy soil in Alabama and in Georgia using several varieties of cotton that differed in Fusarium wilt susceptibility. Soils were naturally infested with Fusarium and Heterodera marioni (Meloidogyne sp.). Using a fumigant that contained 90% EDB (dosages varied from 12.5-37.5 gal/acre; 118-355 liters/ha), he obtained striking control of Fusarium wilt of cotton and increases in the yield of cotton lint (the average increase in yield due to the nematicide ranged for different varieties from 42 to as much as 1312%. Effective control was obtained in one experiment by placing the fumigant (12.5 gal/acre; 118 liters/ha) beneath the row before planting.

Minton and Minton [92] compared the effect of infesting soils of three different soil series with Fusarium oxysporum f. sp. vasinfectum alone and with the root-knot nematode, Meloidogyne incognita acrita, and with the sting nematode, Belonolaimus longicaudatis. Their results were striking in that severe wilt occurred only in the plots in which both Fusarium and either of the nematodes were present. Plants inoculated with both the sting nematode and Fusarium showed severe root rot in addition to the vascular discoloration usually caused by Fusarium wilt on all three soils. They concluded that the relationship of both nematodes with Fusarium was synergistic. They hypothesized that Fusarium alone was a weak pathogen and required predisposition of the roots by the nematodes for invasion. There was no evidence for this assumption, but it might be worth further investigation. In most pathogenicity testing done for plant breeding, test plants are bare rooted and dipped in inoculum composed of a spore suspension. It has been assumed by many that wounds are required for pathogenesis.

Morgan [93] reported results from three field experiments in Maryland with tobacco. The soil was fumigated with EDB at 5 gal/acre (47.3 liter/ha). The percentage of occurrence of Fusarium wilt was markedly reduced from 6 to 0%, 89 to 12%, and 22 to 3% in three consecutive tests. Despite the nematicidal effect of EDB, populations of root-knot nematode Meloidogyne incognita acrita had increased markedly by the end of the season. This may suggest that there is a nematode population threshold for predisposition to Fusarium infection to occur.

Although there is no unequivocal evidence that the root-knot nematode increases the susceptibility of plants to Verticillium wilt [94], the increase in yield of potato tubers and delay in symptom expression of Verticillium wilt of potato by side dressing with the systemic organophosphorous insecticides (phorate, disulfoton, carbamate, aldicarb) [95, 96] may indicate such an interaction.

Hoyman and Dingman [96], in the state of Washington, reported results of an experiment in which disulfoton at 6 lb/acre (6.7 kg/ha) was applied as a split application in a band under the row at planting time (3 lb) and as a side dressing (3 lb). Verticillium wilt and root-knot symptoms were decreased and the yield was increased. However, as evidence against the interpretation that the effect of these chemicals on Verticillium wilt was due to a primary effect on the nematode, they showed data indicating that O,O,diethyl-O-p-(methylsulfonyl)phenylphosphorodithioate (Bayer 2514) had excellent nematicidcal activity but, when used as a side dressing, was one of the least effective treatments in delaying Verticillium wilt [95].

Despite the promising nature of the results presented by Hoyman and co-workers [95, 96], the effects of the systemic insecticides, phorate, disulfoton, or aldicarb on delay of Verticillium wilt of potatoes have not been consistent in other areas in which they have been tested. Easton [97], also in the state of Washington, tested disulfoton and aldicarb at similar dosages and treatments in fields that were infested with Verticillium but not root-knot nematode. He reported neither an increase in yield nor a

delay in Verticillium wilt and suggested that, "it is possible that a synergistic phenomenon existed in Hoyman's field between root-knot and the
Verticillium wilt organism, in which case Verticillium wilt symptoms would
be more severe than if the Verticillium wilt organism were present alone."

In a recent report, Krause et al. [98] found that the systemic insecticide
acephate (O-S-dimethylacetylphosphoramidothioate), applied at the low
dosages of 1.0 and 2.0 oz/100 lb (6 and 12 g/kg), significantly delayed the
appearance of symptoms of Verticillium wilt, but after the plants became
infected, progress of the disease was not affected. They postulated that
control of soilborne insects which perhaps augmented the number of successful infections by Verticillium albo-atrum (DM) and V. dahliae (MS) in
the soil, might be responsible for the observed response.

In summary, the control of root-knot nematodes by fumigation where
they are associated with Fusarium wilt has a dual effect on the nematode
population and the ensuing root-knot disease, and a secondary effect on
Fusarium wilt due to the removal of the predisposing effect of the root-knot
nematode. However, there is little evidence that the root-knot nematode
predisposes plants to Verticillium wilt.

Golden and Van Gundy [99] reported that galled tissue was more attractive than healthy tissue to Rhizoctonia solani, the cause of root rot, and
that the fungus colonized galled tissue more rapidly than healthy tissue.
Golden [100] also found that the galled tissue acted as a biochemical sink
which competed with healthy sinks, roots, and stem apices, for carbohydrates formed from radioactive $^{14}CO_2$.

Wang and Bergeson [101] reported that galled roots of tomato plants
infected with Meloidogyne incognita contained a markedly higher sugar content and a lower amino acid content than healthy tissue. They proposed
that higher than normal sugar content in the xylem sap may be responsible
for the predisposition of plants to Fusarium wilt.

Although the root-knot nematode has a predisposing effect on cotton to
Fusarium wilt, there is little indication that these nematodes predispose
cotton plants to Verticillium wilt. McClellan et al. [94] applied EDB (4.5
gal/acre; 42.7 liter/ha) 1 week before planting and at planting time to a
field on which the root-knot nematode Meloidogyne incognita var. acrita
had been a serious problem for two years. Although excellent control of
the root-knot nematode was obtained and growth was more rapid in fumigated plots, the average proportion of Verticillium wilt (77 to 81%) was not
significantly reduced by any treatment. Likewise no yield differences were
noted. This may have been due to the overriding factor of severe Verticillium wilt.

An interaction between Verticillium wilt caused by V. albo-atrum (MS)
and Pratylenchus penetrans, the root lesion nematode, has been reported
by several researchers. McKeen and Mountain [102] reported that these
two organisms exerted a synergistic effect on eggplant in a growth chamber
study in which both organisms were added to soil before planting. The

inoculum of V. albo-atrum (standard dosage was one plate of a microsclerotial colony blended in 400 ml of water) was tested at four concentrations (standard dosage, 1/4, 1/12, and 1/30 of the standard dosage) with and without nematodes (2,600 per pot of soil). When the experiment was terminated 60 days later, there were no symptoms on roots to indicate that the nematodes were pathogenic, yet the incidence of Verticillium wilt at the lower inoculum concentrations indicated that P. penetrans increased the incidence of Verticillium wilt at all inoculum concentrations except at the standard rate (highest concentration) of Verticillium. In this inoculum treatment the amount of disease was the same with nematodes plus Verticillium as with Verticillium alone.

In a subsequent paper [103], using the same Verticillium fungus (the name was changed to V. dahliae (MS) in this paper because of its production of microsclerotia), Mountain and McKeen reported that the population of P. penetrans increased more rapidly in plants infected with V. dahliae than in noninfected plants. The reason for this was not evident from their research.

Faulkner and Skotland [104] reported that the incidence of Verticillium wilt of mint, caused by V. dahliae var. menthae (MS), was decreased by soil application of DD. In a subsequent paper, they reported that inoculating rooted mint cuttings with both Pratylenchus minyus and V. dahliae (MS) resulted in the increase of both the incidence and severity of Verticillium wilt [105]. As reported by Mountain and McKeen [103], the population of Pratylenchus increased more rapidly on Verticillium-infected plants than on noninfected plants. Indicative of the specificity and intricacy of these interactions, Bergeson [106] inoculated mint plants with combinations of P. penetrans and V. albo-atrum (MS) and with each fungus alone. He noted only insignificant differences or additive effects on growth, but diagnostic symptoms of Verticillium wilt appeared about two weeks earlier than when plants were inoculated with Verticillium alone.

Other nematodes that have been experimentally shown to increase the severity of Fusarium wilt are the sting nematode, Belonolaimus gracilis, which inhibits the development of root tips of cotton and causes root decay [107]. This nematode in pot tests increased the incidence of Fusarium wilt on the wilt-resistant cotton variety Coker 100WR and also on the susceptible variety Hurley's Rowden.

On tobacco, Tylenchorhynchus claytoni added to pots of soil, increased the incidence of Fusarium wilt caused by F. oxysporum f. sp. nicotianae. Miller [108] reported that the tobacco cyst nematode Heterodera tabacum increased the susceptibility of Bonny Best tomato plants only at low inoculum levels of Verticillium albo-atrum (although not designated, probably the MS form) and not at a higher inoculum level, but decreased the disease rating of Fusarium wilt caused by F. oxysporum f. sp. lycopersici.

Corbett and Hide [109] reported an interesting interaction between Heterodera rostochiensis and Verticillium dahliae (MS). When both organisms were added together in pots of soil, symptoms of Verticillium wilt of potato occurred 15 days earlier than by inoculation with only V. dahliae. Chlorocholine chloride (CCC), a growth retardant, at a rate of about 2 g per pot caused severe growth retardation but also decreased the number of cysts produced by H. rostochiensis and the amount of stem blackening (from microsclerotia formation on Verticillium-infected plants).

After many years of research, which have confirmed the interaction of root-knot nematodes and Fusarium wilt, there still remains much to be understood about the actual relationship of pathogenic nematodes and vascular wilt fungi. Certainly control of the nematode by exclusion or fumigation is a first step toward reduction of the severity of Fusarium wilt. Likewise where feasible, resistance to both organisms has a high priority in breeding work.

We have already briefly mentioned the proposal by Wang and Bergeson [101] that increase in xylem sugars in root-knot infected plants might be related to the increase in susceptibility to Fusarium wilt. The work of Orion and Hoestra [110] opened a new area which involves "several systems of host parasite relationships . . . plant-nematode-Fusarium, plant-Fusarium-ethylene, plant-nematode-ethylene." Their work showed that the Fusarium wilt resistance of the tomato variety Moneymaker could be broken by the root-knot nematodes, Meloidogyne incognita and M. javanica, but wilt resistance of the Fortos variety could not be broken by these nematodes. They tested the effect of ethylene (as a soil drench with Ethrel, 2-chloroethyl phosphonic acid) on the interaction since Viglierchio and Yu [111] reported an increase in auxin (indole-3-acetic acid) which is known to induce production of ethylene. Ethrel reduced the severity of Fusarium wilt symptoms in the variety Moneymaker. The nematodes, however, counteracted the therapeutic effect of the Ethrel. The variety Fortos remained resistant in the presence of the nematodes and with Ethrel. Ethrel was nonfungitoxic in vitro to Fusarium. Although the observed effects were still too limited to be put into a logical metabolic scheme, they felt that this was further evidence that the root-knot nematode-Fusarium wilt interaction is caused by metabolic disturbances in the plant.

5. Control of Dutch Elm Disease and Oak Wilt by Root Graft Control

Both Ceratocystis ulmi and C. fagacearum can pass from tree to tree by natural root grafts. Treatment of a band of soil between trees with Vapam (sodium N-methyl dithiocarbamate) killed roots and reduced spread of Dutch elm disease [222, 223]. Smalley applied (20% solution v/v) in holes 30-36 inches (76-91 cm) deep at 9-12 inch (22.8-30.4 cm) intervals [112]. All lateral roots were killed, which prevented root grafts between diseased and healthy trees. Since one or two trees beyond the diseased tree often became infected, he recommended blocking off one additional tree on each side of severely diseased trees and also protecting it with an additional chemical barrier.

6. Control by Suppression of Inoculum with Fungicides on Seeds or on
   Asexually Propagated Plant Material

When plants are vegetatively propagated, the propagated plant part often
carries pathogens. At this point, chemical control can exert an important
effect on the relatively small amount of the plant material, which can usually tolerate phytotoxic effects that might be deleterious if used on the entire
plant. In addition, treatment of the entire plant would be much more expensive than treatment of vegetative propagules. The importance of such a
seed or "tuber seed" treatment control method is more critical if the intention is to grow the plant in noninfested land where background contamination
is not a factor. Examples of vascular wilts that can be at least partially
controlled in this way include Verticillium wilt of potato.

Verticillium wilt of potatoes is caused by both Verticillium dahliae (MS)
and by V. albo-atrum (DM). Since potato is vegetatively propagated by
means of tubers which are produced in the soil, it might be expected that
this factor is extremely important primarily in the initial transmission of
the causal agents to uninfested areas. Powelson indicated that seedborne
inoculum does not play a significant role in the epidemiology of the disease
[113], but that the disease severity is directly proportional to the inoculum
density in the soil. However, Easton reported that seed transmission was
the means of introducing Verticillium into many uninfested areas in the
state of Washington [114] and there is evidence of this in the Negev region
of Israel [115]. Easton et al. [114] reported that 51% of all the certified
potato seed lots from several states and Canada, sampled for Verticillium
by plating soil carried on the tubers in an ethanol-streptomycin-penicillin
medium [80], yielded the MS form of Verticillium. When internal tissue
was plated on the ethanol medium, 30% of all the lots yielded both the MS
and DM forms with the DM type predominating. Several chemicals that had
a possible use as preplant treatment for potato seed tuber were tested by
mixing the compound with 1 g of naturally infested field soil, containing the
MS form of Verticillium, for 2 min. and plating appropriate dilutions of the
soil suspension on the ethanol medium. Semesan Bel (containing hydroxy-
mercurinitrophenols and hydroxymercurichlorophenols) was most effective
in reducing the number of Verticillium propagules from 944/g for the untreated control soil to 0 at 1.08 lb/100 gal (1.3 g/liter). Since the organic
mercury fungicides such as Semesan Bel are no longer registered for use
by the Environmental Protection Agency, this compound is useful only as an
experimental control. Captan, N-[(trichloromethyl)thio]-4-cyclohexane-
1,2,-dicarboximide, used at rates of 0.93-7.5 lb/100 gal of water (1.1-9.0
g/liter); Polyram, containing a mixture of 5.2 parts (w/w) of ammoniates
of [ethylenebis(dithiocarbamate)]zinc with 1 part of ethylenebis[dithiocar-
bamic acid] bimolecular and trimolecular cyclic anhydrosulfides and disul-
fides, at 0.62-1.87 lb/100 gal water (0.7-2.2 g/liter); and sodium hypochlorite bleach (5.25% solution), at 1.0-10.0% solutions prevented germination and growth of almost all of the Verticillium propagules. Mixing captan

or Polyram as dusts with infested soil was much less effective than aqueous suspensions.

Since isolates of Fusarium oxysporum which cause vascular diseases survive for long periods of time in soil, probably as chlamydospores, it would be expected that seeds from a Fusarium wilt affected crop such as bean (Fusarium oxysporum f. sp. phaseoli) might be infested with spores and bits of plant debris containing the fungus, as shown by Kendrick [116] and Kendrick and Snyder [117]. Observational evidence indicated that single plant infections in disease-free fields probably came from infested seeds. The fungus on the seed was then controlled by seed treatment with the mercurial fungicides Semesan, hydroxymercury nitrophenol plus 3.1% hydroxymercury phenol, and Ceresan, N-(ethylmercury)-p-toluene sulfonanilide. Since the organic mercury fungicides are no longer approved for registration, other fungicides capable of killing Fusarium have to be substituted now. About 3% of the seeds from tomato plants infected with Fusarium oxysporum f. sp. lycopersici may also carry spores of the Fusarium wilt organism, and this may be a factor in spread of a new race or in carrying the fungus to noninfested land. The acid soak method commonly used for control of the bacterial canker pathogen (Corynebacterium michiganense) probably serves also to control Fusarium on the seed.

Transmission of vascular wilt diseases on seeds or on asexually propagated plant parts is an important aspect of sanitation in a disease control program. With the loss of the organic mercury compounds which had strong eradicant action, there is a need for safe chemicals that have eradicant as well as protective action.

In a recent paper, Ayers [118] reported that benomyl, and thiophanate methyl as dust formulations were highly effective for the control of ("tubers")-seedborne Verticillium albo-atrum on potato. Whether this isolate was DM or MS was not indicated.

B. Systemic Fungicides

1. Introduction

The use of systemic fungicides to control vascular wilt diseases has been envisioned since the early 1940s when Horsfall and co-workers began a research program at the Connecticut Agricultural Experiment Station to find a chemotherapeutant that was active against Dutch elm disease caused by Ceratocystis ulmi. Dimond and co-workers [119] at the same station also sought a chemotherapeutant to control Fusarium wilt of tomato caused by Fusarium oxysporum f. sp. lycopersici. Some of this work will be discussed in the sections to follow; however, a more complete history of progress in this area is given by Wain and Carter [120]. The early studies defined research goals, contributed methods of testing and application and, as the studies progressed, presented a large amount of theoretical and practical information.

Dimond et al. [119] defined a chemotherapeutant as a chemical that controls the disease after the pathogen has entered the plant. They classified chemotherapeutants into (1) those chemicals that counteracted a toxin produced by a pathogen; (2) those that altered the metabolism of the host in such a way that it became more resistant to infection and disease; and (3) systemic fungicides, which entered the plant and were translocated to the locus of infection and acted directly on the pathogen.

Since none of the currently used systemic fungicides appear to act against toxins (category 1), only chemicals that act directly against the pathogen will be covered in this section. Those chemicals that act indirectly and appear to increase resistance of the host will be covered in a later section.

The only systemic chemicals found in recent times to exert a controlling effect on vascular wilt diseases are the 2-substituted benzimidazoles. These chemicals have a high degree of toxicity to fungi which occur taxonomically in the Ascomycetes and the Fungi Imperfecti. Dark-spored members of the Deuteromycetes are not sensitive [4].

The minimal concentration of benomyl which inhibits Verticillium spores was about 0.4 µg. The assay consisted of placing benomyl in 0.1 ml of water on a cellulose assay disc (12 cm diam.) which was transferred to a plate of potato dextrose agar and sprayed with Verticillium spores. The area of the zone of inhibition of growth (but not germination) that could be detected is directly correlated with concentration [121]. The benzimidazole fungicides are generally considered to be fungistatic [51]. Benomyl hydrolyses to methyl 1-benzimidazolecarbamate in plant tissue [9, 10]. However, up to 20 times as much benomyl penetrated leaf tissue when compared to that of MBC [122]. Thiophanate M was comparable in fungitoxicity to benomyl but only after it hydrolyzed to MBC [14]. Thiabendazole was also fungitoxic at comparably low levels to Verticillium albo-atrum (MS). In shake culture, 0.3 µg/ml of TBZ added to a synthetic liquid medium reduced the number of viable conidia to 13% of the control [6]. For a more detailed account of the mode of action of these chemicals, see the reviews by Erwin [51] and Wain and Carter [120], and the review on structure-activity relationship by Woodcock [123]. A relatively recent finding on the mode of action is by Davidse [124]—not covered in the review by Erwin [51]—who reported that MBC interferes with mitosis. His data indicated that MBC exerted an effect on fungal cells which cytologically resembled that of colchicine on human cells. He presented chemical data that suggested that there was a complex formed between MBC and a subunit of the microtubuli which interfered with normal spindle formation. (For more detail on this phenomenon, see Volume 2, Chapter 10.)

a. Translocation Patterns. The benzimidazole chemicals are readily
taken up by roots and translocated to the upper part of the plants [24, 125]
where they suppress the development of vascular wilt fungi in xylem tissue
[6, 126, 127]; see also the review by Erwin [51].

Systemic fungicides exert different degrees of systemicity. Benomyl
translocates translaminarly [128] from petals and calyxes to the stigma and
style in the flower [129], and from roots to the fully expanded leaves [121].
Translocation of nearly all known systemics follows the transpiration
stream in the apoplast (xylem tissue) [24, 127]. Prasad [125] considered
that a small portion of MBC may move in the symplast. Rarely is there
evidence of translocation from one leaf to another, such as in the example
given by Baron who reported movement of benomyl from one banana leaf to
another [130].

Translocation of the fungitoxic moiety (MBC) of benomyl [24] or of TBZ
[131] appears to be a one-way path, although Prasad [125] considered that
a small portion may be symplastic in movement. The fungicide moves from
the root into the xylem system through the stem and into the leaves. Leaves
with the most transpirational pull accumulate the most MBC. Prasad [125]
showed increased translocation of [^{14}C] MBC with increase in temperature
and light intensity. The chemical usually moves to the periphery of older
leaves but does not recirculate to the upper part of the plant. Apoplastic
movement is a fundamental principle on which all treatment of vascular
diseases must be based, since the chemical which reaches the leaf accumulates in the periphery but does not translocate back to the main stelar
system.

Busch and Hall [132] reported that painting chrysanthemum leaves with
benomyl (20 μg/ml) at weekly intervals prevented symptom expression on
leaves of plants inoculated with Verticillium dahliae (MS) but did not inhibit
development of Verticillium in the stems. Since the stunting effect caused
by Verticillium was not prevented, the indication was that the fungicide had
not migrated back to the stem from leaves and that the factors that cause
leaf symptoms are most likely found in leaves and not stems. More detail
on the physiology of translocation of systemic fungicides is given by Crowdy
[25] and in Volume 2, Chapter 12.

For the control of vascular diseases, the need is not only for translocation of a fungitoxic compound upward in xylem tissue, but also for enough
fixation in xylem tissue to control the pathogen. Dimond [30] stated in 1952,
"Systemic diseases present the most complex situation in chemotherapy ...
The chemotherapeutant must either be systemic, if action is to make the
host resistant; or it must be fixed to xylem tissue, if its toxic action is
direct. The chemotherapeutant that is translocated to leaves through the
stem is of no value against the wilt diseases nor is the material translocated
in phloem." The phrase "the chemotherapeutant that is translocated to
leaves through the stem is of no value," is at first puzzling. Yet, reflecting
on the difficulties in the field with the benzimidazole compounds for control

of vascular diseases, perhaps the conjecture applies to some extent to benomyl. This appears to be the case because benomyl, which is insoluble and not taken up in large quantities from soil, translocates rapidly (as MBC) through stems to leaves [24] where it is deposited, and is then returned to the soil with leaf fall [133]. Siegel and Zabbia [127] reported that in pea plants, fed [^{14}C]MBC via the roots, 99% of the radioactivity moved from the roots to the shoots when the source of [^{14}C]MBC was removed. Peterson and Edgington [24] noted that fungitoxic activity can be detected in the bark of plants treated in the root zone with benomyl. They showed conclusively that by separating the bark from xylem tissue with a nonpervious paper, translocation took place only via xylem tissue but not via phloem tissue in bark.

Thus the properties of a chemical that allow it to become systemic, that is, not to adsorb to xylem tissue or localize, may preclude its being persistent in xylem tissue for more than a short time. The supply of benomyl, which hydrolyzes to MBC, must be continuous from soil or from plant injection. In light of this hypothesis, it is not surprising that foliar application appears to be largely unsuccessful in inducing long-term control in the field.

Fuchs et al. [134] treated pea and several other plants by allowing roots to take up systemic fungicides (400 μM) for two days, after which the roots were washed free of external fungicides and transplanted to soil. The MBC translocated quickly and induced foliar disease control, but the effect did not persist long; when benomyl or thiophanate M was used, the translocated fungicidal effect on the leaves lasted much longer. They interpreted these data to indicate that thiophanate M and benomyl caused a much more gradual release of fungitoxicant into the aerial parts.

Pinkas et al. [135] injected solubilized TBZ (in phosphoric acid) into pear trees and noted an accumulation near the application site. In the course of time they noted that a secondary distribution occurs. In a similar experiment, Shabi et al. [21] noted that TBZ moved more slowly after trunk injection than solubilized MBC, but that later assays indicated that more TBZ could be detected in leaves than MBC. This indicated that all of the available MBC had translocated, whereas TBZ was still moving but more slowly in what they termed in the earlier paper a "secondary translocation" [135]. In studies with movement of [^{14}C]TBZ on cotton, Wang et al. [131] found that TBZ apparently became complexed with higher-molecular-weight compounds in cotton plants. While this appears to impair the efficacy of TBZ as a systemic fungicide (which is less effective than benomyl or MBC), this may indicate a need in the future for compounds that do not move as freely as MBC. If systemics are to be continuously effective, a source of supply must always be available and with tree injection the source cannot be replaced economically very often.

b. Eradication. An early part of this chapter introduced the concept of suppression of initial inoculum, which implies eradication. One of the useful attributes of benomyl is its eradicative capability, as exemplified by suppression of formation of perithecia by Venturia inaequalis, the cause of apple scab [136]. The systemic fungicides have not yet been proved to be useful in this way against vascular diseases in the field. However, there is experimental evidence that benomyl, applied to Verticillium-infected cotton foliage and stems, is capable of suppressing the formation of microsclerotia (the initial inoculum which becomes active during the following year) in plant tissue.

Evans [137] first showed that formation of microsclerotia could be prevented in cotton leaves infected with Verticillium dahliae (MS) by foliar treatment with a suspension of benomyl.

We have confirmed this and found that treatment of infected leaves with a benomyl and a paraffinic oil (20-50%) emulsion [138] prevented microsclerotia formation. Paraffinic oils increase penetration of benomyl [139, 140]. However, in stems up to 1-2 cm in diameter the suppression of microsclerotia by benomyl was less efficient than in leaves. This effect appeared to be due to lack of penetration of benomyl because of the large size of the stems compared to leaves.

Tsai and Erwin found by a method of assaying numbers of microsclerotia in plant tissue by a dilution technique, that there were in the order of 40,000 microsclerotia per gram (green weight) in stems of cotton, 70,000 per gram in petioles, and 50,000 per gram in leaves collected from infected plants that had been allowed to incubate and decompose in unsterilized soil [141]. Theoretically the potential number of microsclerotia that could form in a field of infested cotton and be returned to the soil is immense, and any approach that would reduce this number, especially on fields with relatively light degrees of infestation, should be studied. The main difficulty with use of an eradicative fungicide, such as benomyl, to suppress the formation of microsclerotia is the relative impenetrability of large stems by an externally applied benomyl-oil emulsion.

In one experiment, leaves, petioles, and stems (about 15 mm in diam.) from cotton plants infected with Verticillium dahliae (MS) were soaked in a suspension of benomyl (2,500 μg/ml) for 1 hr and subsequently incubated in unsterilized moist soil for about one month until microsclerotia had formed. When microsclerotia were counted by the dilution technique [141], benomyl caused only a slight reduction in stems (37,000/g in the control) and petioles (50,000/g in the control), but in leaves the number was reduced from 15,000/g in the control to about 1,500/g in the treated. However, a 5-hr soak treatment completely prevented microsclerotia formation in leaves and petioles and reduced the number to about 5,000/g in the stems. Although a 5-hr treatment is not practical, this is an example of the type of eradication that is possible.

In a similar experiment, treated petioles, leaves, and stems were allowed to dry after treatment, and the reduction in microsclerotia was even greater [142]. When infected but still green leaves (79% moisture) were treated with benomyl (100 μg/ml) in a paraffinic oil (10% v/v) and immediately buried in unsterilized soil, the number of microsclerotia was reduced to 12% of the untreated control. When similar leaves were dried for 7 days (14% moisture) after treatment with benomyl, the number of microsclerotia was reduced to 2% of the control. When the effect of drying alone (no benomyl) was tested, the number of microsclerotia was reduced to 20% of the control (untreated infected green leaves buried in unsterilized soil). Similar trends were shown for stems and petioles treated with benomyl and oil and by drying. This experiment showed that desiccation of infected plant material probably plays a role in suppressing mycelium or spores before production of microsclerotia [142].

c. Application of Solubilized Benzimidazoles. Looking back on the progress in systemic fungicide research, it is not surprising that those interested in improving uptake of benomyl or MBC, both of which are virtually insoluble compounds, should turn to ways of making them soluble with the view that water-soluble compounds should be more readily translocated. Indeed, early experiments in which various types of injectors were utilized to force benomyl into trees met with little success due to the fact that benomyl as a suspension translocated poorly [23, 143].

In recent years there has been a considerable amount of research with solubilized benomyl or MBC. Buchenauer and Erwin [15] in 1971, were probably the first to report that benomyl after being hydrolyzed to MBC from an acetone-hydrochloric acid solution (pH 1.7) formed a water-soluble salt (MBC·HCl) which could be applied to leaves and stems of cotton plants. This compound penetrated stems and translocated upward controlling Verticillium wilt caused by subsequent stem inoculation. The MBC was detected by bioassay in leaves and treated stems above the treated area. In comparable plants treated with nonacidified benomyl, little or no MBC could be detected above the treated area. No phytotoxicity due to the acidified MBC was noted in greenhouse experiments. However, in later field experiments some leaf burning was noted [144]. Seldom cited patent applications by Littler et al. [145] and Klopping [146] describe methods of preparing soluble MBC salts with inorganic acids.

After treatment of the stems and foliage with MBC·HCl before inoculation of plants by stem puncture, not only were the symptoms of Verticillium wilt suppressed, but the population of Verticillium albo-atrum (MS) in plant tissue was also reduced [16] (Fig. 2). Population of Verticillium in inoculated plants was determined by blending inoculated stems and assaying by a dilution plate assay on potato dextrose agar. Whether the water solubility or low pH of the solution caused the increase in uptake of MBC through stems or leaves could not be determined. However, increasing the molar ratio of HCl to MBC in the MBC·HCl solution from 1:1 to 1:2, to 1:3 increased the amount of uptake of MBC through stem tissue into the xylem tissue to 8, 10, and 12 μg/g, respectively. In a comprehensive review of the literature

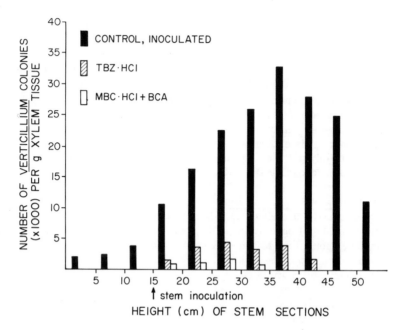

FIG. 2. Effect of three foliar sprays with MBC·HCl (1,650 ppm MBC) and
TBZ·HCl (2,500 ppm TBZ) (3, 5, and 7 days before inoculation by stem
puncture with 10^5 spores per milliliter) on the population of Verticillium
albo-atrum in the xylem of cotton plants 22 days after inoculation. Reprinted
from Ref. 16, p. 129, by courtesy of Verlag Paul Parey, Berlin, West
Germany.

on the possible mode of action for increased uptake of MBC·HCl over MBC
alone, there was theoretical evidence that both the increase in water solubility and the increase in hydrogen ion concentration could be involved in
increasing uptake through the cuticle on leaves [16]. Solel and Edgington
[128] reported later that acidifying MBC increased its uptake through detached cuticle from apple leaves. The cuticle is probably the most important barrier to penetration of MBC through stems or leaves.

Despite the increase in uptake due to use of MBC·HCl, frequent applications (1,500 μg/ml) to foliage of cotton in the field failed to control Verticillium wilt to an extent great enough to increase yield of cotton [51]. The
reason for this failure in the field was probably in part the relatively low
percentage of uptake even of MBC·HCl. Zaki et al. [140] found that, when
[^{14}C]MBC·HCl was applied to stems of cotton plants in a growth chamber
and assayed after 4 days, 70% of the residual MBC remained on the bark,
8.0% in the stele, 9.4% in the leaves, and 3.5% in the petioles (see Table 1).
This study indicated that a large amount of MBC would have to be deposited

TABLE 1

Distribution[a] of [^{14}C]MBC·HCl in Cotton Plants at Various Times
After Treatment of Stems With and Without P795 Oil[b]

Plant part[c]	4 days		7 days		11 days	
	MBC·HCl	MBC·HCl + oil	MBC·HCl	MBC·HCl + oil	MBC·HCl	MBC·HCl + oil
R	0.00	0.33	0.10	0.05	0.05	0.05
B_F	0.10	0.17	0.38	0.28	0.72	0.36
S_F	0.03	0.04	0.38	0.41	0.35	0.25
T_B	70.10	65.40	67.50	57.90	60.00	32.00
T_S	8.00	8.30	8.90	7.40	7.40	9.40
St_A	9.00	11.80	7.70	11.50	9.40	8.90
P	3.50	3.00	3.00	3.40	2.00	2.30
L	9.40	11.30	12.20	17.30	20.30	47.00
Total	100.00	100.00	100.00	100.00	100.00	100.00

[a]Each value represents the counts per minute (c/min) per plant part divided by the total c/min experimentally recovered from all parts of the plant × 100.
[b]P795 is Orchex 795 paraffinic oil, used in 22% concentration.
[c]R = root; B_F = bark of the first internode below the treated internode; S_F = stele of the first internode; T_B = bark of treated internode; T_S = stele of treated internode; St_A = stem above the treated internode; P = leaf petioles; L = leaf blades.
Source: Reprinted from Ref. 140, p. 491, by courtesy of the American Phytopathological Society.

on the stems for the upper petioles and leaves to accumulate a significant amount of the fungicide. Eleven days after treatment 60% of the MBC·HCl applied remained on the bark. However, when [^{14}C]MBC was applied in a paraffinic oil emulsion, movement of MBC from the treated bark to the stele and upper plant parts continued until the last day (11th) of assay, at which time only 32% of the radioactivity remained in the bark and 47% in the leaves [140].

Since certain phytobland paraffinic oils increase the uptake of benomyl [51], we tested the effects of a paraffinic oil, a naphthenic oil and an iso-paraffinic oil on the uptake of benomyl and thiabendazole on control of

Verticillium wilt of cotton. Emulsions containing 10 to 30% concentrations of the paraffinic oil (Orchex N795) increased uptake and fungitoxic activity above the region of application of benomyl (MBC) following application to the stems. The naphthenic oil (Orchex 792) was less effective than the paraffinic oil, but the isoparaffinic oil was ineffective in increasing uptake of benomyl [141].

Field work was also done with benomyl and MBC·HCl in oil emulsions, as foliar sprays. In the field these fungicides did not control Verticillium wilt or increase the yield. The slight phytotoxicity of oil was a complicating factor, reducing the yield slightly [147]. The U.S. patent 3,657,443 contains much information on compatibility and efficacy of applying many adjuvants including oils with MBC to plants [146].

Another factor that may affect control by foliar application is the limited space on the stems, as well as the limited time that the stem remains wet. Once the MBC·HCl, which is a weak acid, becomes neutralized by plant buffers, it undoubtedly precipitates MBC. Dimond reviewed the pertinent physical factors affecting uptake of chemicals through foliage and stated that uptake is reduced once the soluble chemical has crystallized on the leaf [148].

All research on administering fungicides by tree injection is now being done with one of the soluble inorganic acid salts of MBC or TBZ. In Canada in 1974, the group at the Canadian Forestry Service (Sault Saint Marie, Ontario) published a leaflet "Dutch Elm Disease" in which the use of the water soluble fungicides, MBC-phosphate (CFS-1020) and MBC-sulfate (CFS-1021), as tree injections for control of Dutch elm disease was described. In 1973, Kondo et al. [18] reported methods for the preparation of water-soluble MBC salts, MBC-sulfate, MBC-phosphate, MBC-hydrochloride, MBC-nitrate, and an "MBC-sodio derivative." They chose MBC-phosphate for field work because the salt was a necessary plant nutrient and was fungitoxically active but not phytotoxic to xylem of elms. They reported that the acidic solutions of MBC as well as water at similar pH values caused necrotic spots on leaves. This observation was different from the report by Buchenauer and Erwin [15, 16] in which leaves of cotton sprayed with MBC·HCl (1,500 μg/ml) were not injured.

Since there are several methods used to prepare solubilized benzimidazole fungicides, the principal techniques published by several laboratories will be summarized here.

Buchenauer and Erwin [15, 16] prepared soluble MBC by dissolving technical benomyl in acetone, in which it hydrolyzes to MBC and butylcarbamic acid (BCA). When the acetone solution was concentrated in vacuo, MBC precipitated. It was separated from BCA by filtration and washed several times with small amounts of cold acetone. Hydrochlorides of MBC or thiabendazole (TBZ) were prepared by dissolving MBC in 1 or 2 N HCl at 60°C. The salts were then obtained by removing the excess HCl at 40°C under vacuum. When benomyl was used to prepare a crude MBC·HCl solution (which also contained BCA), 10 g of benomyl was added to 100 ml of acetone to which 300 ml of 0.34 N HCl (molar ratio of MBC·HCl was about 1:2) was added and the mixture was heated to 70° C for 1-1.5 hr. The final pH of the MBC·HCl solution was 1.7 and of the TBZ·HCl solution 2.7.

Smalley [22] prepared a solution of MBC by warming benomyl (50 g/liter) in 85% lactic acid at 75°C, but lactic acid formulations were too phytotoxic for practical use. Then MBC·HCl was prepared by dissolving 25 g of benomyl at 24°C in 500 ml of acetone and warmed to 70°C for 30 min in 0.35 N HCl.

Gibbs and Clifford [17] dissolved MBC (9.56 g, 0.05 mole) in 160 ml of hot 1 N HCl and evaporated the solution to dryness in a rotary evaporator. The purple solid was triturated with ethanol and filtered. Another method consisted of adding MBC (40 g) to a solution of 28 ml of conc. HCl plus 472 ml of water which was heated until the MBC dissolved. This solution was diluted to 1 liter with tap water (pH of final solution was 1.7). When this solution was diluted eight times with tap water (pH was then 2.0), no precipitation occurred.

A formulation of MBC-nitrate and lactic acid was made by adding MBC (10 g) to distilled water (150 ml) containing 4.7 ml of HNO_3 (density 1.42). The mixture was heated to boiling and evaporated to dryness. The residue was triturated with ethanol, filtered, washed with ethanol, and dried. The residue was dissolved in lactic acid and diluted with tap water to give a stable concentrate containing 6.6% MBC·HNO_3 and 20% lactic acid (pH 2.3) [17]. As found by Smalley [22], the lactic acid formulation was phytotoxic.

A formulation of MBC-bisulfate and $KHSO_4$ was prepared by mixing MBC (28.2 g) in conc. H_2SO_4 (27 ml) and water (27 ml), diluted further with water to 1.7 liters, and heated until dissolved; $KHSO_4$ (7.5 g) was added and the solution evaporated to dryness in a rotary evaporator. The residue was triturated with ethanol, filtered, and dried. The residue contained 85% MBC·HSO_4 and 15% $KHSO_4$. This solid was dissolved in 2 liters of boiling tap water to give a 0.5% (w/v) solution of MBC (pH 1.4) [17].

Pinkas et al. [135] prepared solutions of TBZ in hypophosphorous acid (pH 1.88); MBC (30 g/liter) was dissolved in 1.0 N HCl (final pH 0.4) [21].

McWain and Gregory [19] prepared MBC·lactic acid by dissolving 100 g benomyl in 1 liter of 42.5% lactic acid by heating. The solution was diluted to 20 g/liter and filtered. The solute did not precipitate.

Van Alfen and Walton [23] made MBC·HCl by dissolving MBC (4.5 g) in 1 liter of 0.05 N HCl.

Kondo et al. [18] prepared MBC-sulfate by suspending MBC (1 g) in methanol (10 ml) which was heated and stirred. H_2SO_4 (50%) (0.5 ml) was added dropwise until the MBC dissolved. MBC-phosphate was made from MBC (1 g) in methanol (25 ml) and water (25 ml) which was heated to boiling. Orthophosphoric acid was added slowly until the MBC was dissolved. MBC·HCl was made from MBC (1 g) in acetone (8 ml) and methanol (2 ml) to which HCl (0.7 ml) was added and heated until the MBC dissolved.

MBC-nitrate was made from MBC (1 g) in acetone (8 ml) and methanol
(2 ml) to which concentrated HNO_3 (0.4 ml) was added. The slurry was
heated until MBC dissolved. An MBC-sodio derivative was made from MBC
(1 g) in methanol (15 ml) which was heated and stirred; NaOH (9 ml of a 1 N
solution) was added to the boiling solution until the MBC dissolved. All of
the salts were recrystallized by appropriate methods and tested for purity
by chemical means.

d. Prospects for Use of Systemic Chemicals. The use of soil fumigation
for control of soilborne vascular wilt diseases is in general successful but
is limited mainly by the economics of application. Control of vascular diseases by use of systemic fungicides is still in the experimental stage.
Although many principles have been learned, there are probably not many
areas where systemic chemicals are now used in practice for the control of
vascular diseases. In intensive fields of agriculture such as production of
carnations in greenhouses [149] or the field [150], systemic fungicides
appear to have a practical use economically. However, for field crops
such as cotton [147] the cost is still too high and the means of application
too inefficient to expect systemic chemicals to make a significant contribution. At present, benomyl as a 50 WP (Benlate) (wettable powder), costs
about $9.00/lb ($20/kg). Similar benzimidazoles such as thiophanate methyl
and thiabendazole are only slightly cheaper.

The Dutch elm disease, caused by Ceratocystis ulmi, is a possible candidate for the practical use of systemic fungicides, especially since it is
important to save large valuable trees and there simply is no other way left
to control this disease. Thus there is a major research effort going on in
eastern Canada and the northeastern United States. The highlights of this
effort will be assessed in Section III.B.3.

2. Methods of Application

In principle, there are three kinds of application: injection into trees, soil
application, and spraying of stems or trunks and foliage. Each method has
advantages and disadvantages.

Campana's theoretical and practical evaluation of systemic fungicides
for control of tree diseases aptly presented the various advantages and disadvantages of application by foliar spraying, or soil or tree injection [151].
Foliar spraying avoids physical injury to the tree, but it does not always
place the chemical in the part of the plant where it is needed. Soil application has the advantage that the chemical is placed near the roots where the
natural uptake of solutes takes place. However, Campana cites "dilution
and decomposition" as complicating factors in soil treatment. Large quantities may be needed to introduce a small amount of chemical into the plant.

The strategy of control by suppressing the pathogen in its active stage
in the living host differs from that of suppression of initial inoculum. In the
former case, we accept the initial inoculum and subsequent infection of the

host, but the intention is to suppress the pathogen after it has entered the host. This strategy is much more complex than suppression of the initial inoculum or protection of a host from a foliar pathogen. The demand is for persistent fungitoxicity, but not phytotoxicity, and the capability of the compound to translocate in the host.

With the vascular wilt diseases, the secondary inoculum within the host either comes from new infections via the roots or from established infected xylem tissue or both. Thus, the fungitoxicant must be in continuous supply in the xylem fluid or be renewed frequently enough to keep the fungus in a suppressed state.

Many studies [52, 152-154] lead to the conclusion that soil application of benomyl is perhaps not an economical method. However, it also can be assumed from other studies that benomyl (or MBC) is more readily taken up and translocated via the roots than via leaves, bark, or stems. Indeed, this is the case with most chemicals since roots rather than leaves are the usual organs of uptake by plants.

Application of the benzimidazole systemic fungicides to the soil has been experimentally successful for controlling vascular diseases [51, 120] of many crops including cotton, elm, strawberry, carnation, potato, and tomato. However, the control achieved has not shown sufficient return to be economical. At least a part of the problem appears to be due to the lack of sufficient uptake of the chemical from the soil rather than inefficiency of the chemical within the plant. Pionnat [150] showed that a concentration of only 0.5 μg of MBC/g of collar tissue of carnation stems was sufficient for control of Phialophora wilt. However, to achieve this level in plant tissue, soil-drench application of 0.37 g of benomyl in 10 liters of water/m^2 every 14 days was required. It is obvious that only a small percentage of the benomyl applied reached the xylem tissue of the carnation plants.

The lack of mobility of benomyl and MBC in soil [155, 156] indicates that any diffusion or movement after the initial placement cannot be expected. Placement of a chemical deep enough in soil to significantly affect the root system requires special equipment and an expenditure of energy; therefore soil application is not an easy task and many problems exist in placement. For instance, in our work with control of Verticillium wilt of cotton [147], chisel blades oriented horizontally or vertically were pulled through the soil at different depths behind a tractor while, by means of nozzles mounted behind the blades, an aqueous suspension of benomyl was sprayed into the soil. However, the width of the actual band of benomyl after the soil had fallen in behind the blade was less than 1 inch. Judging from the data of Leach et al. [53], this was less than the minimal area in the profile which must be treated to allow the maximum uptake.

For control of Verticillium wilt of cotton by preplant (chisel) or postplant (side dressing) applications of benomyl, dosages of 20-40 lb/acre (22.4-56.0 kg/ha) were required for partial control [52]. Yet, in a pot containing 1 kg of soil, as little as 10-20 mg/800 g of soil—which is the

equivalent of about 2 lb/acre (2.4 kg/ha)—controlled Verticillium wilt induced by stem-puncture inoculation.

Another study by Leach et al. [53], reviewed by Erwin [51], elegantly
showed that dilution of benomyl in the soil was a factor in its lack of success
in the field for control of Verticillium wilt of cotton. In a greenhouse pot
test treatment of soil with benomyl (100 μg/g) suppressed Verticillium wilt
of cotton grown in naturally infested field soil for up to 200 days. However,
when the same amount of benomyl used to treat the soil in the pot at 100
μg/g was mixed in smaller fractions of the soil in pots (50, 25, 12, 6, and
3% of the total soil mass in each pot), the degree of uptake of MBC, as
measured by a leaf-agar diffusion bioassay, was reduced in proportion to
the reduction in the fraction of the soil that was treated. In pots in which
only 3% of the soil was treated with benomyl, no uptake could be detected.
In the field they noted a similar effect when they treated different areas of
the soil profile. When 21.7 lb of benomyl per acre (24.3 kg/ha) was mixed
in soil at preplant time by rototilling in a profile 36 square inches (232.2
cm^2) in the row and in a profile 72 square inches (464.5 m^2), the area of
the zone of inhibition by the leaf bioassay 26 days after planting cotton in
the treated zone was 360 mm^2 in the 72 square inch plots but only 178 mm^2
in the 36 square inch plots. The difference 81 days after planting was even
greater. In the 72 square inch plots the leaf bioassay data showed a zone
of inhibition which averaged 18 mm^2 but in the 36 square inch zone there
was no inhibition, indicating no detectable uptake of MBC. By the end of the
season a rating of disease severity indicated that benomyl at 21.7 lb/acre
(24.3 kg/ha) did not exert a perceptible degree of control of Verticillium
wilt.

Thus it appears that for soil and root application, placement of a systemic chemical is a major physical problem. Perhaps soil injection at high
pressure, as used by Biehn [152, 154] and by Stipes [8], forces the suspension of benomyl into a large enough zone to effect control of Dutch elm disease in a large tree. However, too few dosage response tests using high-
pressure soil injection have been reported to determine the relative efficiency
of this method.

Adjuvants, e.g., Tween 20 and Tween 80, increased movement of benomyl in soil cores [157, 158] and its effectiveness as a fungicide [159-161],
but did not induce enough increase in uptake to improve control of Verticillium wilt [8, 220]. Application to soil by pressure injection of an aqueous
suspension of fungicides as reported by Stipes [8] and Biehn [152, 154]
appears to be effective in experimental work with the Dutch elm disease.
This method might be practical for individual plantings, but not for row
crops. Application of fungicides to soil in the root zone therefore poses a
problem.

An interesting historical account of methods of tree injection is given
by May [162]. His paper is still valuable, since most of the "modern"
methods are based on older techniques. Injection simply implies introducing

a substance into the tree through an opening that extends into the xylem tissue, allowing the injected material to be swept upward in the transpiration stream. May's review extends back to the injection of arsenic into trees by Leonardo da Vinci in the 15th century to prevent people from stealing the fruit, and to the classic researches of Hales in 1731, who studied how water rises in trees. Basically, May's paper gives many practical inovations from his own horticultural trials and from the work of others on ways of making injection more effective [162].

Tree injection through the trunk or stem appears to be the most popular method [151]. It has advantages, but Campana [151] and Stipes [163] list several problems, such as wounding and the possibility of secondary infections by weak pathogens. Other possible disadvantages seen by Campana include: thickness of bark on old trees which may be an obstacle to attaching a feeder tube; the physiology of the tree limiting application to certain times of the year; the chemical moving so rapidly through the xylem that its effect may be "nullified"; and finally, distribution may not be uniform in the tree.

In favor of stem injection, Campana [151] stated that it has become easier especially following the introduction of the Mauget injector (J. J. Mauget Co., Burbank, California) used for application of the insecticide Bidrin to elm trees for control of bark beetles. This method is economical in that precise amounts of chemical can be used and none is wasted, movement of the chemical is rapid, and in a woody plant it is the easiest means of widespread distribution. The Mauget injector is a closed device which can be inserted into the xylem tissue and which, when activated, exerts a pressure of 8-10 psi.

Jones and Gregory [164] described a method of tree injection using an air compressor or a nitrogen pressure tank, a pressure-holding tank, and a clamp to hold the pressure line against the tree. The preparation of the site of injection appeared to be an important innovation which would overcome the problem associated with the bark thickness [151]. A hole—about 1-3/8 inches in diameter (3.5 cm)—was cut through the bark of the tree with a bow punch or chisel, taking care not to cut into the sapwood. Then a wedge-shaped hole was cut into the sapwood with a chisel which extended to a depth of two or three annual growth rings. A neoprene washer of the same diameter as the hole in the bark was inserted. The injector head was placed against the neoprene washer and held in place with a jack and a belt pulled tightly around the tree. The solution was then applied under pressure (up to 96 psi), after which the hole was sealed.

Pinkas et al. [135] introduced a novel method that required little equipment. They forced 500 ml of a fungicidal solution (acidified TBZ) into the end of a piece of latex surgical tubing (40 cm long, i.d. 6.3 mm, and wall thickness 2.4 mm) that was closed at one end with a wire. This caused the tubing to expand (pressure 1.4 atm). The end of the tubing was clamped and the syringe removed. The clamped end of the tubing was affixed to a

plastic tube inserted into a hole in the trunk of an apple tree. The clamp
was removed, allowing the fungicide to be forced slowly into the tree. The
movement of TBZ to regions above the application area was monitored by
assaying sawdust collected from borings in different segments of the tree.

The method of "root injection" by Kondo [165] uses a pressure tank
equipped with intake and outgo valves. The main difference from other
methods is that the material is injected directly by connecting the delivery
hose by clamps to carefully severed (not crushed) lateral roots at three or
four locations around the tree (root flare method). The material is delivered
to the roots by gravity flow at first and later with pressure up to 10 psi for
24-48 hr. See also [18] for use of this method in disease treatments.

Kondo listed the advantages of the root flare method as (1) uniform distribution; (2) the large volume of solution delivered permits a lower concentration of chemical, which should reduce phytotoxicity; (3) there is a
reduction in use of excess chemicals (as is necessary for root drenching
and foliage spraying); and (4) wounds created by severing the small roots
(1.9 cm = 3/4 inch in diam.) heal rapidly without ill effect on the tree.

In 1974, Kondo described (166) a large-capacity injection system which
modified the earlier apparatus described in 1972 [165] with trunk, root
flare, and root connectors. The larger system has the advantage of allowing larger quantities of material to be delivered over a relatively long time,
serving to dilute and distribute the material widely. This prevents or
reduces phytotoxic effects, such as that of acid in the solubilized MBC
preparations.

Himelick [167] described a high-pressure system utilizing a spray pump
which developed up to 400 psi, and high pressure hoses attached to a lag
bolt with a hole drilled in the center. The lag bolt was screwed into a
slightly smaller hole drilled in the tree. He injected 12 gal (45 liters) of
soluble material in a 24-inch (30.5 cm) diam. elm tree in 10 min at 125 psi.
One important advantage is that material can be forced into roots as well
as into stems. Also, high-pressure injection is more rapid than low pressure, allowing the operator to be in control of the apparatus at all times.
Certain gravity-flow or low-pressure systems require that the reservoir
be left connected with the tree for a long time.

Prasad and Travnick [20], using a trunk injection system based on that
of Himelick [167], studied the distribution patterns of MBC hydrochloride,
prepared by reacting MBC with HCl [168], and of rhodamine. They found
that large volumes could be injected at 80 psi. The distribution was nearly
complete and mainly followed the xylem pathway.

Filer [169] reported a method for constructing a pressure apparatus at
a cost of about $12.00 for injecting fertilizers, insecticides, or fungicides
in trees. The equipment consisted of a pressure tank designed to hold freon.
It was fitted with a suitable valve and plumbing and was filled with the liquid
chemical; then it was closed and pressurized to 65 psi with air. The plug
to deliver the material to the xylem tissue in the tree was made by drilling

a longitudinal hole in a lag bolt, which was filed on two sides so it could be screwed into the tree. Filer recorded that at 65 psi, 1 liter could be injected into an elm tree in an average time of 30 min.

3. Control of Dutch Elm Disease

Since 1933, Dutch elm disease of the American elm (Ulmus americanum) has moved from Europe to the eastern United States and thence across the Midwest to the Pacific coast. It seems that only time will determine when all elms in the United States will be affected. Until the recent introduction of a more severe form of Ceratocystis ulmi in England, the disease there was mild [170]. According to Lincoln in 1968, the disease destroys 400,000 trees annually at a loss of $100 million [171].

Dutch elm disease differs from Verticillium and Fusarium wilt diseases in that insect vectors, the European bark beetle, Scolytus multistriatus, and the native elm bark beetle, Hylurgopinus rufipes are largely responsible for transmission. The disease can also be transmitted by natural root grafting [112]. The spores of Ceratocystis ulmi are formed in the tunnels made by the bark beetles and are transmitted by them from tree to tree. A life cycle and a general discussion of the disease are found in Agrios [172], and a somewhat briefer treatment in Smith's book on tree pathology [173].

In the case of Dutch elm disease, the initial inoculum is the dying tree and not the soil as in the annual crops affected by Verticillium or Fusarium. The first logical approach made to reduce the inoculum was destruction of the dying trees. This was practiced in the eastern United States until almost 1940 [174]. However, Zentmyer et al. [174], as early as 1946, presented data that showed that the rate of spread of the disease through Connecticut was almost 5.4 miles per year and that this rate did not increase when destruction of diseased trees was stopped in 1940. This lack of response to sanitation may have been due to inadequate methods of detecting diseased trees; for instance, an infected tree which did not yet show symptoms would still be a source of inoculum.

Control of the insect vector by use of insecticides is now used in most areas where there is great enough interest in saving the elm trees. Healthy trees have been treated with DDT and methoxychlor. However, the recent ban on the use of DDT in the United States has renewed interest in control by systemic fungicides [8]. The systemic insecticide Bidrin has been used with the idea of preventing the elm beetles from colonizing trees, but has not been entirely successful in controlling the disease.

Dutch elm disease is highly destructive. Historic and beautiful trees are killed, and the cost of their removal is immense to both the home owner and to cities. It is thus not coincidental that control of this disease by systemic chemicals was the subject of study by Horsfall, Dimond, Zentmyer, and others at the Connecticut Agricultural Experiment Station [26-28, 174, 175]. A history of progress in the control of this disease with systemic chemicals is given by Wain and Carter [120].

In this section only a few of the highlights of the early research on control of Dutch elm disease are given. The emphasis will be on present research on control by the systemic benzimidazole fungicides. The chemical control of Ceratocystis ulmi was reviewed by Stipes in 1973 [224].

Perhaps the earliest comprehensive report on the search for a systemic fungicide for Dutch elm disease was by Zentmyer et al. in 1946 [174]. As indicated by the literature review, there was little background information at that time on which to base a rationale for selection of systemic compounds. A large number of compounds were administered by gravity flow through bored holes in the sapwood of the trunk, through an excised lateral branch, or by drench application to a basin around the tree. Later, the tree was inoculated by puncturing through a drop of spores deposited on the axil of a branch. This method of inoculation is still used by most workers in the field.

Application by injection consisted of boring holes into the sapwood near the base of the tree, and inserting a tapered or threaded piece of brass tubing which was connected by a rubber tube to a reservoir. The material (concentration about 1,000 μg/ml) was readily taken up by gravity flow. Another method was to insert a gelatin capsule containing the dry chemical into a bored hole, cover it with water, and close the hole with a cork. In some cases the chemical was placed directly into the hole, which was then filled with water and sealed with a cork. The advantage of the dry chemical or gelatin capsule method is that moving water in the transpirational stream dilutes the chemical if it is water soluble. A disadvantage is phytotoxicity at the immediate site due to the high concentration of chemical [174].

Although none of the compounds tested completely suppressed the disease when applied before inoculation, 8-hydroxyquinoline sulfate, 8-hydroxyquinoline benzoate, hydroquinone, p-nitrophenol, pyrogallol, and quinone (all at 1,000 μg/ml) delayed the onset of disease symptoms [174].

In a later study, Dimond et al. [119] studied the suitability of the Fusarium wilt assay for selection of compounds active against Dutch elm disease. In this assay, roots of tomato plants were drenched with the chemical for about 9 days; then the roots were removed, washed free of chemical and of sand, and inoculated with a suspension of conidia of Fusarium oxysporum f. sp. lycopersici. The degree of disease was quantitatively evaluated on the basis of the severity of vascular discoloration and leaf necrosis.

Dimond et al. concluded that compounds selected by this method were in general more active locally in the roots than systemically, and that there was little or no correlation of activity against Fusarium wilt activity against Dutch elm disease. Although 8-quinolinol benzoate showed moderate chemotherapeutic activity against Dutch elm disease [175], it had little effect on Fusarium wilt. In studying the possible reason for the lack of correlation between the two systems, they sprayed chemicals in a nonphytotoxic oil-in-water emulsion (percentage of oil was not given) on foliage of elm trees at weekly intervals for 5 weeks before inoculating with Ceratocystis ulmi (apparently by stem inoculation). None of the chemicals that had been effective against Fusarium wilt were effective against Dutch elm disease, except the water-soluble potassium salt of 2-carboxymethyl mercaptobenzothiazole. Quantitative data were not given on the degree of control; however, a photograph contrasting the appearance of the healthy treated trees with the untreated inoculated trees was impressive.

With the advent of the benzimidazoles, the interest in systemic fungicidal control of Dutch elm disease was revived. The first published report of systemic activity of TBZ and benomyl in elms appeared in 1969 [176]. Stipes [177] treated potted 2-year-old elms with suspensions of TBZ and benomyl (1,000 and 4,000μg/ml) by soil drench before inoculation of stems with spores of Ceratocystis ulmi. Benomyl eliminated and TBZ reduced the population of C. ulmi as indicated by reisolation from inoculated trees. When bandages around branches of 8-year-old elms in a nursery were soaked with benomyl in 90% dimethylsulfoxide (DMSO), benomyl in water, and TBZ in 100% DMSO, subsequent appearance of symptoms was substantially reduced. Zaronsky and Stipes [176] showed by a paper-disc bioassay method that a substance fungitoxic to Verticillium albo-atrum could be detected in acetone-ethanol extracts from leaves collected from elm trees that had received benomyl by soil drench. These substances were assumed to be TBZ and MBC.

Stipes later [8] reported that pressure injection of benomyl, captan, and TBZ into the soil at the root zone of 3-year-old elms before inoculation with spores by stem wounding, resulted in marked suppression of symptoms (4%) by benomyl, a less marked suppression by TBZ (23%), and even less by captan (55%). The symptom value on control trees which received only water was 76%. By the use of a paper-disc bioassay of ethanol-acetone extracts of leaf tissue collected three months after soil treatment, evidence was found of translocation of fungitoxic substances in all three treatments, indicating that TBZ, MBC, and the breakdown products of captan had been translocated to the leaves from the soil. The evidence for the translocation of captan or its fungitoxic breakdown products is worthy of note. Bankuti and Thomas [178] reported that both Fusarium and Verticillium wilt diseases of tomato were controlled by root dipping transplants or by applying difolatan as a nursery soil drench before transplanting to an infested field. Both disease control and yield increase were reported. Difolatan is closely related to captan in chemical structure.

Although Tween 80 was shown to increase movement of benomyl by a thin-layer chromatography soil assay [157], its use (1%) with benomyl injected into soil in the field did not increase the degree of disease control [8].

Smalley [179] applied benomyl as a soil drench to planting of elms in trenches 15 cm deep and 15 cm apart. In a dosage series, reduction of disease did not occur until the dose reached 224 kg/ha or higher; TBZ was largely ineffective. He noted that control with benomyl, though effective, was much less successful than injection with the growth regulator

2,3,6-trichlorophenylacetic acid [180]. Smalley commented that a "better way of introducing the chemical into elm trees needs to be discovered" [179].

Stipes and Weinke [181] reported that TBZ (500 and 2,000 μg/g soil) used as a dry soil treatment around each tree, which was subsequently inoculated by trunk wounding and introduction of Ceratocystis ulmi spores, exerted better control of foliar symptoms than benomyl, captan, or difolatan. Captan gave about the same level of control as benomyl, being effective (38% reduction of foliar symptoms) at 2,000 μg/g but not at 500 μg/g of soil.

It does not appear that soil application (by drench or incorporation) is effective with benomyl. This can be explained by its relative immobility in soil [154, 155], and by the data from Leach's experiment on cotton, in which he found a threshold minimal part of the root zone that must be treated with a certain dose to obtain enough uptake for control [53].

In 1971, Biehn and Dimond [152] reported using 260 lb/acre (290 kg/ha) of benomyl with surfactant F and 405 lb/acre (453 kg/ha) with Tween 20 as soil injections in the root zone around elm trees which were subsequently inoculated. Foliar symptoms were reduced 79% by the former treatment and 97% by the latter. In 1973, Biehn reported symptom data of 3 years for trees treated with benomyl in 1970 (405 lb/acre; 453 kg/ha) and subsequently reinoculated with Ceratocystis ulmi each spring. The benomyl treatment suppressed symptoms 97, 96, and 93% in 1970, 1971, and 1972, respectively. Although there was a decline in fungitoxic activity in trees with time, a significant amount was detected each year, indicating that benomyl (or MBC) was persistent in soil [154].

Although 405 lb/acre (453 kg/ha) is a large dose and may not be economical at present prices, the continued activity of benomyl for 3 years is an attribute that might be of immense value in reducing the frequency of injections. Pressure injection of soil appears to give better distribution of benomyl than soil drenching; however, there are few data bearing on this question.

Stipes and Schreiber [182] reported reduction in the severity of symptoms on young elm trees (2.5 inches diameter at breast height, 6 cm dbh) which had been treated with captan (2.4 g/liter or 50% WP in 5% dimethyl sulfoxide) by saturating a layer of absorbent cotton wrapped around the trunk 60 cm above the ground level. In a subsequent paper, Roberts and Stipes [183] reported that captan stimulated root development of elm seedlings when applied by drenching.

Although foliar spraying with systemics does not seem to cause a consistent reduction in vascular disease, in a few cases this method has led to some control. Hart [184] conducted an experiment in 1971 in a natural planting of elms containing a few diseased trees as a source of inoculum. Benomyl (8 lb of 50% WP/100 gal water, 3.6 kg/946 liters) was applied as a mist spray to "run off" in the spring. In the treated trees, only 1.4% of 71 showed symptoms in the fall, but in the untreated trees, 14% of 197 had

Dutch elm disease. Similar results were obtained in 1971. Hart was able to demonstrate translocation of fungitoxic activity (MBC) in the xylem tissue of hand-sprayed trees but not in trees sprayed with a mist-blower machine. Presumably coverage was better with hand spraying. For a bioassay he used the method of Himelick and Neely [185]. A cellophane disc was placed on the end of a cross section of a branch and a drop of C. ulmi spores were incubated 24 hr on the cross section of cellophane. The discs were examined microscopically for growth of fungus. Where MBC was present, spore germination but no growth occurred.

Stipes [186] reported that the disease was not prevented by foliar spraying with benomyl, thiophanate M, or TBZ solubilized with HCl, or by applying a bandage soaked with TBZ·HCl or benomyl·HCl to the trunks of young elm trees [187].

Prasad [188] reported an experiment using [^{14}C]MBC applied as a spray to bark. As reported by Zaki et al. [140], only a small proportion of the activity was detected in the wood (0.32%). Slightly more was found when the bark was wounded. About twice as much activity moved through young bark compared to old bark; however, the total amount was not significant. To date, it does not appear from controlled experiments that application to bark allows enough uptake to obtain sufficient fungitoxic concentration. It might be worthwhile to test the effects of paraffinic oils with benomyl on the bark of elm trees. The slight phytotoxic effect observed on cotton might not be a problem on an ornamental tree from which a yield of fruit is not expected.

Foliar spraying with benomyl (4.8 g/liter) in a natural situation in Wisconsin eliminated the incidence (3%) of Dutch elm disease [22]. In greenhouse trials, Smalley et al. found that a brown discoloration of the cambial layer, which occurred on the inside of the bark, was associated with the use of benomyl·HCl and benomyl-lactic acid formulations. Since the necrotic spots were located near the lenticel openings, they speculated that lenticels may be the major route of uptake of benomyl applied to the trunks of trees.

Judging from the number of papers published, pressure injection application to trees is being used more frequently and is more successful than foliar spray or soil treatment. The degree of control by this means of application has been impressive. As indicated in Section III.B.2, tree injection by various means has been practiced for years. However, technological modifications that increase efficiency and ease of operation usually determine whether the method will be practically useful.

The use of solubilized (acidic) MBC and TBZ has made tree injection a good possibility for the control of Dutch elm disease since benomyl or TBZ, even when micronized, do not move in the xylem system as insoluble materials. These compounds are apparently filtered out by the small pores in the xylem tissue. Van Alfen and Walton [23] showed that passing a suspension of benomyl through a 20-cm elm stem reduced the detectable fungitoxic

activity by 90%. Despite the poor movement through stems, injection of
trees with benomyl (2 g/liter) gave a degree of control slightly better than
that caused by use of the insecticide methoxychlor to control the vector.
Translocation of MBC·HCl, however, was much better than that of benomyl.
When MBC·HCl was forced through stems, none of the fungitoxicity was
lost and the conductance through the elm xylem tissue was the same as that
of water. When benomyl was forced through stems, it reduced the conductance of fluid through the stems and most of the fungitoxic activity was
lost. Data from Van Alfen and Walton are reproduced in Figure 3 [23].
Gregory et al. [189] also found that benomyl injected slowly into trees
(25-80 psi) did not move far from the area of injection, but that MBC·HCl
was readily translocated and controlled Dutch elm disease [168].

In Canada, Kondo et al. [18] reported the preparation of a series of
soluble forms of MBC, summarized in Section III.B.1.c. They tested each
salt by applying solutions (250 ppm) to a sterile filter disc and placing the
disc on a plate of nutrient agar seeded with <u>Penicillium expansum.</u> The

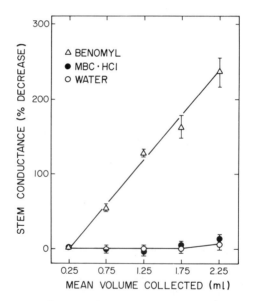

FIG. 3. Conductance of benomyl (4 g Benlate 50 WP/liter), MBC·HCl (4.5
g/liter), and water through elm stems 3-5 mm in diam. and 20 cm in length.
Results expressed as percent decrease in conductance (cm^3/bar sec) of
each 0.5 ml of fluid passed through the stems vs. the mean of each 0.5 ml
volume. The vertical lines represent the range of two times the standard
error. Reprinted from Ref. 23, p. 1233, by courtesy of the American
Phytopathological Society.

size of the zone of inhibition of fungal growth (which is influenced by water solubility as well as fungitoxicity) was greatest for MBC-nitrate, MBC-phosphate, and MBC-sulfate; intermediate for MBC·HCl and MBC-sodio-derivative, and least for MBC and benomyl. When these compounds were applied by the severed-root injection method at 10 psi [165] as much as 50 liters/hr were taken up. When concentrations of MBC-phosphate at 250-1,000 μg/ml were applied to xylem tissue by injection there was no phytotoxicity; however, acidic solutions (of MBC or water) were toxic to leaves. An interesting graph of the rate of uptake as influenced by weather conditions was presented. As might be expected, the uptake on sunny days was better than on foggy days. Using this method of application, a high degree of control of Dutch elm disease was obtained, and careful bioassays indicated that 75-100% coverage of the vascular system in each treated tree was obtained.

Kondo and Huntley [190] reported the use of a piece of transparent plexiglass marked in 1/2 inch (1.27 cm) squares for evaluation of disease severity. The tree was viewed through the transparent plexiglass grid, the outline of the crown and the diseased areas were traced on the plexiglass, and the affected proportion of the crown was calculated. The index ranged from 0-400 for a diseased elm tree. Using their method, there was only a 1-2% difference in evaluations done by different assessors; whereas, in visual assessment without the plexiglass grid, the variation was 5-25%. They suggested that a uniform method of disease assessment should be adopted by different researchers.

Prasad and Travnick [20] acknowledged that Kondo's root flare injection [165] was effective, but pointed out that it may not be feasible for street trees and may be more expensive than trunk injection. Employing a modification of the method described by Himelick [167], using high pressure up to 80 psi for application of benomyl and MBC·HCl (prepared from benomyl dissolved in 6 N HCl at 90°C), benomyl produced no fungitoxic activity detectable in leaves, but there were zones of inhibition of 3.8 mm (diameter) in leaves when MBC·HCl was injected into trees. This is another example of the lack of translocation of the insoluble form of benomyl. They estimated that application of 1,000 μg/ml resulted in 70% coverage of vascular tissue, 4,000 μg/ml in 80%, and 16,000 μg/ml in 90%.

Prasad injected the trunks of four large trees which had symptoms of Dutch elm disease. An improvement in general health of all four trees was observed after high-pressure (80 psi) injection of MBC·HCl. No physiological problems occurred due to phytotoxicity or use of high-pressure injection. The cost varied from $80-475 per tree [191]. Further cost analyses were determined after treatment of several other mature elm trees (7 inches dbh) (17.7 cm). Using MBC·HCl at 1,000 μg/ml, there was 100% suppression of symptoms at a cost of about $24 per tree [133].

Gibbs and Clifford [17] applied MBC·HCl by trunk injection and recorded disease on subsequently inoculated trees. Only 28% of the treated trees showed symptoms compared to all of the control inoculated trees. Injection

of MBC·HCl in trees 2 and 4 weeks after inoculation reduced symptom
expression to less than 76% of the untreated inoculated trees.

Smalley et al. [22] used the Mauget cup trunk injection method as well
as a newer low-pressure injection system to apply MBC·HCl and concluded
that the Mauget system was the "technique of choice," as large volumes
can be introduced with little wounding of the tree. Application by the low-
pressure injector (Elm Research Institute), which utilized pressures of
$2.1-2.8$ kg/cm^2 for application of MBC·HCl to trees already diseased,
appeared to be promising.

4. Experimental Approaches and Problems in Controlling Vascular Wilts
   of Field, Ornamental, and Vegetable Crop Diseases

In the recent book edited by R. W. Marsh on Systemic Fungicides, there is
a small section on wilt diseases of greenhouse crops by Spencer [192] and
on soilborne diseases of vegetable crops by Maude [193]. It is obvious from
the brevity of these two accounts that systemic fungicides are not yet in use
in very many areas for control of vascular wilt diseases.

Spencer [192] states that benomyl is being used in Jersey (England) on
greenhouse tomatoes at a rate of 1 pint of benomyl (4 oz a.i./100 gal; 0.3 g/
liter) per plant at 4-week intervals for control of Fusarium wilt. Maude
[193] cites only the example of Corbett and Hide [194] who showed that
treatment of potato seed tubers with benomyl had no effect on yield; however,
rotovating benomyl (20 lb/acre; 22.4 kg/ha) into soil before planting increased the yield. It is unlikely, however, that the increase in yield was
economical. These examples exemplify the trends in use of benomyl for
the control of vascular wilt diseases of commercial crops.

It is likely that use of benomyl on many greenhouse crops such as carnations [149] for control of Verticillium wilt, Phialophora wilt, or Fusarium
wilt will be practical. However, when rigorous greenhouse sanitation is
practiced (soil sterilization by steam or fumigation with chloropicrin,
methyl bromide, or combinations), it is unlikely that a systemic fungicide
will be needed, except for cases of accidental contamination.

The problem in applying an insoluble compound such as benomyl, thiabendazole, or thiophanate M to soil is partly that of dispersion. Probably
more important is the lack of efficiency of uptake once the compound is
applied to soil. These two problems have already been discussed in Section
III.B.1.c.

Resistance of fungi to the benzimidazoles is another factor to consider.
Since the chapter by Georgopoulos (Volume 2, Chapter 12) deals with this
subject it will not be discussed here. However, it is obvious that where the
use of these compounds is continuous, the likelihood of resistance developing in the fungal pathogens causing wilt diseases will be high. In greenhouse
crops, dependence on a benomyl program would be dangerous. Ideally,
alternation with a second fungicide with a different mode of action should be

practiced; however, none of the other available systemic fungicides are
active on vascular wilt fungi.

C. The Effect of Growth Regulators on Resistance:
An Experimental Challenge

Horsfall and Dimond [26] in their definition of chemotherapy included compounds that act on the metabolism of the host to increase resistance.
The physiological aspects of growth-regulating compounds and their
effect on diseases have been reviewed by Wain and Carter [120], van der
Kerk [195], van Andel [196, 197], Buchenauer [33, 198, 199], and Erwin
[34, 35]. Bockmann [200] described increased resistance caused by chlorocholine chloride to foot rot but decreased resistance to Septoria leaf spot of
wheat. The literature on the physiological effects of growth regulators on
plants was reviewed by Cathey [201]. The text of Leopold and Kriedemann
[202], who have specialized in the physiology of several growth regulators,
is particularly useful. In the past few years research on several growth-
regulating compounds with little or no fungitoxicity indicates that they can
be classified in Horsfall and Dimond's category of compounds that act on
the metabolism of the host to increase resistance [26].
    Davis and Dimond [29] tested the effect of several growth regulators
on Fusarium wilt of tomato using the Fusarium wilt assay [119]. In this
assay, the foliage of tomato plants was immersed in the growth-regulator
solution for 15 sec to simulate foliar spraying. Plants were inoculated by
dipping washed roots in a spore suspension of <u>Fusarium oxysporum</u> f. sp.
<u>lycopersici</u>. Different sets of treated plants were inoculated at 10 and 5
days before treatment and 4 days after treatment. Assay of the severity of
xylem necrosis was made several days after inoculation. The growth regulator 2,4-dichloropenoxyacetic acid (2,4-D) (5 μg/ml) was most effective in
reducing the degree of vascular discoloration. The greatest reduction resulted from treatment 10 days before inoculation (no vascular discoloration);
the next greatest when treatment was 4 days before inoculation (33.0% of
the untreated, inoculated control); and the least, 4 days after inoculation
(78.0%). The compound α-naphthaleneacetic acid (50 μg/ml) was also most
effective when applied 10 days before inoculation [29]. Sinha and Wood [203]
also reported reduction in severity of Verticillium wilt of tomato by 2,4-D
when applied before inoculation but increased severity when applied after
inoculation.
    Neither 2,4-D nor α-naphthaleneacetic acid (100 μg/ml) was fungitoxic
when added to a nutrient solution on which <u>Fusarium</u> was grown. Since the
compounds were not fungitoxic and since they decreased the reducing sugar
content in plants, induced formative effects on plants, and reduced plant
weight, it was concluded that the reduction in disease was due to a change in
plant metabolism.

Despite the fact that such a potent herbicidal compound as 2,4-D would not likely become a practical disease control agent on tomato, these examples show that chemical alteration of metabolic processes could lead to resistance to a vascular wilt and perhaps in some cases [203] to a decrease in resistance.

Sinha and Wood [203, 204] reported that chlorocholine chloride (CCC), as a soil drench on potted tomato plants, delayed and mitigated symptoms caused by subsequent root inoculation with Verticillium albo-atrum (DM). This remarkable finding stimulated much subsequent research. They attributed the delay in production of symptoms to an increase in tylosis formation in xylem tissue caused by CCC, since CCC was nonfungitoxic to Verticillium in a growth medium [204].

A series of studies on the effects of various growth regulators was launched in the laboratory of Prof. Grossmann at the University of Giessen, West Germany, by Buchenauer. A series of growth regulators which were relatively nontoxic to Verticillium albo-atrum or Fusarium oxysporum f. sp. lycopersici was utilized. The morphactins, which included n-butyl 9-hydroxyfluorene-9-carboxylate, n-heptyl 9-hydroxyfluorene-9-carboxylate, methyl 2-chloro-9-hydroxyfluorene-9-carboxylate and n-heptyl 2-chloro-9-hydroxyfluorene-9-carboxylate, were applied at 10^{-6} and 10^{-8} M concentrations to roots or to leaves [31]. All of the morphactins delayed the appearance of both leaf symptoms and vascular browning. When sections of the xylem tissue were blended, diluted, and plated on an agar medium, there was a reduction in the number of propagules of Fusarium in stems of treated plants. Although no antifungal activity was detected in xylem fluid, it appeared, from the reduction of fungal propagules in the xylem and the relative lack of in vitro antifungal effects of the growth regulators, that the disease-delaying effects were due to plant response induced by the growth regulators.

Another series of growth-regulating compounds was also studied for their effects on the Fusarium and Verticillium wilt diseases of tomato [32]. The following fluorenes significantly reduced the severity of Fusarium wilt symptoms on tomato and also the number of propagules in xylem tissue: 2-bromofluorene, 9-bromofluorene, 9-chlorofluorene, 9-hydroxyfluorene, 2-acetylfluorene, and 2-nitrofluorene. Phenanthrene and xanthene compounds acted in a similar way.

In still another study, application of CCC and N-dimethyl-(β-chloroethyl)-hydrazonium chloride (CMH) (BASF Co., West Germany) as a soil drench reduced the severity of Fusarium wilt symptoms. The three compounds CCC, 2,4-dichlorobenzyl-tributylphosphonium (Phosphon D), and N-dimethyl-aminosuccinamic acid (B 995) applied as a soil drench reduced the severity of wilt symptoms caused by Verticillium. None of the growth regulators were fungitoxic to either fungus at 10^{-3} and 10^{-4} M concentrations in a nutrient medium [33].

Following the finding that CCC suppressed Fusarium wilt and Verticillium wilt diseases of tomato, Buchenauer and Erwin showed that two related quaternary ammonium compounds $\underline{N},\underline{N}$-dimethylpyrrolidinium iodide (DPYI) and $\underline{N},\underline{N}$-dimethyl piperidinium iodide (DPII) had growth-retardant properties on corn, tomato, wheat, and cotton [198]. Since the growth-retardant effect of DPII and DPYI on plants was counteracted by exogenously applied gibberellic acid (GA$_3$) and since both DPYI and DPII interfered with the induction of α-amylase activity in barley endosperm, these compounds were considered to have anti-giberellic acid effects.

When the research with DPII and DPYI was conducted, only the iodide form of these compounds was available (obtained from Aldrich Chemical Company, Milwaukee, Wisconsin). In 1973, Zeeh et al. [205] reported the growth-retardant activity on wheat and barley of similar compounds, 1,1-dimethylpiperidinium chloride (BASF 083) and 1,1-dimethylpyrrolidinium chloride; the former was about 40% more active than the latter in retarding the growth of wheat.

Both DPYI and DPII when applied to cotton by soil drench (40 μg per pot) reduced the growth and, when plants were subsequently inoculated by stem puncture with spores of Verticillium dahliae (MS) (5×10^4 spores/ml), delayed the onset of disease 2-3 weeks [199]. To determine whether the interaction of DPII or DPYI was only symptomatic, xylem tissue from stems of treated and inoculated plants were blended in an Omnimixer cup and the diluted suspension was planted on potato dextrose agar 17 days after inoculation. There was a marked reduction in the number of propagules of Verticillium in xylem tissue in plants that had been treated with the growth retardants. This indicated that the effect was on some host system that apparently prevented the increase of propagules of Verticillium. The relationship of propagules of Verticillium to treatment with DPYI and DPII is shown in Figure 4.

In a subsequent experiment, extracts were made of xylem tissue from stems by blending the tissue in ethanol and extracting with ethyl acetate after the method of Zaki et al. [206]. When the extracts from treated and untreated plants were spotted on thin layer silica gel plates and chromatographed, large antifungal spots were noted for extracts of plants treated by soil drench with DPII and DPYI, whereas much smaller spots were noted for extracts from untreated plants [199]. The finding that the growth retardants DPII and DPYI reduced the numbers of propagules of Verticillium in xylem tissue and also induced the production of antifungal compounds strengthens the hypothesis of Horsfall and Dimond [26] that growth regulators can increase the resistance of the plant to disease.

In field work with growth regulators, experiments were done in the San Joaquin valley of California (1967-1969) using CCC as a foliar spray and as a soil drench on cotton. In general, CCC caused some delay in disease onset, but the effects were not marked when foliar symptoms on cotton plants were used to evaluate the effect on disease. Also, CCC at dosages of 200 g/ha, by foliage spray, not only retarded growth of cotton (which is desirable in some agricultural situations where rank, tall cotton lodges, and the dense canopy favors boll rot fungi) but also retarded yield [221]. However,

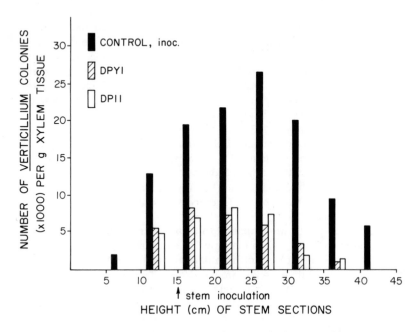

FIG. 4. Effect of root treatments (5 and 2 days before inoculation by stem puncture with 5×10^4 spores per milliliter) with DPYI and DPII (each 40 mg per plant) on the number of propagules of Verticillium albo-atrum in the xylem of cotton plants 17 days after inoculation. Reprinted from Ref. 199, p. 578, by courtesy of Verlag Eugen Ulmer, Stuttgart, West Germany.

in 1970 Singh [207] reported a yield response of cotton in India due to foliar application of CCC but only when low dosages (10 and 40 g/ha) were used. In 1974, we applied CCC to foliage of cotton on June 25 (first flower-buds stage) at dosages of 10, 25, 50, and 75 g/ha on a field in which the incidence of Verticillium wilt always exceeded 75% and caused severe yield losses. Although the degree of symptom expression of Verticillium wilt was not markedly affected by the treatments, the population of Verticillium in plant tissue was reduced. Petioles selected from uppermost fully expanded leaves were blended in an Omnimixer cup and the numbers of Verticillium propagules were assayed per g of petiole tissue. On September 5, there was a marked reduction from 20,000/g in the untreated control plants to a range of 900 to 4,400/g in the treated plants. The reduction of plant height was inversely proportional to the dosage used. Yield of cotton seed and lint was increased 16 and 12% by the 10 and 25 g/ha dosages of CCC, respectively. Yield was unchanged at 50 g/ha and reduced 6% by the 75 g/ha dosage of CCC. In some way, CCC caused a significant reduction in numbers of propagules and increased the yield of cotton at low dosages [35, 208].

Another example of the effect of a yield-enhancing, growth-regulating chemical on Verticillium wilt, is that of tributyl[(5-chloro-2-thienyl)methyl]-phosphonium chloride (TTMP) [34]. When this chemical was applied to soil in pots before inoculation by stem puncture (about 10 spores per plant) of cotton plants in the greenhouse, the number of Verticillium propagules in treated plants was markedly reduced from that of the untreated control. Likewise in the field, propagules were reduced from 20,000/g (control) to 900/g in the treated plants. Also the yield of seed and lint was increased 20% [35].

These data are not yet sufficient to prove that the yield enhancement was due entirely to Verticillium control, since there are other reports of yield enhancement on soybean in the field where a disease was not involved [209]. However, the data strongly encourage further study of the role of growth retardants on the possible increase in resistance or tolerance of plants to vascular wilt diseases.

Beckman [210] noted that elm trees were most susceptible to Cerato-cystis ulmi during the period of rapid spring growth. To test the hypothesis that inhibition of growth might increase disease resistance, he tested a series of chemicals by applying them as powders placed in bore holes oriented circumferentially in the trunks of elm trees. Sodium 4,5-dimethyl-2-thiazolyl mercoptoacetate (4,5-MTMA) was applied, at a rate of 2 g/inch (0.79 g/cm) of the diameter of the tree, once and three times at weekly intervals. The treatments induced a measurable reduction in sapwood thickness and leaf length (growth inhibition) and a corresponding reduction in the incidence of wilt following artificial stem inoculation with Ceratocystis ulmi. Reinoculation after the growth inhibition had passed resulted in 100% symptom expression of all the trees, regardless of previous treatment. He suggested two possibilities to explain this response. Morphologically, inhibition of sapwood growth would mean that the inoculum was deposited in the previous summer wood and that an intermediate growth-inhibition response, such as was produced by only one treatment with 4,5-MTMA, would result in fewer and smaller vessels in the sapwood. Thus the numbers of conidia that could migrate upward would be reduced. A second possibility is a change in the physicochemical state in the host.

Beckman [211] treated elm trees with 2,3,5,6-tetrachlorobenzoic acid (65% in oil) by painting the bark with the chemical at different times of the year, at bud-break (May 9) and at full-leaf (May 22) stages. Subsequently, the trees were inoculated by stem puncture with spores of Ceratocystis ulmi. By September 13 Dutch elm disease occurred in 95% of the untreated trees, in 52% of the trees treated at full leaf, and in only 16% of the trees treated at bud break. This indicated that growth inhibition before inoculation increased disease resistance.

Smalley [180] confirmed the findings of Beckman, but stated that the morphological foliar malformation due to polychlorobenzoic acid would preclude its use for practical control. A related compound, the herbicide

2,3,6-trichlorophenylacetic acid (TCPA) was used as a water soluble sodium salt and applied in several ways including injection at several rates into drilled holes in the trunks of elm trees. Trees were inoculated following treatment. As found by Beckman, little control of Dutch elm disease occurred when trees were treated by bark application after they had reached full leaf, but injections of TCPA (10 μg/ml) in early June gave complete control with no phytotoxicity noted. To test the hypothesis that formation of tyloses in the new wood might impede movement of the Ceratocystis inoculum in plants, histological studies were made with greenhouse grown elms treated by stem injection with TCPA (10 μg/ml). When the longitudinal sections were observed microscopically, both tyloses and gumming had occurred in the new wood. The intensity of occurrence of such occlusions increased with increasing concentration of the chemical, and these morphological changes appeared to impede movement of conidial inoculum.

IV. DISCUSSION

In this chapter several methods have been described which utilize fungicidal control of vascular wilt diseases. For most herbaceous crops, breeding for resistance has been the most actively pursued method of control and will continue to be where there is a source of resistance available.

The old concept of sanitation or the destruction of initial inoculum discussed by Van der Plank [36] was revived here as a model on which to base strategy of control methods. Under ideal conditions with use of long rotations to insusceptible crops, destruction of initial inoculum could be accomplished naturally in the soil. However, in an agriculture which becomes more intensive each year, this ideal ecological approach to disease control cannot be intelligently followed if the production of food and fiber is to continue on an economical basis.

There is need, then, in some cases to destroy the initial inoculum by such means as fumigation with general biocides such as methyl bromide or chloropicrin. With Dutch elm disease, the causal fungus can be destroyed by burning or burying the diseased trees. This practice eventually proved to be ineffectual against the spread of the disease in the eastern United States. This might have been due to incomplete detection of diseased trees or to delay in properly destroying them.

For soil borne vascular diseases, the discovery of volatile chemicals that had the capability of dissemination through soil allowed the grower to reduce the inoculum by killing a large portion of the surviving units of the causal fungus. This practice is expensive, however, and not free of problems.

One of the important problems that has arisen as a consequence of mycorrhizal destruction in fumigated soil is the stunting of plants such as citrus [75] or cotton [72, 73]. The growth of these plants is materially aided by the effect of these fungi on uptake of certain inorganic nutrients from soil. The mycorrhizal fungi previously considered to be Endogone [72, 73, 75] are now included under the new genus Glomus [212].

It is now clearer than ever that mycorrhizal fungi are necessary for growth of many crops, including an annual crop such as cotton, and that inordinate destruction of these "useful" fungi must be either prevented or they must be replaced in the ecosystems after destruction of the pathogen. With a crop such as strawberry, which is asexually propagated each year, the mycorrhizal flora is apparently maintained on transplant roots which are placed in fumigated soil.

Retention of a mycorrhizal population in fumigated soil might require careful time-dosage studies with selected fumigants. So far there has been little research on this aspect of control, although in the field a time-dosage effect might be difficult to determine and more difficult to practice once the relationship is understood.

Much has been written about the "increased growth response" due to fumigation of soil with various fumigants. Although it is beyond the scope of this chapter to cover this interesting and speculative area, a recent paper which deals with separation of the lethal effects of methyl bromide on different components of the soil microflora is interesting. Millhouse and Munnecke [213] fumigated a sandy loam soil with methyl bromide (30,000 μl/liter, v/v) for different periods of time and found that fungi were eliminated in 8 hr and fluorescent pseudomonads in 16 hr but that even after 32 hr, populations of Streptomyces and Bacillus were not reduced. When samples of soil were fumigated in parallel for different periods of time and planted with Nicotiana glutinosa, growth was significantly greater in soils fumigated longer than 32 hr. Plants grown in soil fumigated for 64 and 128 hr flowered sooner than in soil fumigated for lesser periods of time. The correlation of the elimination of certain components of the soil flora by different times of fumigation with increased growth strongly suggests a relationship of growth to components of the soil flora.

The recent development of systemic fungicides has stimulated much research on the control of the vascular wilt diseases. The 2-substituted benzimidazoles such as TBZ, benomyl, and thiophanate M are all efficacious against all fungi that cause vascular wilt disease. However, the problem now remains: how can they be placed in the plant where they can exert their fungistatic effect on these deep seated pathogens? Thus far root, foliar and stem uptake of such compounds as benomyl far exceed that of any previous fungicide, but the uptake is not enough to accomplish both effective and economical control of most vascular wilt diseases.

Systemic fungicides solubilized in inorganic acids and injected into trees seems to be a possibility for control of the Dutch elm disease. Indeed, in the history of control of this disease, this appears to be a last stand before pathologists must admit defeat. This disease, which is caused by a fungus with little persistence outside the host, is transmitted mainly by bark beetles. It has continued to spread despite the practice of eradication of diseased trees (initial inoculum) and attempts to control the vectors with protective and systemic insecticides.

Soil injection with benomyl has been experimentally successful but the dosage required for effective control appears to be much too high for practicality. Foliar sprays have been occasionally effective but it is clear this approach cannot be recommended with much confidence. Tree injection appears to be the best possibility for control. A most important factor with this approach is the development of soluble forms of MBC by use of the inorganic acids HCl, H_2SO_4, H_3PO_4, and HNO_3. Solubilization of MBC with lactic acid and acetic acid is effective, but the organic acids are too phytotoxic for practicality.

Tree injection has been done in two different ways: by the root flare method of Kondo [165, 214] in which tubes are affixed to severed lateral roots and by the trunk injection method in which threaded tubes are inserted into the tree. The use of pressure is a variable factor. Some utilize high pressures (up to 100 psi) and others low (less than 10 psi). The low-pressure method usually utilizes a higher volume of fungicidal solution.

It is too soon to predict whether these methods will mature into practical usage. However, with the high value placed on the American elm, it might be safe to predict that the technology of tree application will advance to the point where treatment will become practical.

Although this chapter does not include the oak wilt disease caused by Ceratocystis fagacearum, the principles of application learned from research on the Dutch elm disease should be immediately applicable to this and to other vascular wilt diseases of trees.

The use of nonfungitoxic growth regulators as an indirect control mechanism has intrigued investigators for years. It is too soon to predict whether the action of growth regulators to delay symptoms of Verticillium wilt and to reduce Verticillium propagules in infected cotton plant tissue will become practical for disease control. In contrast to the phytotoxicity of compounds used by Davis and Dimond [29] to delay Fusarium wilt of tomato, CCC and TTMP [CHE 8728) used on foliage of cotton at the proper dosage have not exerted phytotoxic effects and have increased yields [34, 35]. Likewise, some of these compounds have increased yields of soybeans, presumably in the absence of a vascular wilt disease [209].

REFERENCES

1. J. C. Walker, Amer. Phytopathol. Soc. Monograph No. 6, 1971.
2. S. Wilhelm, R. C. Storkan, and J. M. Wilhelm, Agri. Environ., 1, 227 (1974).
3. B. von Schmeling and M. Kulka, Science, 152, 659 (1966).
4. L. V. Edgington and G. L. Baron, Phytopathology, 57, 1256 (1967).

5. T. Staron, C. Allard, H. Darpoux, H. Grabowski, and A. Kollman, Phytiatr.-Phytopharm., 15, 129 (1966).
6. D. C. Erwin, J. J. Sims, and J. Partridge, Phytopathology, 58, 860 (1968).
7. C. J. Delp and H. L. Klopping, Plant Disease Reptr., 52, 95 (1968).
8. R. J. Stipes, Phytopathology, 63, 735 (1973).
9. G. P. Clemons and H. D. Sisler, Phytopathology, 59, 705 (1969).
10. J. J. Sims, H. Mee, and D. C. Erwin, Phytopathology, 59, 1775 (1969).
11. H. Buchenauer, D. C. Erwin, and N. T. Keen, Phytopathology, 63, 1091 (1973).
12. A. Matta and I. A. Gentile, Mededel. Rijksfac. Landbouwetensch. Gent, 36, 1151 (1971).
13. H. A. Selling, J. W. Vonk, and A. Kaars Sijpesteijn, Chem. Ind., 1625 (1970).
14. J. W. Vonk and A. Kaars Sijpesteijn, Pestic. Sci., 2, 160 (1971).
15. H. Buchenauer and D. C. Erwin, Phytopathology, 61, 433 (1971).
16. H. Buchenauer and D. C. Erwin, Phytopathol. Z., 75, 124 (1972).
17. J. N. Gibbs and D. R. Clifford, Ann. Appl. Biol., 78, 309 (1974).
18. E. S. Kondo, D. N. Roy, and E. Jorgensen, Can. J. Forest Res., 3, 548 (1973).
19. P. McWain and G. F. Gregory, U.S. Dept. Agr. Forest Serv. Res. Paper NE-234, 1971.
20. R. Prasad and D. Travnick, Chem. Control Res. Inst. Ottawa, Ont. Inform. Rept. CC-X-53, 1973.
21. E. Shabi, Y. Pinkas, and Z. Solel, Phytopathology, 64, 963 (1974).
22. E. B. Smalley, C. J. Meyers, R. N. Johnson, B. C. Fluke, and R. Vieau, Phytopathology, 63, 1239 (1973).
23. N. K. Van Alfen and G. S. Walton, Phytopathology, 64, 1231 (1974).
24. C. A. Peterson and L. V. Edgington, Phytopathology, 60, 475 (1970).
25. S. H. Crowdy, in Systemic Fungicides (R. W. Marsh, ed.), Wiley, New York, 1972, pp. 92-115.
26. J. G. Horsfall and A. E. Dimond, Ann. Rev. Microbiol., 5, 209 (1951).
27. J. G. Horsfall, Principles of Fungicidal Action, Chronica Botanica, Waltham, Mass., 1956.
28. J. G. Horsfall, Indian Phytopathol. Soc. Bull., I, 13 (1963).
29. D. Davis and A. E. Dimond, Phytopathology, 43, 137 (1953).
30. A. E. Dimond, Phytopathology, 52, 1115 (1962).
31. H. Buchenauer and F. Grossmann, Deut. Bot. Ges. Neue Folge, 3, 149 (1969).
32. H. Buchenauer and F. Grossmann, Nature, 227, 1267 (1970).
33. H. Buchenauer, Phytopathol. Z., 72, 53 (1971).
34. D. C. Erwin, S. D. Tsai, and R. A. Khan, Phytopathology, 66, 106 (1976).

35. D. C. Erwin, in Crop Protection Agents: Their Biological Activity (N. R. McFarlane, ed.), Academic Press, New York, 1977, pp. 477-490.
36. J. E. Van der Plank, Plant Diseases: Epidemics and Control, Academic Press, New York, 1963.
37. P. M. Halisky, R. H. Garber, and W. C. Schnathorst, Plant Disease Reptr., 43, 584 (1959).
38. R. L. Powelson, in Root Diseases and Soil Borne Pathogens (T. A. Toussoun, R. V. Bega, and P. E. Nelson, eds.), University of California Press, Berkeley, 1970, pp. 31-33.
39. R. H. Stover, N. C. Thornton, and V. C. Dunlap, Soil Sci., 76, 225 (1953).
40. R. H. Stover, Soil Sci., 80, 397 (1955).
41. C. E. Horner and H. L. Dooley, Plant Disease Reptr., 49, 581 (1965).
42. J. L. McIntyre and C. E. Horner, Phytopathology, 63, 172 (1973).
43. R. L. Powelson and A. E. Gross, Phytopathology, 52, 364 (1962).
44. S. Wilhelm, in Biology and Control of Soil-Borne Plant Pathogens (G. W. Bruehl, ed.), The American Phytopathological Society, St. Paul, Minn., 1975, pp. 166-171.
45. I. Isaac, Ann. Rev. Phytopathol., 5, 201 (1967).
46. W. C. Schnathorst, in Verticillium Wilt of Cotton, U.S. Dept. Agr. Res. Serv., ARS-S-19, 1973, pp. 1-19.
47. B. A. Rudolph, Hilgardia, 5, 353 (1931).
48. W. C. Schnathorst and D. E. Mathre, Phytopathology, 56, 1155 (1966).
49. S. Wilhelm, R. C. Storkan, J. E. Sagen, A. G. George, and H. Tietz, Califor. Agr., 26, 4 (1972).
50. J. E. DeVay, O. D. McCutcheon, A. V. Ravenscroft, H. W. Lembright, R. H. Garber, D. E. Johnson, M. Hoover, and J. Quick, in Cotton Disease Research in the San Joaquin Valley 1968-1969, Progress Report, Univ. Calif. Div. Agr. Sci. (mimeographed), pp. 218-226.
51. D. C. Erwin, Ann. Rev. Phytopathol., 11, 389 (1973).
52. D. C. Erwin, Summary Proc. Western Cotton Prod. Conf., March 1-3, 1972, pp. 17-20.
53. L. D. Leach, E. M. Huffman, and W. C. Reische, Proc. Beltwide Cotton Prod. Res. Conf., 29th Cotton Disease Council, 1969, pp. 36-40.
54. S. Wilhelm, J. E. Sagen, and H. Tietz, Phytopathology, 64, 924 (1974).
55. L. J. Ashworth, Jr., O. D. McCutcheon, and A. G. George, Phytopathology, 62, 901 (1972).
56. L. J. Ashworth, Jr., J. E. Waters, A. G. George, and O. D. McCutcheon, Phytopathology, 62, 715 (1972).
57. O. C. Huisman and L. J. Ashworth, Jr., Phytopathology, 64, 1043 (1974).
58. J. E. DeVay, L. L. Forrester, R. H. Garber, and E. J. Butterfield, Phytopathology, 64, 22 (1974).

59. L. J. Ashworth, Jr., D. M. Harper, and H. L. Andris, Phytopathology, 64, 629 (1974).
60. E. J. Butterfield and J. E. DeVay, Proc. Amer. Phytopathol. Soc., 2, 111 (1975).
61. E. J. Butterfield and J. E. DeVay, Proc. Amer. Phytopathol. Soc., 2, 111 (1975).
62. W. C. Schnathorst, Mycologia, 57, 343 (1965).
63. W. C. Schnathorst, Proc. Beltwide Cotton Prod. Res. Conf. 29th Cotton Disease Council, 1969, p. 30.
64. P. H. Tsao, Ann. Rev. Phytopathol., 8, 157 (1970).
65. K. F. Baker, in Root Diseases and Soil Borne Pathogens (T. A. Toussoun, R. V. Bega, and P. E. Nelson, eds.), University of California Press, Berkeley, 1970, pp. 234-239.
66. D. E. Munnecke, Ann. Rev. Phytopathol., 10, 375 (1972).
67. S. Wilhelm, Calif. Agr. Expt. Sta. Circ., 494, 26 pp. (1961).
68. S. Wilhelm, Proc. 16th Intern. Hort. Congr., 1962, pp. 263-269,
69. D. E. Munnecke and D. L. Lindgren, Phytopathology, 44, 605 (1954).
70. S. Wilhelm, R. C. Storkan, and J. E. Sagen, Phytopathology, 51, 744 (1961).
71. A. A. R. Hafez, P. R. Stout, and J. E. DeVay, Agron. J., 67, 359 (1975).
72. J. H. Hurlimann, R. D. Raabe, and S. Wilhelm, Proc. Amer. Phytopathology Soc., 1, 84 (1974).
73. J. H. Hurlimann, Ph.D. Thesis, University of California, Berkeley, 1974.
74. J. W. Gerdemann, Mycologia, 57, 562 (1965).
75. G. D. Kleinschmidt and J. W. Gerdemann, Phytopathology, 62, 1447 (1972).
76. S. Wilhelm, Ann. Rev. Phytopathol., 4, 53 (1966).
77. J. G. Bald and R. N. Jefferson, Plant Disease Reptr., 40, 840 (1956).
78. L. T. Richardson and D. E. Munnecke, Can. J. Bot., 42, 301 (1964).
79. G. D. Easton, M. E. Nagle, and D. L. Bailey, Phytopathology, 62, 520 (1972).
80. G. D. Easton, M. E. Nagle, and D. L. Bailey, Phytopathology, 59, 1171 (1969).
81. G. D. Easton, M. E. Nagle, and D. L. Bailey, Amer. Potato J., 51, 71 (1974).
82. R. A. Young, Plant Disease Reptr., 40, 781 (1956).
83. J. P. Jones, A. J. Overman, and C. M. Geraldson, Phytopathology, 56, 929 (1966).
84. S. N. Smith and W. C. Snyder, Phytopathology, 65, 190 (1975).
85. J. P. Jones and S. S. Woltz, Plant Disease Reptr., 51, 645 (1967).
86. J. P. Jones and A. J. Overman, Phytopathology, 61, 1415 (1971).
87. N. T. Powell, Phytopathology, 53, 28 (1963).
88. N. T. Powell, Ann. Rev. Phytopathol., 9, 253 (1971).

89. F. L. Stark, Jr. and B. Lear, Phytopathology, 37, 698 (1947).
90. I. J. Thomason, D. C. Erwin, and M. J. Garber, Phytopathology, 49, 602 (1959).
91. A. L. Smith, Phytopathology, 38, 943 (1948).
92. N. A. Minton and E. B. Minton, Phytopathology, 56, 319 (1966).
93. O. D. Morgan, Plant Disease Reptr., 41, 27 (1957).
94. W. D. McClellan, S. Wilhelm, and A. George, Plant Disease Reptr., 39, 226 (1955).
95. W. G. Hoyman and E. Dingman, Amer. Potato J., 42, 195 (1965).
96. W. G. Hoyman and E. Dingman, Amer. Potato J., 44, 165 (1967).
97. G. D. Easton, Amer. Potato J., 47, 419 (1970).
98. R. A. Krause, J. P. Huether, P. Sotiropoulos, and L. E. Adams, Plant Disease Reptr., 59, 159 (1975).
99. J. K. Golden and S. D. Van Gundy, Phytopathology, 65, 265 (1975).
100. J. K. Golden, Ph.D. Thesis, University of California, Riverside, 1974.
101. E. L. H. Wang and G. B. Bergeson, J. Nematology, 6, 194 (1974).
102. C. D. McKeen and W. B. Mountain, Can. J. Bot., 38, 789 (1960).
103. W. B. Mountain and C. D. McKeen, Nematologica, 7, 261 (1962).
104. L. R. Faulkner and C. B. Skotland, Plant Disease Reptr., 47, 662 (1963).
105. L. R. Faulkner and C. B. Skotland, Phytopathology, 55, 583 (1965).
106. G. B. Bergeson, Phytopathology, 53, 1164 (1963).
107. T. W. Graham and Q. L. Holdeman, Phytopathology, 43, 434 (1953).
108. P. M. Miller, Phytopathology, 65, 81 (1975).
109. D. C. M. Corbett and G. A. Hide, Ann. Appl. Biol., 68, 71 (1971).
110. D. Orion and H. Hoestra, Neth. J. Plant Pathol., 80, 28 (1974).
111. D. R. Viglierchio and P. K. Yu, Exptl. Parasitol., 23, 88 (1968).
112. E. B. Smalley, Proc. Intern. Shade Tree Conf. Midwest Chapter, 20, 41 (1965).
113. R. L. Powelson, Amer. Potato J., 41, 303 (1964).
114. G. D. Easton, M. E. Nagle, and D. L. Bailey, Amer. Potato J., 49, 397 (1972).
115. J. Krikun and M. Chorin, Israel J. Agr. Res., 16, 177 (1966).
116. J. B. Kendrick, Phytopathology, 24, 1139 (1934).
117. J. B. Kendrick and W. C. Snyder, Phytopathology, 32, 1010 (1942).
118. G. W. Ayers, Can. Plant Disease Surv., 54, 74 (1974).
119. A. E. Dimond, D. Davis, R. A. Chapman, and E. M. Stoddard, Conn. Agr. Expt. Sta. New Haven, Bull. 557 (1952).
120. R. L. Wain and G. A. Carter in Systemic Fungicides (R. W. Marsh, ed.), Wiley, New York, 1972, pp. 7-33.
121. D. C. Erwin, World Rev. Pest Control, 8, 6 (1969).
122. P. M. Upham and C. J. Delp, Phytopathology, 63, 814 (1973).
123. D. Woodcock, Systemic Fungicides (R. W. Marsh, ed.), Wiley, New York, 1972, pp. 34-85.
124. L. C. Davidse, Pestic. Biochem. Physiol., 3, 317 (1973).

125. R. Prasad, Chem. Control Res. Inst. Ottawa, Ont. Inform. Rept. CC-X-35, 1972.
126. D. C. Erwin, H. Mee, and J. J. Sims, Phytopathology, 58, 528 (1968).
127. M. R. Siegel and A. J. Zabbia, Jr., Phytopathology, 62, 630 (1972).
128. Z. Solel and L. V. Edgington, Phytopathology, 63, 505 (1973).
129. D. C. Ramsdell and J. M. Ogawa, Phytopathology, 63, 959 (1973).
130. M. Baron, Fruits, 26, 643 (1971).
131. M. C. Wang, D. C. Erwin, J. J. Sims, N. T. Keen, and D. E. Borum, Pestic. Biochem. Physiol., 1, 188 (1971).
132. L. V. Busch and R. Hall, Can. J. Bot., 49, 1987 (1971).
133. R. Prasad, Chem. Control Res. Inst. Ottawa, Ont., Inform. Rept. CC-X-73, 1974.
134. A. Fuchs, G. A. van den Berg, and L. C. Davidse, Pestic. Biochem. Physiol., 2, 191 (1972).
135. Y. Pinkas, E. Shabi, Z. Solel, and A. Cohen, Phytopathology, 63, 1166 (1973).
136. S. R. Connor and J. W. Heuberger, Plant Disease Reptr., 52, 654 (1968).
137. G. Evans, private communication, 1973.
138. D. C. Erwin, S. D. Tsai, and R. A. Khan, Proc. 1974 Beltwide Cotton Prod. Res. Conf., 34th Cotton Disease Council, pp. 31-32.
139. D. C. Erwin, R. A. Khan, and H. Buchenauer, Phytopathology, 64, 485 (1974).
140. A. I. Zaki, D. C. Erwin, and S. D. Tsai, Phytopathology, 64, 490 (1974).
141. S. D. Tsai and D. C. Erwin, Phytopathology, 65, 1027 (1975).
142. D. C. Erwin, S. D. Tsai, and R. A. Khan, unpublished work, 1975.
143. W. K. Hock and L. R. Schreiber, Plant Disease Reptr., 55, 58-60 (1971).
144. D. C. Erwin, Proc. 1972 Beltwide Cotton Prod. Res. Conf., 32nd Cotton Disease Council, pp. 19-20.
145. C. A. Littler, B. L. Richards, Jr., and H. L. Klopping, Fr. Pat. 1,532,380 (1968).
146. H. L. Klopping, U.S. Pat. 3,657,443 (1972).
147. D. C. Erwin, L. Carter, R. H. Garber, T. A. DeWolfe, and R. Khan, Proc. 1971 Beltwide Cotton Prod. Res. Conf., 31st Cotton Disease Council, pp. 75-76.
148. A. E. Dimond, in Modern Methods of Plant Analysis, Vol. 5 (H. F. Linskens and M. Tracy, eds.), Springer Verlag, Berlin, 1962, pp. 368-382.
149. R. Baker, Colo. Flower Growers Assoc. Bull., 260, 1 (1972).
150. J. C. Pionnat, Ann. Phytopathol., 3, 207 (1971).
151. R. J. Campana, Trees, 30, 11 (1970).
152. W. L. Biehn and A. E. Dimond, Plant Disease Reptr., 55, 179 (1971).

153. W. L. Biehn and A. E. Dimond, Phytopathology, 60, 571 (1970).

154. W. L. Biehn, Plant Disease Reptr., 57, 35 (1973).

155. R. C. Rhodes and J. D. Long, Bull. Environ. Contam. Toxicol., 12, 385 (1974).

156. R. B. Hine, D. L. Johnson, and C. J. Wenger, Phytopathology, 59, 798 (1969).

157. R. J. Stipes and D. R. Oderwald, Phytopathology, 60, 1018 (1970).

158. R. E. Pitblado and L. V. Edgington, Phytopathology, 62, 513 (1972).

159. T. E. Rawlins and J. A. Booth, Plant Disease Reptr., 52, 944 (1968).

160. J. A. Booth and T. E. Rawlins, Plant Disease Reptr., 54, 741 (1970).

161. J. A. Booth, T. E. Rawlins, and C. F. Chew, Plant Disease Reptr., 55, 569 (1971).

162. C. May, Trees, 4, 7, 10, 14, 16 (1941).

163. R. J. Stipes, Gard. Chron., 169, 20, 26 (1971).

164. T. W. Jones and G. F. Gregory, U.S. Dept. Agr. Forest Serv. Res. Paper NE-233 (1971).

165. E. S. Kondo, Can. Forest Serv. Sault Ste. Marie, Ont., Inform. Rept. O-X-171 (1972).

166. E. S. Kondo and G. D. Huntley, Can. Forest Serv. Sault Ste. Marie, Ont. Inform. Rept. O-X-192 (1974).

167. E. B. Himelick, Arborist's News, 37, 97 (1972).

168. G. F. Gregory, T. W. Jones, and P. McWain, U.S. Dept. Agr. Forest Serv. Res. Note NE-176 (1973).

169. T. H. Filer, Jr., Plant Disease Reptr., 57, 338 (1973).

170. J. N. Gibbs, H. M. Heybroek, and F. W. Holmes, Nature, 236, 121 (1972).

171. A. C. Lincoln, Proc. Tree Wardens, Arborists, and Utilities Conf., March 11, 1968, Amherst, Mass, pp. 138-147.

172. G. N. Agrios, Plant Pathology, Academic Press, New York, 1969, pp. 272-280.

173. W. H. Smith, Tree Pathology, Academic Press, New York, 1970, pp. 202-207.

174. G. A. Zentmyer, J. G. Horsfall, and P. P. Wallace, Conn. Agr. Expt. Sta. New Haven, Bull., 498 (1946).

175. A. E. Dimond, G. H. Plumb, E. M. Stoddard, and J. G. Horsfall, Conn. Agr. Expt. Sta. New Haven, Bull., 531 (1949).

176. C. Zaronsky, Jr. and R. J. Stipes, Phytopathology, 59, 1562 (1969).

177. R. J. Stipes, Phytopathology, 59, 1560 (1969).

178. M. M. Bankuti and W. D. Thomas, Jr., Phytopathology, 54, 1431 (1964).

179. E. B. Smalley, Phytopathology, 61, 1351 (1971).

180. E. B. Smalley, Phytopathology, 52, 1090 (1962).

181. R. J. Stipes and K. E. Weinke, Plant Disease Reptr., 56, 604 (1972).

182. R. J. Stipes and L. R. Schreiber, Plant Disease Reptr., 50, 573 (1966).

183. B. R. Roberts and R. J. Stipes, Forest Sci., 15, 87 (1969).

184. J. H. Hart, Plant Disease Reptr., 56, 685 (1972).

185. E. B. Himelick and D. Neely, Plant Disease Reptr., 49, 949 (1965).

186. R. J. Stipes, in Fungic. Nematic. Tests, Results of 1972 Amer. Phytopathol. Soc., Report 242, 1972.

187. R. J. Stipes, in Fungic. Nematic. Tests, Results of 1972 Amer. Phytopathol. Soc., Report 243, 1972.

188. R. Prasad, Chem. Control Res. Inst. Ottawa, Ont. Inform. Rept. CC-X-47, 1972.

189. G. F. Gregory, T. W. Jones and P. McWain, U.S. Dept. Agr. Forest Serv. Res. Paper NE-232, 1971.

190. E. S. Kondo and G. D. Huntley, Can. Forest Serv. Sault Ste. Marie, Ont. Inform. Rept. O-X-201, 1974.

191. R. Prasad, Chemical Control Res. Inst. Ottawa, Ont. Inform. Rept. CC-X-72, 1974.

192. D. M. Spencer, in Systemic Fungicides (R. W. Marsh, ed.), Wiley, New York, 1972, pp. 206-224.

193. R. B. Maude, in Systemic Fungicides (R. W. Marsh, ed.), Wiley, New York, 1972, pp. 225-236.

194. D. C. M. Corbett and G. A. Hide, Proc. 6th Brit. Insectic. Fungic. Conf., Brit. Crop Prot. Council, 1971, p. 258.

195. G. J. M. van der Kerk, Acta Phytopathol., 6, 311 (1971).

196. O. M. van Andel, Phytopathol. Z., 45, 66 (1962).

197. O. M. van Andel, Neth. J. Plant Pathol., 74, 113 (1968).

198. H. Buchenauer and D. C. Erwin, Z. Pflanzenkrankh. Pflanzenschutz, 80, 567 (1973).

199. H. Buchenauer and D. C. Erwin, Z. Pflanzenkrankh. Pflanzenschutz, 80, 576 (1973).

200. H. Bockmann, Euphytica Supp., 1, 271 (1968).

201. H. M. Cathey, Ann. Rev. Plant Physiol., 15, 271 (1964).

202. A. C. Leopold and P. E. Kriedemann, Plant Growth and Development, 2nd ed., McGraw Hill, New York, 1975.

203. A. K. Sinha and R. K. S. Wood, Ann. Appl. Biol., 60, 117 (1967).

204. A. K. Sinha and R. K. S. Wood, Nature, 202, 824 (1964).

205. B. Zeeh, K.-H. König, and J. Jung, Kemia-Kemi, 1, 621 (1974).

206. A. I. Zaki, N. T. Keen, and D. C. Erwin, Phytopathology, 62, 1402 (1972).

207. S. Singh, Indian Farming, 20, 5 (1970).

208. S. D. Tsai and D. C. Erwin, Proc. 1975 Beltwide Cotton Prod. Res. Conf., 35th Cotton Disease Council, pp. 31-32.

209. C. A. Stutte and R. D. Rudolph, Arkansas Farm Res., 23, 3 (1973).

210. C. H. Beckman, Phytopathology, 48, 172 (1958).

211. C. H. Beckman, Phytopathology, 49, 227 (1959).

212. J. W. Gerdemann and J. M. Trappe, Mycologia Memoir No. 5, New York, Botanical Gardens, 1974.

213. D. E. Millhouse and D. E. Munnecke, Proc. Amer. Phytopathol. Soc., 2, 115 (1975).

214. E. S. Kondo and G. D. Huntley, Can. Forest Serv. Sault Ste. Marie, Ont. Inform. Rept. O-X-182, 1973.

215. D. C. Erwin, unpublished work, 1972.

216. S. Wilhelm, private communication, 1975.

217. G. D. Easton, private communication, 1975.

218. C. E. Horner, private communication, 1975.

219. J. P. Jones, private communication, 1975.

220. D. C. Erwin, unpublished work, 1974.

221. D. C. Erwin, unpublished work, 1968.

222. E. B. Himelick, D. Neely, and J. H. Tyndall, Plant Disease Reptr., 47, 85 (1963).

223. D. Neely and E. B. Himelick, Plant Disease Reptr., 49, 106 (1965).

224. R. J. Stipes, in Dutch Elm Disease Proc. of IUFRO Conf., Minneapolis-St. Paul, U.S.A., Sept. 1973, USDA Forest Service Northeastern Forest Expt. Sta., Upper Darby, Pa., 1975, p. 1.

225. E. J. Butterfield and J. E. DeVay, Phytopathology, 67, in press (1977).

226. E. J. Butterfield, Ph.D. Dissertation, University of California, Davis, 1976.

Chapter 8

CONTROL OF FOLIAGE AND FRUIT DISEASES

Zvi Solel

Division of Plant Pathology
Agricultural Research Organization
The Volcani Center
Bet Dagan, Israel

225

I. IMPORTANCE OF FOLIAGE AND FRUIT DISEASES

This chapter is concerned with a great variety of fungal diseases on many different crops. Foliar diseases may cause direct losses when the vegetative part of the plant itself is the marketable product. Examples of destructive diseases on such crops are early and late blights of celery, downy mildew of lettuce, blue mold of spinach and tobacco, Pseudopeziza leaf spot of alfalfa and clover, and rusts and mildews on grasses. Indirect losses are caused when foliar diseases reduce the quantity of fruit or root yield or impair market quality. Thus, grain yield of cereals is affected by diseases like rusts, mildews, Septoria, and Helminthosporium leaf spots. Fruit yield of many trees and of vegetable crops such as tomatoes, peppers, cucumbers, and legumes is lost due to foliar diseases such as blights, rusts, mildews, and various leaf-spotting diseases. Sugar production is greatly reduced by Cercospora leaf spot of sugarbeet.

Direct infection of fruit inflicts severe losses of yield and quality. Examples of such diseases are scab of apples and pears, brown rots of stone and citrus fruit. Vegetable fruits are subject to diseases like late blight and anthracnose of tomatoes, anthracnose of pepper, Ascochyta blights of pea and chick-pea. Smuts cause severe damage to grain cereals, and corn is also subject to losses due to ear rots. Fruit infections are often interrelated with foliar infections since they are incited by airborne inoculum, the source of which is spores formed in foliar lesions. For example, late infections of scab of apple fruit occur if leaf lesions have been permitted to develop. Infection of tomato fruit by late blight occurs after build up of inoculum on foliage.

Under optimal conditions for disease development (when inoculum is plentiful, the cultivar is susceptible, and the weather conditions are favorable for the pathogen) epidemics may be devastating. The disastrous famine in Ireland in the middle of the 19th century was due to late blight epidemics of potatoes which occurred under conditions ideal for disease development. Epidemics of powdery mildew of vines, which spread in Europe at that time, were also catastrophic. Nowadays such occurrences are less common, but diseases still take their share of the yield. Many are debilitating rather than devastating.

Farmers throughout the world produce food for mankind. However, most of them, whether they cultivate a small garden or a huge farm, are interested mainly in producing a profitable crop. Therefore, when discussing chemical measures to minimize losses of yield due to leaf and fruit diseases, one should bear in mind that the ultimate goal of a farmer is to gain the best profit from his crop. Maximum profit does not necessarily coincide with maximum yield, as a result of best disease control. In their review "Economic bases for protection against plant disease," Ordish and Dufour [1] put forward the following economic principles: (1) Disease causes a loss of crop at a certain value. (2) A remedy overcomes the disease and prevents all or some of the loss. (3) If the cost of 2 is less than 1, all is well. If 2 is equal or more than 1, the work should not be undertaken. Obviously, economic considerations differ between countries according to prices of components of crop production, their protection and yield value.

Amongst the most destructive enemies of mankind are rusts of wheat. In India, fungicides were evaluated in the field for control of leaf and stem rusts, and five sprays of maneb either alone or supplemented by nickel sulfate, applied with a knapsack sprayer, were economically beneficial [2]. In Canada, oxathiin systemic fungicides were evaluated for control of the same diseases [3]. Cost-benefit values were calculated for various methods of application and prices of chemicals (which were not commercially available at that time). The most economical treatment was two sprays of oxycarboxin; three treatments were, on the averages, less profitable. In Israel, Septoria leaf blotch of wheat is quite severe; three sprays with chlorothalonil or maneb increased grain yield by 23 and 17%, and profits by 15 and 13%, respectively [4].

II. INTRINSIC FACTORS INFLUENCING THE ACTIVITY OF FOLIAGE AND FRUIT FUNGICIDES

A. Introduction

Fungicides may be classified according to their function as:

1. Protectants, applied to protect the plant from an expected infection.
2. Therapeutants, applied soon after onset of infection process to suppress development of the pathogen and thus cure the plant.
3. Eradicants, applied to diseased organs to destroy inoculum, or prevent its formation.

The most common fungicides are protectants which are applied topically to the surface of foliage or fruit for local protection. Most systemic compounds act in part as surface protectants and in this sense, they need not be distinguished from conventional protectants. The systemic fungicidal activity of these compounds will be discussed separately.

The prime requirement for a useful fungicide is that it exert a toxic effect on certain plant pathogens. Various methods have been devised to assess fungitoxicity in vitro [5, 6], but none simulate the many factors encountered under natural conditions. Laboratory tests may provide indications as to the spectrum of fungi affected by a toxicant, and the "specific activity" of the chemical, which is often expressed as a dosage-response curve [7]. While the innate toxicity to fungi is the basis for its action, the chemical must be safe to plants and to humans or animals utilizing the plants. In recent years, more attention is paid to the effect of chemicals on the environment, especially on animals, beneficial insects, worms, and microorganisms. Phytotoxic effects may be manifested by discoloration, necrosis, defoliation, deformation, or inhibition of growth. There may be physiological negative effects resulting in reduced yield without obvious symptoms. Thus copper fungicides are known to have such toxic effects on certain crops [8], and recently systemic fungicides have been shown to cause symptomless phytotoxic effects [9].

Although fungitoxic properties and lack of phytotoxicity are prerequisites for a fungicide, there are many more features that a chemical must possess in order to function under field conditions. Furthermore, the chemical must be formulated into a particulate or liquid formulation to suit application equipment and distribution on foliage and fruit. Formulation has a tremendous effect on the success or failure of a chemical. Factors affecting the field performance of fungicides from the moment of their deposition on the plant will now be analyzed.

B. Surface Protectants

A protectant fungicide is applied to the surface of foliage or fruit to form a barrier between the host organ and the inoculum in order to prevent infection. Best protection is conferred by a continuous film of a fungicide completely covering the surface during the entire infection season. This goal has never been achieved, because in practice a chemical is deposited on the surface in the form of particles and never forms a complete layer. It would be worthwhile to state that not only dusts and wettable powders, but also emulsified or solubilized formulations ultimately yield a deposit of solid particles, sometimes natural crystals, distributed on the host surface. Thus, an initial deposit is the amount of solid particles which is retained by a dusted or sprayed surface after evaporation of the liquid carrier or solvent. Any single particle will be capable of protecting a certain sphere around itself according to its characteristics. However, since we are interested in the organ, the plant, and the field as a whole, we should consider the mass of particles as a population. Thus, the perfection of distribution, or coverage, is of paramount importance, since an uncovered spot is also unprotected. The perfection of the coverage depends on the method of

application, which will be discussed in the next section of this chapter, and
on the properties of the chemical and the way it is formulated.

1. The Initial Deposit

Most foliar fungicides are formulated as particles suitable for dispersal in
water and spraying on the target. Hence, surfactants are added in order
to suspend the particles in the water and to improve spreading of the spray
droplets on the waxy surface of leaves and fruits. The type and quality of
the surfactant may have a remarkable effect on coverage of the fungicide
through its effect on the spread of droplets. The size and number of droplets are also important; numerous small and dense droplets may confer
better protection than large and scattered droplets containing the same
number of fungicide particles. Usually, the initial deposit is measured as
the amount of fungicide per unit of surface [10, 11]. This quantitative data
do not provide any information on the quality of coverage, which is the
consequence of density and randomness of particles. Quality of coverage
might explain why different values are obtained in the laboratory and in the
field under practical conditions. Large et al. [12] found in field trials that
a copper residue of $4 \mu g/cm^2$ of leaf was needed to assure protection from
potato blight (Phytophthora infestans). Courshee [13] showed in the laboratory that potato leaves covered thoroughly with as little as $0.01 \mu g/cm^2$
of leaf were completely protected. This gap suggests that coverage under
field conditions is far from perfect.

Several large particles on a leaf would fail to protect the whole surface,
while the same dose subdivided into numerous small particles might be
adequate to protect any locus of potential infection. This was demonstrated
with sulfur dusts where powders of different particle size differed in toxicity
when compared on an equal weight basis; the more finely divided dusts
inhibited spore germination at lower doses [14]. However, on the basis of
equal number of particles per unit area there was no major differences in
toxicity. Obviously, this would be true only if the particles are uniformly
distributed on the surface. Activity might be related to the number of particles only in a certain range of sizes, because under a certain size the
zone protected by one particle would be proportional to its size. This was
illustrated when fractions of different sizes of dichlone particles were
tested in the greenhouse for control of late blight of tomatoes [15]. When
particle size was reduced from a mean radius of $24.5 \mu m$ to $0.45 \mu m$, fungicidal efficiency was increased proportionally to the logarithm of the number of particles per unit area of surface. The use of extremely small particles is limited by two factors, namely the price of their production and
their unsatisfactory persistence. For example, the sublimation of sulfur
is proportional to the external surface of a particle [16]. One might expect
that compounds with high intrinsic toxicity might be better suited to extreme
subdivision.

2. Redistribution

In the preceding paragraphs, considerable importance was ascribed to quality of coverage. On the farm, unfortunately, uniform and dense coverage is seldom reached. Yet, fortunately, disease control is quite satisfactory. The process which corrects uneven coverage is called redistribution, in which the deposit is redistributed on surfaces not protected by the initial deposit. The driving forces for redistribution are dew, rain, and sprinkler irrigation. For many diseases these factors are also essential for onset of infection. The dual interrelated effects of water may result in good disease control.

The efficiency of redistribution may be demonstrated by reviewing experiments with new application methods devised in Japan for citrus-disease control by Yamada et al. [17-19]. The authors have come up with the idea that in the case of many rainborne pathogens sufficient control could be obtained by applying a fungicide deposit to the upper surface of the leaves in the top of the plants. Fungicides were applied either as concentrated suspensions sprayed to the top of citrus trees, or by the "netting" method. With this method, a net made of cotton-staple fiber yarn or hemp string was impregnated with fungicides and placed over the canopy of the trees. The authors studied redistribution of several fungicides and adjuvants to lengthen the period of their effectiveness and their efficiency for controlling two main citrus diseases, melanose (Diaporthe citri) and scab (Elisnoe fawcetti). Fungicides could be traced on leaves in the inner parts of the trees, and some of them were effective in controlling the diseases. Certain protectants such as Bordeaux mixture and other inorganic copper derivatives were not suitable for application based on the described concept, because redistribution was too limited to inhibit pycnospore germination of D. citri or conidia formation by E. fawcetti. On the other hand, captafol was excellent. It should not be concluded that redistribution of copper fungicides is ineffective, as rains improved the protection of potato foliage by Bordeaux mixture against infection by Phytophthora infestans [20]. An excellent redistribution of dodine was reported [21]; when the fungicide was deposited to the upper surface of apple leaves and the leaves were held in a moist environment, it moved to the undersurface and effectively controlled apple scab (Venturia inaequalis).

3. Weathering

From the moment a deposit is formed it is subjected to weathering. Redistribution might be considered as a positive aspect of weathering. However, weathering gradually reduces the amount of the residue on the surface until it drops below the threshold of effectiveness [12]. For that reason, and also when new growth is developing, protective treatments must be repeated at certain time intervals. The factors for weathering are many. Rain, sprinkler irrigation, and dew are of major importance due to their leaching effect. Any chemical, even those often designated as practically insoluble, has some degree of solubility and may be leached. Rain, especially in storms and in the tropics, and sprinkler irrigation remove deposits also by mechanical erosion. Tenacity, which designates the retention of fungicide particles to the plant's surface, is an intrinsic factor of the fungicidal formulation. It confers resistance to mechanical removal by water, wind, and scraping. Tenacity is extremely important in preventing erosion by such common factors as rain and sprinkler irrigation. Bordeaux mixture is superior to numerous other fungicides and has been used extensively for almost a century, primarily because of its superb tenacity. Thus, Bordeaux mixture applied to citrus fruit in autumn persists throughout the rainy winter [22]. Miller [23] compared the performance of many copper fungicides and showed that Bordeaux mixture was superior in control of cherry leaf spot (Coccomyces hiemalis) as well as in retention to leaves (or Pyralin plates) and in LD$_{50}$ value. Burchfield and Goenaga [24] attributed the outstanding tenacity of Bordeaux mixture to its hydrogel consistency which affects its spread over the plant surface to form more or less continuous films.

Tenacity is influenced by the spreading properties of the spray on the plant's surface. Thus even Bordeaux mixture, when applied to extremely waxy leaves of bananas, collects as discrete droplets at the leaf vein [24]. On drying, these form friable pellets of Bordeaux which are easily dislodged mechanically by the first rain. The addition of a surfactant to the mixture reduces interfacial tension so that the droplets spread and the deposit thus formed is more tenacious. Obviously, the surfactant would improve coverage as well. On the other hand, surfactants also tend to reduce tenacity because they enable water to wet particles. Furthermore, when spraying at high volume, surfactants promote runoff to occur at an earlier stage and thus a thinner layer of spray remains; consequently a smaller amount of fungicide is deposited. Therefore, the amount of surfactant added to a fungicide should be balanced so as to improve spreading without too much loss due to runoff.

Rich has compared the dynamics of deposition and tenacity of Bordeaux mixture and zineb formulations [10]. He found that during the weathering of dried Bordeaux deposits, the larger the initial deposit, the smaller is the percentage lost. He conjectured that dried particles of Bordeaux mixture cohere more strongly to each other than they adhere to the leaf surface. Dried zineb particles weather in an opposite fashion; the larger the deposit of zineb, the greater the proportion of its mass lost by weathering, apparently because of stronger adherence to leaf surface than coherence of particles to each other. Both fungicides weathered slower on hairy bean leaves than on smooth celery leaves.

Stickers might be the answer for low tenacity. In the past, natural products like casein or linseed oil were tested without much success or even with failure due to inactivation of copper fungicides [25, 26]. Recent "cocktails" like Biofilm or Citowet (alkyl and aryl ethers of polyglycol) are adjuvants designated as spreader-sticker, which seem to enhance performance of fungicides, especially on waxy plants.

Dusting is the best application method to achieve good coverage of foliage and fruit. Yet it is rarely used with protectants because of the poor adherence of dust particles. Only with sulfur is dusting still largely used, because of its low price. A device which imparts a positive charge to dusted particles greatly improves their deposition on the negatively charged leaves. However, it does not improve their retention afterwards.

Persistence of protective fungicides depends on their chemical and physical stability. For some, instability is essential for their fungicidal action. Physical factors of disappearance are sublimation and volatility. Sulfur deposits are attenuated by sublimation, and in warm climates protection may disappear quickly [27]. Fine sulfur persists less than coarse sulfur, due to the difference in the total surface exposed to sublimation. The organic compounds, even if regarded as nonvolatile, have some degree of volatility which affects their persistence. For many, fungitoxic activity by vapor phase is evident. This mode of action is not necessarily one of practical importance. Vapor-phase action was shown to be involved in control of powdery mildews by drazoxolon, oxythioquinox, binapacryl, dinocap [28], alkyldinitrophenols [29], triarimol, pyrazophos, and methyrimol [30]. Several other fungicides show vapor phase action. Benomyl, thiabendazole, thiophanate, fentin acetate, and maneb are active against Cercospora beticola [31]; phenyl mercury chloride, maneb, mancozeb, dichlofluanid, and oxythioquinox are active against Botrytis fabae [32]. Usually the chemical nature of the vapor was not identified and modifications from the original structure cannot be excluded. The vapors emanating from a solution of nabam were identified as carbon disulfide and ethylene diamine [33].

Fungicides exposed to the environment are subjected to oxidation, reduction, hydrolysis, and other chemical reactions. Sometimes the reaction is positive while in other cases fungicides are inactivated. Light, including UV irradiation, has a catalytic effect on decomposition. Measurements of the photolytic decomposition rates of chloranil and dichlone showed that dichlone was more stable [15]. This agrees with practical experience that both are good for seed treatment, but for foliar protection chloranil has poor persistence. While irradiation generally has a negative effect, the thiophanates may be activated by photochemical transformation. The authentic structure of this group is nonfungitoxic [34], and hence nonprotective. On sprayed leaves subjected to sunlight and dew, residues are transformed to fungitoxic alkyl benzimidazolecarbamates [35].

C. Systemic Protectants

To protect foliage, systemic fungicides may be applied either directly to the target for local distribution, or elsewhere for systemic translocation.

In the first case, most fungicides would act primarily as surface protectants by virtue of their fungitoxic properties. In addition, some part of the chemical enters into the leaf and either moves through the leaf (translaminar protection) or is distributed within it. The pattern of distribution depends on the nature of systemicity. All systemic fungicides used at present are basically apoplastic, and therefore move via the transpiration stream toward the tip and margins of leaves [36, 37]. Carbendazim was also shown to have a negligible movement in the symplast toward the root and growing tip of the plant [37, 38]. The systemic fungicidal activity of foliar treatment has often been demonstrated but rarely quantified. In bean and tomato leaves sprayed with benomyl, less than 1% of the chemical enters the leaves [37]. Penetration into the leaves can be enhanced by the addition of a surfactant [37], and also by solubilizing the chemical in acidic medium [39].

The superior performance of a foliar systemic fungicide, especially in rainy periods, might stem also from the resistance to leaching of that part absorbed by the leaf. This is true only if the chemical is persistent within the plant.

Sometimes the host plant has a very aggressive effect on a fungicide [40, 41], while in other combinations fungicides persist for quite some time. The decay of a fungicide might be characterized by its "half life" measured by days or hours. The use of a term borrowed from the world of radioisotopes does not imply that inactivation of a fungicide has an exponential curve. This type of decay might occur when the degrading forces are in relative "excess" and hence degradation would depend largely on the amount of chemical left at each moment.

Systemic fungicides can best exhibit their unique feature if administered to the soil, thus being absorbed by the plant via its natural uptake organ, namely the root system. When a reservoir is available in the soil, continuous uptake and translocation of the fungicide toward the leaves can occur via the transpiration stream for a prolonged period. Thus the toxicant can be distributed there in a pattern and concentration that can protect foliage from within. Furthermore, the fungicide can constantly protect newly formed leaves and also inner foliage which is hardly accessible to spray treatment. The fungicide is not likely to accumulate in fruit [42, 43] since their transpiration is negligible or absent. Most fruit diseases occur from inoculum built up on foliage. If this phase is checked then fruit infection is not in real hazard.

While soil application has proved to be effective with herbaceous plants [44, 45], it is less successful with trees [46, 47]. A substitution for soil treatment might be trunk injection. Recent studies demonstrated that by pressure injection of benzimidazole fungicides a reservoir is formed in the trunk with subsequent secondary distribution into branches and leaves [48, 49]. When pear trees were injected before leaf fall, most of the fungicide was reserved in the trunk, but some had translocated to the leaves [49]. In the next spring and summer, both thiabendazole and carbendazim had translocated and accumulated in the leaves, and protected them from scab infections (Venturia pirina).

D. Therapeutants

A chemical useful in therapy is termed a chemotherapeutant [50] or simply a therapeutant. Such a chemical can be applied after the onset of infection to interfere with the development of disease, and may either prevent development of symptoms or limit their severity. A therapeutant must have systemic properties in order to be taken up by the host and act within the infected organ. Speed of fungicide penetration into the infected tissue is of paramount importance since it has to compete with fungus development. The fungicides available for therapy are actually the same as those listed for systemic protection. Their main disadvantage is their fungistatic rather than fungicidal nature [51]. Therefore, if a fungicide is metabolized and the concentration drops below the effective threshold, the fungus can resume its pathogenic activity.

E. Eradicants

The purpose of eradication is to suppress inoculum formation at the source. Eradicants are fungicides applied to infected organs to prevent sporulation or to destroy inoculum already developed in the lesions. If the treatment is applied to nonliving organs, e.g., fallen leaves, the choice of chemical is wide because phytotoxicity is not a limiting factor. When applied to fruit trees while dormant, phytotoxicity is still not so limiting, and chemicals like monocalcium arsenite [52] or sodium pentachlorophenate [53] may be used. Antisporulation agents are mainly selected from the list of systemic protectants. They manifest such action in addition to their main protectant activity.

F. Fungicides Used for Control of Foliage and Fruit Diseases

The most commonly used fungicides are listed in Tables 1 and 2. For rough characterization, distinction has been made between systemic and conventional fungicides, although some of the latter might possess a certain degree of systemicity. The fungicides are grouped according to their chemical structure. Generally, chemicals of the same group act similarly, but exceptions cannot be excluded. Only the most common uses are indicated, and chemicals employed on a large variety of crops are designated as general.

TABLE 1

Fungicides Commonly Used for Control of Foliar and Fruit Diseases

Chemical group	Generic name	Chemical name	Formulations[a]	Main uses	
				Diseases	Crops
Inorganic sulfur		Sulfur	WP, D	Powdery mildews	General
	Lime sulfur	Calcium polysulfide	S, D	Powdery mildews	General
Inorganic copper		Cuprous oxide	WP, D, SC	Broad range	General
	Basic copper sulfate	Cupric sulfate monohydrate	WP, D	Broad range	General
	Copper oxychloride	Basic cupric chloride	WP, D, SC	Broad range	General
	Bordeaux mixture	Complex of copper sulfate and calcium hydroxide		Broad range	General
Inorganic arsenic		Monocalcium arsenite	D	Brown rot	Stone fruits
Organocopper compounds	Copper oxinate	Copper-8-hydroxyquinoline	S, SP	Broad range	Vegetables, field crops
		Copper salts of fatty and rosin acids	SC	Brown rot, mildews	Citrus, vegetables
Organomercury compounds		Phenylmercuric acetate	SP, S	Scab	Fruit trees, ornamentals

235

TABLE 1 (Cont.)

Chemical group	Generic name	Chemical name	Formulations[a]	Main uses	
				Diseases	Crops
		Phenylmercurimonoethanol ammonium acetate	S	Scab	Fruit trees, ornamentals
		Phenylmercuritriethanol ammonium lactate	S	Scab	Fruit trees, ornamentals
Organotin compounds	Fentin acetate	Triphenyltin acetate	WP	Broad range	Field crops, vegetables
	Fentin chloride	Triphenyltin chloride	WP	Broad range	Field crops, vegetables
	Fentin hydroxide	Triphenyltin hydroxide	WP	Broad range	Field crops, vegetables
Carbamates	Ferbam	Ferric dimethyldithiocarbamate	WP, D	Broad range	General
	Maneb	Manganese ethylenebisdithiocarbamate	WP, D	Broad range	General
	Propineb	Zinc propylenebisdithiocarbamate	WP, D	Broad range	General
	Zineb	Zinc ethylenebisdithiocarbamate	WP, D	Broad range	General
	Ziram	Zinc dimethyldithiocarbamate	WP, D	Broad range	General
	Mancozeb	Complex of maneb and zinc	WP	Broad range	General
Dinitrophenols	Binapacryl	2-(1-Methyl-n-propyl)-4,6-dinitrophenyl-2-methylcrotonate	WP, EC, SC	Powdery mildews	General
	Dinocap	2-(1-Methyl-2-heptyl)-4,6-dinitrophenyl-crotonate	WP, D, EC	Powdery mildews	General

Class	Common name	Chemical name	Formulation	Target	Crops
Chlorinated compounds	Captan	N-(Trichloromethylthio)-3a,4,7,7,7a-tetrahydrophthalimide	WP, D	Broad range	General
	Folpet	N-(Trichloromethylthio) phthalimide	WP, D	Broad range	General
	Captafol	N-(1,1,2,2-Tetrachloroethylthio)-3a,4,7,7a-tetrahydrophthalimide	WP, SC	Broad range	General
	Chloranil	2,3,5,6-Tetrachloro-1,4-benzoquinone	WP, D	Various	Vegetables, turf
	Chlorothalonil	Tetrachloroisophthalonitrile	WP, D, tablets	Broad range	General
	Dichlone	2,3-Dichloro-1,4-naphtoquinone	WP, D, SC	Various	Vegetables, ornamentals
	Dichloran	2,6-Dichloro-4-nitroaniline	WP, D	Rhizopus, Botrytis, Sclerotinia	Vegetables, fruit trees
	Drazoxolon	4-(2-Chlorophenylhydrazono)-3-methyl-5-isoxalone	EC	Powdery mildews	General
	Anilazine	2,4-Dichloro-6-(2-chloroanilino)-1,3,5-triazine	WP, D	Various	Vegetables, field crops
Miscellaneous compounds	Dithianon	2,3-Dicyano-1,4-dithiaanthraquinone	WP, SC	Various	Fruit trees, vegetables
	Dodine	Dodecylguanidine acetate	WP, D, SC	Scab	Apple, pear
	Oxythioquinox	6-Methyl-2,3-quinoxalinedithio cyclic S,S-dithiocarbonate	WP, D, SC	Powdery mildews	General

TABLE 1 (Cont.)

| Chemical group | Generic name | Chemical name | Formulations[a] | Main uses | |
				Diseases	Crops
	Ditalimfos	O–O–Diethyl phthalimidophosphonothioate	WP	Powdery mildews	General
		Sodium pentachlorophenate	S	Brown rot	Stone fruits

[a]Key: D = dust, EC = emulsifiable concentrate, G = granular, S = solution, SC = suspendable concentrate, SP = soluble pow–der, WP = wettable powder.

TABLE 2

Systemic Fungicides Commonly Used for Control of Foliar and Fruit Diseases

| Chemical group | Generic name | Chemical name | Formulations[a] | Main uses | |
				Diseases	Crops
Benzimidazoles	Carbendazim	Methyl 2–benzimidazolecarbamate	WP, SC	Broad range	General
	Benomyl	Methyl (1–butylcarbamoyl)–2–benzimida– zolecarbamate	WP	Broad range	General
	Thiabendazole	2–(4'–Thiazolyl)–benzimidazole	WP, SC	Broad range	General
	Cypendazol	1–(5–Cyanopentylcarbamoyl)–2–methoxy– carbonylamino)–benzimidazole	WP	Broad range	General
Thiourea based[b]	Thiophanate	1,2–Di–(3–ethoxycarbonyl–2–thioureido) benzene	WP	Broad range	General

238

Group	Common name	Chemical name	Formulation	Broad range	General
	Thiophanate methyl	1,2-Di-(3-methoxycarbonyl-2-thioureido)benzene	WP	Broad range	General
Oxathiins	Carboxin	5,6-Dihydro-2-methyl-1,4-oxathiin-3-carboxanilide	WP, SC	Smuts, rusts	Cereals, vegetables
	Oxycarboxin	5,6-Dihydro-2-methyl-1,4-oxathiin-3-carboxanilide-4,4-dioxide	WP	Rusts, smuts	Cereals, vegetables
Pyrimidines	Ethirimol	5-Butyl-2-(ethylamino)-4-hydroxy-6-methylpyrimidine	SC	Powdery mildews	General
	Dimethirimol	5-Butyl-2-(dimethylamino)-4-hydroxy-6-methylpyrimidine	S	Powdery mildews	General
	Triarimol[c]	α-(2,4-Dichlorophenyl)-α-phenyl-5-pyrimidinemethanol	WP, D, EC	Powdery mildews	General
Piperazine	Triforine	N,N'-bis(1-formamido-2,2,2-trichloro-ethyl)-piperazine	EC	Powdery mildews	General
	Chloraniformethan	1-(3,4-Dichloroanilino)-1-formylamino-2,2,2-trichloroethane	EC	Powdery mildews	Cereals, vegetables
Morpholines	Dodemorph	4-Cyclododecyl-2,6-dimethylmorpholine	EC	Powdery mildew	Roses
	Tridemorph	2,6-Dimethyl-4-tridecylmorpholine	EC	Powdery mildews	Cereals
Organophosphorus compounds	Edifenphos	O-Ethyl-S,S-diphenyl phosphorodithioate	EC, D	Various	Rice
	Kitazin[d]	O-O-Diisopropyl-S-benzyl thiophosphate	EC, D, G	Various	Rice

239

TABLE 2 (Cont.)

Chemical group	Generic name	Chemical name	Formulations[a]	Main uses	
				Diseases	Crops
	Pyrazophos	2-(O,O-Diethylthionophosphory)-5-methyl-6-carbethoxy-pyrazolo-[1, 5a] pyrimidine	EC	Powdery mildews	General
Antibiotics	Blasticidin[d]	Benzylaminobenzene sulfonate	WP, D, S	Blast	Rice
	Cycloheximide	β[2-(3, 5-Dimethyl-2-oxocyclohexyl)-2-hydroxyethyl]-glutarimide	WP	Various	Ornamentals, turf
	Griseofulvin	7-Chloro-4, 6-dimethoxycoumaran-3-one-2-spiro-1'-(2'-methoxy-6'-methylcyclo-hex-2'-en-4'-one)	WP	Various	General
	Kasugamycin[d]		WP, D, S	Blast	Rice
	Polyoxins	3-Ethylidene-L-azetidine-2-carboxylic acid	WP, D	Various	Rice, fruit trees

[a]For key, see Table 1.
[b]Transformable to benzimidazoles.
[c]The fungicide has been rejected for toxicological considerations.
[d]Trade name, no generic name has been accepted.

240

III. METHODOLOGY OF FUNGICIDE APPLICATION

When Bordeaux mixture was developed by Millardet in 1885, it was applied
to vine foliage by means of a brush made of heath straw. The brush was
dipped into the thick mixture (of 6% copper sulfate and 11% calcium hydroxide in water) and then swung at the vines, thus splattering the foliage. The
coverage produced by this method was uneven and inadequate by present
standards. Despite this, at that time application of a fungicide to plant foliage was a breakthrough in disease control, although the application technique
was primitive. At present, more effective methods are used to apply fungicides more uniformly to the target, and it has been shown that the application
method may have considerable influence on field performance of a fungicide.

To control foliar diseases, fungicides are applied either topically to the
foliage or indirectly, in the case of systemic toxicants, e.g., to the soil or
into the trunks of trees. Each method requires a suitable formulation for
the specific technique. Most spray application techniques utilize water as
a carrier, with formulation aids, such as solvents, emulsifiers, dispersants,
wetting agents, stickers, stabilizers, and safeners added to the active ingredient by the manufacturer.

A. Spraying of Foliage and Fruit

Spraying techniques may be characterized by the following variables: (1) volume of liquid applied per unit of area; (2) runoff from the sprayed surface;
(3) concentration of fungicide in the spray; and (4) type of sprayer used for
application. All these variables are interrelated, so that by selecting one
of them the others must comply with it. In the following discussion, spraying techniques will be classified according to volume of spray applied per
unit area. It is not the volume per se which is so important but the wetting
of foliage and subsequent density of coverage that are affected by it. Classification of volumes per unit of area are arbitrary because they do not relate
specifically to the type of crop and its various development stages, which
determine its total foliar surface per field area. Some guidelines have been
given by Maas [54] for field crops:

1. High volume, more than 400 liters/ha. Usually much more is used,
   e.g., in orchards volumes of 2,000-20,000 liters/ha are common.
2. Low volume, 5-400 liters/ha. Volumes of 15-75 liters/ha are commonly encountered in aerial applications.
3. Ultralow volume, less than 5 liters/ha.

1. High-Volume Spraying

In this application method all the foliage and fruit are thoroughly sprayed to
cause runoff. The impacted droplets spread and coalesce, often forming a
liquid film on the sprayed surfaces. Excessive droplets overlap and coalesce

into large globules of spray. Since the sprayed surfaces tend to be inclined
rather than horizontal, globules larger than some critical size become
gravitationally unstable and roll down the surface to discharge from the
lowest edge, marking the onset of runoff [55]. When the liquid has drained,
a thin film is retained to the surface by physicochemical forces. Thickness
of this film is related to the interaction of the characteristics of both the
formulated toxicants and the nature of the plant's surface, and depends
largely on the angle of tilt [55]. The liquid volume retained by a crop can
be calculated from the initial deposit left on its surface after the carrier
liquid evaporates. With high-volume spraying this deposit has also been
termed as "runoff deposit" [56]. When Somers [56] sprayed particulate
copper fungicides containing 0.25% copper, a runoff deposit of about 5 μg
Cu/cm^2 was measured. This is equal to 0.002 ml/cm^2 or 200 liters/ha of
leaf surface. Obviously foliage surface does not equal ground surface.
Usually the fungicide drained in the runoff process is wasted. In practice,
the proportion of the fungicide wasted by high-volume spraying may be high,
as much as 50 [57] or even 90% [58].

Somers studied some factors influencing the level of runoff deposits of
copper fungicides [56]. He indicated that the deposit level was not increased
by spraying beyond incipient runoff, and that spray retention depends on the
wettability of the surface. He showed that representative surfactants of
anionic, cationic and nonionic types, when used at low concentrations,
slightly lowered the amount of deposit; at concentrations greater than those
used in practice, serious reductions in deposit level occurred. Supple-
menting thickening agents which increase the viscosity of the spray did not
increase deposits. Although surfactants decreased deposit we should not
neglect their necessity for suspending powders in water and their positive
effect on the quality of coverage. It would be worthwhile to mention that a
general characteristic of runoff spraying is accumulation of deposit at the
tips and edges of leaves and fruits.

Generally, the runoff deposit is proportional to the concentration of the
chemical in the spray [56, 59]. Rich [10] studied the effect of fungicide
concentration in the spraying tank on the runoff deposit of two fungicides.
He found that while zineb is deposited in direct proportion to the concentration, Bordeaux mixture is deposited according to the logarithm of the concentration. This was ascribed to the negative and positive electrokinetic
charges of the two compounds, respectively. Rich hypothesized that the leaf
area acts as an adsorbing surface. Since Bordeaux particles have a negative charge they are adsorbed to the leaf. At low concentrations the adsorptive surfaces have room for all the Bordeaux particles and the runoff liquid
contains less particles than the original spray. However, when spray concentration increases, there are fewer chances for all the spray particles to
find an uncovered adsorptive spot. All those particles which are not held
are lost in the runoff liquid. In the case of zineb the electrokinetic conditions necessary for adsorption are not present. Hence, the initial size of

zineb deposit would be strictly proportional to the volume of spray retained on the leaf surface and the concentration of the spray used. This would account for the arithmetic relation between initial retention and spray concentration.

2. Low-Volume Spraying

In this method the spray is distributed on the surface as discrete droplets, and no runoff occurs. The main advantages are savings in labor required to spray a crop unit, and often the dose of fungicide may be reduced by 20-25% [60]. Decrease in spray volume can be obtained either by reducing droplet density per unit surface while maintaining droplets larger than 200 μm (sizes common in high-volume spraying), or, if dense coverage is desired, by utilizing droplets smaller than this size. Johnstone [61] calculated the relation between droplet density per unit sprayed area (number/ cm^2), droplet diameter, and volume of spray. These relations are graphically illustrated in Figure 1. As an example, if a cover of 50 droplets/cm^2 is necessary and the appropriate droplet size is 100 μm, the required volume will be 2.5 liters/ha, provided no losses occur. Obviously, the ratio of foliar to ground surface should be borne in mind, as well as efficiency of spraying, when estimating the required spray volume per area of plot. In practice, spraying equipment emits droplets within a range of sizes rather than of a uniform size. Statistical expression of the droplet sizes within the range is frequently characterized by parameters derived from the mass or volume size distribution. Thus, the mass or volume median diameter is the droplet diameter dividing the mass or volume of the spray into equal parts; the 16% and 84%, or 25% and 75% diameter are also used as indicators of the spectrum [55]. Having chosen droplet size and spray volume, the concentration of the fungicide in the spray can be calculated according to the required deposit (μg/cm^2).

Selection of droplet size for spraying must take into consideration the energy necessary to assure impaction of the droplets to the sprayed target. Droplets too fine usually do not have the energy necessary to pass through the air barrier which is at the plant leaf surface. The choice of density and fineness of droplets would depend on the fungicide in the formulation, the redistribution capacity, and on the mode of action on one hand, and the nature and epidemiology of the disease on the other.

For the control of brown rot of citrus fruit (<u>Phytophthora citrophthora</u>) high- and low-volume spraying with different sprayers were compared. Klotz et al. [62] found no difference in control performance when copper fungicides were sprayed with either boom sprayer or hand guns at about 6,000 liters/ha or with Turbo Mist sprayer at about 600 liters/ha. Likewise, Pappo et al. [63] did not observe any significant difference in protection when spraying with hand guns at 2,180 liters/ha or with a blower at either 2,820 or 350 liters/ha.

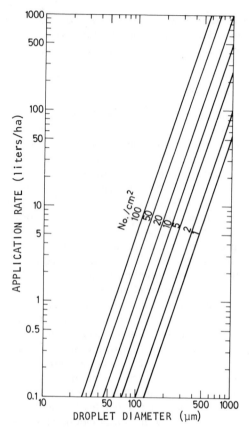

FIG. 1. Relation between droplet density, diameter, and volume application rate. Reprinted from Ref. 61, p. 78, by courtesy of the Society of Chemical Industry.

In spraying potatoes the use of a blower may confer some advantage, apparently due to the air-blast used as a carrier for droplet deposition. Fulton [59] reviewed the work of Goosen, who compared the distribution of copper sprayed by a hydraulic sprayer, with droplets greater than 150 μm on one hand, and by a blower generating droplets at the range of 50-150 μm on the other. He found that distribution, canopy penetration, lower surface coverage, and resistance to weathering were superior when applied by the blower.

Low-volume spraying by aircraft has several specific advantages in addition to diminution of cost. The epidemiological advantage is that large areas can be treated in a relatively short period, especially when weather conditions favor disease development. It is important that no untreated locations are left when obstacles interfere with a thorough treatment. By aircraft it is possible to treat immediately after rain or irrigation, when damp soil restricts the use of ground equipment. Furthermore, soil compaction and damage to dense crops are precluded.

3. Ultralow-Volume Spraying

This method, though basically similar to low-volume spraying, has several advantages. The further lowering of labor cost per unit of treated area, independence on water availability due to use of ready-to-use formulations, and use of ground types of sprayers relatively small and lightweight, which can be operated easily under difficult ground conditions. The main disadvantage is that when using very fine spray a wind velocity of more than 5 m/ sec greatly increases the drift hazard.

The ultralow-volume (ULV) application techniques originated in the control of locust in East Africa. Vast areas had to be treated rapidly, and any reduction in volume without loss of effect was of great importance. Very small droplets were emitted (about 70 μm) and even the lightest wind carried the spray and dispersed it over a large area. This technique has been called drift-spraying. For agricultural crops on limited areas, spray drift is an environmental hazard as well as a waste of chemical. Therefore, spraying techniques have been modified to minimize drift, and ULV aircraft spraying is carried out from low altitudes between 2 and 4 m, the usable swath width being limited to 15-25 m.

The control of plant diseases demands denser coverage than that generally called for in insect control. If the droplet spectrum commonly used in low-volume technique is maintained, then at best a sparse coverage may be reached if application is performed under ideal conditions. Hence special ULV equipment emitting a fine spray has been developed, which, if properly used, can be adjusted to give satisfactory density of coverage. Droplets too fine (apparently under 70 μm volumetric median diameter) are undesirable because they are difficult to direct to the target and are liable to drift even in light air currents. In addition, such small droplets of water-based spray evaporate quickly. This drawback can be overcome by using nonvolatile solvents. If suspensions are sprayed, the size of the particles must be very fine.

Maas [54] has reviewed the work of several investigators for disease control using ULV or very low volume, and mentioned the following examples: Vine diseases were successfully controlled with 10-15 liters/ha of oil-base suspensions of fungicides. Potato late blight (Phytophthora infestans) was controlled by applying 1.6 kg of maneb per ha, sprayed in a volume of 15 liters/ha using a mineral oil as the carrier liquid. An ULV formulation containing 26% glyodin and 6.5% dodine gave outstanding control of apple scab (Venturia inaequalis). A comparison was made of ULV and conventional applications of dinocap to control powdery mildew of roses (Sphaerotheca pannosa). The ULV treatment with 0.06 kg/ha in a volume of 5 liters/ha was superior to 0.15 kg/ha in 1,100 liters/ha.

Methods of ULV spraying for disease control deserve intense research and cooperation of the chemical industry to provide suitable formulations of fungicides.

B. Equipment for Foliar Spraying

The spraying process may be divided into two phases: atomization of the liquid into droplets, and launching the droplets toward the target. Atomization is achieved by using nozzles, rotating disks, and other devices. The common type of nozzle contains a whirl chamber and an orifice. The spray liquid is forced into the whirl chamber where it rotates, and issues from the nozzle orifice as a jet of fine droplets in the form of a hollow cone. The size of the droplets thus formed is related mainly to pressure and flow rate. In some types of airblast sprayers twin-fluid atomizers are used, where air is introduced under pressure and shatters the liquid into fine droplets. Fine spray generators consist of a rotating disk or cup fed centrally upon the surface with the spray liquid. A thin sheet of liquid is formed which breaks up into fine droplets. Their size range is related mainly to rotating speed and capacity.

The common sprayer is of the hydraulic type and consists of a pump forcing the liquid through the nozzle. In knapsack sprayers the pump is operated by hand producing a limited pressure. In machine-operated equipment, high pressure is employed enabling higher rates of work, and the jet of droplets may be launched a greater distance. The distance is critical when high trees are to be sprayed. Knapsack sprayers also include a pneumatic version where gas under pressure is used to force the liquid out of the tank into the nozzle.

Blower sprayers are devices by which the spray liquid is hydraulically atomized and then carried to the target by an air stream which passes the source of the spray. They can be classified into those having axial-flow blowers and those having centrifugal fans. A modification commonly used as shoulder mounted equipment is the mist blower, in which an air stream is used not only to carry the spray but also to atomize the spray liquid into fine droplets.

The various types of ULV sprayers have been reviewed [54, 57]. The ULV spraying can be accomplished with mist blowers by fitting special nozzles or adapting a "minispin" on the sprayer which is operated by the air blown from the sprayer. Both modifications produce very fine droplets, ranging from 40 to 100 μm. Special ULV ground sprayers produce very fine droplets by a spinning disk, often operated electrically (connected to batteries). In the most simple equipment no directional force is present, and the distribution of the spray takes place by means of prevailing winds, spraying being carried out from the upwind position in a wind velocity ranging from 1 to 5 m/sec. In spite of this drawback, this instrument is

popular because of its light weight. Improved types also include a fan
forming a directional air stream. The unit may be mounted on a skid to
which a small wheel has been attached. More elaborate equipment is
mounted on a tractor; in some a big blower produces the directing air
stream. Generally, ground-spraying equipment is still in the developmental
stage.

Aerial spraying obviously utilizes the lower range of low volume, 15-75
liters/ha and ULV, under 5 liters/ha. The principles of droplet generation
are similar to those used in ground equipment.

C. Application via Sprinklers

This experimental method has been tried in sprinkler-irrigated fields with
the purpose of saving the work of spraying. The chemical is introduced
into the irrigation system and sprinkled onto the foliage, and though thorough
coverage cannot be expected, redistribution might correct this. Some
wastage of fungicide which falls to the ground cannot be avoided. The treatment can be applied just at the end of an irrigation cycle, when ground
sprayers cannot be operated. Good control of potato late blight (Phytophthora
infestans) has been reported in Israel [64]. In Japan [19] the method has
proved effective against citrus melanose (Diaporthe citri).

D. Application via "Netting"

The "netting method" was experimentally devised in Japan for treatment of
fruit trees [17-19]. The idea is that a fungicide available at the top-of-the-
tree canopy will redistribute over the whole foliage under the influence of
rain. For placing fungicides over the canopy, the chemical is impregnated
in the fibres of a net, and the net is located over the top of the tree. Nets
made either of cotton-staple fiber yarn or of hemp string were found adequate [18]. The method was devised with the purpose of saving the cost of
the spraying operation.

E. Foliar Dusting

Dusting of a fungicide confers a thorough coverage of the foliage. However,
dusts are generally weak in tenacity and therefore are limited in usefulness.
The main fungicide broadly dusted is sulfur, due to its excellent control of
powdery mildews and other diseases, but with recent developments of other
fungicides very effective against mildew the use of sulfur has declined.

Dusting is based on the principle of emitting a blast of air in which the
dust particles are borne. In hand dusters the air is pumped by a plunger,

the compression stroke of the piston expelling a blast of dust-laden air. Knapsack dusters are operated by a bellows; a hand lever operates the bellows and agitates screens to allow the dust to flow into a hose and a pipe tipped with a sieve-type nozzle. Power dusters, driven by motors, may have a long boom with many delivery pipes. Sometimes a hood is attached to the boom to reduce the amount lost by drift. For orchards, manifolds with eight nozzles and strong blowers are used. Electrostatic dusters impart very high positive charges to particles to increase their attraction to the negatively charged leaves. Dusting 50 kg/ha of 70% sulfur by aircraft [65] effectively controlled powdery mildew of apples (Podosphaera leucotricha).

F. Application via Vaporization

This method is applicable only in enclosed structures. The fungicide is thermally vaporized, usually by an electric heating element, and the hot vapor condenses due to the thermal difference to form minute particles dispersed in the air. Finally the particles settle on plant surfaces. Labor saving is the primary merit of this technique over conventional application methods. Often a system of vaporizers is distributed in the greenhouse which can be operated by turning of a knob or even automatically. Usually the treatment is made at the end of the day when ventilation systems are closed.

This application method is suitable in principle for chemicals which are temperature stable. A fungicide suitable for vaporization is sulfur. It vaporizes quite rapidly at 112°C, and is used to control powdery mildew. Chlorothalonil is available as tablets for thermal vaporization. It is suitable for control of snapdragon rust (Puccinia antirrhini) at 2g/1000 ft^3 (7 g/ 100 m^3) [66], and is also effective for control of Botrytis.

G. Nonfoliar Application of Systemic Fungicides

Systemic fungicides intended for foliage may be applied to the soil or to the trunk, from which they are translocated into the foliage.

1. Soil Treatment

Various application methods have been used:

1. Mixing of dry powder or granules with soil, used mainly with cereals and vegetables [3, 67, 68].
2. Seed dressing, employing seed as carriers, used mainly with cereals [69].
3. Drench of solutions or suspensions of fungicides, either as a special treatment or via the irrigation system, used mainly in greenhouses [44, 70].
4. Injection of suspensions into the rhizosphere [71, 72].

Effectiveness of the treatment is determined by the physiochemical nature of the fungicide and its interaction with soil constituents which may influence movement in the soil and availability for root uptake. The application method must match the nature of the root system in the specific type of soil. Persistence of fungicides in soil is influenced by soil microorganisms [73]. The physiological activity of the plant, especially the transpiration rate, also plays an important role in the system.

2. Trunk Injection

In a recent paper, Shabi et al. [49] suggested that injection of systemic fungicides into the trunk of trees may lead to control of foliar diseases. The benzimidazole fungicides, initially accumulated in the trunk of pear trees, gradually translocated and accumulated in the leaves and protected them from scab infection (Venturia pirina).

Injection of fungicides into the trunk may be performed either by gravity flow or with the aid of pressure. With either method, holes are drilled into the trunk to a considerable depth. With the gravity-flow method [74], a tube is connected to a container of the fungicide liquid and a seal is made at the opening of the bore hole. The liquid then slowly diffuses into the xylem of the trunk. Insufficient control encountered with vascular diseases suggests that the fungicides did not reach desirable concentrations and distribution in the tree.

Injections aided by pressure forcing of the liquid into the trunk have several advantages. Larger doses of the fungicide can be injected in a shorter period and distributed more uniformly in the trunk. This modification enables the formation of desired concentrations in infection sites over all the tree's crown.

Various techniques to perform pressure injection, most of which utilize home-made equipment, have been described [48, 75-77]. The principle of all instruments is that the liquid is pressurized in a container connected to a tube sealed into the bore hole in the trunk. The Maujet Company has been supplying commercial units. Fungicides for injection must be solubilized in a suitable nonphytotoxic medium.

IV. TIMING OF FUNGICIDAL TREATMENT

A. Introduction

Time is an important parameter in the progress of many diseases after their initial outbreak. Primary infections produce lesions where the pathogen multiplies and gains inoculum to initiate secondary disease cycles. The shorter the life cycle, the more time is a critical factor in the development of an epidemic. With many diseases such as mildews or blights, lesion size increases with time, thus increasing the inoculum producing area and damage to the plant. Van der Plank [78] has developed a concept that control of plant diseases is part of the study of populations of pathogens, which is the subject of epidemiology. Fungicides control a disease by either reducing the initial population of the pathogen or by retarding its subsequent increase. For many foliar diseases, at any time during the course of an epidemic, the amount of disease is determined by the amount of initial inoculum and the subsequent rate of disease development. A schematic equation is then $X = X_0 \, e^{rt}$. The value X is the proportion of disease at any time, and this is determined by the initial inoculum X_0, the average infection rate r, and the time t during which infection has occurred. The equation applies only to the logarithmic phase of an epidemic, thus excluding the extreme phases when most of foliage is either healthy or infected, at the beginning or at the end of an epidemic. Most methods of control reduce X_0 or r or both. With this concept, optimal timing of fungicidal treatment is aimed to interfere with inoculum build up at certain phases in the progress of the disease.

In practice, a single fungicidal treatment would seldom eradicate inoculum since some fungal spores survive to initiate new disease cycles. When a source of inoculum is evident or expected, the extent to which a farmer would be interested in reducing it would generally be based on economic considerations. A disease does not cause economic losses until it passes a certain threshold. Hence another approach might be that only when this stage in the progress of disease is reached should chemical control measures be applied.

The onset and development of a disease caused by airborne pathogens depend largely on weather conditions during the growing season of the crop. Quite often treatments that are applied prophylactically or in response to initial disease outbreak prove later to be unnecessary owing to lack of disease progress. Development of epidemics of diseases incited by airborne propagules is highly dependent on environmental factors that affect spore formation, dissemination, and infection. It is an old wish of plant pathologists to forecast the necessity of fungicidal treatment according to weather factors and host proneness [79]. Such predictions are designed to warn growers of expected significant development in the plant disease picture. In particular, short-range disease forecasts issued at intervals during the growing season give guidance on the best seasonal tactics, e.g., on whether or not chemical control measures are necessary, on the optimum dates to apply such treatments, and on similar day-to-day decisions in the fight against diseases [80]. Disease forecasting is especially important for diseases that are hard to detect and develop rapidly. Success is quite rare because of varying combinations of climatic conditions and difficulties in precise monitoring of conditions in the field. This last drawback applies especially to dry climates, which may have a decisive effect on disease development. For several diseases in areas of temperate climate and where the epidemiology of certain diseases has been extensively studied, regional warning services are operated [79, 80]. Forecasting has been used for potato late blight (Phytophthora infestans); apple scab (Venturia inaequalis); downy mildews of vine (Plasmopara viticola), cucurbits (Pseudoperonospora cubensis), lima beans (Phytophthora phaseoli), and tobacco (Peronospora tabacina); rusts of wheat and barley; and rice blast (Pyricularia oryzae).

Timing of treatments must take into account the potency of the fungicide to be applied as well as the nature of the disease. Types of treatment may be classified according to the disease phase at the moment of application, as follows: (1) prophylactic treatment; (2) responsive treatment with protective fungicides; (3) therapeutic treatment; and (4) eradicative treatment.

B. Prophylactic Treatment

Fungicidal treatment applied to a crop before infection has occurred in the specific field or orchard is prophylactic treatment. Such treatment is a routine procedure in some crops grown in areas known to harbor the causal organism. The treatment is aimed at preventing primary infections which would incite an epidemic.

Rowell [81] suggested that treatment which would restrict the initiation of primary rust infection when inoculum arrives early in the season in the spring wheat areas of North Central states of the United States could also reduce the initial rate of increase of the rust after infection is established. Dressing seed with a systemic fungicide, such as oxycarboxin, would suppress primary infection in the early stage of crop growth and could have considerable value for the control of rust. Indeed, field trials [82, 83] revealed that for 2 months or more, wheat plants were protected against stripe and leaf and stem rusts (Puccinia striiformis, P. recondita, and P. graminis, respectively). Oxycarboxin applied as powder drilled between the sown rows proved to be less effective [3]. A similar goal was attained by a single spray of spring wheat with 4-n-butyl-1,2,4-triazole (RH-124), a fungicide specifically effective against leaf rust. A spray that was applied 21 days after planting, preceding primary infection, was superior to single treatments started when primary infections were apparent [84].

In Europe, rusts of grain cereals are less critical than in North America, while powdery mildew (Erysiphe graminis) is more significant. Prophylactic application of ethirimol to the soil, which serves as a reservoir for prolonged activity, is quite rewarding. For barley, different application methods were tried, including seed dressing, pelleted seed, liquid sprayed into the furrow, broadcast granules and combined drilling and placing of granules [85]. The most consistent success has been obtained with ethirimol used as a seed dressing. Almost complete control of barley mildew can be expected for at least 15 weeks. Yield responses vary much between experiments and years; on the average, yield increase ranged between 7 and 18%. On wheat, mildew control is much less effective, and yield increases have rarely exceeded 5%.

The key to successful control of apple scab (<u>Venturia inaequalis</u>) is to prevent primary infections [86]. One approach to achieve effective control is to keep the rapidly expanding organs covered with protective fungicide during the period of their highest susceptibility and of weather conditions generally favorable for infection. Protection may be achieved either by a series of sprays at intervals depending on growth rate, or by applying a single massive dose of captafol at bud opening, which assures protection of new growth for an extended period [86].

In Florida, melanose (<u>Phomopsis citri</u>) is assumed to attack fruit and foliage of citrus. Prophylactic treatment with a copper fungicide is recommended soon after the fruit is set in April, and under severe conditions a second application 3 or 4 weeks later [87].

Quite often prophylactic treatments by themselves are not sufficient to prevent disease outbreak, and they must be followed-up by responsive treatments. However, their role is to inhibit disease propgress. This is well known for late-blight of potato and tomato (<u>Phytophthora infestans</u>) which is one of the most destructive diseases known. Treatment should begin at least 10 days before blight usually attacks. Therefore a grower should know the blight date for his own district.

For the control of onion downy mildew (<u>Peronospora destructor</u>), two prophylactic and one responsive schedule of treatments with a maneb preparation, at 10-day intervals, were compared [88]. Large plots (250 m^2) were planted in early winter and disease outbreak was expected in February. The results (Table 3) show that both prophylactic schedules were superior to the responsive one in disease control during a part of the season, and this is reflected in a 20% increase in yield. In this case, a delay in the progress of the disease might well pay for one or two prophylactic sprays.

In citriculture, the most widespread disease is brown rot of fruits caused by species of <u>Phytophthora</u> which survive in the soil. In winter, zoosporangia formed on the surface of the ground are splashed by rain and contaminate lower fruits on the tree. After incubation, infected fruits fall off the tree and yield is lost. Furthermore, when fruits are picked and packed during the incubation period, brown rot develops during shipment and contaminates other fruits in the box. Usually one spray with Bordeaux mixture toward winter will protect the fruits during the picking months [89]. Only in a very rainy season might a second spray be needed. In Israel, where all citrus fruit is marketed by a Marketing Board, the growers are obliged by legislation to apply prophylactic treatment for brown rot.

C. Responsive Treatment with Protective Fungicides

Responsive treatment with a protective fungicide is applied when a certain level of disease has developed. The treatment is aimed at protecting the healthy foliage or fruit from infection, and thus prevents the initiated disease from developing into an epidemic. Often the responsive treatment is

TABLE 3

Effect of Maneb Treatments Applied at 10-day Interval on Infection
and Yield of Onions Naturally Infected with Downy Mildew
Treatments Were Started Prior to or After
Appearance of Symptoms

| Date of assessment | Percent of infected leaves after treatment | | Responsive |
| | Prophylactic | | |
	from 1 Feb.	from 10 Feb.	from 20 Feb.
17 Feb.	0	0	0
20 Feb.	0	0	0.1
26 Feb.	1.6	1.6	6.0
17 March	30.0	39.0	63.0
17 April	70.0	67.0	74.0
Yield kg/0.1 ha			
	2590	2650	2201

Source: Reprinted from Ref. 88, p. 20.

complementary to a prophylactic one when the latter fails to prevent primary infections. Responsive treatment is most commonly used in disease control practice. From the epidemiological point of view, responsive treatment is intended to check development of the inoculum as well as the disease, and to maintain them below a certain threshold which is specific for each case. The success of the treatment depends on the potency of a protectant fungicide to prevent infection under pressure of the available inoculum potential. Repetitions of treatment might be needed because application techniques are seldom perfect, and erosion of the protective cover as well as new growth of the plant expose tissues to infection.

For disease control in crops of relatively low value, any single treatment must be timed for maximum effect. Therefore, the number of responsive treatments and the need of prophylactic ones have to be evaluated. Chickpea (Cicer arietinum) is an ancient cultivated crop in Israel. Nowadays its cultivation is limited by anthracnose (Ascochyta rabiei) which attacks foliage and pods. Solel and Kostrinski [90] studied anthracnose control with a zineb preparation; the results are summarized in Table 4. A single responsive treatment applied shortly after the first symptoms were recorded was quite effective in controlling the disease and had a considerable effect on the yield. A further moderate improvement in both disease reduction and yield was

TABLE 4

Effect of Treatments with Zineb on Anthracnose Infection
and Grain Yield of Chick-Pea

Treatment	Infection grade[a]	Grain yield, kg/1000 m^2
Single prophylactic spray on 22 Jan.	3	390
Single responsive spray on 20 Feb.	2	820
Six responsive sprays at 10-days interval from 20 Feb.	1	1180
Control	3.5	320

[a]The infection was recorded on 18 March according to the severity of the
disease on a 0-5 scale.
Source: Reprinted from Ref. 90, p. 120.

obtained with a schedule of six sprays. However, the economic desirability
of this schedule is questionable. One prophylactic treatment, which hap-
pened to be applied 4 weeks before disease outbreak, had only a slight effect.

For the control of grape powdery mildew (Uncinula necator), responsive
or prophylactic biweekly treatments with sulfur dust were equally effective
[91]. Performance at different time intervals between applications was
clearly dependent on conditions for disease development. Biweekly treatments were highly effective even under conditions of severe infection, while
treatments at intervals of 19 days were satisfactory only where conditions
were not very favorable to the mildew.

Solel [92] studied timing programs and performance of three different
fungicides at two dosage levels for control of Cercospora leaf spot (Cercospora beticola) of sugarbeet. Infection of winter leaves of sugarbeet planted
in autumn has a negligible effect on the yield, but might play a major role
in development of disease epidemics. Since the season is rather long, a
schedule of treatments applied at 3-week intervals from early winter on was
compared to weekly and biweekly responsive treatments commenced after
the disease was established in the field. The results (Table 5) reveal that
efficacy of the schedule started early was good early in the season but
declined later. Weekly treatments that began later were the most satisfactory, but were also more expensive. The number of biweekly sprays was
close to that of triweekly sprays, but less effective. One might conclude
that early triweekly treatment up to May, followed by several weekly sprays

TABLE 5

Severity of Infection of Cercospora Leaf Spot[a] on Sugarbeet Treated with Fungicides at Different Timing Schedules[b]

Date of assessment and dosage, g/ha	Fixed copper, 16%			Copper oxinate, 20%			Zineb, 65%			Control
	Weekly	Biweekly	Triweekly	Weekly	Biweekly	Triweekly	Weekly	Biweekly	Triweekly	
1 March										2.2
3,000	1.6	2.5	0.2	1.8	1.9	0.2	1.8	2.1	0.5	
6,000	1.8	1.5	0.4	1.7	2.0	0.1	1.8	2.1	0.1	
24 May										2.7
3,000	1.7	2.4	1.4	0.8	1.2	1.2	0.7	1.6	1.0	
6,000	0.9	1.9	1.0	0.7	0.9	0.9	0.5	1.1	0.6	
1 June										3.7
3,000	3.0	3.6	3.4	1.4	3.0	2.5	0.9	2.7	2.3	
6,000	2.9	3.4	2.4	1.1	2.2	1.7	0.6	1.4	1.6	

[a]Severity of infection assessed on Kleinwanzlebener 0–5 scale.
[b]Seventeen weekly and nine biweekly sprays started on 12 February; ten triweekly sprays started on 18 December of the previous year.
Source: Reprinted from Ref. 92.

would be the optimal schedule. The difference in performance of the three fungicides was clear. Doubling the dose improved performance of all three fungicides, but not to a degree that would compensate for longer intervals between treatments. In a later study the newly developed fentin derivatives gave adequate disease control with only three triweekly applications started in the spring [93]. Stallknecht and Calpouzos [94] investigated the basis of superiority of fentin compounds over copper and carbamate fungicides. Three protective characteristics were tested: fungitoxicity, distribution on leaves, and tenacity under simulated rain. For all three parameters they found superiority of fentin hydroxide. In a later study, Solel [95] demonstrated that in addition to protective activity, fentin acetate possesses a moderate systemic activity, chiefly by translaminar action. Recently, benzimidazole fungicides were shown to be superior to the fentin derivatives [96].

In Tunisia, the limiting factor for development of disease caused by Leveillula taurica is spore formation. Timing for control has been based on periodic field observations to discover the first appearance of lesions on the lower side of leaves of selected plants. For artichokes, Laudanski et al. [97] recommend initiation of treatments when conidiophores have appeared on 4-5 lesions on the lower leaves of marked plants. On pepper, Guillerm [98] recommends the first treatments when formation of conidiophores is evident on three leaves per marked plant. These lesions are marked at each inspection and further treatments are applied as further lesions with conidiophores appear. This method of timing has made it possible to reduce the number of treatments applied to a crop from twelve to nine.

D. Therapeutic Treatment

Therapeutic treatment is applied after the pathogen has entered the host and established incipient infection. In this chapter, therapy refers to the use of chemicals and is synonymous to chemotherapy [50] as well as to curative action. Therapeutic treatment is generally scheduled intentionally when the chance of infection cannot be forecast and thus prophylactic actions may be saved. With therapeutic treatment, timing precision is critical and must match both the characteristics of the fungicide and the nature of the disease. This interaction may be exemplified by apple scab control.

In order for scab infection to occur, the apple leaves must be wet for a variable period of time, depending on the temperature and the kind of inoculum as is indicated by selected data presented in Table 6. Once infection has been established, no further wetting is required, and lesion appearance will depend solely on the temperature (Table 6). Some conventional fungicides are able to arrest the development of the lesions if applied soon enough after penetration of the fungus into the host. This after-infection therapy

TABLE 6

Approximate Hours of Wetting at Indicated Temperature Required for
Leaf Scab Infection, and Days Required for Lesions to Appear

| Average temperature °C | Hours wetting required | | Days required for lesion to appear |
	From primary inoculum (ascospores)	From secondary inoculum (conidia)	
6	25	16.5	
9	15	9.9	17
11	12	7.9	15
14	10	6.6	12
17	9	5.9	9
20	9	5.9	9

Source: Ref. 86, p. 10.

has been termed "kickback action." Properties of conventional fungicides recommended in New York State for this type of therapy are summarized in Table 7. Other fungicides tested in England showed significant therapeutic action when sprayed 42 hr after inoculation. These include phenylmercury chloride, dichlofluanid, and didecyldimethylammonium bromide [99]. To control apple scab, transcuticular movement of a fungicide would suffice for it to encounter the pathogen, whose habitat is confined to the subcuticular zone. Several of the fungicides mentioned were found to be capable of penetrating through cuticles of the apple leaf. These were captafol, captan, thiram, and phenylmercury monoethanol ammonium acetate [100].

In recent years true systemic fungicides such as the benzimidazole derivatives [101, 102] and triforine [103] have enhanced our capacity for therapeutic scab control. Extreme effectiveness was shown by triarimol, which, when applied to foliage up to 72 hr after inoculation, prevented any symptom development. Even when applied 96 hr after inoculation, lesions were smaller than normal and necrotic [104]. The actions of the fungicide were attributed to the arresting of development of the stromatic hyphae [105].

Mal secco disease of citrus (Phoma tracheiphila) often starts from infections in foliar wounds. Postinoculation therapy of leaves was attempted with systemic fungicides fortified with 2% spray oil to enhance activity [106]. Efficacy of benomyl, carboxin, and thiabendazole declined rapidly when the interval from inoculation to treatment increased from 2 to 4 and to 6 days.

Several systemic fungicides were compared for timing effect on therapeutic action against Cercospora leaf spot (Cercospora beticola) of

TABLE 7

Properties of Fungicides for Apple Scab Therapy

Fungicide	Amount per 100 gallons	Hours for after-infection therapy[a]
Captan 50% WP	2 lb	18–24
Captafol 4F	5 qt	Several days
Dichlone 50% WP	1/2 lb	36–48
Dichlone 50% WP	1/4 lb	30–36
Dodine 65% WP	3/8 lb	18–24
Ferbam 76% WP	2 lb	None
Glyodin 30% sol	2 pt	None
Maneb 80% WP	1-1/2 lb	18–24
Phybam–S	4 lb	24–36
Polyram 80% WP	2 lb	18–24
Sulfur	5 lb	None
Thiram 65% WP	2 lb	15–20

[a]The data are given for wet infection period at 50–60°F (10–15°C). The lower time period is for 60°F (15°C).

Source: Ref. 86, p. 10.

sugarbeet [107]. When applied 5 days after inoculation, benomyl and thiabendazole markedly reduced incidence of lesions that normally develop in a minimum of 10 days. Thiophanate and fentin acetate were less effective. The latter was ineffective when applied 6 or 7 days after inoculation, while thiabendazole at that stage reduced lesion incidence by two-thirds. Microscopic examination of the leaves showed that the percentage of penetrated germ tubes which colonized the mesophyll was significantly lower with thiabendazole than with fentin acetate, which differed only slightly from the control. Varietal susceptibility proved to have an impact on therapeutic performance. On a tolerant sugarbeet cultivar, fentin acetate as well as thiabendazole successfully controlled the disease even 5 days after inoculation [107].

Almost complete control of Cladosporium leaf mold (Cladosporium fulvum) of tomato was achieved by a foliar spray of benomyl applied 10 days after inoculation, which anticipated the end of incubation period by 6 days [108].

Triforine has shown therapeutic activity for the control of wheat leaf rust
[109]. While rust pustules normally appear 7-9 days after inoculation, post-
inoculation leaf treatment within 3 days after inoculation completely controls
leaf rust. Application one day later is still very effective, but a further
delay in application results in loss of effectiveness. Flax rust was effectively controlled in greenhouse tests by application of oxycarboxin 5 days
after inoculation [110].

E. Eradicative Treatment

The aim of eradicative treatments is to arrest the advance of a disease.
Eradication may either stop the development and expansion of an established
lesion in a disease which otherwise may indefinitely spread on the infected
organ, or it may act to destroy inoculum and thus preclude new infections.
   Potted cucumber plants, infected with powdery mildew (Sphaerotheca
fuliginea), were treated with the systemic fungicides benomyl, triarimol,
tridemorph, and dimethirimol by soil drench. The mildew was eradicated,
and microscopic examination revealed that hyphae, conidiophores, and
conidia had shriveled [111]. When tridemorph was applied to an established
infection of powdery mildew of barley (Erysiphe graminis), it rapidly killed
the mildew, and the crop sometimes regained its green color [85].
   A single spray intended to suppress sporulation in overwintering lesions
has been attempted successfully with several diseases, mainly of fruit crops.
For the control of brown rot (Sclerotinia laxa) of almonds and other stone
fruits, a dormant spray of monocalcium arsenite [52], sodium pentachlorophenate [53], or benomyl [112] is applied to twigs to reduce sporodochia
formation.
   Several approaches to control apple scab have already been discussed.
An additional one has been used for eradication of primary inoculum. Sodium arsenite was added to Bordeaux mixture and sprayed on apple leaves in
the late autumn to kill the pathogen (Venturia inaequalis) or to prevent formation of the ascigerous stage [113, 114]. Similar success has been
achieved with urea spray [115], and recently, with systemic fungicides [116,
117].
   Another version of this approach is the ground spraying of orchards
with sodium dinitro-o-cresylate to reduce sporulating inoculum by preventing
the maturation of ascospores and the discharge of those which have formed
at the time of application [114]. Likewise, benomyl spray on the ground in
the spring markedly reduced spore discharge [118].
   Various chemicals not in practical use exhibit antisporulant action in
vitro [119, 120]. Most fungicides if applied directly onto a foliar lesion
suppress sporulation on that area, but not on the opposite leaf surface.
Fentin acetate was effective as a "contact" antisporulant for Cercospora
beticola [107]. However, only systemic benzimidazole fungicides were

capable of antisporulant action on both surfaces of sugarbeet leaves when applied to one surface only. Systemic fungicides have been reported as antisporulants for various diseases [121, 122], but in practice this specific mode of action is obscured by the protective and therapeutic action of these fungicides. In Japan, Kasugamycin, a systemic antibiotic, has been used successfully for control of rice blast (Pyricularia oryzae). It has the eradicative action of preventing sporulation, as well the therapeutic action of limiting lesion enlargement. Interestingly, this fungicide has no protective action on the leaf surface [85].

F. Strategy of Selecting Timing Schedule

The key to profitable disease control is knowledge of the epidemiological features of the disease and its effect on yield. In cucumbers grown in summer, an effective control of powdery mildew (Sphaerotheca fuligenea) demands spraying at short-time intervals. Yet mildew covering up to 20% of foliage area does not reduce yield [30], and hence would not justify fungicidal treatments. It is desirable to know what increment of yield, either weight or grade, would offset the cost of a certain treatment.

Epidemiological study provides curves of disease dynamics during the season in relation to environmental conditions and susceptibility of the host cultivar. The farmer can then plan his strategy according to disease progress and treat the crop at the stage when he expects optimal results from each treatment, selecting the fungicide with the proper action. He might need to adjust his tactics according to developments over the season. Tactics should comply with weather conditions which affect both disease development and weathering of fungicides. For certain diseases a regional warning service promotes correct timing. When the time interval between treatments is affected mainly by the loss of the fungicide, monitoring the decay of the fungicide deposit might provide the best tool for precise timing [123]. The time interval between protective treatments has to coincide also with growth rate, since newly formed tissues are unprotected and are exposed to disease attack.

For high-value crops one might start with prophylactic treatments to prevent primary infections, and in the case of failure, follow up with responsive protective treatments to arrest secondary infections. If the farmer's arsenal includes effective systemic fungicides he might add this ammunition to the battle. The real challenge is to achieve economical control of diseases of relatively low value crops. Here the correct selection of fungicide, application method, and precise timing are indispensable. A single prophylactic treatment, if untimely applied, might miss the target with no chance for correction. Such treatment would be successful if preceded by a thorough study of dates of expected pathogen appearance. Preferably spore traps should be used. The fungicide applied at the correct time might inhibit build up of inoculum to a significant degree and thus prevent or delay disease progress. Often it would be safer to reserve this single treatment until a later stage, when primary infections had started. A systemic fungicide if available, combining protection with therapy and eradication, might be superior to a conventional one.

V. IDEAS FOR THE FUTURE

This chapter may be biased in favor of the systemic fungicides. Since it is not an historical review of the use of fungicides to combat foliage and fruit diseases, an effort was made to demonstrate principles. The space given to systemic fungicides is disproportionate to their practical use at present. One should remember that it is hardly a decade since the modern true systemic fungicides were introduced. Esteemed scientists such as Dimond and Horsfall [124] recognized the high potential of systemic fungicides, but only recently have really potent chemicals been developed. Systemic fungicides provide not only conventional surface protection, but also internal protection, therapy of infected organs and eradication of inoculum. It seems that these several modes of action are not yet fully harnessed by us to match their potential. If the fungicides are applied by conventional methods as foliar spray repeated at fixed time intervals, they act mainly as surface protectants, while therapy and eradication occur as side effects. When applied to foliage, the currently used systemic fungicides, which are apoplastically transported, do not move out of a treated leaf into newly formed foliage in any significant quantity. The apoplastic nature of fungicide transport might be advantageous if the compounds were used properly. Thus, if systemic fungicides enter the xylem system at the base of the plant they move via the transpiration stream to the transpiring foliage. When a reservoir is available, continuous translocation of the fungicide toward the leaves would occur over prolonged periods, and the chemical would act by internal protection, therapy, and eradication. It is true that fruits, which transpire very little, would not be protected. However, if the foliage is effectively protected, there would be no inoculum to threaten fruit. The production of fruit uncontaminated with fungicides is a desirable goal in itself.

The logical way of treatment with systemic fungicides would be to apply the toxicants to the soil for root uptake. Several successful examples were mentioned earlier. However, there are many failures indicating the complexity of this approach, and much research is still needed to advance economical control of foliage diseases via soil treatment. To escape from these difficulties the idea was introduced to inject the fungicide directly into the trunk of trees to form a reservoir, from which the fungicide would gradually migrate to foliage during the whole season [48, 49]. The balance between fast movement needed to protect actively growing trees, especially deciduous trees after bud burst, and a long-lasting effect, is quite delicate.

Although preliminary results are encouraging [49], the full potential of this approach needs to be explored.

Ultralow-volume (ULV) spraying and application via sprinkler irrigation were the two methods introduced to substitute for conventional spraying methods. The economic, operational and epidemiological advantages have already been mentioned. ULV spraying is widely adopted for insecticidal treatments, especially using aircraft sprayers, but rarely for fungicide application. While coverage might be more critical for fungicide performance than for other pesticides, the low toxicity and low phytotoxicity of most fungicides are of great advantage in ULV spraying. The fungicide industry has not developed suitable formulations of fungicides and only limited field research has been done. Both aspects should be further developed. The cost for fungicide application via sprinkler irrigation is very low, and it might prove particularly effective in cases where irrigation itself on the one hand stimulates disease development as an ecological factor [125], and on the other hand decreases protection by leaching of fungicide deposits. Thus the idea deserves a trial whenever sprinkler irrigation is used.

Though methodology of application is an important component of success, it is the fungicide that controls the disease. After the breakthrough in developing systemic fungicides, there is an urgent need for new systemic compounds to meet several demands. We might need to develop systemic fungicides and formulations to suit specific modes of application, as discussed previously.

The present commercial systemics do not cover the whole range of pathogens. They are completely inactive against Phycomycetes, and weak against some other groups like Porosporae, including important pathogens like Alternaria and Helminthosporium. We are threatened now by rapid development of resistance to our leading systemic fungicides, although this is not the rule. We may hope that new fungicides will not be eliminated by this hazard. Developments still to come are systemic fungicides which move in the symplast. Such compounds could be applied to foliage and migrate to growing tips to protect new foliage.

Obviously if new fungicides are introduced, some old ones will be rejected. It appears that one group is quite immune, namely the copper fungicides. These conventional fungicides, especially Bordeaux mixture, are distinguished by excellent protective performance primarily due to resistance to weathering and broad spectrum of activity. These features together with the relatively low price are good reason not to discard them from our arsenal.

REFERENCES

1. G. Ordish and D. Dufour, Ann. Rev. Phytopathol., 7, 31 (1969).
2. D. V. Singh, B. M. Khanna, and A. N. Khanna, Plant Disease Reptr., 53, 352 (1969).

3. W. A. F. Hagborg, Can. J. Plant Sci., 50, 631 (1970).
4. G. Elal and A. Dinoor, Hassadeh, 53, 885 (1973) (in Hebrew).
5. American Phytopathological Society, Comm. on Standardization of Fungicidal Tests, Phytopathology, 33, 627 (1943).
6. D. C. Torgeson, in Fungicides: An Advanced Treatise (D. C. Torgeson, ed.), Vol. 1, Academic Press, New York, 1967, pp. 93-123.
7. J. G. Horsfall, Principles of Fungicidal Action, Chronica Botanica, Waltham, Mass., 1956, p. 279.
8. J. G. Horsfall and N. Turner, Amer. Potato J., 20, 308 (1943).
9. D. Prusky, A. Dinoor, and Y. Eshel, Phytopathology, 64, 812 (1974).
10. S. Rich, Phytopathology, 44, 203 (1954).
11. E. Somers and W. D. E. Thomas, J. Sci. Food Agr., 7, 655 (1956).
12. E. C. Large, W. J. Beer, and J. B. E. Paterson, Ann. Appl. Biol., 33, 54 (1946).
13. R. J. Courshee, in Fungicides: An Advanced Treatise (D. C. Torgeson, ed.), Vol. 1, Academic Press, New York, 1967, pp. 239-286.
14. F. Wilcoxon and S. E. A. McCallan, Contrib. Boyce Thompson Inst., 3, 509 (1931).
15. H. P. Burchfield and G. L. McNew, Contrib. Boyce Thompson Inst., 16, 131 (1950).
16. E. F. Feichtmeir, Phytopathology, 39, 605 (1949).
17. S. Yamada, H. Tanaka, M. Koizumi, and S. Yamamoto, Bull. Hort. Res. Sta. Japan, Ser. B., 5, 75 (1966).
18. S. Yamada, H. Tanaka, Y. Homma, and M. Koizumi, Bull. Hort. Res. Sta. Japan, Ser. B., 5, 89 (1966).
19. S. Yamada, Japan. Agr. Res. Qu., 7, 98 (1973).
20. K. Björling and K. A. Selgren, Kgl. Landbruks-Hogskol. Ann., 23, 291 (1957); through Rev. Appl. Mycol., 37, 675 (1958).
21. B. M. Jones and M. Szkolnik, Phytopathology, 57, 341 (1967).
22. S. Pappo, I. Baumann, and Y. Oren, Hassadeh, 48, 45 (1968) (in Hebrew).
23. H. J. Miller, Phytopathology, 33, 899 (1943).
24. H. P. Burchfield and A. Goenaga, Contrib. Boyce Thompson Inst., 19, 141 (1957).
25. J. W. Heuberger and J. G. Horsfall, Phytopathology, 32, 370 (1942).
26. E. Somers, J. Sci. Food Agr., 7, 160 (1956).
27. R. W. Tatcher and L. R. Streeter, N.Y. State Agr. Expt. Sta. Tech. Bull. 116, Geneva, N.Y., 1925.
28. K. J. Bent, Ann. Appl. Biol., 60, 251 (1967).
29. D. R. Clifford, E. C. Hislop, and M. E. Holgate, Pestic. Sci., 1 18 (1970).
30. G. Elal, M.Sc. Thesis, Hebrew University of Jerusalem, Rehovot, 1973 (in Hebrew).
31. Z. Solel, Pestic. Sci., 2, 126 (1971).
32. E. C. Hislop, Ann. Appl. Biol., 60, 265 (1967).
33. C. E. Cox, H. D. Sisler, and R. A. Spurr, Science, 114, 643 (1951).

34. T. Noguchi, K. Ohkuma, and S. Kasaka, Second International IUPAC Congress of Pesticide Chemistry, Tel-Aviv, Israel, Vol. V, 1972, p. 263.

35. H. Buchenauer, L. V. Edgington, and F. Grossmann, Pestic. Sci., 4, 343 (1973).

36. C. A. Peterson and L. V. Edgington, Phytopathology, 61, 91 (1971).

37. P. M. Upham and C. J. Delp, Phytopathology, 63, 814 (1973).

38. Z. Solel, J. M. Schooley, and L. V. Edgington, Pestic. Sci., 4, 713 (1973).

39. H. Buchenauer and D. C. Erwin, Phytopathology, 61, 433 (1971).

40. M. Snel and L. V. Edgington, Phytopathology, 60, 1708 (1970).

41. A. Ben-Aziz and N. Aharonson, Pestic. Biochem. Physiol., 4, 120 (1974).

42. P. M. Smith and D. M. Spencer, Pestic. Sci., 2, 201 (1971).

43. M. R. Siegel and A. J. Zabbia, Phytopathology, 62, 630 (1972).

44. D. Netzer and I. Dishon, Phytoparasitica, 1, 33 (1973).

45. M. H. Ebben and D. M. Spencer, Proc. Brit. Insectic. Fungic. Conf., Brighton 1973, p. 211.

46. B. D. Cavell, R. J. Hemingway, and G. Teal, Proc. Brit. Insectic. Fungic. Conf., Brighton 1971, p. 431.

47. W. K. Hock and L. R. Schreiber, Plant Disease Reptr., 55, 58 (1971).

48. Y. Pinkas, E. Shabi, Z. Solel, and A. Cohen, Phytopathology, 63, 1166 (1973).

49. E. Shabi, Y. Pinkas, and Z. Solel, Phytopathology, 64, 963 (1974).

50. A. E. Dimond, in Plant Pathology, Problems and Progress 1908-1958 (C. S. Holton et al., eds.), The University of Wisconsin Press, Madison, 1959, pp. 211-228.

51. D. C. Erwin, J. J. Sims, and J. Partridge, Phytopathology, 58, 860 (1968).

52. E. E. Wilson, J. Agr. Res., 64, 561 (1942).

53. E. E. Wilson, Phytopathology, 40, 567 (1950).

54. W. Maas, ULV Application and Formulation Techniques, N.V. Philips-Duphar, Amsterdam, 1971, p. 64.

55. D. R. Johnstone, in Pesticide Formulations (W. Van Valkenburg, ed.), Marcel Dekker, New York, 1973, pp. 344-386.

56. E. Somers, J. Sci. Food Agr., 8, 520 (1957).

57. F. N. Matthee and A. C. Thomas, Decid. Fruit Grow., 24, 27 (1974).

58. H. Frankel, unpublished data, 1968.

59. R. H. Fulton, Ann. Rev. Phytopathol., 3, 175 (1965).

60. F. H. Lewis and K. D. Hickey, Ann. Rev. Phytopathol., 10, 399 (1972).

61. D. R. Johnstone, Pestic. Sci., 4, 77 (1973).

62. L. J. Klotz, E. C. Calavan, T. A. DeVolfe, J. E. Pehrson, M. P. Miller, and H. S. Elmer, Calif. Citograph, 56, 520 (1971).

63. S. Pappo, Y. Oren, I. Baumann, and N. Zamir, Hassadeh, 49, 370 (1969) (in Hebrew).

64. M. Merchav, M.Sc. Thesis, Hebrew University of Jerusalem, Rehovot, 1973 (in Hebrew).
65. S. Pappo and Z. Shoham, Hassadeh, 49, 620 (1969) (in Hebrew).
66. J. L. Peterson, C. B. Redington, and S. H. Davis, Jr., Plant Disease Reptr., 53, 886 (1969).
67. J. B. Rowell, Plant Disease Reptr., 51, 336 (1967).
68. K. R. W. Hammet, Plant Disease Reptr., 52, 754 (1968).
69. P. Leroux and M. Gred, Phytiatr.-Phytopharm., 21, 45 (1972).
70. E. W. Ryan, W. P. Staunton, and T. Kavanagh, Proc. Brit. Insectic. Fungic. Conf. Brighton 1971, p. 355.
71. R. J. Stipes, Gadners' Chron., 169, 20 (1971).
72. N. Aharonson and Z. Solel, Proc. 4th Israel Congress of Plant Pathology, Rehovot, 1975, p. 14 (in Hebrew).
73. M. R. Siegel, Phytopathology, 65, 219 (1975).
74. L. R. Schreiber, Plant Disease Reptr., 54, 240 (1970).
75. T. H. Filer, Jr., Plant Disease Reptr., 57, 338 (1973).
76. E. B. Smalley, C. J. Meyers, R. N. Johnson, B. C. Fluke, and R. Vieau, Phytopathology, 63, 1239 (1973).
77. R. Prasad and D. Travnick, Information Report CC-X-53, Chemical Control Research Institute, Ottawa, 1973, p. 28.
78. J. E. Van der Plank, Plant Diseases: Epidemics and Control, Academic Press, New York, 1963, p. 349.
79. P. R. Miller, in Plant Pathology, Problems and Progress, 1908-1958, (C. S. Holton et al., eds.), The University of Wisconsin Press, Madison, 1959, pp. 557-565.
80. P. M. A. Bourke, Ann. Rev. Phytopathol., 8, 345 (1970).
81. J. B. Rowell, Ann. Rev. Phytopathol., 6, 243 (1968).
82. R. L. Powelson and G. E. Shaner, Plant Disease Reptr., 50, 806 (1966).
83. J. B. Rowell, Plant Disease Reptr., 57, 567 (1973).
84. J. B. Rowell, Plant Disease Reptr., 57, 653 (1973).
85. D. H. Brooks, in Systemic Fungicides (R. W. Marsh, ed.), Wiley, New York, 1972, pp. 186-205.
86. G. H. Oberly, P. A. Arneson, and J. L. Brann, Jr., 1973 Tree-Fruit Production Recommendations, Cornell University, Ithaca, N.Y., 1973, p. 48.
87. R. M. Pratt, Florida Guide to Citrus Insects, Diseases and Nutritional Disorders in Color, Agricultural Experiment Station, Gainesville, Fla., 1958, p. 191.
88. S. Brosh and J. Palti, Onion Downy Mildew, Ministry of Agriculture, Tel Aviv, 1970, p. 23 (in Hebrew).
89. M. Schiffmann-Nadel and E. Cohen, Israel J. Agr. Res., 18, 4 (1968).
90. Z. Solel and J. Kostrinski, Phytopathol. Mediterranée, 3, 119 (1964).
91. I. Reichert, J. Palti, S. Moeller, and N. Hochberg, Agr. Res. Sta. Rehovot, Bull. 38, 1946, p. 4.

92. Z. Solel, unpublished data, 1961.

93. Z. Solel and G. Minz, Hassadeh, 43, 409 (1963) (in Hebrew).

94. G. F. Stallknecht and L. Calpouzos, Phytopathology, 58, 788 (1968).

95. Z. Solel, Phytopathology, 61, 738 (1971).

96. Z. Solel, J. Amer. Soc. Sugar Beet Technol., 16, 93 (1970).

97. F. Laudanski, S. Mehani, and R. Cassini, Rapp. trav. rech. effect. en 1956, Serv. bot. agron. de Tunis., 2, 213 (1957).

98. J. L. Guillerm, Rapp. trav. rech. effect. en 1956, Serv. bot. agron. de Tunis., 2, 193 (1957).

99. M. H. Moore, Ann. Appl. Biol., 57, 451 (1966).

100. Z. Solel and L. V. Edgington, Phytopathology, 63, 505 (1973).

101. C. J. Delp and H. L. Klopping, Plant Disease Reptr., 52, 95 (1968).

102. K. Ishii, Summaries of Papers, VIIth Intern. Congr. Plant Prot., Paris, 200 (1970).

103. K. G. Adlung and C. A. Drandarevski, Proc. Brit. Insectic. Fungic. Conf., Brighton 1971, p. 577.

104. I. F. Brown, Jr., H. R. Hall, and J. R. Miller, Phytopathology, 60, 1013 (1970).

105. I. F. Brown, Jr. and H. R. Hall, Phytopathology, 60, 1286 (1970).

106. Z. Solel, Y. Pinkas, and G. Loebenstein, Phytopathology, 62, 1007 (1972).

107. Z. Solel, Phytopathology, 60, 1186 (1970).

108. R. E. Pitblado and L. V. Edgington, Can. J. Plant Sci., 52, 459 (1972).

109. C. Ebenebe, H. Fehrmann, and F. Grossmann, Plant Disease Reptr., 55, 691 (1971).

110. G. F. Froiland and L. J. Littlefield, Plant Disease Reptr., 58, 737.

111. Y. Esheth and A. Dinoor, Hassadeh, 51, 111 (1971) (in Hebrew).

112. D. C. Ramsdell and J. M. Ogawa, Phytopathology, 63, 959 (1973).

113. G. W. Keitt and D. H. Palmiter, J. Agr. Res., 55, 397 (1937).

114. G. W. Keitt, C. N. Clayton, and M. H. Langford, Phytopathology, 31, 296 (1941).

115. R. T. Burchill, K. E. Hutton, J. E. Crosse, and C. M. E. Garett, Nature, 205, 520 (1965).

116. S. R. Connor and J. W. Heberger, Plant Disease Reptr., 52, 654 (1968).

117. D. L. McIntosh, Plant Disease Reptr., 53, 816 (1969).

118. P. M. Miller, Plant Disease Reptr., 54, 27 (1970).

119. J. G. Horsfall and R. J. Lukens, Bull. Conn. Agric. Expt. Sta. 694, New Haven, 1968.

120. B. Sneh, D. R. Clifford, and A. T. K. Corke, Pestic. Sci., 3, 433 (1972).

121. J. O. Whiteside, Plant Disease Reptr., 54, 865 (1970).

122. J. R. Hardison, Phytopathology, 62, 609 (1972).

123. C. H. Blazquez, Phytopathology, 61, 885 (1971).
124. A. E. Dimond and J. G. Horsfall, Ann. Rev. Plant Physiol., 10, 257 (1959).
125. J. Rotem and J. Palti, Ann. Rev. Phytopathol., 7, 267 (1969).

Chapter 9

CONTROL OF POSTHARVEST DISEASES

Joseph W. Eckert

Department of Plant Pathology
University of California
Riverside, California

I. INTRODUCTION

Mature fruits and vegetables are susceptible to attack by a variety of microorganisms to which they were resistant during the period of their development on the plant. Ripening plant products should not, however, be regarded
merely as a culture medium for the growth of a wide array of saprophytic
microorganisms. Each variety of fruit and vegetable may be attacked by
only a relatively small and unique group of parasitic fungi, and possibly
bacteria, that have nutritional requirements and enzymatic capabilities
which permit them to develop extensively in the tissues of their host.
Penicillium digitatum causes postharvest disease of citrus fruits only,
whereas P. expansum is a serious pathogen of apples and pears, but not of
citrus fruits. Monilinia fructicola incites brown rot of peaches, cherries,
apples, and pears but does not produce a disease in tropical fruits. On the
other hand, Rhizopus stolonifer, Botrytis cinerea, Alternaria tenuis, and
Geotrichum candidum cause diseases of a wide variety of fruits and vegetables. Bacterial diseases induced by species of Erwinia are common on
vegetables but rare on fruit crops, with the notable exceptions of tomatoes,
peppers, cucumbers, and pears.
   The "soft rots" of fruits and vegetables caused by Rhizopus, Geotrichum,
Sclerotinia, and Erwinia are a serious group of postharvest diseases because
they progress rapidly under optimum environmental conditions and the pathogens develop extracellular enzymes which macerate the fleshy tissues of the
host into an incoherent, watery mass within a few days. In addition, these
pathogens may spread by contact to adjacent units creating "pockets" of
decayed produce. The "brown rots," exemplified by Monilinia fructicola on
stone fruits, species of Gloeosporium on apples, and the stem-end rots of
citrus fruits incited by Diplodia natalensis, Phomopsis citri, and Alternaria
citri, may become a serious problem when a substantial portion of the crop
is infected at the time of harvest, but usually the diseased fruit remain firm
and the pathogen does not spread readily during storage and shipment.
Sometimes, microorganisms may adversely affect the appearance of fruits
and vegetables without actually invading the edible portion of the product.

"Soilage" of citrus fruits by spores of <u>Penicillium digitatum</u> and <u>P. italicum</u>, moldy carpels of dessert apples, and mold on the surface of melons are examples of conditions which devaluate the product without influencing its palatability. The major postharvest diseases of fruits and vegetables and their causal microorganisms are listed in Table 1. The comprehensive series of monographs on postharvest diseases of crop groups published by the U.S. Department of Agriculture should be consulted for detailed information on the symptoms and development of specific diseases [1-8].

TABLE 1

Major Postharvest Diseases of Fresh Fruits and Vegetables

Crop	Disease	Pathogens
Apple and pear	Lenticel rot	Cryptosporiopsis malicorticis (Condl.) Nannf. (= Gloeosporium perennans Zeller and Childs)
		Phlyctaena vagabunda Desm. (= Gloeosporium album Osterw.)
	Eye rot	Nectria galligena Bres.
	Blue mold rot	Penicillium expansum Thom
	Gray mold rot	Botrytis cinerea Pers. ex Fr.
Banana	Crown rot	Colletotrichum musae (Berk. and Curt.) Arx (= Gloeosporium musarum Cke. and Mass.)
		Fusarium roseum Link emend. Snyd. and Hans.
		Verticillium theobromae (Turc.) Hughes
		Ceratocystis paradoxa (Dade) Moreau (= Thielaviopsis paradoxa [de Seynes] Höhn.)
	Anthracnose	Colletotrichum musae (Berk. and Curt) Arx (= Gloeosporium musarum Cke. and Mass.)
Carrot	Gray mold rot	Botrytis cinerea Pers. ex Fr.
	Centrospora rot	Centrospora acerina (Hartig) Newhall

TABLE 1 (Cont.)

Crop	Disease	Pathogens
Carrot	Watery soft rot	Sclerotinia sclerotiorum (Lib.) de Bary
Citrus fruits	Stem-end rot	Phomopsis citri Fawc.
		Diplodia natalensis P. Evans
		Alternaria citri Ell. and Pierce
	Green mold rot	Penicillium digitatum Sacc.
	Blue mold rot	Penicillium italicum Wehmer
Grape	Gray mold rot	Botrytis cinerea Pers. ex Fr.
Papaya	Anthracnose	Colletotrichum gloeosporioides (Penz.) Sacc.
Peach and cherry	Brown rot	Monilinia fructicola (Wint.) Honey (= Sclerotinia fructicola [Wint.] Rehm)
	Rhizopus rot	Rhizopus stolonifer (Ehr. ex Fr.) Vuill.
Pineapple	Black rot	Ceratocystis paradoxa (Dade) Moreau (= Thielaviopsis paradoxa [de Seynes] Höhn.)
Potato	Bacterial soft rot	Erwinia carotovora (Jones) Holland and other species
	Dry rot	Fusarium spp.
	Gangrene	Phoma exigua var. foveata (Foister) Boerema
	Skin spot	Oospora pustulans Owen and Wakef.
	Silver scurf	Helminthosporium solani Dur. and Mont.
Strawberry	Gray mold rot	Botrytis cinerea Pers. ex Fr.
	Rhizopus rot	Rhizopus stolonifer (Ehr. ex Fr.) Vuill.
Sweet potato	Black rot	Ceratocystis fimbriata Ellis and Halst. [= Endoconidiophora fimbriata (Ell. and Halst.) Davidson]

TABLE 1 (Cont.)

Crop	Disease	Pathogens
Sweet potato	Rhizopus soft rot	Rhizopus stolonifer (Ehr. ex Fr.) Vuill.
Leafy vegetables	Bacterial soft rot	Erwinia carotovora (Jones) Holland and other species
	Gray mold rot	Botrytis cinerea Pers. ex Fr.
	Watery soft rot	Sclerotinia sclerotiorum (Lib.) de Bary

Fresh fruits and vegetables are often produced in areas distant from population centers and sometimes they mature at a time of the year when consumer demand for the product is weak. These circumstances may necessitate a period of several weeks or months for storage and shipment before the product reaches the consumer. Substantial decay losses may occur during this period if the product is not treated with an effective inhibitor of microbial growth and stored in an environment unfavorable to disease development. Surveys in tropical countries have revealed that postharvest losses of 25-50% are commonplace in situations where refrigerated facilities are not available and suitable chemical treatments are not utilized [9, 10]. The value of postharvest treatments in technologically advanced areas is more difficult to assess since statistics on postharvest disease losses are often compiled from shipments of produce which were treated with fungicides or bactericides after harvest. Decay losses in the "controls" of large-scale trials and in commercial shipments of untreated produce are given in Table 2 to provide an estimate of the losses which could be anticipated in certain crops if effective postharvest treatments were not utilized. Statistics based on the loss of individual commodity units do not reveal the entire economic impact of postharvest diseases. Important aspects of the problem which are ignored by this treatment are: (1) the loss of a large investment in harvesting, packaging, storage, and transportation, the combined costs of which may be several-fold greater than the total value of the raw product in the field; (2) the partial or total loss of consumer packages of produce which contain only one or several diseased units; and (3) the accelerated or uncontrolled ripening of fruit in a storage or shipping compartment due to the ethylene evolved from diseased fruits. Losses from all sources combined may result in economic failure of an agricultural enterprise despite months of the best agronomic practices in crop production. The utilization of postharvest fungicide treatments is essential to world trade in citrus fruits, bananas, and grapes and the potential

TABLE 2

Postharvest Losses of Fruits and Vegetables
due to Pathological Diseases

Commodity	Origin	Loss, %	Reference
Storage[a]			
Apples, McIntosh	Nova Scotia	2–7	11
Apples, Cox's Orange Pippin	England	20–50	12
Carrots	England	38	13
Carrots	New Jersey	6–12	14
Lemons	California	3–6	15
Lemons	Florida	30–35	16
Oranges, Sanguinello	Italy	52	17
Marketing[b]			
Apples	New York	29	18
Lemons	California	4–13	19
Oranges	Florida	0–25	20
Lettuce	California	10–15	21
Peaches	California	15–24	22
Strawberries	California	25–35	23

[a]Loss during low-temperature storage near the production area.
[b]Loss during transport and marketing.

value of these treatments for other crops as well has been gaining recognition in recent years.

This chapter will emphasize the principles of postharvest disease control and strategies which have been developed to reach this goal. The discussion of specific fungicides and bactericides will be limited for the most part to compounds in current practical use, those which show considerable promise at this time, and chemical treatments that illustrate a principle or a unique approach to the control of postharvest disease. Other chemical treatments that have been evaluated or used in earlier times have been discussed in previous reviews of this subject [24–26].

II. DEVELOPMENT OF POSTHARVEST DISEASES

Knowledge of the time and mechanism of infection is essential to the development of an effective program for the control of a postharvest disease. The eradication of a wound infection becomes difficult, if not impossible, after several days of incubation, whereas a latent infection arising in the field may be treatable with a systemic fungicide several months later. Postharvest diseases may be initiated by two distinct types of infection: (1) the active infection which progresses continuously as long as the environment is favorable for the growth of the pathogen, and (2) the quiescent infection which is initiated before maturity of the host, but which becomes active only after harvest when the fruit begins to ripen. The phenomenon of quiescent or latent infections of both fungal and bacterial pathogens has been reviewed recently [27, 28].

A. Infection Before Harvest

1. Direct Penetration

Conidia of Botrytis cinerea are abundant in the atmosphere of strawberry and raspberry plantations during flowering in the spring. These spores may germinate in a water film on the petals and other parts of the flower. The pathogen then moves from the diseased flower parts into the proximal end of the receptacle and therein forms a latent infection [29-32]. This infection becomes the site of disease development after the berries are harvested several months later. Similarly, spores of Colletotrichum gloeosporioides germinate in moisture on the surface of avocado, banana, citrus, mango, and papaya fruits at any time during the development of the fruit on the plant [33-39]. Appressoria begin to form at the end of the germ tubes within several hours after spore germination. These appressoria are considerably more resistant to an unfavorable environment than spores and, in addition, they adhere strongly to the surface of the fruit for mechanical support during the infection process [40, 41]. On bananas and mangoes a slender infection hypha grows from the bottom of some of the appressoria, pierces the cuticle and outer wall of an epidermal cell, and forms a small mass of hyphae. Further development of the pathogen is prevented by the resistance of the immature host tissue [35, 41, 42]. These "latent" infections remain viable for months and may give rise to disease after harvest when the fruit ripens. In citrus, avocado, and papaya fruits, latent infections are not frequently observed on immature fruit, but the appressorium may function as the quiescent stage, giving rise to an infection hypha after the fruit have been harvested [37, 39, 43].

Stem-end rots of citrus fruits caused by Diplodia and Phomopsis are the major postharvest diseases of this crop produced in humid, subtropical areas. These diseases arise from quiescent infections in the stem button (calyx plus disc) of the fruit. These infections are initiated at any stage of development of the fruit when rainfall is adequate for spore dispersal and germination [44, 45]. Species of Alternaria may incite a serious stem-end rot of citrus fruit grown in arid as well as humid climates, since the conidia of this pathogen are airborne and not dependent upon rainfall for dispersal [46-48]. Quiescent infections of Diplodia, Phomopsis, and Alternaria in the button of the fruit do not become active until this organ becomes senescent and begins to separate from the fruit.

Postharvest diseases of several fruits arise from infections initiated late in the growing season, if moisture conditions are favorable to the infection process. Gray mold (Botrytis cinerea), which is the most serious disease of grapes during low-temperature storage, originates from late season infections in the vineyards [1, 49]. In Australia, quiescent infections of Monilinia fructicola established in peaches and nectarines before maturity are regarded to be a significant factor in the control of this disease after harvest [50-52]. Brown rot of citrus fruits, which develops in storage after harvest, arises from incipient infections initiated by zoospores of species of Phytophthora which splash onto low-hanging fruit several days before harvest [53]. Fruits infected by the fungus earlier in the season usually fall from the tree or show obvious symptoms and are discarded at the time of harvest. Harvested carrot roots may carry incipient infections of Botrytis cinerea which arise from the crown of the plant in the field [54]. These infections may cause serious decay losses of the crop during low-temperature storage. Incipient infections of Sclerotinia sclerotiorum on snap beans at the time of harvest are also an important factor in the development of postharvest decay of this crop [55].

2. Infection Through Lenticels

Lenticel rotting of apples, a major storage disease of fruit grown in Europe and other humid areas, arises from latent infections of Phlyctaena vaga-bunda (Gloeosporium album) and Cryptosporiopsis curvispora (G. perennans) which develop in the lenticels of the fruit during periods of relatively high temperature and humidity late in the summer [56-61]. The severity of lenticel rotting during storage of apples in Nova Scotia has been correlated with rainfall during the three months before harvest [11]. The pathogenic fungi cannot develop further until the fruit reaches a certain degree of ripeness in storage.

The lenticels of potato tubers are a common site of initiation of bacterial soft rot after harvest. The lenticels of most tubers are contaminated with cells of the bacterium Erwinia carotovora at the time of harvest [62-65], but the bacteria remain quiescent in the lenticels until the tubers are subjected to environmental conditions, such as free moisture and low oxygen, which increase their susceptibility to decay [66-69]. Evidently, free soil water causes proliferation of the filling cells of the lenticels, thereby increasing their susceptibility to infection and also transports the bacterial cells into the lenticels [70-72]. The surface of potato tubers are contaminated with virulent strains of E. carotovora which may be moved into the lenticels during the washing operation after harvest [65].

3. Control of Preharvest Infections

Infections that are well established at the time of harvest are difficult to eradicate with conventional fungicides or bactericides applied after harvest because most of these compounds do not penetrate into the host tissue efficiently. Postharvest treatment with heat, irradiation, and systemic fungicides has proved effective in some instances, but the principal approach to the control of preharvest infections is by the application of protectant or eradicant fungicides to the crop in the field.

Application of a fungicide spray at intervals of 7 to 14 days during the period of flowering has given good control of Botrytis rot of strawberries and raspberries after harvest. More frequent treatment may result in even better disease control by providing better fungicide coverage on the internal parts of the flowers which open between spray applications [31]. Field sprays of thiram and captan have given good control of Botrytis rot of strawberries after harvest [32, 73], but chlorothalonil, dichlofluanid, and benomyl are the most effective materials currently available for this application [31, 74-77].

Several tropical fruits are susceptible to infection by Colletotrichum gloeosporioides at all stages of their development and, therefore, it is essential that these fruits be protected continuously with a fungicide deposit from the time of fruit set until harvest. Postharvest decay (anthracnose) of Florida mangoes by Colletotrichum may be controlled by weekly applications of an organic fungicide during bloom, followed by a monthly application of a copper fungicide throughout the remainder of the season [78]. More recently, the systemic fungicide benomyl applied every 14 days has provided excellent control of anthracnose on harvested fruit, probably because this compound can eradicate latent infections as well as protect against new infections. Equally effective control of anthracnose on bananas in the Windward Islands has been obtained by spraying of benomyl in the plantation [34]. In Hawaii, the incidence of anthracnose on papayas after harvest was reduced substantially by application of maneb every 10 days during the growing season [79]. This treatment also reduced the development of stem-end rots after harvest [80].

Lenticel rotting of apples (species of Gloeosporium) during storage has been controlled in England and France by late summer sprays of captan, benomyl and thiabendazole which protect against or eradicate lenticel infection at this time of the year [12, 81, 82]. Benomyl sprays applied to oranges more than one month before harvest prevent postharvest development of stem-end rots arising from quiescent infections of Diplodia and Phomopsis in the button of the fruit [83-85]. Brown rot of peaches after harvest caused by <u>Monilinia fructicola</u> has been controlled by orchard sprays of benomyl applied 2 to 3 weeks before harvest [51, 86], a treatment which eradicates latent infections.

Before the early 1960s, the control of latent infections in fruits was limited to providing a protective coating on the fruit during the period when infection was likely to occur. In recent years, it has been shown that systemic fungicides such as benomyl and thiabendazole can eradicate latent infections as well as protect against their establishment [34, 81, 83, 84, 86-88]. The control of postharvest diseases is improved considerably by the application of preharvest fungicides to reduce the number of latent infections since postharvest treatments are most effective if the level of infection at harvest is minimal [87]. However, caution should be exercised in choosing a preharvest fungicide in order to avoid the selection of strains of the pathogen which will be resistant to the postharvest treatment. For example, the use of benomyl both pre- and postharvest would be unwise since field sprays of benomyl may lead to the development of resistant strains of postharvest pathogens [34, 75, 77, 89-92].

B. Infection During and After Harvest

Some of the most devastating postharvest diseases arise from infections initiated in mechanical and physiological injuries on the surface of the fruit [93]. A certain degree of random mechanical injury is inevitable in the course of harvesting, processing, and packaging fruits and vegetables, even when these operations are carried out with the utmost care. Mechanical harvesting results in a far greater number of injuries than hand harvesting [13, 94, 95]. It is almost impossible to avoid surface skinning of some varieties of Irish potato and sweet potato during harvesting and handling, due to the immature cork layer and active cambium of the tubers at this time of their development [69, 96].

The injury created by severing the product from the plant is a frequent point of initiation of postharvest diseases by wound pathogens. Familiar examples of diseases which develop in this fashion are crown rot of banana [97, 98], pedicel rot of pineapple [99, 100], and stem-end rots of mango [101], papaya [80], avocado [102], green pepper [103], and pear [4]. Excessive pressure on the surface of apples and potatoes damages the lenticels and predisposes them to infection by pathogenic microorganisms. Bruising usually results in the development of <u>Penicillium</u> in lenticels of apples [104, 105] and also induces the early development of species of <u>Gloeosporium</u> in lenticels bearing latent infections of this fungus [56, 58, 60]. Bruising the surface of bananas stimulates the development of latent infections of <u>Colletotrichum.</u> Finger-stem rot results from twisting the individual banana fingers on the hand and a lenticular rot develops at points where the ridges of the fingers have rubbed against the fiberboard shipping carton [36, 97].

Physiological injuries caused by cold, heat, oxygen deficiency, and other environmental agents predispose fresh agricultural products to postharvest diseases [93]. Fruits and vegetables, particularly those of tropical origin, may be injured by storage at temperatures much below 10°C. This injury frequently results in an increase in postharvest diseases, even though the symptoms of chilling injury are not apparent. Storage at chilling temperatures has been reported to increase stem-end rot of grapefruit [106], Alternaria black rot and bacterial soft rot of tomatoes [107, 108], surface rot of yams [109], and Nigrospora rot of bananas [110]. Low temperatures may also increase decays of white potatoes and sweet potatoes by inhibiting periderm formation [111].

Moderately high temperatures also predispose fresh produce to decay. Exposure of harvested potatoes, yams, and sweet potatoes to direct sunlight for more than a few hours is known to increase the severity of storage rots [112, 113]. It was demonstrated that soaking lemons in water at 48°C for 4 minutes, a recommended treatment for eradication of incipient infections of Phytophthora, predisposed some lots of lemons to excessive decay by Penicillium and other fungi during storage without producing symptoms of heat damage on the fruit [15]. Tomatoes have been predisposed to Alternaria rot by a similar hot water treatment [114].

Poor ventilation, especially combined with surface water, has been recognized as a major factor in the development of bacterial soft rot of potatoes [64-66, 115]. An inadequate supply of oxygen predisposes potatoes to bacterial soft rot by inhibiting the development of periderm over injuries and by causing a leakage of water and solutes from turgid cells into the intercellular spaces around the lenticels [67].

Propagules of pathogenic fungi and bacteria are abundant in the atmosphere and on the surface of fruits and vegetables as they approach maturity in the field [29, 38, 116, 117]. Many of these pathogens, e.g., Penicillium, Rhizopus, Geotrichum, and Thielaviopsis, are not capable of penetrating the surface of the host, but if they gain entry through injuries or natural openings they may cause devastating rots of mature produce. Other pathogens such as Monilinia fructicola and Colletotrichum gloeosporioides may penetrate directly through the epidermis of peaches and bananas, respectively, but mechanical injuries are also important in the practical development of these diseases [2, 36, 97]. In addition to preharvest contamination with pathogenic microorganisms, harvested produce may become infested by contact with decayed fruits or contaminated equipment in the packinghouse. The atmosphere of most packinghouses is heavily laden with the spores of pathogenic fungi and fruit may be contaminated by airborne inoculum subsequent to disinfection by chemicals [118].

Water used to clean, transport, cool, or chemically treat produce after harvest is an important source of disease inoculum. Dumping of potatoes, citrus fruits, bananas, and apples into water can lead to inoculation of injuries and lenticels unless appropriate measures are taken to maintain the water in a sanitary condition [65, 98, 119, 120]. Immersion of apples after harvest in suspensions of diphenylamine or ethoxyquin for control of scald during storage has been implicated as a factor in increasing decay of these fruits [121-124].

Most of the usual treatments and practices for controlling postharvest decay are aimed at preventing infection at injuries created during harvesting and handling of the crop. The greatest degree of success has been achieved in the treatment of mechanical injuries with fungicides. Infections originating in bruises, lenticels, and from latent infections are much more difficult to control by postharvest treatment.

III.   STRATEGY OF POSTHARVEST DISEASE CONTROL

The use of fungicide sprays in the field to control latent infections initiated before harvest was considered in Section II.A.3. The remainder of this chapter will emphasize treatments applied after harvest for the purpose of reducing postharvest disease losses.

From a strategical standpoint, postharvest disease losses may be considered to arise from two sources, which although related, can be treated separately. The most obvious consequence of postharvest disease is that it makes a single unit of produce inedible. The second source of loss arises from the fact that an entire package of produce may be devalued, or even discarded, because of unsightly superficial contamination, even though the palatability of the individual units is not affected. Diseases caused by fungi which spread from fruit to fruit, such as Rhizopus or Botrytis, or which sporulate heavily on decayed fruit, such as Penicillium, fall into this latter category. In addition, some fungi grow on the surface of melons, citrus fruit, and apple sepals, thereby adversely affecting the appearance of these products without actually reducing their palatability or nutritive value.

A.   Prevention of Infection

Many postharvest diseases are initiated by the invasion of pathogens through injuries in the host surface which arise at the time of harvest and during handling of the crop. Incidence of such diseases can be decreased by substantially reducing the number of injuries or the number of pathogen propagules on the surface of the host. Recommended measures for reducing mechanical injuries during harvesting and handling are discussed in detail in the U.S. Department of Agriculture publications on postharvest diseases of specific crops [1-8]. This subject is beyond the scope of this chapter. Another approach to reducing surface injury is to retard maturation of the surface tissues of the product or to make the surface slightly flaccid by loss

of small amounts of water. Preharvest treatment of citrus fruits with gibberellin retards senescence of the peel and reduces mechanical injury and
decay [125]. "Wilting" citrus fruits before handling in the packinghouse
has been a standard practice in California for years.

1. Chemical Treatments to Reduce the Level of Inoculum on the Host
    and in the Environment

The principal sources of inoculum for contamination of fruits and vegetables
after harvest are (1) containers used to transport or store the harvested
product; (2) water used to clean, cool, transport, or apply a chemical treatment after harvest; (3) the atmosphere of the packinghouse; and (4) brushes
and conveyor belts which come in contact with the product as it flows through
the packinghouse. Recycled fruit bins or boxes may become heavily contaminated with pathogenic microorganisms, especially if the containers are
used for storage or for holding the product under circumstances which may
result in a substantial decay. Examples are boxes used for storage of
lemons at 15°C or for degreening of oranges at 22°C. Formaldehyde and
nitrogen trichloride have been used to disinfest lemon storage boxes and
sulfur dioxide has been recommended for picking boxes for figs [126]. These
three treatments have the disadvantage that they are corrosive to the box
materials. Lemon boxes are currently surface sterilized by deluging them
with a hot borax solution before the boxes are refilled with a new lot of
lemons for storage (see Section V.A).
    Fruits and vegetables are frequently cleaned, cooled, and conveyed in
water as well as treated with aqueous formulations of chemicals for purposes other than disease control. Familiar examples are cleaning citrus
fruits, apples, and potatoes; removing latex from the cut stems of bananas
and mangoes; hydrocooling peaches, tomatoes, and peppers; and applying
antiscald treatments (diphenylamine or ethoxyquin) to apples and pears before storage. In many instances, the water is recycled for several days
and may become heavily contaminated with propagules of pathogenic fungi
and bacteria unless an effective concentration of a broad-spectrum antimicrobial agent is maintained in the water. Fungi and bacteria in untreated
water readily infect fresh produce through lenticels, stomates, and injuries
created during the harvesting and handling operations [65, 98].
    Hypochlorous acid, added to water as chlorine gas or a hypochlorite
salt, rapidly kills microbial cells in the process water and reduces the hazard of inoculating the product with pathogenic microorganisms during operations involving recycled water [40, 119, 127-129]. Sodium o-phenylphenate
has been used occasionally to reduce the number of pathogenic microorganisms in water used for produce treatment (see Section V.F.1.). Some
possible advantages of sodium o-phenylphenate over hypochlorite for special
applications are noncorrosiveness, stability of the solution, and compatibility with chemicals which react with chlorine. Sodium o-phenylphenate
(0.1% pH 11.5) has been added to hydrocooler water to prevent inoculation

of peaches with brown rot [130] and 100 ppm has been added to solutions of sec-butylammonium phosphate (2-aminobutane) used to disinfect oranges before ethylene degreening (see Section V.D.2).

The population of pathogenic fungi on the surface of citrus fruits and peaches arriving from the orchard may be greatly reduced by surface treatment with a solution of hypochlorite or sodium o-phenylphenate [118, 119, 129]. In California, lemons are brushed in a solution of sodium o-phenylphenate after storage to remove spores and other decay debris from the fruit surface [131].

Washing with water alone has reduced the incidence of Fusarium dry rot of potatoes [111, 132], anthracnose and stem-end rot of oranges [37, 133], and decay of carrots in long-term storage [13]. Water alone may physically remove propagules of pathogenic microorganisms from the surface of the produce.

The atmosphere of citrus packinghouses, especially lemon houses, is heavily contaminated with spores of Penicillium digitatum and P. italicum. In recent years it has become common practice to fumigate the packinghouses each night by atomizing a solution of 1-3% formaldehyde into the atmosphere. The conveyors and waxer brushes of packinghouses for tomatoes and peppers may become contaminated with soft-rot bacteria and pathogenic fungi within a few hours after the operation commences and serve as an efficient inoculator for all produce which then passes across the equipment [103]. Fruit conveying and treating equipment should be disinfested regularly with quaternary ammonium compounds, hypochlorite, sodium o-phenylphenate, or formaldehyde.

2. Chemical Treatments to Combat Infection after Harvest

Two major strategies have been developed for the treatment of produce to combat infection by pathogens: (1) cover the surface of the product with a fungicide deposit before harvest with the expectation that superficial injuries created during or subsequent to harvest will contain an inhibitory concentration of the fungicide; and (2) treat the product with a fungicide after harvest with the intention of depositing an inhibitory quantity of the fungicide in those superficial injuries which were formed during harvesting and handling of the crop.

a. Preharvest Protectant Treatment. Several investigators have demonstrated that Rhizopus rot of peaches after harvest may be substantially reduced by spraying the fruit 1 week before harvest with dicloran (2,6-dichloro-4-nitroaniline) [134-137]. Similarly, orchard sprays of benomyl before harvest greatly reduced the incidence of brown rot (M. fructicola) developing on peaches after harvest, but had little effect upon the development of Rhizopus rot [51, 86]. Oranges sprayed with 300-1,000 ppm benomyl more than 30 days before harvest showed substantially less stem-end rot and Penicillium mold 2 weeks after harvest [83-85]. Orchard sprays of benomyl and thiabendazole on pears produced a reduction in decay due to <u>Penicillium</u> <u>expansum</u> and <u>Botrytis cinerea</u> [138, 139]. As a means of preventing infection of harvest injuries, the application of fungicides in the field is a less desirable practice than treatment of the fruit after harvest because (1) only a fraction of the fungicide applied in the field is bound to the harvested product where it can protect subsequent injuries; and (2) the deposit required for postharvest disease control may be removed by brushing or waxing the fruit after harvest [135, 137]. Preharvest treatments appear justifiable in situations where substantial harvest injury is anticipated, but handling practices make postharvest treatment difficult to carry out. The application of preharvest sprays of fungicides to control Diplodia stem-end rot of oranges during ethylene degreening [83-85] and Rhizopus rot of peaches during controlled ripening or storage before processing, appear to be situations which justify the preharvest treatment [135, 140]. Mechanical harvesting invariably results in a substantial increase in decay arising at superficial injuries and preharvest treatments may be a desirable approach to this problem, especially if the crop cannot be treated soon after harvest [94, 95, 135]. The fungicide chosen for preharvest application should be selected with care since a significant risk exists that residues from the preharvest treatment will select fungicide-resistant strains of the pathogen that would nullify any potential benefits of a postharvest treatment with the same fungicide. The usefulness of preharvest sprays to control postharvest diseases has been questioned by some investigators because of this potential problem.

b. Postharvest Treatments. After inoculation of an injury with a pathogen, it is still possible to prevent disease development by interrupting the infection process with a fungicide treatment, provided that development of the pathogen in the host is still quite limited. Since the usual solution treatment of a fungicide contacts the product for only a brief period of time, it is often assumed that postharvest fungicide treatments are lethal to the pathogen during the time of treatment. This does not seem to be the normal situation, however, since most treatments that are highly effective in preventing decay of inoculated fruit are not lethal to the pathogen in vitro under the same conditions. Examples of chemicals which are only fungistatic under the usual conditions of fruit treatment are sodium o-phenylphenate, sodium tetraborate, sodium carbonate, sec-butylamine, biphenyl, and thiabendazole. On the other hand, sodium hypochlorite is lethal to propagules of bacteria and fungi in vitro, but it is not effective in preventing decay of inoculated fruits and vegetables. It is not necessary to kill a pathogen present in an injury in order to prevent disease development, but rather only to maintain a fungistatic concentration of the chemical at the injury site for as long as the injury is receptive to growth and infection by the pathogen. Stability of the fungicide deposit in the injury may be desirable, but not essential. Superficial injuries in the peel of apples and oranges remain susceptible to

infection by species of <u>Penicillium</u> for only a few days [104, 141]. Susceptibility of the crown tissue of a banana hand to infection by several fungi
decreases with time after the hand is cut from the stem [98]. Potatoes and
yams form a periderm layer over injured surfaces which is resistant to
invasion by pathogens. The tubers are, therefore, susceptible to infection
by some wound pathogens for only a few days after harvest. In each of
these cases, infection can be prevented by maintaining an effective concentration of a fungicide at the injury site or by placing the produce in an
environment that is unfavorable to infection during the period of wound
susceptibility.

Many postharvest treatments result in the accumulation of fungistatic
residues in superficial injuries which prevent infection at these sites.
However, ammonia and aliphatic amines appeared to prevent infection of
oranges by temporarily increasing the pH of the tissues associated with
superficial injuries beyond the limits for growth of the pathogens [141].
The pH returned to the normal value for orange peel tissues after several
days, but the injuries were no longer receptive to infection, despite the
fact that the amine residues were not fungitoxic.

c. <u>Timing of the Fungicide Treatment to Prevent Infection</u>. The length of
the post-inoculation period when a fungicide may be effectively applied depends
upon the type of inoculation or infection, growth rate of the pathogen, susceptibility of the host to active invasion by the pathogen, humidity and temperature, and the depth to which an inhibitory concentration of the fungicide
penetrates into the host tissues. In the case of latent and quiescent infections, systemic fungicide treatments may produce excellent results even
when applied weeks or months after inoculation of the host. In these cases,
the pathogen is not able to develop extensively because of the host resistance
and, therefore, it remains vulnerable to the postharvest fungicide treatment
until such time as ripening of the fruit permits the development of the disease. Thus, postharvest treatment with benomyl and thiabendazole have
given good control of stem-end rot of citrus fruits [83-85], anthracnose of
bananas [34, 88] and mangoes [142], and lenticular rots of apples [56, 57, 87],
despite the fact that inoculation and perhaps infection occurred weeks before
the treatment was applied.

As a general principle, treatments to prevent infections at injuries created during harvesting and handling of the crop should be applied as soon as
possible after harvest. However, treatments which are marginal in effectiveness may be improved by delaying their application for a few hours after
inoculation (Fig. 1). Observations of this nature suggest that the sensitivity
of the pathogen increases with hydration of the spores or with increased
physiological activity as a result of being in an environment favorable for
infection. Germinated spores of <u>Monilinia fructicola</u>, <u>Rhizopus stolonifer</u>,
and <u>Alternaria tenuis</u> are more easily killed by heat treatments than are
dormant spores of these fungus species [114, 143]. Germinated spores of

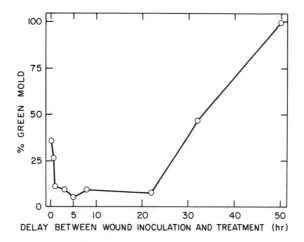

FIG. 1. Influence of time between wound inoculation and treatment upon the effectiveness of sodium o-phenylphenate (SOPP) in controlling green mold (Penicillium digitatum) on oranges. The oranges were submerged in a 2.0% SOPP solution pH 12 for 2 min. and then rinsed with water. Reprinted from Ref. 152, p. 1313.

Rhizopus are killed by exposure to temperatures near zero for 7 days, whereas over 50% of the dormant spores survive this treatment [144]. Couey and Uota [145, 146] reported that the toxicity of sulfur dioxide to spores of Botrytis and Alternaria was greatly increased if the spores were hydrated before the exposure period or if the fumigation was carried out in an atmosphere of high relative humidity. Conidia of Penicillium digitatum and Aspergillus niger became more sensitive to fungicidal phenol derivatives after germination was initiated [147, 148]. In practice, however, a few hours will almost certainly elapse between harvest and treatment of the crop; therefore, the primary goal should be to apply the treatment as soon as possible before the pathogen can penetrate deeply into the host tissues.

Rhizopus rot of stone fruits, which develops rapidly at temperatures prevailing during the harvest season, is controlled best if the fungicide treatment is applied within 12 hr after inoculation [149], although dicloran will retard the development of this disease even when applied after the lesion is visible [150]. Penicillium digitatum and P. expansum infect citrus fruits and apples, respectively, at a much slower rate and an effective fungicide will satisfactorily control decay of these fruits if applied within 24-48 hr (at ambient temperatures) after inoculation [151-153]. If the prevailing temperature is below optimum for the development of the pathogen, the period of incubation may be extended considerably without adversely influencing the effectiveness of the fungicide treatment [151]. The specific

fungicide treatment can have a profound effect upon the maximum delay
between inoculation and effective treatment. Long and Roberts [151] reported that a delay of as much as 48 hr at 16°C after inoculation with
P. digitatum did not diminish the effectiveness of a sodium o-phenylphenate
solution treatment; whereas sodium tetraborate had to be applied within a
few hours after inoculation for maximum effectiveness. Edney [87] observed
that both thiabendazole and benomyl could prevent decay of apples incubated
13 days after lenticel inoculation with Gloeosporium perennans, but the
effectiveness of captan was markedly reduced if the treatment was delayed
for 2 days. Meredith [154] found that sodium salicylanalide would effectively
control decay of bananas inoculated with Colletotrichum 6 hr before treatment; whereas nystatin had to be applied immediately after inoculation for
the same degree of effectiveness. More recent investigations with bananas
have shown that thiabendazole and benomyl treatments may be delayed for
2-3 days after inoculation with only slight loss of effectiveness [36, 155-157].
Long [156] showed that the benomyl treatment could be delayed for 48 hr and
still provide quite effective control of Gloeosporium rot at the cut end of the
fruit. Fumigation of seed potatoes with sec-butylamine to control gangrene
(Phoma exigua), skin spot (Oospora pustulans), and silver scurf (Helminthosporium solani) may be delayed for 14 days after lifting, whereas organomercury dips for control of these diseases must be applied within 3 days
after lifting, reflecting the relative penetration of the two fungicides into the
suberized periderm [158]. Treatment of potatoes with thiabendazole for
control of these diseases also may be delayed for several days without
diminishing the beneficial effects of the treatment.

B. Eradication or Attenuation of Established Pathogens

The eradication of a pathogen in an established infection is difficult because
a large number of propagules may be involved and contacting all of them
with an inactivating dose of the toxicant is a difficult task. Chemical treatments have retarded the development of a pathogen and perhaps eradicated
infections in a few instances, but most of the past successes in the therapy
of established infections has been by heat and irradiation treatments. More
recent investigations describing the inhibition of latent infections by systemic
fungicides have provided a basis for renewed optimism regarding the chemotherapy of established infections on harvested crops.

1. Chemical Treatments

Most fungicides that have been applied after harvest are very limited in their
ability to penetrate host tissues and thus are ineffective against established
infections. However, the development of established lesions of Botrytis in
grapes and Rhizopus in peaches has been retarded by high concentrations of
sulfur dioxide [159] and dicloran [150, 160], respectively. Benomyl

also inhibits the expansion of visible lesions of Monilinia fructicola on harvested peaches. Benomyl penetrated the epidermis of the peach and fungistatic concentrations of this compound were detected in the flesh several mm below the surface of the fruit [161]. Several investigators have reported that benomyl prevents, or at least delays, the development of latent infections of species of Gloeosporium on apples [81, 87], bananas [34, 88, 156], and mangoes [142]. Benomyl is also able to penetrate the disc and the calyx of the button of orange fruits, thereby halting the development of latent infections of Diplodia and Phomopsis [84, 85]. Thiabendazole can penetrate at least 5 mm into the flesh of the potato tuber to inhibit development of 48-hr infections of Fusarium [162]. Tisdale and Lord [163] detected thiabendazole in the flesh of potatoes treated with high dosages of thiabendazole, but observed that the major portion of applied thiabendazole was bound to the periderm.

2. Heat Treatments

In some instances, incipient infections of pathogenic fungi may be eradicated or attenuated by heating the surface of the host to a few degrees below the injury threshold for several minutes. Heat treatments have the advantage of low cost, relatively simple equipment, and no chemical residues on the treated product. Water is the most efficient heat transfer medium for most applications, but moist air may be advantageous because the product can then be treated without removing it from the original container. This could be a significant advantage for crops which are stored or cured in field containers before grading and packing.

The purpose of a heat treatment is to increase the temperature of the surface cells of the product to a point above the thermal death point of the pathogen which may be present in incipient or latent infections. However, heat treatments approaching the injury threshold for the plant tissue may alter its physiology, often resulting in an increase in disease susceptibility or an increase in the rate of ripening of the host. Brief hot water treatments do not raise the average temperature of the product significantly, so additional refrigeration is not needed. On the other hand, heated moist-air treatments are applied for several hours and do increase the pulp temperature of the product significantly, so that additional refrigeration is required to bring the temperature of the product down to the desired storage temperature.

Rosenbaum [164] reported in 1920 that incipient infections of species of Phytophthora in mature green tomatoes could be inactivated by immersing the fruit in water at 60°C for 1.5 min, providing that the fungus had not penetrated too deeply into the fruit tissues at the time of treatment. The only effective postharvest treatment for eradication of species of Phytophthora on citrus fruit consists of immersing the fruit in water at 48°C for 2-4 min within several days after infection [53]. However, this treatment may physiologically weaken some lots of lemons, increasing their susceptibility to other pathological diseases during storage [15].

The utility of heat treatments for reducing decay of peaches and other stone fruits has been actively investigated over the past decade. An effective and practical treatment consists of submerging peaches in water at 51.5°C for 2-3 min or at 46°C for 5 min [129, 165, 166]. A shower of water at 51.5°C for 3 min has given equivalent results [167]. Hot-water treatments have provided a 70-80% reduction in decay under commercial conditions in California and in the southeastern United States [167, 168]. The hot-water treatment inactivates spores and hyphae located in the skin and in the outer 3 mm of the flesh, and is more effective against Monilinia brown rot than Rhizopus rot [168, 169]. Fungus cells which survive the heat treatments are inhibited, but begin to grow again in a few days [165, 169, 170]. The hot-water treatment (52°C) was less effective in controlling brown rot in fruit stored at low temperatures after treatment [169, 171]. Hot water may cause injury to peaches, observed when stored at low temperature after treatment, but not when stored at ambient temperature. This injury is reflected in an increase in Penicillium, Alternaria, and Botrytis on fruit stored at low temperature after treatment [169, 171].

Australian peaches ripen well at 32-35°C and 90% RH (relative humidity) and the development of brown rot is inhibited at this temperature [172]. Exposure of peaches and sweet cherries for 24 hr to an environment of 40°C and 85-90% RH reduced brown rot to about the same extent as the best hot-water treatment [166]. Shorter exposure to higher temperatures tended to injure the fruit. Black rot (Ceratocystis) of sweet potatoes was reduced significantly by holding the tubers for 24 hr in an atmosphere at 41-43°C and high RH [173]. Exposure of Irish potatoes after harvest to 43°C for 2 hr can greatly reduce infection by Oospora pustulans [174].

Lenticel rot of apples caused by Gloeosporium perennans (Cryptosporiopsis) was reduced substantially by submersion of apples for 10 min in water at 45°C, but the treatment tended to increase a physiological disorder of the fruit known as "core flush" during storage [175, 176]. Similarly, tests in the United States have shown that water at 54°C reduced Penicillium decay of apples, but the treated fruit developed internal breakdown and scald during storage [177]. Hot-water treatments have provided control of Penicillium and Diplodia on oranges [178], Colletotrichum on papayas [179], crown rot of banana [180], stem scar and rind mold on cantaloupes [181, 182], and Colletotrichum on mangoes [142]. Mangoes, like peaches and apples, may show symptoms of heat damage if stored at 13°C after treatment in hot water at 54°C. Heat damage was not apparent on fruit which were ripened at 21°C immediately after treatment [142].

Immersion of bell peppers in water at 53.5°C for 1.5 min significantly reduced bacterial soft rot of the stem [103]. Treatment of tomatoes with water at 46°C for 4 min reduced Alternaria rot [114], but bacterial soft rot was not influenced by heat treatment at 60°C [183]. Decay of snap beans caused by Pythium and Sclerotinia was significantly reduced by dipping the beans for 30 sec in water at 52°C [55].

The conditions of heat treatment for maximum disease control are usually quite close to the limit that can be tolerated by the crop, although injury may be evident only under a special set of environmental conditions such as low temperature storage. Even if heat damage is not visible, the product may show an increased susceptibility to wound pathogens such as Alternaria and Penicillium. Considerable interest has developed in recent years in treatments combining heat and fungicides. The temperatures employed are usually borderline for decay control but well below the injury threshold. The elevated temperature increases the effectiveness of the fungicide, resulting in acceptable disease control with lower concentrations of the fungicide and little risk of heat damage to the crop. The applications of these combination treatments for specific fungicides will be described in Section V.

3. Irradiation

Gamma rays are capable of penetrating fruits and vegetables and destroying pathogens in established lesions deep in the host tissue. Mature fruits and vegetables are relatively resistant to radiation damage because cell division is a comparatively rare event at this stage of development. These facts, coupled with the potential for treatment of packaged products have encouraged extensive research over the past 20 years on the application of γ radiation to the control of postharvest diseases. The current status of the most promising applications of irradiation in this field has been summarized in a recent review [184].

Although attractive in theory, the control of established infections by γ radiation has two serious limitations. First, deep-seated infections often consist of relatively large populations of pathogen cells. Radiation dosages sufficiently high to inactivate the most resistant individuals of the population generally exceed the tolerance of the host tissues, resulting in radiation damage to the crop [185, 186]. Second, species of pathogens which are present as deep-seated, quiescent infections, e.g., Diplodia and Alternaria on citrus fruits, are relatively resistant to radiation, whereas fungi which invade through superficial injuries and are controllable by fungicide treatments, are relatively sensitive to radiation treatments. At radiation dosages tolerated by the host, the principle benefit of a radiation treatment is most likely to be a delay in the onset of decay rather than eradication of the pathogen, as was visualized by early investigators. Despite the wave of enthusiasm which accompanied the initial proposals for the application of irradiation for control of postharvest diseases, sober reevaluation in recent years of all data and experiences has revealed that γ radiation has few, if any, promising applications in the field of postharvest pathology. Maxie et al. [187] have summarized the principal reasons for the infeasibility of γ radiation treatments for the majority of fresh fruits and vegetables: (1) many commodities are damaged by the radiation dosages which are required to substantially retard the development of decay; and (2) for those fruits and vegetables which do respond favorably to γ radiation treatment, there are alternate procedures available for decay control which are less expensive, noninjurious, and more effective.

4. Low Temperature Storage

Low temperature storage is the most effective and practical method for delaying the development of decay in fruits and vegetables with deep-seated infections that cannot be eradicated by other postharvest treatments [26]. With rare exceptions, however, low temperature does not have a permanent effect upon the pathogen cells in an incipient infection but rather it merely delays their development until later in the postharvest period. Pathogens with relatively high temperature optima, such as Geotrichum and Erwinia, may be inhibited almost indefinitely by storage near 0°C, but will rapidly produce a disease when the product is brought to a higher temperature for ripening or marketing. Other microorganisms such as Botrytis, Penicillium, Sclerotinia, Centrospora, and Cladosporium grow slowly at near-freezing temperatures and limit the storage life of the crop if they are not controlled by some other treatment [13, 188, 189]. Low temperature delays the development of postharvest diseases by two mechanisms: (1) inhibition of ripening of the host thereby prolonging the disease resistance associated with immaturity; and (2) by direct inhibition of the pathogen at a temperature unfavorable for growth. An optimum storage temperature exists for each variety of fruit and vegetable which will maintain the desired quality of the product for the maximum period of time. This optimum is near 0°C for many fruits and vegetables, e.g., apples, oranges, carrots, and grapes. Storage at these low temperatures greatly retards the development of diseases as well as maintains the quality of the product. Other crops, typically of tropical origin such as bananas, lemons, tomatoes, grapefruit, and pineapples, suffer chilling injury at storage temperatures below 10°C and become more susceptible to postharvest disease if exposed to such temperatures for more than a few hours. These fruits are harvested in a slightly immature state, and if the fruit pulp temperature is reduced to 10°C soon after harvesting, they can be held for 2 weeks or longer without serious disease problems in storage. After storage, the fruit must be ripened for a few days at a higher temperature, and during this period previously attenuated infections develop rapidly and may involve a substantial portion of the fruit.

Rhizopus rot is an interesting exception to the generalization that low temperature does not "cure" postharvest diseases, but merely delays their development. Germinated sporangiospores of several species of Rhizopus are killed by exposure to near freezing temperatures although dormant spores are tolerant of these same conditions [144, 190]. The lethal effect of low temperature on Rhizopus spores provides an explanation for the

observation that peaches and strawberries which were stored for 1 week at
0°C showed a lower incidence of Rhizopus rot after ripening at a higher
temperature than fruit which were stored for 1 week at ambient temperature
before ripening [191].

5. Maintenance of Host Resistance

Several practical methods for retarding ripening and senescence, in addition to low temperature, have been investigated for reducing postharvest
disease losses. Most lemons grown in California are treated with the isopropyl ester of 2,4-dichlorophenoxyacetic acid (2,4-D) before storage to
delay senescence of the button (calyx plus disc) which is the usual point of
attack by the fungus Alternaria citri [46, 192, 193]. Oranges and grapefruit
are treated similarly in South Africa to reduce Diplodia and Alternaria stem-
end rots during shipment to markets in northern Europe [194]. Gibberellic
acid may reduce Penicillium decay of navel oranges by delaying senescence
of the peel which renders it more susceptible to injury [125].

Storage atmospheres, modified by altering the levels of oxygen and carbon dioxide, can influence the development of postharvest disease either by
direct inhibition of the pathogen or by altering the resistance of the host.
Decay of strawberries and sweet cherries can be reduced significantly by
raising the CO_2 level of the storage atmosphere to 20-30% [195, 196]. These
levels of CO_2 reduce decay by retarding the physiological deterioration of
the berries [197, 198] and by directly inhibiting the growth of the pathogen
[199]. High CO_2 storage for strawberries is most beneficial at temperatures above 5°C which do not completely inhibit the growth of Botrytis. The
high CO_2 treatment is superfluous if the pulp temperature of the berries is
rapidly reduced to less than 5°C and maintained at this point for the duration
of the shipment [197, 198]. The high CO_2 treatment has been utilized extensively in shipments of strawberries from California to distant markets.
The CO_2 is generated by sublimation from solid carbon dioxide (dry ice)
placed in the shipping compartment or in pallet stacks covered with polyethylene film [200].

Modified atmospheres composed of O_2 and CO_2 levels less than 10% have
been evaluated extensively as a means of extending the physiological life of
fruits and vegetables in cold storage. With a storage temperature close to
0°C, it has been observed that the incidence of decay is somewhat less in
modified atmosphere storage than in air [171, 201-203]. However, when
these fruits are removed to air for ripening at ambient temperatures, decay
develops to the same extent regardless of the previous storage atmosphere
[171]. These results are explainable by the effect of modified atmospheres
on the resistance of the host to disease development. The usual modified
atmosphere for fruit storage is unlikely to have a pronounced effect upon
pathogens since most fungi grow well in atmospheres containing more than
1% O_2 [204, 205].

Florida avocados stored for 45 and 60 days at 10°C in an atmosphere of 2% O_2 and 10% CO_2 showed a remarkable reduction in anthracnose compared to fruit stored in air at the same temperature. This atmosphere also prevented chilling injury, thereby permitting storage of the fruit at 7.2°C. Under these conditions avocados could be stored for 6 weeks, a period that is three times the maximum storage time in air [206-208]. Storage at reduced pressures (hypobaric storage) has a similar effect upon the development of postharvest diseases [209-211].

Modified atmosphere storage does not always result in a reduction of postharvest diseases. Parsons and Spalding [212] reported that mature green tomatoes stored in an atmosphere of 5% CO_2 and 3% O_2 had a higher incidence of bacterial soft rot than fruit stored in air. They attributed this to the fact that green tomatoes are uniquely more susceptible to this disease than are riper fruit. Low oxygen (less than 5%) and high CO_2 greatly increased the development of bacterial soft rot on newly harvested potatoes by inhibiting the development of a periderm over injured areas [213]. The severity of rot produced by Geotrichum candidum on tomatoes is increased by storage in low O_2 and high CO_2 atmospheres because the growth of the fungus is stimulated by an elevated CO_2 level [214]. Swinburne [215] observed that the development of Nectria in apples during storage was less at 4% CO_2 than at higher concentrations or in air. The accumulation of benzoic acid, a natural inhibitor of Nectria infections, is stimulated by concentrations of CO_2 up to 2.5% in the atmosphere.

C. Moderation of Disease Symptoms

The value of fresh fruits and vegetables may be reduced significantly by superficial fungal growth or by visible contamination with spores or other debris arising from adjacent diseased units, even though the palatability of the product is not altered. "Sooty mold" on oranges, superficial fungal growth on the cut surface of banana crowns or on melon rinds, moldy sepals on apples and soilage of citrus fruits with Penicillium spores are all examples of this condition. Often a postharvest fungicide treatment will also control these conditions, and in some cases, may be the sole purpose of the treatment.

1. Inhibiting Growth and Sporulation of the Pathogen
   on the Surface of the Diseased Host

In his first report in 1936 on the potential usefulness of biphenyl for control of postharvest diseases of citrus fruits, Tomkins [216] noted that vapors of this compound inhibited sporulation of Penicillium on the surface of decayed oranges. This property has been utilized in long distance shipment of unwrapped citrus fruits to control "soilage" of many sound fruit by a few diseased fruit in a shipping container [19]. Lemons may be fumigated repeatedly with nitrogen trichloride during storage to control Penicillium sporulation as well as to reduce decay. As a result of this treatment, fewer fruit are rejected after storage because of visible contamination with Penicillium spores.

The superficial growth of Cladosporium, Alternaria and other fungi on the rind and stem tissues of a number of fruits during storage under high humidity conditions significantly detracts from the appearance of these fruit. Saprophytic growth of Alternaria on the buttons of oranges can be prevented by vapors of dibromotetrachloroethane [217], but not by biphenyl or o-phenylphenol [218]. Latent infections of Colletotrichum on the peel of bananas may give rise to brown lesions on the peel when the fruit is ripe, but these do not penetrate into the pulp. The appearance of these brown spots can be delayed or prevented by postharvest application of benomyl [34, 88, 219]. Superficial growth of Alternaria and other fungi on melon rinds is controlled by postharvest treatment with a dithiocarbamate, tetraiodoethylene, sodium o-phenylphenate, and heat treatments [181, 182].

2. Preventing the Spread of the Pathogen
    from Diseased to Healthy Produce

Rhizopus, Botrytis, Trichoderma, Sclerotinia and other fungi can spread by contact from diseased to sound fruit under conditions encountered in storage rooms and during long distance transport to market. In fact, contact spread frequently accounts for the major losses caused by these pathogens.

The spread of Botrytis on pears can be prevented by oiled paper wraps containing copper sulfate [220], a treatment used extensively today for this purpose. Copper impregnated wraps were not successful in controlling Rhizopus on tomatoes [221], but wraps impregnated with organic compounds which generated active chlorine and active bromine retarded the spread of Rhizopus in tomato packs experimentally [222]. Paper wraps impregnated with 2, 6-dichloro-4-nitroaniline are highly effective in preventing the spread of Rhizopus on peaches [135, 223]. Trichoderma rot was confined to initially diseased fruit by wrapping lemons in paper impregnated with o-phenylphenol [218]. More recently, esters of o-phenylphenol have been evaluated for this purpose since they are less injurious to produce than the parent phenol [224]. It must be stressed that fungicide-treated wraps are usually not very effective in preventing decay of fruit which are infected at the time of packaging, but rather they prevent secondary infection of neighboring fruits by pathogens which are capable of spreading by contact.

Incipient infections of Botrytis on grapes at the time of harvest may develop into "nests" of decay during long-term cold storage. The standard commercial practice is to fumigate grapes in storage with sulfur dioxide at 10-day intervals to prevent the growth of Botrytis away from diseased berries [1]. Sodium bisulfite is added to shipping containers to generate sulfur dioxide to prevent the spread of Botrytis during storage and long distance shipment of grapes [225-227].

IV. METHODS FOR APPLYING POSTHARVEST TREATMENTS

A. Gases and Airborne Particulates

1. Fumigants

Fumigation is potentially one of the most versatile methods for applying a fungicide to a harvested crop. It presents a means for treating crops such as strawberries, grapes, grains, and other crops which, for various reasons, cannot be treated in any other fashion after harvest. Citrus and peaches may be treated with aqueous solutions of fungicides in the packing-house, but fumigation can be carried out immediately after harvest to prevent infection of injuries on fruit which must be transported long distances to the packing plant, degreened before processing, or must be held for several days at the cannery before processing. Furthermore, fumigation is the only practical means of periodically treating products such as grapes and lemons which are held for long periods of time in storage. Finally, fumigation is the only convenient means of treating film-wrapped or bagged consumer units of fruits and vegetables. Despite these promising applications, relatively few fumigants have been developed for the control of postharvest diseases. A suitable material must be reasonably volatile, possess the desired antimicrobial spectrum, and either have an intrinsic low order of phytotoxicity or not penetrate the surface of the product.

The sole objective of the fumigation treatment is to deliver an effective dosage of the active agent to the potential infection site on the product. Any of the fumigant that does not reach these sites is not only wasted, but may appear as undesirable nonfunctional residues on the commodity. Fumigant fungicides absorbed by the uninjured surface of the fruit are often inactivated [228, 229] and thus do not protect against future infection. Polar fumigants such as amines and sulfur dioxide can be applied at relatively low vapor concentrations with reasonable expectation that these fungicides will accumulate at potential infection sites (moist injuries) with little extraneous absorption by areas of the product that are covered with an intact, waxy cuticle. Unfortunately, water-soluble fumigants are also absorbed by other moist and hydrated surfaces in the fumigation chamber and this phenomenon makes it very difficult to fumigate fruits and vegetables with water condensed on their surfaces. The effect of free moisture and relative humidity on sorption losses have been extensively evaluated for sulfur dioxide fumigation of grapes [230-232].

The effectiveness of a fumigation treatment is a function of both time and concentration (Fig. 2), indicating that a certain amount of the toxicant must be absorbed by the pathogen cells and by the tissues surrounding the infection site in order to inactivate the pathogen. An increase in temperature increases the rate of diffusion of the fumigant and its activity directly against the pathogen. A high relative humidity increases the losses of polar fumigants by sorption and the quantity of fumigant required for a given level

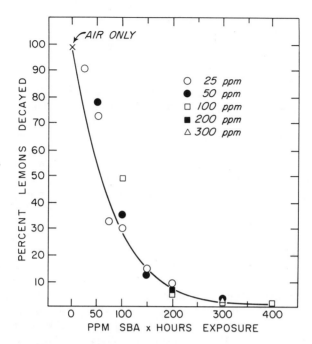

FIG. 2. Effect of dosage of sec-butylamine (SBA) gas on decay of lemons inoculated with Penicillium digitatum. The abscissa scale is the product of the gas concentration (indicated by the plotted symbols) and the time (in hours) of exposure. Reprinted from Ref. 314, p. 547.

of disease control, and accentuates the corrosion problem inherent in the use of chemically reactive fumigants such as sulfur dioxide and nitrogen trichloride. Many fresh fruits and vegetables, e.g., grapes, must be stored at temperatures near 0°C and high humidity to maintain quality. In these cases, the dosage of a fumigant must be increased to compensate for the lower efficiency under these environmental conditions.

The distribution of fumigants in large chambers can be greatly improved by fans and other mechanical devices that distribute and agitate the atmosphere during the fumigation treatment. However, molecular diffusion is the principle means whereby the fumigant ultimately moves into the interstices between commodity units and penetrates to potential infection sites. Moreover, the distribution of in-package fumigants such as biphenyl and sulfur dioxide (see Section V) depends almost entirely upon molecular diffusion. The physical behavior of fumigant gases has been considered in detail in another review [233]. The properties and applications of postharvest fumigants will be discussed in Section V.

2. Smokes, Fogs, and Dusts

The application of a postharvest fungicide in the form of airborne particu-
lates is a rather inefficient process. Smoke and fog treatments cannot
approach the efficiency of a gas in penetrating into containers filled with
fresh produce. Smoke and fog applications of thiabendazole have been evaluated recently as a convenient means for treating potatoes in storage to
control postharvest diseases [234]. Thiabendazole is thermally stable and
sublimes at 310°C. By means of appropriate formulations and specialized
equipment, a smoke consisting of very fine particles of thiabendazole may
be generated and passed through the piles of potatoes in the storage room,
ideally producing a uniform deposit of thiabendazole on the surface of each
potato [235, 236]. This is a common method for applying the sprout inhibitor isopropyl phenylcarbamate. For small storage operations, a tablet
consisting of thiabendazole blended with a combustible mixture is ignited to
generate the smoke. Trials in Europe have demonstrated that a thiabendazole smoke can control potato storage diseases, but this application method
is inferior to a mist or low-volume spray applied to the potatoes as they
are transported into the storage room [132, 234-242]. Mist (ultralow-
volume spray) applications of thiabendazole are less effective than low-
volume spray treatments which uniformly wet the surface of the tuber [132,
237-239, 241]. Storage rooms and recycled fruit containers may be disinfested conveniently with a fog (aerosol) of a dilute solution (1-3%) of formaldehyde. This technique is essentially a means for generating formaldehyde
gas (boiling point, -20°C) rather than an aerosol treatment per se.
　　Dust formulations of fungicides have been used for treatment of tubers
and bulbs to prevent diseases which develop in storage or in the field the
following season. Thiabendazole and benomyl dusts controlled gangrene
and Fusarium on potato tubers, but were less effective than spray or dip
treatments which resulted in much greater absorption of the fungicide by the
tuber [132, 235, 236, 240-242]. A dust formulation must contain a concentration of thiabendazole 10 times greater than an aqueous suspension for
equivalent protection of the potatoes against disease [163, 241, 242]. Dust
treatments have appeal because they are easy to apply and, unlike aqueous
treatments, do not present the potential hazard of developing bacterial soft
rot in treated potatoes. In addition to their reduced effectiveness, dust
formulations have the disadvantage that the treated product is unpleasant to
handle.

B. Solutions, Suspensions, and Emulsions

Water is the most popular vehicle for application of postharvest fungicides
because the formulations are easy to prepare and apply, the entire surface
of the product may be coated uniformly with the fungicide, and water application techniques are compatible, in most instances, with mechanical fruit

handling operations. Fungicides and bactericides may be added to water
used for applications of other treatments, such as cleaning citrus fruits,
applying antiscald agents to apples, removing latex from bananas and mangoes, and hydrocooling a variety of fruits and vegetables. However, if a
fresh product is not treated with water in the usual handling operation, the
application of a fungicide in water should be carefully considered lest the
treatment increase decay compared to the dry operation. In fact, it is
customary to add microbiocides to water used for washing, hydrocooling,
and applying antiscald treatments to prevent excessive decay due to these
practices alone. The probable role of water in increasing decay is through
the hydration of injuries, water-soaking of stomates and lenticels, and as
a reservoir of pathogen propagules.

True solutions of fungicides and bactericides are simple to apply and
usually the antimicrobial agent penetrates well into injuries on the surface
of the fruit and possibly into natural openings such as stomates and lenticels.
which may function as sites of infection. True solutions do not require
agitation to remain homogeneous and they may be pumped, sprayed, and
recirculated as required. Emulsions tend to be somewhat less stable than
solutions and may require some agitation during application. Dispersions
of wettable powders must be continuously agitated during application and
still may settle in the pipes and clog up spray nozzles unless due care is
exercised.

Several factors may influence the effectiveness of a postharvest fungicide applied in water, such as fungicide concentration, surfactants, temperature, pH, contact time, and presoaking the fruit before and rinsing it
after treatment. All of these factors influence the quantity of the fungicide
which penetrates and accumulates at the infection site. An increase in
temperature will increase (1) the concentration of a slightly soluble fungicide in true solution; (2) the movement of the solution into channels and
pores which might be associated with infection sites; and (3) the diffusion of
fungicide molecules through waxy coverings and the cuticle. Several investigations have demonstrated that the amount of the fungicide absorbed by the
treated commodity is related to the fungicide concentration, temperature,
and time of contact, and that these parameters are also correlated in a
positive fashion with disease control [147, 243-245]. Fungicides which
interact with the surface of the product by hydrophobic forces are likely to
be taken up to a maximum value quite rapidly. The absorption of o-phenylphenol by oranges and thiabendazole by potatoes are examples of uptake by
hydrophobic mechanisms [147, 163, 245]. These investigations have related
gross residues on the treated commodity with the treatment parameter
(temperature, pH, time) being studied. The gross residue, however, may
not be a measure of the amount of fungicide which accumulated in the infection court. From such data, one might conclude that virtually all fungicide
uptake was complete in a fraction of a minute, whereas, in fact, fungicide
uptake by the injuries was continued for a considerably longer period of time.

The effectiveness of an aqueous fungicide treatment may be influenced by pH in two ways: (1) the intensity of its action against the pathogen, and (2) the amount of the fungicide taken up by superficial tissues of the host. The pH of the solution determines the relative concentrations of the ionic and nonionic (neutral) forms of weakly acidic and weakly basic fungicides. Invariably, the nonionic form is more fungitoxic and more phytotoxic than the ionic form. The effect of pH on toxicity of a fungicide or bactericide is most evident in a range of about 2 pH units on either side of the pK_a value.

Several investigators have demonstrated that the total residues of o-phenylphenol on oranges is greatest at pH values below the pK_a of this fungicide ($pK_a = 10.01$) [147, 151, 245], presumably indicating that the intact cuticle of the orange is more permeable to the neutral form of o-phenyl-phenol than to the anionic form (Fig. 3). This effect on uptake is clearly evident by the degree of injury to oranges resulting from treatment with o-phenylphenol at different pH values [147, 151]. sec-Butylamine, a weak base, is more strongly absorbed as the nonionic form at pH values greater than 10, the pK_a of this amine [246]. Thiabendazole may exist in either an

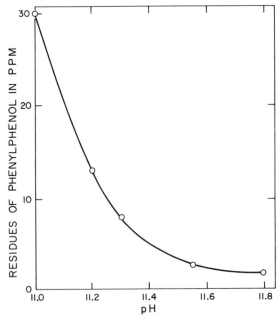

FIG. 3. Effect of pH of sodium o-phenylphenate (SOPP) solution upon residues of o-phenylphenol in treated oranges. The fruit were flooded with a 2.0% SOPP solution at indicated pH at 32°C for 3 min. and then rinsed with water. Reprinted from Ref. 245 with permission from J. Agric. Food Chem. Copyright by the American Chemical Society.

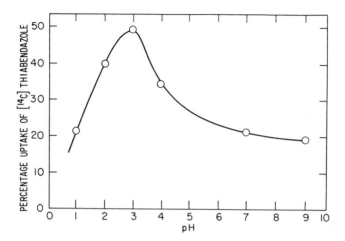

FIG. 4. Effect of solution pH on the uptake of thiabendazole by potato tubers. The tubers were exposed to an agitated solution of 50 ppm thiabendazole at indicated pH for 60 min. Reprinted from Ref. 163, p. 125, by courtesy of Blackwell Scientific Publications Ltd.

anionic, neutral, or cationic form depending upon the pH of the solution. Tisdale and Lord [163] showed that potatoes took up thiabendazole most strongly from a solution at pH 3, which is the pK_a of this compound (Fig. 4). These investigators believed that uptake was greatest at this pH because thiabendazole was in the neutral form and ionization of carboxyl groups on the surface of potatoes was suppressed. Both factors favored the partition of thiabendazole into the waxy periderm of the tuber. The influence of pH on the effectiveness of a fungicide treatment may be quite different when dealing with a pathogen which invades the host exclusively at wounds. The microenvironment at these sites is most likely to be hydrophilic in nature in comparison with the intact cuticle of a citrus fruit or the suberized periderm of a potato. In the case of the wound pathogen, the lipophilic barrier separating the fungicide and the pathogen should be minimal and diffusion should be the major force driving the fungicide into wounded tissue. The rate of diffusion is a function of the concentration gradient; hence, the pH which brings about the greatest solubility of slightly soluble organic compounds is likely to be the pH at which the treatment is most effective. This may be a factor in the greater activity of thiabendazole and methyl benzimidazolecarbamate applied in slightly acid solutions [247-249] and o-phenylphenol in alkaline solutions [147]. Figure 5 shows that the quantity of o-phenylphenol taken up by injuries in the peel of oranges is proportional to the concentration of o-phenylphenate ion in the treating solution and that the quantity in the injuries is directly related to the control of the wound pathogen

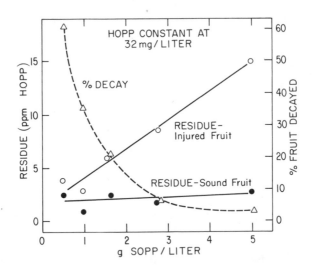

FIG. 5. Relationship between residue of o-phenylphenol (HOPP) accumulated in superficial peel injuries and the control of green mold (Penicillium
digitatum) on Valencia oranges. The fruits were submerged for 4 min. in
solutions of sodium o-phenylphenate tetrahydrate at the indicated concentrations and rinsed with water. The concentration of HOPP in the solutions
was held constant by slight adjustments in the pH of the solutions. Reprinted
from Ref. 147, p. 1102.

Penicillium digitatum [147]. Furthermore, the quantity of o-phenylphenol
in the injuries is independent of the total residue of this fungicide on treated
oranges.

The sequence of fruit treatments in many packinghouses includes the
application of water or another aqueous treatment prior to the application of
the fungicide. This practice tends to reduce the effectiveness of the fungicide treatment by diluting the concentration of fungicide in the infection
court. This occurs because the first aqueous solution saturates the injuries
on the fruit surface and, even if the fruit is partially dried, the fungicide
applied thereafter must penetrate the water-saturated injuries by molecular
diffusion. This should result in a lower concentration of toxicant at the
infection site than if the fungicide solution had been applied first and flooded
the injuries without dilution. The adverse influence of a water pretreatment
on the efficiency of a fungicide treatment can probably be offset by increasing
the concentration or the temperature of the fungicide solution.

The effectiveness of an aqueous fungicide treatment may be increased by
certain additives which alone are not active. Agents which increase the
water solubility of benzimidazole fungicides, such as dilute mineral acids,
lactic acid, and organic phosphate surfactants have increased the effectiveness

of these fungicides against certain diseases [247-249]. Surfactants which increase penetration into injuries, lenticels, under the calyx, etc., should markedly improve the effectiveness of the treatment. Brown [244] has recently shown that the penetration of benomyl into the flavedo and albedo layers of Valencia oranges was increased by the addition of 1% spray oil to the benomyl suspension. The improved uptake of benomyl may have been due to better distribution of the fungicide on the surface of the fruit or to increased penetration through the intact cuticle. The effectiveness of solution treatments may also be improved by certain additives such as wax solids and colloidal substances which tend to increase the surface residues and perhaps the uptake of the fungicide [250, 251].

Fungicides and bactericides in solution or suspension have been applied to fruits and vegetables by several techniques, such as immersion, flooding, spraying, misting, and wet brushing. Application of a fungicide treatment by immersion of the harvested crop in a bath filled with a solution or suspension of the active material has several advantages over other methods of applying aqueous formulations. First of all, the temperature of the solution can be accurately regulated to improve the effectiveness of the treatment, especially against disease such as Phytophthora brown rot which cannot be controlled by the direct action of the fungicide itself (see Section II.A). Furthermore, the bath treatment provides the maximum opportunity for the fungicide to penetrate to potential infection sites since the fruit is totally submerged in the solution for 1-4 min. In the case of borax, sodium carbonate, or sodium o-phenylphenate treatments, all of which must be followed by a water rinse to prevent phytotoxicity or reduce the fungicide residue, those factors which improve penetration and retention of the active compound in injuries will improve the effectiveness of the treatment. In the case of treatments with sec-butylamine and thiabendazole, the fruits do not have to be rinsed after treatment and, therefore, the application temperature and contact time are less important because rather large residues of the fungicide remain on the treated fruit. Figure 6 shows that in the treatment of oranges with a solution of sec-butylamine factors which increase the rate of penetration into superficial injuries, such as contact time, solution pH, and temperature, are important only when the oranges are rinsed following immersion in the solution of sec-butylamine. Rinsing is always required when the fungicide and the cleaning operations are combined, e.g., a bath containing borax and soap or sodium o-phenylphenate.

The principal disadvantage of a bath treatment is that a relatively large quantity of fungicide is required and the treating solution must be used over a period of several weeks for reasons of economy. The bath treatment is best suited for fungicides which are inexpensive, chemically stable, and preferably possess antibacterial as well as antifungal properties so that the bath does not become unsanitary after a few days of use. The concentration of the fungicide in a bath must be measured frequently, and additions made to maintain the concentration at an effective level. Also, the pH of the

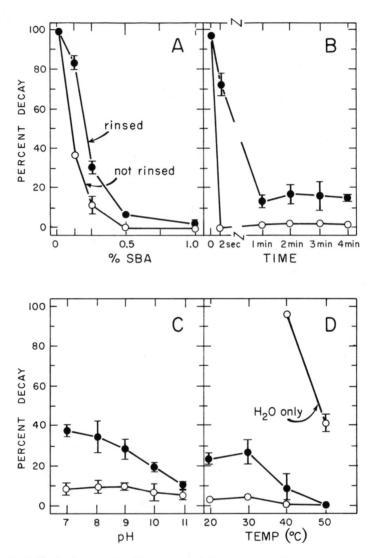

FIG. 6. Effect of concentration, contact time, pH, and temperature on the effectiveness of solutions of sec-butylamine hydrochloride against green mold decay (Penicillium digitatum) on oranges. Unless specified otherwise, the oranges were submerged for 4 min. in solutions at pH 8.5 and 23°C. In plot B, the sec-butylamine concentration was 0.5%, whereas in C and D, the concentration was 0.25%. Reprinted from Ref. 246, p. 983.

solution must be frequently checked and readjusted as in the case of sodium o-phenylphenate which is hydrolyzed by weakly acidic materials brought into the bath along with the fruit. In addition, the bath treatment imposes a greater limitation on the volume of fruit which can be handled during a given time period compared to the other systems described below. Bulk dipping is a variation of the bath treatment which is currently recommended in Australia for application of fungicides within a few hours after harvest [252]. Bins of fruit direct from the orchard are immersed by forklift truck into a suspension of thiabendazole or benomyl for 30 sec. In California, oranges which require ethylene degreening or storage for a few days before washing are drenched with a recirculated aqueous solution of sec-butyl-amine soon after harvest (Fig. 7).

A common commercial practice today is the application of postharvest fungicides in the form of a spray, flood, or some modification of these techniques, to the fruit as they are transported on a horizontal conveyor. These techniques are better suited for the application of the newer organic fungicides which have been introduced over the past decade. Relatively small volumes of formulation may be applied without recycling, thereby conserving expensive compounds while minimizing problems of stability, pH control, and sanitation. These fruit-treating systems are uncomplicated in operation, and product throughput is usually substantially higher than in the case of the packing line with a bath treatment. The disadvantages of flood and spray treatments are that they are difficult, and in some cases impossible, to apply at an elevated temperature to improve the effectiveness of the treatment. Furthermore, the flood- and spray-type treatments are usually less effective than a bath treatment in disease control since the time the product is in contact with the fungicide solution is substantially reduced. Those application techniques which apply a substantial volume of the fungicide solution to the fruit are invariably more effective than very low volume treatments. The fungicide solution must wet the fruit thoroughly.

In Florida, recirculated solutions of sodium o-phenylphenate are flooded onto citrus fruit as they are carried along a roller conveyor or transverse brushes [253-256]. This method eliminates the large volume of fungicide formulation required, but the temperature and pH of the solution are difficult to control accurately. However, the solution is not heated above 32°C, if at all, and pH control is less critical because hexamethylenetetraamine is added to the formulation to prevent phytotoxicity (see Section V.F.1). A flood application of thiabendazole is an effective means of controlling citrus fruit decay, but the treatment is not recommended for commercial use in Florida because of the practical difficulty of maintaining the desired concentration of thiabendazole in the recycled suspension which is flooded onto the fruit [253, 256]. Suspensions of benomyl and dicloran have been flooded onto peaches and nectarines for control of brown rot and Rhizopus rot [135, 257].

FIG. 7. A. Recently harvested oranges in field bins being drenched with a solution of sec-butylamine phosphate before degreening with ethylene gas. B. Oranges being cleaned and treated with a solution of 2.0% sodium o-phenyl-phenate tetrahydrate in a foam washer. Photograph by FMC Corp. Reprinted from Ref. 25, pp. 128 and 193, by courtesy of Fisons, Ltd.

Low-volume sprays of benomyl and thiabendazole are applied to citrus fruit on roller conveyors and brushes. This method of application is somewhat less effective than the flood application method, but is nonetheless recommended in Florida because the suspensions are not recirculated, eliminating the necessity for monitoring the fungicide concentration. This method works well with suspensions containing 500-1,000 ppm thiabendazole or benomyl which are effective in reducing fruit infection. At concentrations of 2,000-6,000 ppm required for sporulation control, problems may arise in maintaining a homogenous suspension and with blocking of the spray nozzles. Spray applications of dicloran and benomyl have been used effectively for control of postharvest decay of peaches, nectarines, and sweet cherries [135, 257]. Sprays of chlorinated water have been applied to a variety of fruits and vegetables to inactivate surface contamination. Low-volume and ultralow-volume sprays (mists) of thiabendazole are applied to potatoes entering storage. Designs of equipment suitable for this purpose have been published [237, 239]. The mist treatments are less effective than dip and spray treatments for control of Phoma and Fusarium on potatoes during storage [132, 237]. The advantage of the mist treatment is that the volume of water applied is so small that drying is not necessary before the tubers enter storage. The spray application is more effective than the mist application for control of Fusarium, but the wet tubers must pass over a bed of foam rolls in order to partially dry the tubers before storage [239].

The preferred method for applying thiabendazole and benomyl to oranges in South Africa is to brush the fruit on a roller conveyor with an overhead transverse brush with long filaments which have been wetted with a suspension of the fungicide [258, 259]. The brush is supplied with the fungicide by means of a flood trough or by sprays. This method results in a uniform wetting of the fruit with the fungicide suspension and does not require the low-volume spray nozzles which may be blocked by heavy suspensions. A similar technique has been used to apply dicloran to peaches [135].

The foam washer is conventional equipment for simultaneously washing and treating citrus fruits with sodium o-phenylphenate (see Fig. 7B). A solution of sodium o-phenylphenate and a high-foam surfactant are whipped into a foam which descends as a curtain on the fruit being transported on brushes [131]. The procedure is a convenient method of washing lightly soiled fruit, but the foam washer is inferior to the bath treatment for control of Penicillium mold [131]. Phytophthora brown rot may be a serious problem in citrus fruit which are treated only with a foam washer since there is no means of heating the fruit to eradicate incipient infections of this fungus.

Many fruits and vegetables are coated with wax formulations to prevent excessive water loss from the product during marketing. The addition of one or more fungicides to a wax formulation is a convenient means to treat produce for disease control and, for some crops, the wax formulation may be the only liquid applied after harvest. Water-emulsion waxes are usually applied to fruit by brushes which are fed the formulation by spray nozzles.

Emulsion wax formulations containing sodium o-phenylphenate, thiabendazole, benomyl, and sec-butylamine have been used extensively on citrus
fruits throughout the world [194, 251, 253, 255, 256, 258-260]. Application
of thiabendazole as a water suspension is preferred in Israel because the
wax treatment was slightly less effective despite the greater thiabendazole
residues which resulted from wax applications [251]. Water suspensions of
thiabendazole may wet the surface of the fruit and superficial injuries more
efficiently than the wax treatment. Water-emulsion wax formulations are
often alkaline in reaction and may result in inactivation of some fungicides,
such as benomyl, which are unstable under alkaline conditions [261] (see
Section V.I). o-Phenylphenol, thiabendazole, and benomyl have been dissolved in a formulation of a synthetic coating polymer in a mixture of
hydrocarbon solvents. Benomyl is probably not stable in formulations of
this type [262]. The solvent formulation containing the fungicide is misted
(ultralow volume) onto the fruit. This treatment is convenient to apply and
provides good control of Penicillium decay initiated in superficial injuries,
but is less effective for control of Diplodia stem-end rot of oranges [259].
Apparently, the coverage of the stem end of the fruit with the fungicide is
inadequate for preventing the development of quiescent infections of Diplodia
at this site.

C. Barriers to the Spread of Disease

Certain postharvest diseases, such as those incited by Penicillium italicum,
Rhizopus, Trichoderma, and Geotrichum may spread from diseased to
healthy fruits and vegetables even at the recommended temperature for
storage and transport. Contact spread of these diseases can be reduced by
wrapping the individual fruit in plain tissue paper or, better, in paper wraps
impregnated with a fungistat such as biphenyl, sodium o-phenylphenate,
dicloran, or copper compounds (see Section III.C.2). Plastic and wax-paper
cups and trays are also effective as mechanical barriers to prevent the
spread of postharvest diseases.

V. PROPERTIES AND APPLICATIONS OF
SPECIFIC FUNGICIDES

A. Alkaline Inorganic Salts

Solutions of sodium tetraborate (borax) and sodium carbonate have been
used for several decades to reduce decay of citrus fruits. Early experiments with oranges revealed that a 6-8% borax solution was required for
effective control of Penicillium molds and stem-end rots incited by Diplodia
natalensis and Phomopsis citri [263]. The effectiveness of the treatment

could be improved by heating the solution to 43°C or by not rinsing the borax residues from the fruit after treatment. A 6% borax solution at 38°C
is only weakly fungicidal in vitro to conidia of Penicillium digitatum, but
at 43.5°C the treatment is lethal after 5 min contact [264]. These results
indicate that the effectiveness of the borax fruit treatment, applied at a
temperature lower than 43°C, must be due to the fungistatic action of borax
residues remaining in injuries on the surface of the fruit after rinsing.
Since the borate ion is the active agent of the borax treatment, several
other borate salts of greater cold-water solubility have been suggested as
more convenient than borax for the preparation of fruit-treating solutions.
Solutions of sodium metaborate, sodium pentaborate, lactic acid complexes
of boric acid, and mixtures of borax and boric acid have been shown experimentally to be effective for the control of citrus fruit decay. The physical
and chemical properties of these compounds have been discussed in detail
elsewhere [24]. The borax treatment widely employed in citrus packinghouses uses a solution of 4% borax and 2% boric acid (pH ca. 9) at a temperature of 38-43°C. The oranges are immersed in the solution for 2-4
min, rinsed with sprays of water, and waxed. The water rinse reduces
the effectiveness of the treatment, but is required to prevent the boron content of the fruit from exceeding the U.S. tolerance of 8 ppm. The application of a wax coating to prevent water loss is essential since the warm alkaline borax treatment removes a substantial portion of the natural wax from
the fruit. The borax-bath treatment is little used today because of the problem of disposing of large volumes of rinse water containing a level of boron
which could contaminate underground water supplies. Borax has been
replaced to a large extent, beginning in the 1950s, by sodium o-phenylphenate
which is somewhat more effective against both Penicillium molds and stem-
end rot fungi and constitutes a less serious water pollution problem than
borax. More recently, 3% borax has been added to water-emulsion waxes
applied to lemons before storage, and a 2% borax solution at 66°C is flooded
onto lemon storage boxes to kill mold spores before reuse. Borax solutions
have been reported to reduce decay of melons [265] and sweet potatoes [266],
but the treatment apparently has not been used extensively on these products.

California lemons are immersed in a solution of 3-5% sodium carbonate
at 40-48°C for 2-4 min, followed by a water rinse. The solutions are considerably less fungicidal in vitro than a borax solution of the same strength
and temperature; therefore, control of fruit decay by sodium carbonate
appears to be due almost entirely to the accumulation of alkaline residues in
potential infection sites on the surface of the fruit. These alkaline residues
increase the pH of the infection site environment to a level which does not
permit the germination and growth of Penicillium digitatum and P. italicum.
The effectiveness of the sodium carbonate treatment is improved by heating
the solution to 43-48°C, but this practice should be approached with caution
since it has been shown that this treatment may predispose mature fruit to
decay without causing visible symptoms of injury [15]. Sodium bicarbonate

has also been suggested as a treatment to control <u>Penicillium</u> on citrus
fruits [267] but this compound is inferior to sodium carbonate.

B. Positive Halogen Compounds and
Other Oxidizing Substances

The chemical reactivity of the elemental halogens decreases in the order
chlorine, bromine, iodine. Gaseous chlorine is extremely fungicidal, but is
injurious to fresh produce at dosages required to control decay [268]. Elemental iodine is more selective in its action and has been suggested as a
fungicide for use in fruit wrappers to control postharvest diseases [269].

1. Hypochlorous Acid

Solutions of hypochlorous acid and its salts (hypochlorites) have been utilized extensively to reduce microbiological contamination of water used to
wash, transport, and hydrocool fruits and vegetables and to bleach superficial mold growth and scars on the surface of oranges and melons. Although
solutions of hypochlorous acid have been used as postharvest treatments
for decades, the published reports have, in many instances, disregarded
the principles affecting the properties of aqueous solutions of chlorine.
Elemental chlorine hydrates in water solution to form hypochlorous acid
(pKa = 7.46) which upon dissociation yields the hypochlorite ion. The concentration of elemental chlorine is negligible in a solution of pH 5 or greater
and the relative concentration of hypochlorous acid and hypochlorite ion are
governed by the pH of the solution (Fig. 8). It has been demonstrated repeatedly that both the antimicrobial effectiveness and the bleaching action of
chlorine solutions are a function of the undissociated hypochlorous acid in
the solution [40, 270-272]. That is, hypochlorous acid oxidizes organic
molecules more rapidly than hypochlorite ions; therefore, solutions containing most of the chlorine in the form of hypochlorous acid are more reactive,
whereas solutions with most of the chlorine in the form of hypochlorite are
the more stable in the presence of organic matter. The concentration of
active chlorine (Cl^+) in a solution is frequently expressed in terms of ppm
or % "available chlorine," which is the oxidizing power of the solution determined by titration with iodide ion. Since 1 gram atom of positive chlorine
liberates 1 mole (2 gram atoms) of I_2, a solution containing 100 ppm "available chlorine" actually contains only 50 ppm of Cl^+. The practice of determining the available chlorine iodometrically is highly desirable since it
provides an exact measure of the potential chemical activity of the hypochlorite solution. Unfortunately, many investigators in postharvest pathology
have reported the concentrations of hypochlorite solutions in terms of ppm
chlorine, presumably derived by calculation and dilution from a concentrated
solution of commercial bleach or calcium hypochlorite. Due to the unstable
nature of solutions of Cl^+, this procedure results in treating solutions which

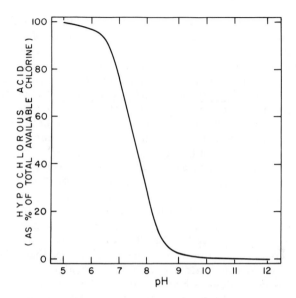

FIG. 8. Relationship between pH and the relative concentration of undissociated hypochlorous acid (HOCl) in a solution of chlorine.

vary greatly in oxidizing power. Statements on the antimicrobial effectiveness of chlorine solutions should specify the concentration of available
chlorine and the pH of the solution.

Hypochlorous acid, introduced as chlorine gas or hypochlorite salts,
rapidly kills microorganisms in recycled water used for handling and treating produce, thereby reducing the hazard of inoculating the produce with
pathogens during these operations [40, 119, 127, 272]. Although several ppm
of active chlorine may be sufficient to kill microbial cells in clear water
[119, 127], 50-100 ppm active chlorine at pH 7-8.5 is frequently employed
in commercial practice with solutions carrying a substantial amount of soil
and organic matter [129, 273]. The pH selected is a compromise between
activity against microorganisms and stability of the solution. Despite the
rapid lethal action of active chlorine against pathogenic microorganisms
suspended in water, chlorine treatments are rarely effective in preventing
infection by propagules of a pathogen imbedded in injured host tissue or in
natural openings such as lenticels or stomates. Hypochlorous acid is a
highly reactive and nonspecific oxidizing agent and is reduced by organic
constituents of the host tissue before it inactivates the pathogen cells [271].

Hypochlorite solutions have been evaluated extensively for control of
Monilinia fructicola and Rhizopus stolonifer on peaches, with major emphasis
on the chlorination of water in hydrocoolers [129, 137]. The usual concentrations have been between 100 and 200 ppm active chlorine, although peaches

can tolerate at least ten times this concentration without injury. Control of Monilinia on peaches has been erratic and the incidence of Rhizopus has been little affected by the hypochlorite treatment. Lipton and Stewart [274] observed that the treatment of melons with hypochlorite solutions had no appreciable influence upon the development of Rhizopus rot, but did reduce superficial growth of Alternaria sp. and Cladosporium sp. on the rind. The superficial growth of Stomiopeltis citri (sooty blotch) on oranges was elimi-nated by soaking the fruit in a hypochlorite bath, but this is principally a matter of bleaching the dark mycelium of this fungus which is growing as a saprophyte on the surface of the fruit [271]. Treatment of inoculated spin-ach [275], peppers [103], and potatoes [69] with hypochlorite solutions has resulted in little or no reduction of bacterial soft rot. This treatment also did not control fungus diseases of carrots and beans after harvest [55, 276].

The use of hypochlorite for reducing the population of microorganisms in water for cleaning, cooling, and treating fruits and vegetables is unques-tionably a beneficial practice because it prevents inoculation of the produce during these operations. However, there is little hope that hypochlorite solutions, even at concentrations of 5,000 ppm or higher, can eradicate pathogenic microorganisms situated in injured host tissue or other environ-ments which bring about a rapid reduction of active chlorine [271].

2. Chloramines

Inorganic and organic compounds containing positive chlorine bonded to a nitrogen atom have attracted interest as decay control agents because they are generally more stable than hypochlorous acid in neutral and slightly acidic environments. These compounds are apparently able to penetrate in an active state into injured host tissue where they react with the pathogen directly or slowly hydrolyze to give hypochlorous acid.

Nitrogen trichloride has been used as a fumigant to control postharvest diseases of citrus fruits, melons, and certain vegetables [268, 277]. Pure nitrogen trichloride is a dense yellow liquid which boils around 71°C and is explosively unstable when concentrated. Nitrogen trichloride is freely soluble in organic solvents but practically insoluble in water. For fumiga-tion of fruits and vegetables, nitrogen trichloride is generated in the pack-inghouse by chlorination of an aqueous solution of ammonium chloride, and immediately volatilized and diluted with a stream of air which carries the dilute nitrogen trichloride to the fumigation chamber. Lemons in storage may be fumigated three times weekly with vapor concentrations of 18-26 mg nitrogen trichloride per m^3. The treatment is not only effective in reducing the number of decayed fruit, but also accelerates the drying of decayed fruit and inhibits the sporulation of Penicillium. Oranges may be fumigated 3-4 hr at dosages of 54-106 mg nitrogen trichloride per m^3. This treatment may be repeated if the oranges are held for several days before packing. The review of Pryor [277] provides the specific details for the fumigation of vegetables with nitrogen trichloride. It is not a common fumigant for fruits and vegetables today because the products of its reactions in a moist environment (hypochlorous acid and HCl) cause a severe problem of deterioration of metal fixtures and wood boxes in the storage room. Papers impregnated with dichloroisocyanuric acid and its salts, which react with water to form nitrogen trichloride, are effective in reducing decay of packaged fruits [278]. Papers impregnated with N-bromo- or N-chlorohydantoins are also effective for this application [222]. The efficiency of nitrogen trichloride in preventing decay of inoculated fruits may be explained by its poor water solubility and lower chemical reactivity compared to hypochlorous acid, properties which could lead to selective action against the pathogen imbedded in host tissues. Attempts have been made to increase the stability of neutral and slightly acidic solutions of chlorine through the use of chloramines [277]. Since low concentrations of hypochlorous acid exist in equilibrium with chloramines in solution, the latter compounds function as reservoirs of highly reactive chlorine at low pH values where the hypochlorite ion does not exist. A satisfactory chloramine for this purpose should be reasonably stable (not subject to decomposition by chlorine), nonvolatile, and have a low combining weight with respect to chlorine. Solutions of melamine (2, 4, 6-triamino-1, 3, 5-triazine) or sulfamic acid have been chlorinated to give more stable solutions of active chlorine.

3. Ozone

Ozone has long been advocated as a fumigant for the control of postharvest diseases. Several independent investigations have conclusively demonstrated that concentrations of ozone tolerated by fresh fruits and vegetables will not prevent the development of decay after inoculation with pathogenic microorganisms. An effective concentration of ozone in the atmosphere will prevent growth and sporulation of fungi on the surface of produce, reduce the viable spore count in the atmosphere of the storage room, and destroy offensive odors in the storage room. Ozone, like chlorine, is rapidly inactivated by reducing substances present at potential infection sites [268, 279-282].

C. Sulfur Compounds

1. Sulfur Dioxide and Bisulfite Salts

The usefulness of sulfur dioxide (SO_2) in the preservation of foodstuffs has been recognized since antiquity and its many applications in this field have been reviewed by Joslyn and Braverman [283]. The use of SO_2 to control postharvest diseases has been limited to grapes and, to a lesser extent, raspberries [284, 285]. They are among the few fresh fruits and vegetables which can tolerate the concentrations of SO_2 required to control postharvest diseases. Liquid SO_2 boils at $-10°C$ at atmospheric pressure and its vapor density is 2.3 times that of air. Sulfur dioxide is soluble in water to the extent of 10.5% (by weight) at 20°C. Aqueous solutions of SO_2 contain molecular SO_2, sulfurous acid, bisulfite ion, and sulfite ions in accordance with the equilibria:

$$[SO_2 \xrightleftharpoons[]{H_2O} H_2SO_3] \xrightleftharpoons[]{pK_1=1.8} [HSO_3^-] \xrightleftharpoons[]{pK_2=7.0} [SO_3^{2-}]$$

Only a few percent of the SO_2 in water solution exists as sulfurous acid. The concentration of SO_2 and ionic species in a solution may be calculated at any given pH (see Section V.F.1) and these values have been presented graphically for the entire pH range by Joslyn and Braverman [283]. The antimicrobial properties of solutions of SO_2 and bisulfites have been attributed to the concentration of undissociated molecules (SO_2 + H_2SO_3) with very little contribution from the ionic species present [286]. Sulfur dioxide in aqueous solutions is highly toxic to spores and vegetative tissue of fungi [145, 146, 287]. Sulfur dioxide gas was much more toxic to spores of Botrytis cinerea and Alternaria sp. after their hydration and the relative humidity and temperature of the fumigation atmosphere were increased [145, 146]. These observations suggest that sulfurous acid is the ultimate toxicant, combining irreversibly with some essential cell constituent. The fumigation of grapes with SO_2 during storage and in railroad cars has been a standard practice in California since 1928 [288]. Fumigation during low-temperature storage makes it possible to market California grapes throughout the year. Certain California varieties (e.g., Emperor and Tokay) can be stored for months at 0°C without serious loss of quality, but Botrytis cinerea slowly develops in field-infected berries and forms "nests" of decayed berries if no remedial action is taken. These grape varieties withstand repeated fumigation with concentrations of SO_2 which prevent the outgrowth of Botrytis from infected berries, but grapes cannot tolerate the levels of SO_2 required to eradicate field infections of this pathogen [287]. The usual program for fumigation of grapes in California consists of an initial application of 1% SO_2 (by volume) for 20 min as soon as possible after harvest, followed in storage every 7-10 days with a fumigation of 0.25% SO_2 for 30-60 min [1, 230]. An effective SO_2 fumigation program is regarded as a compromise between decay control and fruit injury [231]. Injury to the berries, expressed as bleaching around the cap stem and injuries, is accumulative so that the SO_2 dose applied in each fumigation is critical for grapes intended for long-term storage. Nelson and Baker [230] reported that relatively low concentrations (0.05-0.1% SO_2) applied at 7-day intervals gave better control of Botrytis rot with less fruit injury than higher concentrations applied at less frequent intervals.

A commercial practice in Australia, California, and South Africa is the fumigation of grapes during shipment by SO_2 evolved by the hydrolysis of sodium bisulfite or sodium metabisulfite in the shipping container [227, 289, 290]. A small quantity of sodium bisulfite or metabisulfite is mixed with the sawdust or granulated cork in which the grapes are packed. Moisture in the sawdust and the grapes is absorbed by the bisulfite, resulting in its hydrolysis to gaseous SO_2. Ideally, it should be generated slowly over a period of weeks and should permeate around and among the grapes to prevent the outgrowth of Botrytis cinerea from each infected berry. In practice, however, the rate of SO_2 fumigation is a function of the moisture level of the package material which may vary from shipment to shipment. Various techniques have been described for controlling the rate of evolution of SO_2, such as adjusting the pH of a bisulfite solution sprayed onto the packaging material [291], enclosing a bisulfite solution in small polyethylene bags [226], and by enclosing solid sodium bisulfite with a barrier which retards the ingress of water vapor or the rate of diffusion of SO_2 from the formulation into the atmosphere around the grapes [292, 293]. Some interesting formulations have been designed to release SO_2 in two stages: the first burst of SO_2 within 1 or 2 days after the package is closed in order to kill spores on the surface of the berries and perhaps eradicate some incipient infections, and a second-stage release of SO_2 to maintain a concentration of several ppm in the atmosphere, in order to prevent growth of hyphae out from infected berries for several months storage at $0°C$ [293,294]. Formulations designed for multistage release of SO_2 in a package stored at constant low temperature may injure grapes if used under fluctuating temperature conditions such as might be encountered in long-distance shipments [225].

2. Organic Sulfur Compounds

A paste of sodium dimethyldithiocarbamate and sodium 2-mercaptobenzo-thiazole in a synthetic latex was applied to the cut end of a banana main stem to prevent entry of pathogenic fungi at this site [295]. This practice became obsolete in the late 1950s when the Central American banana industry converted to the shipment of individual hands severed from the main stem. The standard practice in the early 1960s was to dip the cut hand stems (cushions) in a suspension of maneb. This treatment has been described as moderately effective but with a tendency to fail when the inoculum potential was high or the shipment conditions were less than ideal [295]. Dithiocarbamate fungicides are less effective than thiabendazole and benomyl for control of crown rot and they have replaced the dithiocarbamates for this application [97, 295, 296].

The coordination product of zinc and maneb (Dithane M-45) reduced Fusarium rots of potatoes [240] and melons [297]. An emulsion wax containing sodium dimethyldithiocarbamate is a standard commercial treatment for control of postharvest diseases of melons. Sodium dimethyldithiocarbamate applied in hot water controlled stem-end rot, anthracnose and Aspergillus rot on mango fruits [298], and maneb, thiram, and ziram have provided some degree of protection against fungi attacking tomatoes after harvest [299].

D. Ammonia and Aliphatic Amines

1. Ammonia

Ammonia has been recognized as a potentially valuable postharvest fungicide for many years [300-302]. Research on its utilization for the control of decay of citrus fruits was revived in the 1950s. Renewed activity in this area was prompted by marketing problems associated with other chemical treatments utilized at that time and by the laboratory observation that a low concentration of ammonia gas (50-200 ppm, v/v) would effectively control Penicillium decay, but was not injurious to fruit exposed to this atmosphere for many hours [303, 304]. The practical application of this finding required the development of a stable formulation which would generate ammonia at a controlled low rate when placed in a fruit container. These investigations were further encouraged by the fact that ammonia is a natural constituent of the juice of citrus fruits.

The two basic ammonia-generating formulations developed were (1) based upon the hydrolysis of ammonium salts of weak acids, and (2) consisted of a mixture of a stable ammonium salt with an alkaline carbonate or oxide. The formulations were prepared as compressed tablets or paper sheets impregnated with the reactants [304-306]. Under constant environmental conditions the release of ammonia from the formulations could be controlled by choice of ingredients or by the physical nature of the formulation. Promising formulations were tested extensively with fruit, both in controlled environments and in commercial shipments with the following conclusions: (1) the rate of ammonia generation by moisture-activated systems is very difficult, if not impossible, to control under the variable humidity conditions which prevail in fruit containers during commercial shipment [307, 308]; (2) atmospheres of 100-400 ppm ammonia surrounding the fruit, which were required for prevention of Penicillium sporulation, could not be sustained by an ammonia-evolving formulation for more than a few days [304]; (3) ammonia formulations that controlled decay effectively also caused darkening of injuries and of the stem buttons in sensitive lots of fruit [304, 309, 310]. Ammonia-generating formulations have been evaluated also for the control of postharvest diseases of small fruits without very promising results [311].

Ammonia gas has been evaluated as a space fumigant for oranges during degreening with ethylene gas at 20-25°C. An applied dosage of ammonia in the range 500-3,000 ppm-hr (ppm ammonia by volume × hours exposure) reduced decay substantially without damaging the fruit [303]. An effective and convenient fumigation schedule consisted of 100 ppm ammonia for

9-10 hr. Ammonia gas could be supplied from a tank of liquid ammonia or from ammonium hydroxide [312]. Despite the effectiveness of ammonia fumigation, the treatment was never utilized on an industry-wide scale because of the slight inconvenience involved and the lack of adequately trained people in commercial packinghouses. In addition, ammonia gas may cause darkening of the calyx and of injuries on lemons [304, 310]. Ammonia fumigation has been suggested also for retarding Rhizopus rot in peaches [313] although this fruit is much more sensitive to ammonia than are citrus fruits. sec-Butylamine is much more effective than ammonia and should be the choice over ammonia for all of the applications described above.

2. sec-Butylamine (SBA)

A number of aliphatic amines have been evaluated for the control of Penicillium decay of citrus fruits. Injection of liquid amines of relatively high vapor pressure into fruit containers produced a substantial reduction in decay although slight injury was sometimes evident [141]. Volatile amines, analogous to ammonia, are absorbed by moist injuries on the fruit surface and temporarily increase the pH of potential infection sites beyond the range which supports germination and development of Penicillium digitatum spores [141]. Amines at high concentrations may be lethal to pathogen propagules, but at concentrations tolerated by fruit the treatment is only fungistatic, preventing infection during the period of susceptibility of the host tissues. Neutral salts of aliphatic amines which remain in the fruit injuries are not fungistatic; therefore, the fumigation treatment has no residual effect upon future growth of the pathogen in the injured host tissue.

   sec-Butylamine (2-aminobutane) is unique in that the sec-butylammonium ion is fungistatic to several species of fungi which cause postharvest diseases [246]. Thus, sec-butylamine may be applied to the harvested crop as a fumigation treatment or as a solution of a salt of sec-butylamine. The fumigation treatment is effective not only because the injuries on the surface of the fruit become alkaline by absorption of the amine vapor, but also because the fungistatic residues of neutral sec-butylamine salts persist in injuries following the fumigation treatment [246, 314].

   Partially neutralized solutions of 0.5-2% sec-butylamine have controlled Penicillium digitatum and P. italicum on citrus fruits (Fig. 6), P. expansum on pome fruits, Monilinia fructicola on peaches, and Gloeosporium in wounds on bananas [246]. The effectiveness of sec-butylamine in solution for the control of Penicillium rots of citrus fruit has been confirmed in several laboratory experiments [315-318] and in large-scale commercial trials [153]. Other investigators have reported that sec-butylamine controls stem-end rots of citrus fruit [319], Penicillium expansum on apples [320], Botrytis on cut flowers [321], and Monilinia fructicola on peaches [322]. sec-Butylamine controlled Nectria galligena on apples, a disease which is not controlled by thiabendazole [12, 323].

In California, oranges are drenched with a solution of 1% sec-butylamine (phosphate salt, pH ca. 9) after harvest to protect the fruit injuries against infection by Penicillium digitatum and P. italicum during the subsequent degreening period or during a holding period of several days between harvesting and processing for shipment as fresh fruit (Fig. 7A). The degreening operation is highly conducive to the development of Penicillium decay since it involves exposure of the fruit to a low concentration of ethylene gas for several days in a warm (23-30°C), humid environment. sec-Butylamine is added also to emulsion wax formulations applied to lemons before storage to control species of Penicillium, but the treatment has little effect upon diseases caused by Alternaria or Geotrichum which are resistant to sec-butylamine as well as to the 2-substituted benzimidazoles. The sec-butylamine treatment may be followed by a thiabendazole or benomyl treatment after storage since isolates of Penicillium rarely exhibit resistance to both sec-butylamine and the 2-substituted benzimidazole fungicides.

Pure sec-butylamine is a liquid with a vapor pressure of 135 mm at 20°C and a boiling point of 63°C. Liquid sec-butylamine may be volatilized in a storage room or in a sealed package of fruit for fumigation action. Exposure of citrus fruits to an atmosphere of 100 ppm sec-butylamine (v/v) for 4 hr reduced Penicillium decay to a very low level [314]. Several equivalent combinations of amine concentration and exposure time gave similar results (Fig. 2). The fumigation treatment resulted in the deposit of a fungistatic residue of neutralized sec-butylamine in the injuries on the surface of the fruit, as revealed by chemical analysis and by the resistance of the injuries to infection 24 hr after the fumigation treatment. The effectiveness of the sec-butylamine fumigation treatment has been confirmed in tests with several varieties of Italian citrus fruits [324]. Grierson and Hayward [325] developed a simple and effective method for fumigating citrus fruits in the grove after harvest for use in developing countries where the fruits are not accumulated in a central packinghouse, but are delivered directly from the grove to the market. An extensive series of trials in Florida have shown that a sec-butylamine fumigation of packaged citrus fruits (35-70 ml/m^3) before and during shipment is highly effective in reducing decay of fruit during marketing [326].

Graham and Hamilton [327, 328] demonstrated that fumigation of seed potatoes with sec-butylamine (200 mg SBA/kg potatoes) gave excellent control for several months of gangrene (Phoma) and skin spot (Oospora) and partial control of silver scurf (Helminthosporium) during storage. They found that sec-butylamine penetrated the suberized periderm more effectively than methoxyethylmercuric chloride applied as a dip treatment and, therefore, the fumigation treatment could be delayed for 14 days after harvest, without a serious loss in disease control.

E. Aldehydes, Epoxides, and Organic Acids

Formaldehyde is used extensively as an inexpensive and effective fumigant to decontaminate picking boxes, packing facilities, and storage rooms.

Formaldehyde in water (1-2%) is usually applied as an aerosol spray or added to the air washers used to humidify storage rooms. Formaldehyde gas volatilizes from the spray and is highly effective in killing spores of pathogenic fungi. The direct application of formaldehyde to living fruits has not been successful because dosages required for decay control usually injure the host. Attempts to utilize low concentrations of formaldehyde and organic compounds which generate formaldehyde for the treatment of fresh fruits and vegetables has been considered in an earlier review [24].

Acetaldehyde is better tolerated by fruits and vegetables than formaldehyde. Tomkins and Trout reported in 1932 that diseases of several fruits could be controlled by fumigation with low concentrations of acetaldehyde [329]. In the early 1970s, several publications reported the control of Botrytis and Rhizopus on strawberries and raspberries, and Penicillium expansum on apples by exposure of the fruits to 0.25-1.0% (v/v) acetaldehyde for 1-2 hr [330-332]. A main attraction of the acetaldehyde treatment for postharvest use is that this compound is a natural volatile constituent of fruits and vegetables and is permitted as a food additive in the United States.

Ethylene oxide and propylene oxide have been injected directly into sealed packages of partially dehydrated dates, figs, and prunes in California for the control of yeast and saprophytic fungi [333]. This practice was discontinued after residues of ethylene chlorohydrin were identified in treated fruits [334].

The successful use of organic acids as microbial inhibitors in the processed food industry inevitably led to the evaluation of these compounds for control of postharvest diseases of fresh fruits and vegetables. Sorbic acid (2,4-hexadieneoic acid) and dehydroacetic acid (3-acetyl-6-methyl-2,4-pyrandione) possess greater antifungal activity in vitro than the common aliphatic and aromatic carboxylic acids and have been widely tested for control of postharvest diseases. Carboxylic acids, like phenols, exhibit greatest antifungal activity in the undissociated form (see Section V.F). The carboxylic acids, propionic and sorbic, have pK_a values in the range 4-5, and therefore exhibit greatest antifungal activity in media at pH < 4 [335]. Their activity is greatly diminished at pH > 6. The antifungal activity of dehydroacetic acid persists, in contrast, at pH 7, but is weak at pH 9 [335]. The pH of the solution of organic acids for treatment of fruits and vegetables may not be critical, however, since the pH of the fruit tissue is acid enough in many cases to hydrolyze residues of salts of organic acids to their more active undissociated form [336].

The relatively few reports on the effectiveness of the salts of simple aliphatic acids indicates that these compounds, although effective in processed foods, have little place in postharvest disease control [337, 338]. However, a great deal of interest has developed in the past few years in the treatment of high-moisture grains with propionic and acetic acid to inhibit the growth of saprophytic species of Penicillium, Aspergillus, and Fusarium on the grain during storage [338-341]. Application rates of 0.5-1.0% of different acids have controlled growth of fungi on 22-24% moisture grains for 8 months at ambient temperature. The treatment destroys the germination

ability of the grains but does not alter their usefulness as feed for livestock.
Propionic and methylenebispropionic acid were superior to other simple
acids tested; the salts were less effective than the free acids [338, 339].
This may indicate that the volatile free acids become better distributed and
penetrate to the potential infection sites more effectively than nonvolatile
salts. On the other hand, nonvolatile sorbic acid was highly effective when
applied in alcohol solution to the grain but was ineffective when applied as
a dry powder or as potassium sorbate [338].

The development of Alternaria sp. and Cladosporium sp. on fresh figs
has been retarded by the application of sorbic acid [126], but this compound
is ineffective in preventing decay of peaches and strawberries [337]. Sodium dehydroacetic acid, applied as a 0.1-0.5% dip treatment, retarded the
development of pathogenic fungi on fresh strawberries and raspberries [342].
The treatment has not been accepted as a commercial practice because of
the possible deleterious effect of water alone on the market life and appearance of the berries. Postharvest dip treatments of 0.5-1.0% sodium dehydroacetic acid have been reported to retard Monilinia fructicola and
Rhizopus stolonifer on peaches [130, 149], but the results have been less
beneficial in other trials. Dehydroacetic acid may injure fruit and vegetables as shown by delayed ripening and injury to the skin of peaches, darkening of the skin of carrots, and bleaching of strawberries.

F. Phenols

1. o-Phenylphenol

In a search for volatile fungicides, Tomkins [343] found that postharvest
diseases of several varieties of fruits could be reduced by wrapping them in
paper impregnated with o-phenylphenol. Unlike biphenyl, which he had
tested earlier, o-phenylphenol caused injury to the fruit. This injury could
be reduced considerably by formulating the o-phenylphenol with high-
molecular-weight bases, such as hexamethylenetetraamine (hexamine), or
with oils [218, 343]. These substances interact with o-phenylphenol, reducing the vapor concentration of the free phenol and thereby reducing phytotoxicity of the treatment. The vapor-phase application of o-phenylphenol
has generated little interest over the years because of the phytotoxicity
problem and because biphenyl has successfully fulfilled this role for citrus
fruits. More recently, the acetate and isobutyrate esters of o-phenylphenol
have been proposed as effective and safe fungicides for impregnation into
wrappers for oranges and tomatoes [224]. These phenol esters also controlled postharvest diseases of grapes and peaches [344, 345].

o-Phenylphenol is soluble in most organic solvents, but is soluble in
water only to the extent of about 800 mg/liter at 25°C [151]. The solubility
in water is reduced considerably by dissolved salts. o-Phenylphenol is a
weak acid with pKa = 10.01 [346]. Sodium o-phenylphenate tetrahydrate

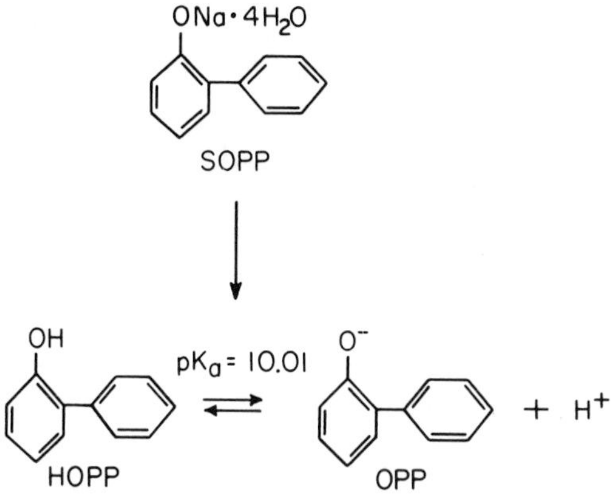

FIG. 9. Equilibrium established between o-phenylphenate (OPP) and o-phenylphenol (HOPP) upon dissolution of sodium o-phenylphenate tetra-hydrate (SOPP) in water. Reprinted from Ref. 147, p. 1097.

(SOPP) is very soluble in water and is hydrolyzed extensively in its aqueous solutions (Fig. 9); a 0.5% (w/v) solution of SOPP is approximately pH 11.2 and a 2.0% solution is pH 11.5. The concentrations of undissociated o-phenylphenol and of the o-phenylphenate anion are dependent upon the strength of the solution (i.e., the concentration of SOPP) and the pH. Their values at any pH may be calculated from the expression,

$$pH - pK_a = \log \frac{[\text{phenate}]}{[\text{phenol}]} = \log \frac{c - x}{x}$$

where x and c - x are the molar concentrations of o-phenylphenol and o-phenylphenate ion, respectively, and c is their sum. Figure 10 shows that a solution of 5 g sodium o-phenylphenate tetrahydrate per liter at pH 12 contains 32 mg undissociated o-phenylphenol per liter and this value increases exponentially to 646 mg/liter at pH 10.6. The graph also reveals that the o-phenylphenol concentration increases, ca. 9.2 fold, as the pH is decreased from pH 12 to pH 11, whereas the concentration of the o-phenyl-phenate anion decreases only about 8% over the same pH range.

   o-Phenylphenol is a broad-spectrum biocide in comparison to most other organic fungicides used for control of postharvest diseases. The undisso-ciated phenol is lethal to microorganisms and is injurious to fresh produce at concentrations of 200-400 mg/liter depending upon the temperature of the solution and the period of contact. The dissociated form (o-phenylphenate ion) is not phytotoxic and a solution of sodium o-phenylphenate (SOPP), con-taining excess alkali to suppress hydrolysis of the salt, is quite safe for treatment of several fruits and vegetables [347, 348]. The intact waxy

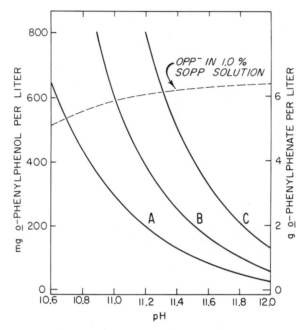

FIG. 10.   Relationship between pH and concentration of undissociated
o-phenylphenol (calculated, $pK_a = 10.01$) in solutions of 0.5% (A), 1.0% (B),
and 2.0% (C) sodium o-phenylphenate tetrahydrate (w/v).   Reprinted from
Ref. 147, p. 1098.

surface of citrus fruits, apples, and other fruits appears to be relatively
impermeable to the o-phenylphenate anion as indicated by low residue levels in fruit treated at pH 12 (Fig. 3).   However, the o-phenylphenate anion
diffuses selectively into injuries where the surface of the fruit is ruptured
and is soon hydrolyzed to o-phenylphenol in the presence of fruit acids and
metabolic carbon dioxide.   Rinsing the treated fruit with water removes
most of the sodium o-phenylphenate residue from the surface of the fruit,
but a substantial amount remains associated with injuries and prevents the
development of pathogens at these sites [147].   The selective accumulation
of o-phenylphenol in injuries and its role in the prevention of fruit decay is
revealed in Figure 5.

    The undissociated o-phenylphenol penetrates into fruit and fungus cells
much more readily than the anionic form [147, 151, 245].   Therefore, the
concentration of o-phenylphenol determines the acute fungitoxicity and phytotoxicity of a solution of SOPP as well as the residue level in treated fruit.
Fruit injury may be anticipated when the concentration of o-phenylphenol
reaches 200-400 ppm; whereas the concentration of o-phenylphenate anion

may be 10-20 times greater without any adverse effect upon the fruit. The pH and temperature of the SOPP solution and the time the fruit are exposed determine fruit injury and residue levels [245]. The principal cause of fruit damage in commercial practice is a drop in pH of the treating solution caused by accidental dilution of the solution with pure water and accumulation of acidic substances such as fruit debris, CO_2, and soil in the treating solution [151].

The benefits of the SOPP treatment when properly administered are twofold: (1) spores of fungi and bacteria on the surface of the fruit or introduced into the cleaning solution are killed; and (2) a residue of o-phenylphenol is deposited in harvest injuries which prevents infection at these sites during storage or marketing. Solutions of SOPP have been used extensively for control of postharvest decay of citrus fruits, apples, pears, peaches, sweet potatoes, and other perishables [24, 26]. The solution recommended for soaking or flooding oranges in Florida contains 2% sodium o-phenylphenate tetrahydrate, 1% hexamine, and 0.2% sodium hydroxide. The pH must be maintained in the range 11.5-12 and the solution should not be heated [253, 254]. Hexamine precipitates free o-phenylphenol as the slightly soluble hexamine phenate (3:1) complex before the concentration of undissociated phenol reaches the threshold for phytotoxicity, and also exerts a buffering effect upon the solution at about pH 11.8 [151]. In addition, the hexamine- o-phenylphenol precipitate reduces the concentration of o-phenylphenol in solution, thereby preventing excessively high residues which are inevitable in fruit treated with solutions of SOPP (without hexamine) at pH values below 11.5 [349]. In other citrus-producing areas the bath for the treatment for oranges usually contains 0.5% SOPP at pH 11.5-12.0, and is heated to 32-43°C. Hexamine is not included in this formulation, but the pH is carefully maintained in the desired range. In the bath treatment the fruit is submerged for 2-4 min in the solution of SOPP, brushed, and finally rinsed with water sprays. Within the past 10 years, the foam washer has become a popular means of treating citrus fruit with SOPP. A solution containing 2% SOPP, 0.1% sodium dodecylbenzenesulfonate, and 0.2% sodium carbonate is agitated into a foam which drops as a curtain over fruit moving beneath it on a horizontal conveyor (Fig. 7B). The foam-covered fruit are brushed for about 15 sec before being rinsed in a spray of water. The foam treatment is less effective in decay control than the bath treatment, but has the advantage that pH control is not a problem because the foam is not recycled [131]. The residues of o-phenylphenol are considerably lower on fruit which have been treated with the foam formulation compared to fruit which have been submerged in a solution of SOPP [350], explaining why the foam treatment is less effective than the bath treatment. Water-emulsion wax formulations may contain 1% SOPP and solvent wax formulations 1% o-phenylphenol, providing an alternate means of applying this fungicide to citrus fruit. The specific recommendations for the use of sodium o-phenylphenate on citrus fruits in different production areas are available in local publications [194, 253, 351].

Apples and pears in the northwestern United States are cleaned in an unheated solution of 0.6% SOPP, pH 11-12 [320]. The treatment kills microorganisms in the washing water, but is not very effective in preventing decay of fruit which are inoculated with Penicillium expansum before the bath treatment. Sodium o-phenylphenate has been added to water used to hydrocool peaches to provide a concentration of 0.1% at pH 11.5. The treatment is fairly effective in controlling brown rot (Monilinia fructicola), but has little effect upon the incidence of Rhizopus rot [130]. The conditions of treatment are critical because peaches are more sensitive to o-phenylphenol than are citrus fruits. Solutions of SOPP have been used to control black rot (Ceratocystis fimbriata) and soft rot (Rhizopus sp.) on sweet potatoes but may cause some injury to bruised tubers [352, 353]. Sodium o-phenylphenate cannot be used safely on Irish potatoes as it injures the periderm and may cause an increase in bacterial soft rot [354]. Treatment of carrots with a solution of about 0.1% SOPP has resulted in good control of decay during long-term cold storage [14, 276, 355], but the same treatment did not prevent the development of incipient infections of Sclerotinia on snap beans [55].

2. Salicylanilide

Salicylanilide was developed initially as a textile preservative but found some applications as a postharvest fungicide for oranges and bananas. Salicylanilide, a weak acid (pK$_a$ = 7.45), is practically insoluble in water but forms soluble sodium and ammonium salts [356]. A solution of 1% sodium salicylanilide effectively reduced Penicillium decay and Phomopsis stem-end rot of oranges [357]. Addition of 1% sodium salicylanilide to an emulsion wax containing SOPP has been recommended as a treatment for oranges when Phomopsis stem-end rot is a problem [351]. Certain hydroxyamine salts of salicylanilide are claimed to be superior to sodium and ammonium salts for control of Diplodia and Phomopsis stem-end rots of oranges [358].

Sodium salicylanilide is highly effective for control of squirter disease of bananas (Nigrospora spherica) [359]. Salicylanilide is much less effective against Gloeosporium rot although it has been tested extensively for control of this disease on bananas in Jamaica [154]. Although small-scale tests there were encouraging, the treatment did not perform well in large-scale shipping trials, presumably because of an excessive delay (incubation period) between natural inoculation and application of the treatment.

G. Hydrocarbons and Halogenated Derivatives

1. Biphenyl

Biphenyl (diphenyl) is a crystalline solid which melts at 71°C and has a vapor pressure of approximately 0.01 mm Hg at 25°C. The vapor pressure curve from 0-40°C is presented graphically in the review of Rajzman [360]. Biphenyl is soluble in many organic solvents, but a saturated aqueous solution at 25°C contains only 10 mg biphenyl per liter.

Biphenyl cannot be considered a broad-spectrum fungicide although its vapors are strongly inhibitory to the mycelial growth of Penicillium digitatum, P. italicum, Diplodia natalensis, Phomopsis citri, Aspergillus sp., Monilinia fructicola, Rhizopus sp., and other plant pathogenic fungi [361]. Bacteria, yeast, Phycomycetes, and resistant strains of Penicillium and Diplodia are not inhibited. The vapor of biphenyl is not essential for fungistatic activity since sensitive strains are also inhibited in aerated liquid cultures.

Tomkins [216] reported in England in 1936 that fruit wraps impregnated with biphenyl controlled Penicillium decay of oranges and reduced sporulation of this fungus on the surface of decaying fruit. He also recognized that a faint odor of biphenyl was detectable on treated fruit, but that the odor was lost after the fruits were exposed to fresh air for several days. The effectiveness of biphenyl in controlling Penicillium decay of citrus fruits has been borne out by several large-scale trials [19, 309] and by more than 30 years' experience of successful commercial use. Biphenyl also reduces stem-end rot caused by Diplodia and Phomopsis but is not effective against fruit diseases caused by Geotrichum, Alternaria, Trichoderma, or Colletotrichum, or biphenyl-resistant strains of Penicillium [362]. Biphenyl has been tested for control of Botrytis on grapes, Rhizopus on peaches, and Fusarium on potatoes with favorable preliminary results, but in each case the odor or flavor of the commodity was adversely affected by the treatment [24]. Biphenyl vapors are injurious to apples and bananas [216].

Citrus fruits destined for distant markets frequently are treated with biphenyl to control decay during transit and storage periods. Biphenyl is impregnated into individual fruit wraps or into paper sheets which are placed at the top and bottom of the fruit containers. The latter method of application has become most common in recent years as the labor of wrapping individual fruit has become prohibitively expensive and mechanical packing of fruit has become a common practice in some areas. Treated tissue wraps are usually 25 × 25 cm and contain 40-50 mg biphenyl. Treated paper sheets are approximately 25 × 40 cm and contain approximately 2.35 g biphenyl.

After the fruit container is closed, biphenyl sublimes into the atmosphere surrounding the fruit and inhibits the development of decays caused by Penicillium digitatum, P. italicum and to a lesser extent Diplodia natalensis. The principal value of the treatment is that it inhibits sporulation of species of Penicillium on decayed fruits and thereby prevents "soilage" of adjacent fruits with mold spores (see Fig. 11).

Oranges and mandarins absorb at least two times more biphenyl per unit of surface area than do lemons or grapefruit [363]. This phenomenon has two important practical consequences: (1) residues of biphenyl in oranges and mandarins may exceed the legal tolerance (U.S., 110 ppm; European EEC, 70 ppm) in poorly refrigerated shipments of fruit in unventilated

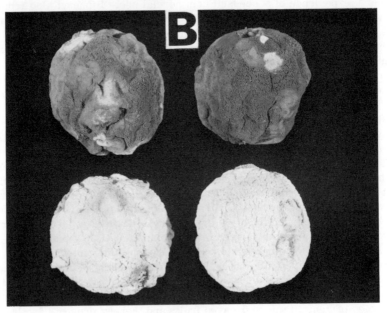

FIG. 11. A. Oranges stored for 3 weeks at 10°C showing soilage of at least eight sound fruits with spores from a single orange decaying with blue mold (Penicillium italicum). B. Inhibition of sporulation of Penicillium digitatum (green mold) on oranges. Lower (white) oranges were taken from a shipping carton with the biphenyl treatment. Reprinted from Ref. 25, p. 120, by courtesy of Fisons, Ltd.

containers, whereas there is little hazard of excessive residues on lemons or grapefruit; (2) the biphenyl treatment is more effective in inhibiting sporulation of Penicillium on lemons than on oranges because the high rate of absorption by oranges tends to reduce the concentration of biphenyl in the atmosphere surrounding the fruit [364-366]. Biphenyl absorbed into the peel of the fruit has a negligible influence upon the development of decay [228]. The usefulness of biphenyl for mandarin-type citrus fruits is highly questionable because of the rather ineffective decay control and high residue levels in fruit which are held at ambient temperature for more than a few days [363].

The biphenyl treatment has been used extensively by citrus exporters over the past three decades and has been a major factor in the development of world trade in these fruits. Despite the practical successes of the treatment, commercial utilization of biphenyl has been beset by several problems. The slight "hydrocarbon" odor associated with treated fruit has resulted in consumer resistance to such fruit, especially in markets where untreated fruits are available for comparison. The odor is due to biphenyl which is loosely absorbed by the surface of the fruit and is no indication of the total biphenyl residue in the fruit. This transient odor dissipates from the fruit within a few days after the fruit are removed from the presence of the biphenyl formulation [366].

Early in the development of biphenyl as a treatment of citrus fruits, the existence of tolerant strains of Penicillium and Diplodia was noted. These strains had no commercial significance until the early 1960s when certain lots of lemons packed with biphenyl papers in California arrived in European markets with high levels of Penicillium decay and sporulation. Before shipment, California lemons are stored for one to four months at 5°C to improve internal quality of the fruit, and to stabilize demand. It was observed that many of the problem shipments originated in packinghouses in which the lemons were treated with sodium o-phenylphenate before storage. Harding [362] found that spores of biphenyl-resistant isolates of Penicillium usually constituted only a small fraction of the total spore population in the atmosphere of packinghouses. However, the atmospheres of those houses where sodium o-phenylphenate had been used for some time were heavily contaminated with spores of resistant isolates. Harding also demonstrated [367] that residues of o-phenylphenol on lemons in storage suppressed the development of Penicillium strains which were sensitive to o-phenylphenol (p-hydroxybiphenyl) whereupon the resistant strains decayed the fruit and sporulated heavily in the absence of competition from the sensitive strains. This selection process resulted in a high level of resistance to o-phenylphenol and biphenyl in the spore population in these packinghouses. The biphenyl treatment was completely ineffective in controlling Penicillium decay in fruit shipped from these establishments. The problem of biphenyl resistance may be corrected by fumigation of the packinghouses with formaldehyde and replacement of sodium o-phenylphenate as a prestorage treatment with a chemically-unrelated fungicide such as sodium carbonate, thiaben-dazole, or sec-butylamine.

The applications, problems, and residue aspects of the biphenyl treat-ment have been discussed in considerable detail in other reviews [360, 366].

2. Halogenated Hydrocarbons

Several reports have described the effectiveness of chlorinated ethanes and ethylenes against postharvest diseases of stone fruits, strawberries, citrus, and grapes, but none of these treatments are in widespread commercial use today. Vandemark and Sharvelle [368] reported that vapors of several liquid chlorinated ethanes and ethylenes reduced decay in peaches, but only tri-chloroethylene was not injurious to the fruit at concentrations required for control of decay. Smith et al. [369] reported that both trichloroethylene and tetrachloroethylene were highly effective for control of Rhizopus rot, but that control of Monilinia fructicola was best from a combination of sulfur dust and tetrachloroethane. Dibromotetrachloroethane (DBTCE) was shown to be effective as a vapor treatment for control of dry rot of potatoes, Botrytis rot of grapes, and Penicillium decay of citrus fruits [217, 370, 371]. Compared to biphenyl for the treatment of citrus fruits, DBTCE is more inhibitory to species of Penicillium, has a broader antifungal spectrum and its odor dissipates more rapidly from treated fruit [217]. Compared to SO_2 for treatment of grapes, DBTCE has the major advantage that evolution of its vapor is independent of moisture, in contrast to the generation of SO_2 from bisulfite salts. Thus, DBTCE is less injurious to grapes than SO_2 and can be used in completely closed packages of grapes, thereby reducing transpiraton from the berries during storage [371]. It was never adopted commercially because it was too volatile for use on citrus fruits and its vapors were faintly lacrimatory, creating a problem for workers in fruit storage rooms. Tetraiodoethylene, 0.3-0.8% in a water emulsion wax, has been used for years as a treatment to control Alternaria and Cladosporium on the rind of melons. Little, if any, experimental data on the efficiency of this treatment have been published.

H. 2,6-Dichloro-4-nitroaniline (Dicloran)

Dicloran is of major importance as a postharvest fungicide for the control of Rhizopus stolonifer, a wound-invading pathogen of many fruits and vege-tables, which is not controlled by any other chemical treatment. Dicloran is essentially insoluble in water and must be applied to crops as a suspen-sion in water or in a wax emulsion. Although the vapor pressure of dicloran is low, it is, nonetheless, sufficient for strong inhibition of the aerial growth of Rhizopus stolonifer on peaches. Trials carried out in California and New Jersey in the early 1960s [372, 373] revealed that Rhizopus rot of sweet cherries and peaches could be controlled effectively by dipping the fruit after harvest in a suspension of 1,000 ppm dicloran. It was less effective against other decays of stone fruits caused by Monilinia fructicola (brown rot) and Penicillium sp. [137, 169, 374] and was inactive against Rhizopus arrhizus and resistant strains of Gilbertella persicaria [375]. Ogawa et al. [150] demonstrated that the treatment was effective in suppressing lesion development in infected peaches, even when the inoculated fruits were incubated for 10-34 hr before treatment. Ravetto and Ogawa [160] observed that dicloran penetrated to a depth of 11 mm into peaches, inhibiting the growth of Rhizopus in the flesh at that depth. This phenomenon undoubtedly explains the action of the dicloran treatment against established infections of Rhizopus rot on peaches. Dicloran impregnated into fruit wraps or in suspensions sprayed upon the fruit inhibits surface growth of Rhizopus and, therefore, the spread of the disease to adjacent fruit [135, 150].

Preharvest sprays of dicloran applied one week before harvest have provided good control of Rhizopus rot of peaches inoculated after harvest [134-136]. The preharvest treatment may be advantageous for peaches destined for local markets, but a postharvest treatment is more effective when peaches are brushed or waxed before shipment as fresh fruit. Ogawa et al. [140] reported that a postharvest treatment with dicloran was superior to a preharvest spray treatment for control of Rhizopus and Monilinia on free-stone peaches during controlled ripening at 20°C. The preharvest spray of dicloran should be necessary only for those special situations where a postharvest treatment is not feasible. A summary of several years' trials comparing different methods of applying dicloran to peaches has been published by Luepschen et al. [135].

The effectiveness of dicloran against postharvest decays of stone fruits is primarily a function of the quantity of dicloran persisting on the fruit after treatment. Both pre- and postharvest treatments which result in residues of at least 2-3 ppm dicloran on the fruit provide adequate control of Rhizopus rot [135, 136, 167]; whereas, the optimum deposit of dicloran for brown rot control appears to be at least 10 ppm and possibly higher under some circumstances [136, 243, 376]. To improve the effectiveness of dicloran for the control of brown rot of stone fruits after harvest, two approaches have been taken: (1) increase the residues of dicloran on the fruit by increasing the concentration of this fungicide in the formulation, raising the treatment temperature, and extending the treatment time; and (2) combine dicloran with another fungicide such as captan, thiabendazole, or benomyl which has greater activity against Monilinia fructicola. Treatment of stone fruits with a suspension of 100-600 ppm dicloran in water at ambient temperature was only slightly effective against brown rot, whereas the same suspension at 46°C greatly reduced the incidence of this disease [168-170]. Wells and Harvey [170] reported that treatment of stone fruits with a suspension of dicloran at 51.5°C delayed the development of brown rot for a longer period of time than treatment with pure water at this temperature. Dicloran treatment at 51.5°C may injure the fruit, whereas application at

46°C provides adequate brown rot control without endangering the condition of the fruit [168-170]. Smith [168] suggested that the dicloran treatment at 46°C might be substituted for the hot water treatment at 51.5°C to lessen the danger of heat injury.

Thiabendazole and benomyl are highly effective against brown rot but have no activity against Rhizopus rot [168]. Treatments combining 450-900 ppm dicloran with 100-1,000 ppm benomyl or thiabendazole have provided outstanding control of both Rhizopus rot and brown rot [86, 167, 169, 257, 377]. Captan also has been combined with dicloran, but the control of brown rot was less than that obtained with the benzimidazoles. The combination of two fungicides to suppress Rhizopus and Monilinia seems a more efficient method for controlling postharvest decay of peaches than application procedures intended to increase the residues of dicloran to overcome brown rot.

Rhizopus rot of sweet potatoes has been controlled by spray and dip applications of dicloran [352, 353, 378]. The treatment is not effective against black rot (Ceratocystis) [378] or a Penicillium surface rot on tubers injured by chilling [109]. Burton [379] reported that a combination of dicloran and sodium o-phenylphenate in a wax formulation provided good control of shrinkage and decay of several tropical root crops. The incidence of Botrytis on carrots during storage was reduced by dipping the roots in a suspension of dicloran [14] and the "nesting" of Botrytis on cabbage was reduced by dusting dicloran on the heads before storage at 0-2°C [189]. Dipping snap beans for 10 sec in 1800 ppm dicloran did not control Sclerotinia and Pythium during storage [55].

Several investigators have evaluated dicloran for control of postharvest decays of strawberries and raspberries since Rhizopus and Botrytis, the principal causative agents, are sensitive to dicloran. Field sprays of dicloran have not been very successful in controlling postharvest decay, although dipping the berries in a dicloran suspension after harvest reduced decay significantly but left a visible residue on the berries [380, 381]. Cohen and Dennis [382] did not consider the use of dicloran on strawberries and raspberries to be a promising practice because of the possible involvement of dicloran-resistant Botrytis strains and Mucor mucedo (a resistant fungus) in the complex of organisms associated with postharvest rotting of soft fruits in England.

I. Benzimidazoles and Related Compounds

Thiabendazole is stable under normal conditions of formulation and in postharvest use. It is soluble in pure water to the extent of approximately 25 μg/ml at 23°C [383], but published values range from 2 μg/ml [384] to 50.6 μg/ml [163]. Thiabendazole is fairly soluble in dilute acid and alkali owing to the amphoteric nature of the imidazole nitrogen atoms. Thiabendazole exists in solutions below pH 3 mainly as a cation, and above pH 14 as an anion. However, these solutions are too acid and too alkaline for treatment of fresh produce. The solubility of thiabendazole is increased by complexing with lactic acid and is slightly more soluble in some complexes with metals and gluconic acid [384]. Other metal complexes are reported to have approximately the same water solubility as pure thiabendazole [385]. All of the present-day postharvest applications are made as suspensions of pure thiabendazole in water. The thermal stability of thiabendazole makes it possible to sublime thiabendazole at 310°C for a "smoke" application.

Benomyl is highly insoluble in water and it decomposes slowly, especially in acidic medium, to methyl benzimidazolecarbamate (MBC) which is only slightly less fungitoxic than benomyl. In an alkaline formulation or environment (pH > 8) the substituents attached to the benzimidazole nucleus of benomyl react to give 1,2,3,4-tetrahydro-3-butyl-2,4-dioxo-s-triazino[a]-benzimidazole [261, 386], which has very weak, if any, antifungal properties (Fig. 12). For this reason, benomyl should never be incorporated into a formulation or application mixture of pH > 8. Benomyl is unstable also in organic solvents, especially benzene, in which it decomposes rapidly to MBC [262]. This compound (MBC) is soluble in pure water to the extent of about 6-7 μg/ml at 23°C [383]. The solubility increases in slightly acidic or basic solutions due to the ionization of the imidazole nitrogen atoms (pK_1 = 3.95; pK_2 = 10.55) [387]. MBC is quite stable in weak acid solution, but in alkaline solution the carbamate ester may hydrolyze slowly. Benomyl appears to penetrate through the plant cuticle and into the plant tissue more rapidly than MBC and this behavior usually leads to better control of pathogen

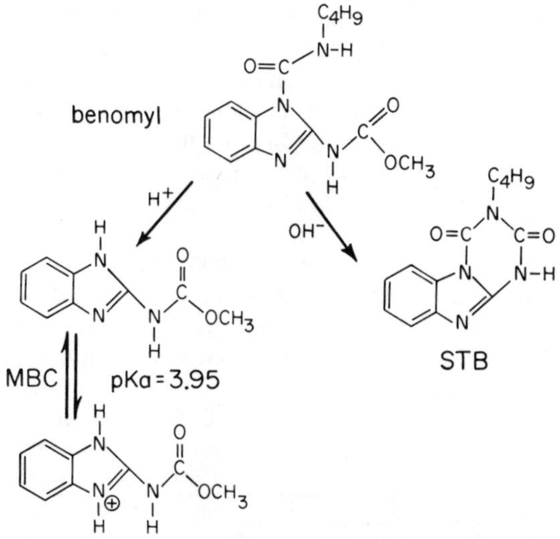

FIG. 12. Reactions of benomyl under acidic and basic conditions.

propagules buried in the host tissues [388]. Therefore, benomyl should be formulated and applied in a manner which avoids decomposition to MBC. A substantial portion of the fungicide residue on oranges treated with a near-neutral suspension of benomyl was identified as benomyl per se [386].

The control of Penicillium digitatum on citrus fruits by thiabendazole was reported first by Crivelli [389] and subsequent investigations have confirmed that dipping or spraying oranges with a suspension containing 1,000 ppm or less thiabendazole greatly reduces Penicillium decay as well as Diplodia and Phomopsis stem-end rots [153, 315, 390-393]. Suspensions of 500-1,000 ppm benomyl applied in the same manner have provided equivalent control of these diseases [392-395]. Benomyl is usually more effective than thiabendazole when the two are compared at equivalent concentrations [396, 397]. Benomyl appears to be capable of limited penetration into superficial tissues of the host to inhibit the development of latent infections of Diplodia and Phomopsis and also to confer resistance to infection by Penicillium for several weeks after treatment [85, 398-400]. The addition of oil to a suspension of benomyl increased the residues of benomyl (benomyl + MBC) in the fruit and resulted in improved decay control [244].

Although dipping fruit in an aqueous suspension of thiabendazole or benomyl provides the most effective control of decay, this method has several major drawbacks as a commercial procedure. Therefore, the principal methods of application are sprays or floods of suspensions of the fungicide in water or in wax emulsions [153, 256, 401, 402]. The more effective application methods are those which thoroughly wet the fruit with the suspension. In Israel, suspensions of thiabendazole in water have provided better control of Penicillium decay than suspensions in wax emulsions [251, 403]. In South Africa, the incorporation of thiabendazole in a solvent wax gave poorer control of Diplodia stem-end rot than a water suspension of the fungicide [258, 259], presumably because the solvent evaporated immediately on contact with the fruit before the fungicide could move under the button of the fruit where the fungus was latent. Trials comparing different methods of applying benzimidazole fungicides and recommendations for different citrus production areas have been published [194, 253, 255, 256, 258-260].

Application of a suspension of 4,000-6,000 ppm thiabendazole in a wax emulsion to citrus fruit resulted in a surface deposit of the fungicide which inhibited sporulation of Penicillium on decaying fruit [153, 251, 402]. Benomyl produced this same effect at a somewhat lower application rate. The significance of this observation is that the benzimidazole fungicides potentially offer an acceptable alternative to the biphenyl treatment for the control of "soilage" of citrus fruits (Section V.G). Thiabendazole and benomyl do not control Geotrichum, Alternaria, or Colletotrichum on citrus fruits [47, 390, 400, 404]. Penicillium italicum usually is more tolerant than Penicillium digitatum to benzimidazole fungicides; therefore, blue mold caused by P. italicum is the more prevalent decay on oranges treated with

low concentrations of these compounds [403] and on lemons in storage rooms
where benzimidazole-treated fruits are being recycled all season. The most
serious problem in this connection is the selection and proliferation of
strains of P. italicum and P. digitatum which are resistant to both thiabendazole and benomyl. Harding [405] reported that this was the reason for
the failure of the thiabendazole treatment in certain packinghouses of
southern California. The occurrence of benzimidazole-resistant strains of
species of Penicillium has now been reported from Florida, Israel, and
Australia [406-408]. Most of the benzimidazole-resistant strains do not
appear to be resistant to sodium o-phenylphenate or sec-butylamine.

Thiabendazole and benomyl are reported to reduce chilling injury on
grapefruit stored at temperatures which normally cause cold-pitting of the
fruit [409-412]. The mechanism of this interesting effect upon the host is
unknown.

Treatment of pineapples with 3,000 ppm suspensions of thiabendazole or
benomyl reduced decay by Thielaviopsis paradoxa at the cut stem but not at
injuries on the side of the fruit [99, 100]. The benzimidazole treatments
were less effective and more costly than salicylanilide which is the standard
postharvest fungicide for pineapples in Martinique and the Ivory Coast.
Investigations carried out over the past 8 years have demonstrated the
effectiveness of thiabendazole and benomyl, applied as dip or spray treatments, for the control of several postharvest diseases of bananas, such as
crown rot (Colletotrichum, Verticillium, Fusarium, and Thielaviopsis),
finger stalk rot (Colletotrichum), nonlatent anthracnose (Colletotrichum)
and "squirter disease" (Nigrospora). These diseases have been controlled
by treatment of the harvested bananas with suspensions of 400-1,000 ppm
thiabendazole [110, 155, 359, 413-415] or 100-500 ppm benomyl [88, 156, 157,
219, 416, 417]. Benomyl appears to penetrate into the cut crown tissues
and the cuticle to a limited extent, therein inhibiting the development of
36-hr wound infections and latent infections of Colletotrichum [34]. Griffee
and Burden [34] found that four preharvest sprays with 1,000 ppm benomyl
controlled postharvest development of latent infections of Colletotrichum,
but had no effect upon the development of this fungus at mechanical injuries
or upon the development of crown rot. The preharvest treatment was not
recommended, however, because of the danger of selecting benomyl-resistant
strains of Colletotrichum. More important, these investigators found that
a single postharvest treatment with 250 ppm benomyl provided excellent
control of both wound and latent Colletotrichum infections as well as crown
rot [34].

Anthracnose (Colletotrichum) on mangoes was controlled by dipping the
harvested fruit in a 1,000 ppm suspension of thiabendazole or benomyl at
55°C before storage for 2-4 weeks at 10-13°C [142, 418]. This treatment
was not effective against stem-end rot in Florida and was slightly injurious
if the treated fruits were stored at 10-13°C before ripening [142]. The same
fungicide suspensions at ambient temperatures did not control anthracnose

for this storage period. In contrast, mangoes treated in India with suspensions of 500 ppm benomyl or 900 ppm thiabendazole showed significantly less anthracnose after storage at room temperature for 7 days [419]. Trials in the Philippines revealed that Diplodia stem-end rot of mangoes was not controlled by treatment with 200 ppm thiabendazole [397].

Treatment of harvested apples and pears with suspensions containing 250-500 ppm benomyl or 500-2,000 ppm thiabendazole controlled decay caused by Penicillium expansum and Botrytis cinerea [120-123, 177, 421-426]. Benomyl or thiabendazole suspensions at 500 ppm and ambient temperatures controlled Penicillium and Botrytis at puncture injuries in apples, but the fungicide suspension had to be heated to 29-45°C to control bruise inoculations [120, 177]. However, the treatment at ambient temperature gave good control of bruise inoculations in pears [426], suggesting that the skin of the pear is more easily penetrated by benzimidazole fungicides than that of the apple. Fungistatic residues of both benomyl and thiabendazole were found in the flesh of pears following treatment of the fruit with 500-1,200 ppm of these fungicides [427]. Benomyl appears to penetrate the skin of the pear and apple more readily than thiabendazole, although the latter is as effective as benomyl in controlling Penicillium expansum and Botrytis cinerea on pears [139, 427]. Preharvest sprays of thiabendazole and benomyl gave good control of Penicillium and Botrytis on pears in storage [138, 139] but the treatment is often superfluous because most pears are dipped after harvest in a bath of the antiscald agent ethoxyquin, and a benzimidazole fungicide could also be applied at this time. In fact, the widespread practice of treating apples with the scald-control agents, diphenylamine and ethoxyquin, may increase decay of fruit during storage. This problem may be alleviated by adding thiabendazole or benomyl to the antiscald treatment. Several investigators have reported that the benzimidazole fungicides are compatible with diphenylamine and ethoxyquin [120, 122, 422], but it was observed that Stamen apples treated with a combination of diphenylamine and thiabendazole or benomyl showed severe injury after storage [123].

Investigations in Europe have shown that the benzimidazole fungicides applied as preharvest orchard sprays or postharvest treatments are effective in controlling lenticel rots of apples, which are the major pathological diseases of this fruit during storage in Europe [56, 57, 60, 61, 81]. These diseases are initiated by preharvest infection in the lenticels late in the growing season and previously hot water was the only effective postharvest treatment. However, the heat treatment had an adverse effect on the storage life of the apples in some cases [105, 176]. Benomyl and thiabendazole orchard sprays applied in July control lenticel rotting by Gloeosporium on the fruit during storage [12]. Excellent control of lenticel rotting is obtained also by dipping unsprayed apples in 300-500 ppm thiabendazole or benomyl [12, 87]. Preharvest orchard sprays may be necessary, however, since high levels of lenticel contamination reduce the effectiveness of the postharvest treatment [87]. Pre- and postharvest treatments with benomyl

controlled lenticel rotting in storage caused by <u>Nectria galligena</u>; whereas
thiabendazole was ineffective against this disease [12, 87, 323, 428]. The
benzimidazole fungicides have no effect upon the development of Alternaria
rot in storage [120, 139, 426]. In fact, some investigators believe that the
benomyl treatment makes this storage disease of apples more serious [426].

Treatment of peaches, nectarines, and cherries after harvest with dips,
sprays and wax formulations containing 100–300 ppm benomyl or 1,000–
2,000 ppm thiabendazole have provided excellent control of Monilinia brown
rot but have been ineffective against Rhizopus rot or Alternaria rot [86, 166,
168, 169, 420]. The treatments are improved by heating the suspensions to
46–52°C or by incorporating the fungicide in a wax emulsion [168, 169, 171,
257, 429]. The concentration of the benzimidazole fungicide in the heated
suspension may be reduced as much as 25% of the recommended dosage
without sacrificing the effectiveness of the treatment [429]. Both brown rot
and Rhizopus rot can be controlled by a single treatment combining dicloran
with a benzimidazole fungicide in the same treating suspension [86, 257,
377, 430].

Trials conducted in the late 1960s in Maine [240] and in the United Kingdom [242] demonstrated that dip and dust treatments of thiabendazole and
benomyl applied to seed potatoes after harvest gave control of the major
potato storage diseases which was equivalent to that provided by organomercurial fungicides. The benzimidazole fungicides controlled dry rot
(<u>Fusarium</u>), gangrene (<u>Phoma</u>), skin spot (<u>Oospora</u>), and silver scurf
(<u>Helminthosporium</u>) [236, 240–242]. Meredith [234] has reviewed the etiology
and control of potato diseases by thiabendazole and benomyl. Thiabendazole
appeared to be somewhat more effective than benomyl for control of dry rot
[132, 162] and gangrene [242]. Fuberidazole provided control of gangrene
equivalent to thiabendazole and benomyl [241].

Several methods for application of thiabendazole to potatoes have been
evaluated, including dip, mist, spray, fog, dust, and thermal smoke. Dip
treatments are the most effective, but are commercially not practical [132,
240–242]. The dust treatment is effective, but the concentration of the
active material must be 10–100 times greater than in an aqueous suspension
which provides equivalent protection [241, 242]. Potatoes take up ten times
more thiabendazole from a dip treatment than from a dust treatment [163].
A low-volume spray of a thiabendazole suspension applied as the potatoes
are conveyed on the bin filler appears to be the most practical and effective
means of treating potatoes before storage [235–239]. Fog and thermal
smoke applications do not give adequate coverage of the tuber with thiabendazole [234–236]. Both benomyl and thiabendazole penetrate the intact
periderm of the tuber, making the underlying tissue inhibitory to pathogenic
fungi [162, 163].

Field sprays of benomyl have reduced postharvest decay of strawberries
caused by <u>Botrytis</u> in some trials [74, 76], whereas the degree of control
has been disappointing in other instances [431]. The treatment has failed

in cases where Rhizopus sp. and Mucor sp. are involved in the disease complex and where populations of benomyl-resistant strains of Botrytis have built up [75, 77, 89, 91]. In fact, field applications of benomyl increased the severity of decay by benomyl-resistant strains of Botrytis and by Rhizopus and mucors. In strawberries artificially inoculated with Botrytis, control of postharvest decay was obtained by dipping the berries in a benomyl suspension 24 hr before inoculation [432]. Application of the treatment 24 hr after inoculation was ineffective, indicating that benomyl provides protective action only in the control of Botrytis rot of strawberries. The postharvest dip treatment was ineffective with naturally inoculated berries [434].

Thiabendazole and benomyl have given control of postharvest diseases of a diverse variety of vegetables and root crops including yams [432], casava [433], carrots [14, 188, 276], celery [435], tomatoes [299], cabbage [189], snap beans [436], squash [437], and cantaloupe [181, 182]. Benomyl possesses a higher degree of antifungal activity in vitro and appears to penetrate the outer layers of fresh produce better than thiabendazole in the few comparative studies that have been made [427]. Therefore, it is interesting that thiabendazole has been reported to be superior to benomyl in the treatment of some crops, such as potatoes for control of dry rot (Fusarium) [132, 162] and gangrene (Phoma) [242]; squash for control of Fusarium [437]; and snap beans for control of Sclerotinia [436]. In these cases, surface binding may be more significant than inherent activity or penetration into the host tissues.

Methyl benzimidazolecarbamate (MBC), a decomposition product of benomyl (Fig. 12) with strong antifungal properties, has been evaluated as a postharvest treatment for control of anthracnose and crown rot of bananas [438, 439], Penicillium decay and Diplodia stem-end rot of oranges [440], and Monilinia brown rot of peaches [441] and sweet cherries [442]. All investigators agree that MBC is slightly inferior or equal to benomyl in preventing infection and is more effective at a lower concentration than thiabendazole. However, MBC is distinctly inferior to benomyl for control of Penicillium sporulation on decaying oranges [443].

Thiophanate methyl gives rise to MBC in vivo by spontaneous reaction of substituents to form a benzimidazole ring [444]. It is about equal to benomyl for control of Fusarium dry rot of potatoes [132] and Penicillium and Botrytis on apples [424], and to thiabendazole for control of anthracnose and crown rot of bananas, but higher concentrations are required for an effectiveness equivalent to benomyl [439, 445]. Thiophanate methyl is distinctly inferior to benomyl for control of brown rot on peaches and cherries [430, 446]. It compares favorably with benomyl and thiabendazole for control of Penicillium decay and Diplodia stem-end rot of citrus fruits in most trials [440, 447, 448], but one investigation rated it as possessing only 10% of the activity of benomyl [441]. It appears that benomyl or thiabendazole should be the fungicides of choice for most postharvest treatments, subject to cost considerations. Strains of fungi which are tolerant to benomyl and thiabendazole are not inhibited by thiophanate methyl or MBC.

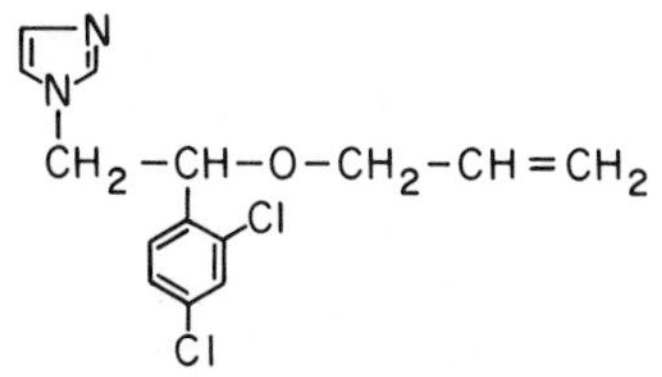

$$CH_2-CH-O-CH_2-CH=CH_2$$

FIG. 13. Structure of Imazalil.

Laville reported in 1973 [449] that Imazalil (R-23979, Fig. 13) possessed protective, curative and antisporulant properties which were equal to benomyl in controlling Penicillium decay of citrus fruits. Furthermore, Imazalil forms water soluble salts that are more convenient to apply than benomyl which must be formulated as a suspension in water. Subsequent tests [450, 452] confirmed the effectiveness of Imazalil for control of Penicillium on citrus and also revealed that Imazalil was fully active against benomyl-resistant strains of the pathogen.

J. Miscellaneous Organic Compounds

Captan has been extensively evaluated as both a pre- and postharvest treatment for control of various diseases during storage and marketing. Some effective preharvest applications of this fungicide have been considered in Section III. The effectiveness of captan as a postharvest treatment is not impressive compared to fungicides which have been considered earlier in this chapter. Captan exhibits a broad antifungal spectrum and it is sometimes combined with one of the newer fungicides in order to extend the range of antifungal activity of the treatment. Captan has been combined with benomyl to control Alternaria on mechanically harvested sweet cherries [446]; chlorothalonil has been combined with benomyl for this same purpose [451].

Preharvest sprays of dichlofluanid, which is structurally related to captan, have given control of Botrytis decay of strawberries after harvest equivalent to that provided by benomyl [74, 75]. Dichlofluanid is considered more reliable than dicloran or benomyl for preharvest treatment of strawberries because strains of fungi resistant to this fungicide have not been observed; furthermore, dichlofluanid shows in vitro activity against Rhizopus and mucors, both of which are resistant to benomyl [382].

Dipping sweet cherries in a suspension of triforine after harvest gave outstanding control of brown rot when applied at concentrations one-fifth to one-tenth of those required for equivalent control by benomyl [442, 451]. Isopropyl carbamoyl-3(3,5-dichlorophenyl)-hydantoin gave good control of brown rot on sweet cherries at double the concentration required for benomyl [446]. This compound was highly effective against Alternaria.

VI. CURRENT PROBLEMS AND TRENDS IN
POSTHARVEST DISEASE CONTROL

The fresh fruit and vegetable industry is now in an era of dynamic innovation in the areas of mechanical harvesting, consumer packaging, and bulk handling and transportation. Many of the proposals that have been made in these areas will tend to intensify postharvest disease problems either by increasing the opportunity for wound infections or by creating an environment favorable for disease development. Mechanical harvesting, in its current version, invariably results in a high frequency of injury to fruit crops which then must be processed, refrigerated, or treated with a fungicide within 24 hr after harvest to avoid serious decay losses. The decreased emphasis on wrapping individual fruits and the ever-increasing interest in consumer packaging appears to limit the future possibilities for in-package generators of fungistatic gases such as biphenyl. These developments do not, however, rule out the direct injection of rapid-acting fumigants, exemplified by ethylene oxide and sec-butylamine, provided that the residues of the treatment are not evident to the consumer. In-line packinghouse treatments to disinfest, disinfect, and protect existent and subsequent wounds against infection appear to be compatible with all current concepts of packaging and marketing fresh produce. Special treatments applied in a wax formulation to control fungus sporulation, surface growth, and disease spread would be of great economic importance in unwrapped and bulk shipments of citrus and perhaps other crops as well.

The materials and technology are available now to satisfactorily control the principal postharvest diseases of the major fresh fruit crops. The development of dicloran in the early 1960s provided, for the first time, a highly effective means for controlling Rhizopus rot, a major source of postharvest loss in peaches, nectarines, cherries, and sweet potatoes. sec-Butylamine, developed in the same years, found major application in the treatment of citrus fruits to prevent infection of harvest injuries by species of Penicillium during degreening and storage. Thiabendazole and benomyl, introduced a few years later, solved the problem of crown rot of banana hands shipped in fiberboard containers. These compounds also provided important breakthrough in the inactivation of latent infections, a major factor limiting the storage and shelf life of apples, citrus, bananas, and other tropical fruits. The combination of dicloran and benomyl offers a means for reducing the major postharvest diseases of stone fruits to an insignificant level. Thiabendazole has equalled or surpassed the effectiveness of organomercury fungicides in controlling the important diseases of seed potatoes in storage and should be useful in the storage of tuber and root food crops as well. Surface deposits of thiabendazole and benomyl within the legal residue tolerance suppress the sporulation of Penicillium molds on the surface of citrus fruits, thereby providing an alternative to the use of biphenyl which imparts a hydrocarbon odor to treated fruit.

Enthusiasm for the benzimidazole fungicides was dampened somewhat by the realization that fungi can develop resistance to this class of compounds with comparative ease. This problem was encountered earlier with the biphenyl treatment and can be surmounted if the situation is recognized in time and dealt with intelligently. Fungicide resistance is being controlled today in the citrus industry by a combination of several measures: (1) Fruit is treated after storage with a fungicide which is structurally unrelated to the fungicide applied before storage, e.g., sec-butylamine and thiabendazole. (2) The premises are disinfested with formaldehyde as often as necessary or practical. (3) The buildup of the fungicide resistant population is monitored and the fungicide program altered when the resistant strains reach a predetermined level. (4) Repacking stored fruit which have received the terminal fungicide treatment is discouraged to avoid contaminating the packinghouse with fungicide-resistant pathogens.

Despite the substantial progress over the past 15 years in the development of highly effective fungicide treatments, there is as yet no chemical treatment which directly acts against several important postharvest diseases. Bacterial soft rot of vegetable crops, Geotrichum sour rot of citrus fruits and tomatoes, and Alternaria rot of several fruits and vegetables are all examples of diseases which cannot be controlled by treatment of the host with an antimicrobial agent. Control measures are limited to sanitation, refrigeration, and retarding the onset of host senescence. This situation appears to result from the insensitivity of the pathogens responsible for these diseases to existing biocides rather than a unique problem associated with the etiology of the disease. Thus, benomyl provides excellent control of Diplodia stem end rot of citrus fruit but not Alternaria stem end rot, apparently because the pathogen is not sensitive to benomyl. Indications that bacterial soft rot could be controlled by a strongly antibacterial agent may be derived from earlier reports on the effectiveness of medicinal antibiotics against this disease [24]. If this line of reasoning is correct, then we may anticipate the development of effective fungicide and bactericide treatments for the remaining postharvest disease problems within our generation.

ACKNOWLEDGMENT

The competent and dedicated editorial assistance of Mrs. Kathleen Eckard is gratefully acknowledged.

REFERENCES

1. J. M. Harvey and W. T. Pentzer, U.S. Dept. Agr., Agr. Handb., 189, 1960, p. 37.
2. J. M. Harvey, W. L. Smith, Jr., and J. Kaufman, U.S. Dept. Agr., Agr. Handb., 414, 1972, p. 64.

3. L. P. McColloch, H. T. Cook, and W. R. Wright, U.S. Dept. Agr., Agr. Handb., 28, 1968, p. 74.
4. C. F. Pierson, M. J. Ceponis, and L. P. McColloch, U.S. Dept. Agr., Agr. Handb., 376, 1971, p. 112.
5. G. B. Ramsey, J. S. Wiant, and M. A. Smith, U.S. Dept. Agr. Misc. Publ., 98, 1949, p. 60.
6. G. B. Ramsey, B. A. Friedman, and M. A. Smith, U.S. Dept. Agr., Agr. Handb., 155, 1959, p. 42.
7. G. B. Ramsey and M. A. Smith, U.S. Dept. Agr., Agr. Handb., 184, 1961, p. 49.
8. J. J. Smoot, L. G. Houck, and H. B. Johnson, U.S. Dept. Agr., Agr. Handb., 398, 1971, p. 115.
9. D. G. Coursey, Biodeterior. Mater., Proc. Intern. Biodeterior. Symp., 2nd (A. H. Walters and E. H. Hueck-Van Der Plas, eds.), Wiley, New York, 1971, pp. 464-471.
10. D. G. Coursey and R. H. Booth, Proc. Brit. Insectic. Fungic. Conf., 6th, 3, 673 (1971).
11. C. L. Lockhart, C. A. Eaves, and F. R. Forsyth, Can. Plant Disease Surv., 50, 90 (1970).
12. R. T. Burchill and K. L. Edney, Ann. Appl. Biol., 72, 249 (1972).
13. D. M. Derbyshire, Chem. Ind., 22, 1052 (1973).
14. J. M. Wells and F. L. Merwarth, Plant Disease Reptr., 57, 697 (1973).
15. P. R. Harding, Jr. and D. C. Savage, Plant Disease Reptr., 48, 808 (1964).
16. W. Grierson, Proc. Amer. Soc. Hortic. Sci., 92, 797 (1968).
17. G. C. Pratella, G. Tonini, and A. Cessari, Proc. Intern. Citrus Symp., 1st, 3, 1317 (1969).
18. M. J. Ceponis and J. Kaufman, U.S. Dept. Agr. Market. Serv., 494, 1963, p. 9.
19. G. L. Rygg, C. W. Wilson, and M. J. Garber, U.S. Dept. Agr. Market. Res. Rept., 500, 1961, p. 14.
20. W. Grierson, Proc. Intern. Citrus Symp., 1st, 3, 1389 (1969).
21. M. J. Ceponis, Plant Disease Reptr., 54, 964 (1970).
22. U.S. Dept. Agr. Agr. Res. Serv., Market Quality Res. Div., Progr. Rept., 1968, p. 22.
23. J. M. Harvey, H. M. Couey, C. M. Harris, and F. M. Porter, U.S. Dept. Agr. Market. Res. Rept., 774, 1966, p. 9.
24. J. W. Eckert, in Fungicides: An Advanced Treatise (D. C. Torgeson, ed.), Vol. 1, Academic Press, New York, 1967, Chap. 9.
25. J. W. Eckert, World Rev. Pest Control, 8, 116 (1969).
26. J. W. Eckert and N. F. Sommer, Ann. Rev. Phytopathol., 5, 391 (1967).
27. K. Verhoeff, Ann. Rev. Phytopathol., 12, 99 (1974).
28. A. C. Hayward, Ann. Rev. Phytopathol., 12, 87 (1974).
29. W. R. Jarvis, Ann. Appl. Biol., 50, 569 (1962).

30. W. R. Jarvis and H. Borecka, Hortic. Res., 8, 147 (1968).
31. O. Kamoen and G. Jamart, Proc. Brit. Insectic. Fungic. Conf., 8th, 1, 347 (1975).
32. R. L. Powelson, Phytopathology, 50, 491 (1960).
33. D. B. Adam, J. McNeil, B. M. Hanson-Merz, D. F. McCarthy, and J. Stokes, Australian J. Sci. Res., Ser. B, 2, 1 (1949).
34. P. J. Griffee and O. J. Burden, Ann. Appl. Biol., 77, 11 (1974).
35. D. S. Meredith, Nature, 201, 214 (1964).
36. D. S. Meredith, Trop. Agr. (Trinidad), 48, 35 (1971).
37. G. E. Brown, Phytopathology, 65, 404 (1975).
38. R. E. D. Baker, Ann. Bot. (London), 2, 919 (1938).
39. M. E. Stanghellini and M. Aragaki, Phytopathology, 56, 444 (1966).
40. G. L. Greene, Phytopathology, 56, 1201 (1966).
41. J. H. Simmonds, Proc. Roy. Soc. Queensland, 52, 92 (1941).
42. T. Chakravarty, Trans. Brit. Mycol. Soc., 40, 337 (1957).
43. N. Binyamini and M. Schiffmann-Nadel, Phytopathology, 62, 592 (1972).
44. G. E. Brown and W. C. Wilson, Phytopathology, 58, 736 (1968).
45. M. Nadel, Palest. J. Bot., Rehovot Ser., 4, 166 (1944).
46. E. T. Bartholomew, Calif. Agr. Expt. Sta. Bull., 408, 1926, p. 39.
47. G. E. Brown and A. A. McCornack, Plant Disease Reptr., 56, 909 (1972).
48. P. Joly, Fruits, 22, 89 (1967).
49. K. E. Nelson, Phytopathology, 41, 319 (1951).
50. P. T. Jenkins and C. Reinganum, Australian J. Agr. Res., 16, 131 (1965).
51. P. F. Kable, Phytopathol. Z., 70, 173 (1971).
52. G. C. Wade, Australian J. Agr. Res., 7, 504 (1956).
53. H. S. Fawcett, Calif. Citrograph, 7, 233, 254 (1922).
54. J. P. Goodliffe and J. B. Heale, Ann. Appl. Biol., 80, 243 (1975).
55. J. M. Wells and T. N. Cooley, Plant Disease Reptr., 57, 234 (1973).
56. G. Bompeix, Rev. Gen. Froid, 10, 1385 (1966).
57. P. Bondoux, C. R. Hebd. Seances, 53, 1314 (1967).
58. K. L. Edney, Ann. Appl. Biol., 46, 622 (1958).
59. K. L. Edney, Ann. Appl. Biol., 44, 113 (1956).
60. C. Moreau, M. Moreau, G. Bompeix, and F. Morgat, Bull. Inst. Intern. Froid, Annexe, 1, 473 (1966).
61. G. C. Pratella and G. Tonini, Rev. Gen. Froid, 60, 3 (1969).
62. M. C. M. Pérombelon, Ann. Appl. Biol., 71, 111 (1972).
63. M. C. M. Pérombelon, Ann. Appl. Biol., 74, 59 (1973).
64. M. C. M. Pérombelon, Plant Disease Reptr., 56, 552 (1972).
65. M. A. Smith and G. B. Ramsey, Phytopathology, 37, 225 (1947).
66. B. M. Lund and J. C. Nicholls, Potato Res., 13, 210 (1970).
67. M. C. M. Pérombelon and R. Lowe, Potato Res., 18, 64 (1975).
68. G. D. Ruehle, Fla. Agr. Expt. Sta., Bull. 348, 1940, p. 36.
69. J. Scholey, C. Marshall and R. Whitbread, Plant Pathol., 17, 135 (1968).

70. M. J. Adams, Ann. Appl. Biol., 79, 265 (1975).

71. M. J. Adams, Ann. Appl. Biol., 79, 275 (1975).

72. R. T. V. Fox, J. G. Manners, and A. Myers, Potato Res., 14, 61 (1971).

73. K. L. Edney, Plant Pathol., 13, 87 (1964).

74. M. Bennett, J. Hortic. Sci., 47, 321 (1972).

75. C. Dennis, Ann. Appl. Biol., 81, 227 (1975).

76. J. A. Freeman and H. S. Pepin, Can. Plant Disease Surv., 48, 120 (1968).

77. V. W. L. Jordan and D. V. Richmond, Proc. Brit. Insectic. Fungic. Conf., 8th, 1, 5 (1975).

78. R. A. Conover, Proc. Fla. State Hortic. Soc., 78, 364 (1966).

79. R. D. Raabe and O. V. Holtzmann, Hawaii Farm Sci., 13, 1 (1964).

80. J. E. Hunter and I. W. Buddenhagen, Trop. Agr. (Trinidad), 49, 61 (1972).

81. G. Bompeix, F. Morgat, B. Poirnet, and J. P. Planque, C. R. Hebd. Seances, 55, 776 (1969).

82. R. W. Marsh, H. B. S. Montgomery, and K. L. Edney, Plant Pathol., 6, 39 (1957).

83. G. E. Brown and A. A. McCornack, Proc. Fla. State Hortic. Soc., 82, 39 (1969).

84. E. G. Brown and L. G. Albrigo, Proc. Fla. State Hortic. Soc., 83, 222 (1970).

85. E. G. Brown and L. G. Albrigo, Phytopathology, 62, 1434 (1972).

86. J. M. Ogawa, B. T. Manji, and E. Bose, Plant Disease Reptr., 52, 722 (1968).

87. K. L. Edney, Plant Pathol., 19, 189 (1970).

88. J. M. Ogawa, H. J. Su, Y. P. Tsai, S. S. Chen, and C. H. Liang, FAO Plant Protect. Bull., 10, 1 (1968).

89. C. Dennis and J. Mountford, Ann. Appl. Biol., 79, 141 (1975).

90. V. W. L. Jordan and D. V. Richmond, Plant Pathol., 23, 81 (1974).

91. V. W. L. Jordan, Plant Pathol., 22, 67 (1973).

92. P. J. Griffee, Trans. Brit. Mycol. Soc., 60, 433 (1973).

93. B. A. Friedman, Econ. Bot., 14, 145 (1960).

94. J. M. Ogawa, J. L. Sandeno, and J. H. Mathre, Plant Disease Reptr., 47, 129 (1963).

95. R. L. Rackham and W. Grierson, HortScience, 6, 163 (1971).

96. L. L. Morris and L. K. Mann, Hilgardia, 24, 143 (1955).

97. R. H. Stover, Banana, Plaintain, and Abaca Diseases, Commonwealth Mycological Institute, Kew, Surrey, England, 1972.

98. G. L. Greene and R. D. Goos, Phytopathology, 53, 271 (1963).

99. P. Frossard, Fruits, 23, 207 (1968).

100. P. Frossard, Fruits, 25, 785 (1970).

101. V. N. Pathak and D. N. Srivastava, Plant Disease Reptr., 51, 744 (1967).

102. M. Schiffmann-Nadel, Y. Cohen, and T. Arzee, Israel J. Bot., 19, 624 (1970).
103. H. B. Johnson, U.S. Dept. Agr. Market. Res. Rept., 738, 1966, p. 6.
104. T. R. Wright and E. Smith, U.S. Dept. Agr. Circ., 935, 1954, p. 14.
105. D. H. Spalding, H. C. Vaught, R. H. Day, and G. A. Brown, Plant Disease Reptr., 53, 738 (1969).
106. M. Schiffmann-Nadel, Proc. Intern. Citrus Symp., 1st, 3, 1295 (1969).
107. R. H. Segall, Plant Disease Reptr., 51, 151 (1967).
108. L. P. McColloch and J. T. Worthington, III, Phytopathology, 42, 425 (1952).
109. R. H. Daines, Plant Disease Reptr., 54, 486 (1970).
110. O. J. Burden, Queensland Agr. J., 95, 621 (1969).
111. A. E. W. Boyd, Rev. Plant Pathol., 51, 297 (1972).
112. L. W. Nielsen, Amer. Potato J., 23, 41 (1946).
113. L. W. Nielsen, Phytopathology, 55, 640 (1965).
114. R. Barkai-Golan, Phytopathol. Mediterranée, 12, 108 (1973).
115. B. M. Lund, J. Appl. Bacteriol., 34, 9 (1971).
116. F. L. Lukezic, W. J. Kaiser, and M. M. Martinez, Can. J. Bot., 45, 413 (1967).
117. G. Roth, Phytopathol. Z., 58, 383 (1967).
118. R. Barkai-Golan, Israel J. Agr. Res., 16, 133 (1966).
119. D. I. Murdock, Proc. Fla. State Hortic. Soc., 84, 266 (1972).
120. R. E. Hardenburg and D. H. Spalding, J. Amer. Soc. Hortic. Sci., 97, 154 (1972).
121. G. D. Blanpied and H. Purnasiri, Plant Disease Reptr., 52, 865 (1968).
122. C. A. Cargo and D. H. Dewey, HortScience, 5, 259 (1970).
123. R. H. Daines and R. D. Snee, Phytopathology, 59, 792 (1969).
124. R. E. Hardenburg, J. Amer. Soc. Hortic. Sci., 99, 236 (1974).
125. C. W. Coggins, Jr. and L. N. Lewis, Proc. Amer. Soc. Hortic. Sci., 86, 272 (1965).
126. J. M. Harvey, Proc. Calif. Fig Inst. Ann. Res. Conf., 10th, 1956, p. 27.
127. P. A. Arneson, Phytopathology, 61, 344 (1971).
128. R. H. Segall, Proc. Fla. State Hortic. Soc., 81, 212 (1969).
129. W. L. Smith, Jr. and W. H. Redit, U.S. Dept. Agr. Market. Res. Rept., 807, 1968, p. 9.
130. W. L. Smith, Jr. and W. H. Redit, Plant Disease Reptr., 46, 221 (1962).
131. P. R. Harding, Jr. and D. C. Savage, Plant Disease Reptr., 49, 332 (1965).
132. S. S. Leach, U.S. Dept. Agr., Agr. Res. Serv. Rept., ARS-NE-55 1975, p. 17.
133. J. J. Smoot and C. F. Melvin, Citrus Ind., 54, 8 (1973).

134. N. S. Luepschen, Plant Disease Reptr., 50, 565 (1966).
135. N. S. Luepschen, K. G. Rohrbach, A. C. Jones, and C. L. Peters, Colo. Agr. Expt. Sta., Bull., 547S, 1971, p. 24.
136. J. M. Ogawa, G. A. Boynack, J. L. Sandeno, and J. H. Mathre, Hilgardia, 35, 365 (1964).
137. R. H. Daines, Plant Disease Reptr., 49, 300 (1965).
138. D. L. Coyier, Plant Disease Reptr., 54, 647 (1970).
139. R. Ben-Arie and S. Guelfat-Reich, HortScience, 8, 181 (1973).
140. J. M. Ogawa, S. Leonard, B. T. Manji, E. Bose, and C. J. Moore, J. Food Sci., 36, 331 (1971).
141. J. W. Eckert and M. J. Kolbezen, Phytopathology, 53, 1053 (1963).
142. D. H. Spalding and W. F. Reeder, Plant Disease Reptr., 56, 751 (1972).
143. W. L. Smith, Jr. and M. Blomquist, Phytopathology, 60, 866 (1970).
144. W. L. Smith, Jr., W. H. Miller, and R. D. Bassett, Phytopathology, 55, 604 (1965).
145. H. M. Couey and M. Uota, Phytopathology, 51, 815 (1961).
146. H. M. Couey, Phytopathology, 55, 525 (1965).
147. J. W. Eckert, M. J. Kolbezen, and B. A. Kramer, Proc. Intern. Citrus Symp., 1st, 2, 1097 (1969).
148. T. Yanagita and S. Yamagishi, Appl. Microbiol., 6, 375 (1958).
149. D. A. Luvisi and N. F. Sommer, Proc. Amer. Soc. Hortic. Sci., 76, 146 (1960).
150. J. M. Ogawa, J. M. Mathre, D. J. Weber, and S. D. Lyda, Phytopathology, 53, 950 (1963).
151. J. K. Long and E. A. Roberts, Australian J. Agr. Res., 9, 609 (1958).
152. J. A. Seberry, Proc. Intern. Citrus Symp., 1st, 3, 1309 (1969).
153. J. W. Eckert, M. J. Kolbezen, and A. J. Kraght, Proc. Intern. Citrus Symp., 1st, 3, 1301 (1969).
154. D. S. Meredith, Ann. Appl. Biol., 48, 824 (1960).
155. C. Beaudoin, J. Champion, and R. Mallessard, Fruits, 24, 89 (1969).
156. P. G. Long, Trop. Agr. (Trinidad), 47, 9 (1970).
157. P. G. Long, Plant Disease Reptr., 54, 93 (1970).
158. D. C. Graham, G. A. Hamilton, C. E. Quinn, and A. D. Rothven, Potato Res., 16, 109 (1973).
159. K. E. Nelson, Proc. Amer. Soc. Hortic. Sci., 71, 183 (1958).
160. D. J. Ravetto and J. M. Ogawa, Phytopathology, 62, 784 (1972).
161. D. G. Phillips, Phytopathology, 65, 255 (1975).
162. A. W. Murdock and R. K. S. Wood, Ann. Appl. Biol., 72, 53 (1972).
163. M. J. Tisdale and K. H. Lord, Pestic. Sci., 4, 121 (1973).
164. J. Rosenbaum, Phytopathology, 10, 101 (1920).
165. W. L. Smith, Jr. and R. D. Bassett, U.S. Dept. Agr. Market. Res. Rept., 643, 1964, p. 14.
166. W. L. Smith, Jr., R. W. Penny, and R. Grossman, U.S. Dept. Agr. Market. Res. Rept., 979, 1972, p. 13.

167. J. M. Wells, U.S. Dept. Agr. Market Res. Rept., 908, 1971, p. 10.
168. W. L. Smith, Jr., Plant Disease Reptr., 55, 228 (1971).
169. R. H. Daines, Plant Disease Reptr., 54, 764 (1970).
170. J. M. Wells and J. M. Harvey, Phytopathology, 60, 116 (1970).
171. W. L. Smith, Jr. and R. F. Anderson, J. Amer. Soc. Hortic. Sci., 100, 84 (1975).
172. G. B. Tindale, P. T. Jenkins, and I. D. Peggie, J. Agr. (Victoria), 56, 107 (1958).
173. L. J. Kushman and J. S. Cooley, J. Agr. Res. (Washington, D.C.), 78, 183 (1949).
174. E. Forsund, Norsk Landbruk, 21, 20 (1968).
175. R. T. Burchill, Plant Pathol., 13, 106 (1964).
176. K. L. Edney, J. C. Fidler, G. Mann, and C. J. North, Bull. Inst. Intern. Froid, Annexe, 1, 243 (1966).
177. D. H. Spalding, H. C. Vaught, R. H. Day, and G. A. Brown, Plant Disease Reptr., 53, 738 (1969).
178. J. J. Smoot and C. F. Melvin, Proc. Fla. State Hortic. Soc., 76, 322 (1964).
179. E. K. Akamine and T. Arisumi, Proc. Amer. Soc. Hortic. Sci., 61, 270 (1953).
180. O. J. Burden, Queensland J. Agr. Animal Sci., 25, 135 (1968).
181. R. E. McDonald and W. R. Buford, Plant Disease Reptr., 55, 183 (1971).
182. J. K. Stewart and J. M. Wells, J. Amer. Soc. Hortic. Sci., 95, 226 (1970).
183. J. A. Bartz and J. P. Crill, Proc. Fla. State Hortic. Soc., 86, 153 (1974).
184. R. A. Dennison and E. M. Ahmed, in Postharvest Biology and Handling of Fruits and Vegetables (N. F. Haard and D. K. Salunkhe, eds.), Avi, Westport, Connecticut, 1975, Chap. 12.
185. N. F. Sommer and R. J. Fortlage, Advan. Food Res., 15, 147 (1966).
186. N. F. Sommer, E. C. Maxie, R. J. Fortlage, and J. W. Eckert, Radiat. Bot., 4, 317 (1964).
187. E. C. Maxie, N. F. Sommer, and F. G. Mitchell, HortScience, 6, 202 (1971).
188. D. M. Derbyshire and A. F. Crisp, Proc. Brit. Insectic. Fungic. Conf., 6th, 1, 167 (1971).
189. A. C. Brown, R. W. Kear, and J. P. Symons, Proc. Brit. Insectic. Fungic. Conf., 8th, 1, 339 (1975).
190. T. T. Matsumoto and N. F. Sommer, Phytopathology, 57, 881 (1967).
191. C. F. Pierson, Phytopathology, 56, 276 (1966).
192. W. S. Stewart, J. E. Palmer, and H. Z. Hield, Proc. Amer. Soc. Hortic. Sci., 59, 327 (1952).
193. T. A. DeWolfe, L. C. Erickson, and B. L. Branaman, Proc. Amer. Soc. Hortic. Sci., 74, 367 (1959).

194. P. du T. Pelser, Recommendations for the Control of Post-harvest
     Decay of Citrus Fruits, South African Co-operative Citrus Exchange
     Ltd., Pretoria, South Africa, 1975, p. 12.
195. C. Brooks, E. V. Miller, C. O. Bratley, J. S. Cooley, P. V. Mook,
     and H. S. Johnson, U.S. Dept. Agr. Tech. Bull., 318, 1932, p. 59.
196. H. W. Smith, Advan. Food Res., 12, 95 (1963).
197. C. M. Harris and J. M. Harvey, Plant Disease Reptr., 57, 44 (1973).
198. N. F. Sommer, R. J. Fortlage, F. G. Mitchell, and E. C. Maxie,
     J. Amer. Soc. Hortic. Sci., 98, 285 (1973).
199. J. M. Wells and M. Uota, Phytopathology, 60, 50 (1970).
200. J. M. Harvey, C. M. Harris and F. M. Porter, U.S. Dept. Agr.
     Market. Res. Rept., 920, 1971, p. 10.
201. K. L. Edney, Ann. Appl. Biol., 54, 327 (1964).
202. L. van den Berg and C. P. Lentz, Can. Inst. Food Sci. Technol. J.,
     7, 260 (1974).
203. J. J. Smoot, Proc. Intern. Citrus Symp., 1st, 3, 1285 (1969).
204. M. N. Follstad, Phytopathology, 56, 1098 (1966).
205. N. A. Littlefield, B. N. Wankier, D. K. Salunkhe, and J. N. McGill,
     Appl. Microbiol., 14, 579 (1966).
206. T. T. Hatton and W. F. Reeder, J. Amer. Soc. Hortic. Sci., 97,
     339 (1972).
207. D. H. Spalding and W. F. Reeder, Phytopathology, 65, 458 (1975).
208. D. H. Spalding and W. F. Reeder, Proc. Fla. State Hortic. Soc.,
     85, 337 (1972).
209. D. K. Salunkhe and M. T. Wu, in Postharvest Biology and Handling
     of Fruits and Vegetables (N. F. Haard and D. K. Salunkhe, eds.),
     Avi, Westport, Conn., 1975, Chap. 13.
210. S. P. Burg, in Postharvest Biology and Handling of Fruits and Vege-
     tables (N. F. Haard and D. K. Salunkhe, eds.), Avi, Westport,
     Conn., 1975, Chap. 14.
211. D. H. Spalding and W. F. Reeder, Proc. Amer. Soc. Hortic. Sci.
     (Trop. Reg.), 18 (in press).
212. C. S. Parsons and D. H. Spalding, J. Amer. Soc. Hortic. Sci., 97,
     297 (1972).
213. W. J. Lipton, Amer. Potato J., 44, 292 (1967).
214. J. M. Wells and D. H. Spalding, Phytopathology, 65, 1299 (1975).
215. T. R. Swinburne, Ann. Appl. Biol., 78, 39 (1974).
216. R. G. Tomkins, G.B. Dept. Sci. Ind. Res., Food Invest. Board,
     Rept., 1935, 1936, p. 129.
217. J. W. Eckert and M. J. Kolbezen, Phytopathology, 53, 755 (1963).
218. J. E. Van der Plank, J. M. Rattray, and G. F. van Wyk, J. Pomol.
     Hortic. Sci., 18, 135 (1940).
219. O. J. Burden, Australian J. Exptl. Agr. Animal Husbandry, 9, 655
     (1969).
220. J. S. Cooley and J. H. Crenshaw, U.S. Dept. Agr., Circ., 177,
     1931, p. 10.

221. E. E. Butler, Plant Disease Reptr., 41, 474 (1957).
222. J. W. Eckert and M. J. Kolbezen, Phytopathology, 52, 180 (1962).
223. N. S. Luepschen, Phytopathology, 54, 1219 (1964).
224. R. G. Tomkins, Nature, 199, 669 (1963).
225. K. E. Nelson and M. Ahmedullah, Amer. J. Enol. Viticult., 24, 75 (1973).
226. A. Paulin, Fruits, 21, 589 (1966).
227. E. G. Hall, Australia CSIRO, Div. Food Res., Food Preserv. Q., 15, 42 (1955).
228. S. M. Norman, D. C. Fouse, and C. C. Craft, Plant Disease Reptr., 53, 780 (1969).
229. M. J. Kolbezen, J. W. Eckert, and B. Bretschneider, Proc. Intern. Citrus Symp., 1st, 2, 1071 (1969).
230. K. E. Nelson and G. A. Baker, Amer. J. Enol. Viticult., 14, 13 (1963).
231. K. E. Nelson and H. B. Richardson, Proc. Amer. Soc. Hortic. Sci., 77, 337 (1961).
232. K. E. Nelson, G. A. Baker, and J. P. Gentry, Amer. J. Enol. Viticult., 15, 93 (1964).
233. W. B. Sinclair and D. L. Lindgren, J. Econ. Entomol., 51, 891 (1958).
234. D. S. Meredith, Proc. Brit. Insectic. Fungic. Conf., 8th, 2, 581 (1975).
235. J. B. Henriksen, Proc. Brit. Insectic. Fung. Conf., 8th, 2, 603 (1975).
236. C. P. Meijers, Proc. Brit. Insectic. Fungic. Conf., 8th, 2, 597 (1975).
237. C. Logan, R. B. Copeland, and G. V. Little, Ann. Appl. Biol., 80, 199 (1975).
238. C. Logan and R. B. Copeland, Proc. Brit. Insectic. Fungic. Conf., 8th, 2, 589 (1975).
239. G. Gray and S. S. Leach, Maine Life Sci. Agr. Expt. Sta., Bull., 23, 1975, p. 12.
240. S. S. Leach, Plant Disease Reptr., 55, 723 (1971).
241. R. B. Copeland and C. Logan, Potato Res., 18, 179 (1975).
242. G. A. Hide, J. M. Hirst, and R. L. Griffith, Proc. Brit. Insectic. Fungic. Conf., 5th, 2, 310 (1969).
243. D. J. Phillips and M. Uota, Blue Anchor, 48, 20 (1971).
244. G. E. Brown, Phytopathology, 64, 539 (1974).
245. F. W. Hayward and W. Grierson, J. Agr. Food Chem., 8, 308 (1960).
246. J. W. Eckert and M. J. Kolbezen, Phytopathology, 54, 978 (1964).
247. R. O. Magie, HortScience, 6, 351 (1971).
248. H. Buchenauer and D. C. Erwin, Phytopathology, 61, 433 (1971).
249. R. E. Pitblado and L. V. Edgington, Phytopathology, 62, 513 (1972).
250. J. I. Mason, J. M. McDougald, and B. G. Drought, HortScience, 9, 122 (1974).

251. Y. Gutter, U. Yanko, M. Davidson, and M. Rahat, Phytopathology, 64, 1477 (1974).
252. L. E. Rippon and B. L. Wild, Agr. Gaz., N. S. Wales, 84, 133 (1973).
253. A. A. McCornack and W. F. Wardowski, Fla., Univ., Agr. Ext. Serv. (Gainesville) Circ., 359, 1972, p. 4.
254. A. A. McCornack, Proc. Fla. State Hortic. Soc., 81, 290 (1969).
255. A. A. McCornack, Proc. Fla. State Hortic. Soc., 83, 229 (1971).
256. J. J. Smoot and C. F. Melvin, Proc. Fla. State Hortic. Soc., 83, 225 (1971).
257. J. M. Wells, Phytopathology, 62, 129 (1972).
258. P. du T. Pelser, Citrus Sub-Trop. Fruit J., 474, 12, 16 (1973).
259. P. du T. Pelser, Citrus Sub-Trop. Fruit J., 463, 12 (1972).
260. P. du T. Pelser, Citrus Sub-Trop. Fruit J., 473, 4, 7, 16 (1973).
261. E. R. White, E. A. Bose, J. M. Ogawa, B. T. Manji, and W. M. Kilgore, J. Agr. Food Chem., 21, 616 (1973).
262. M. Chiba and F. Doornhos, Bull. Environ. Contam. Toxicol., 11, 273 (1974).
263. J. R. Winston, U.S., Dept. Agr., Tech. Bull., 488, 1935, p. 32.
264. L. Hwang and L. J. Klotz, Hilgardia, 12, 1 (1938).
265. W. R. Barger, J. S. Wiant, W. T. Pentzer, A. L. Ryall, and D. H. Dewey, Phytopathology, 38, 1019 (1948).
266. W. J. Martin, J. M. Lutz, and G. B. Ramsey, Phytopathology, 39, 580 (1949).
267. W. R. Barger, Calif. Citrograph, 13, 164, 172 (1928).
268. L. J. Klotz, Hilgardia, 10, 27 (1936).
269. R. G. Tomkins, J. Pomol. Hortic. Sci., 12, 311 (1934).
270. B. P. Ridge and A. H. Little, J. Textile Inst., Trans., 33, 33 (1942).
271. J. E. Van der Plank, S. African Dept. Agr. Tech. Serv., Bull., 241, 1945, p. 60.
272. R. H. Segall, Phytopathology, 58, 1412 (1968).
273. H. L. Link and H. M. Pancoast, Food Ind., 21, 737 (1949).
274. W. J. Lipton and J. K. Stewart, Proc. Amer. Soc. Hortic. Sci., 78, 324 (1961).
275. B. A. Friedman, Phytopathology, 41, 709 (1951).
276. C. L. Lockhart and R. W. Delbridge, Can. Plant Disease Surv., 52, 140 (1972).
277. D. E. Pryor, Food Technol., 4, 57 (1950).
278. J. W. Eckert, M. J. Kolbezen, F. O. Whipple, B. F. Loehr, and R. J. Fuchs, Phytopathology, 50, 83 (1960).
279. J. D. Ridley and E. T. Sims, S. Carolina, Agr. Expt. Sta., Tech. Bull., 1027, 1967, p. 24.
280. H. A. Schomer and L. P. McColloch, U.S., Dept. Agr., Circ., 765, 1948, p. 24.

281. P. R. Harding, Jr., Plant Disease Reptr., 52, 245 (1968).
282. D. H. Spalding, U. S. Dept. Agr., Market. Res. Rept., 801, 1968, p. 9.
283. M. A. Joslyn and J. B. S. Braverman, Advan. Food Res., 5, 97 (1954).
284. R. A. Cappellini, A. W. Stretch, and G. S. Walton, Plant Disease Reptr., 45, 301 (1961).
285. J. T. Worthington, III and W. L. Smith, Jr., Plant Disease Reptr., 49, 783 (1965).
286. W. V. Cruess, P. H. Richert and J. H. Irish, Hilgardia, 6, 295 (1931).
287. K. E. Nelson, Proc. Amer. Soc. Hortic. Sci., 71, 183 (1958).
288. H. E. Jacob, Calif., Agr. Expt. Sta., Bull., 471, 1929, p. 24.
289. K. E. Nelson and J. P. Gentry, Amer. J. Enol. Viticult., 17, 148 (1966).
290. B. B. Beattie and N. L. Outhred, Australian J. Exptl. Agr. Animal Husbandry, 10, 124 (1970).
291. J. Reyneke and J. E. Piaget, II, Farming S. Africa, 27, 477 (1952).
292. J. E. Van der Plank and G. F. van Wyk, S. African Dept. Agr. Forest. Ann. Rept. Low Temp. Res. Lab., Capetown, 1938-1939, 1940, p. 43.
293. K. E. Nelson and J. P. Gentry, Amer. J. Enol. Viticult., 17, 290 (1966).
294. K. E. Nelson and M. Ahmedullah, Amer. J. Enol. Viticult., 23, 78 (1972).
295. D. S. Meredith, Trop. Agr. (Trinidad), 38, 205 (1961).
296. C. A. Phillips, Trop. Agr. (Trinidad), 47, 1 (1970).
297. J. M. Wells and J. K. Stewart, Plant Disease Reptr., 52, 262 (1968).
298. H. Subramanyam and N. V. N. Moorthy, Pestic. Sci., 4, 25 (1973).
299. J. A. Domenico, A. R. Rahman, and D. E. Westcott, J. Food Sci., 37, 957 (1972).
300. A. Grasovsky and M. Shiff, Hadar, 7, 168 (1934).
301. R. G. Tomkins and S. A. Trout, J. Pomol. Hortic. Sci., 9, 257 (1931).
302. E. Bottini, Ann. Chim. Appl., 17, 163 (1927).
303. C. N. Roistacher, I. L. Eaks, and L. J. Klotz, Plant Disease Reptr., 39, 202 (1955).
304. J. W. Eckert, M. J. Kolbezen, and M. J. Garber, Phytopathology, 53, 140 (1963).
305. F. A. Gunther, R. C. Blinn, J. H. Barkley, M. J. Kolbezen, and E. A. Staggs, J. Agr. Food Chem., 7, 489 (1959).
306. C. N. Roistacher, L. J. Klotz, M. J. Kolbezen, and E. A. Staggs, Plant Disease Reptr., 42, 1112 (1958).
307. P. R. Harding, Jr., Plant Disease Reptr., 43, 893 (1959).
308. M. J. Kolbezen and J. W. Eckert, Phytopathology, 53, 129 (1963).

309. C. N. Roistacher, L. J. Klotz, and M. J. Garber, Phytopathology, 50, 855 (1960).
310. I. L. Eaks, Proc. Amer. Soc. Hortic. Sci., 73, 237 (1959).
311. P. R. Harding, Jr., Plant Disease Reptr., 41, 564 (1957).
312. D. Leggo and J. A. Seberry, Australian J. Exptl. Agr. Animal Husbandry, 4, 173 (1964).
313. I. L. Eaks, J. W. Eckert, and C. N. Roistacher, Plant Disease Reptr., 42, 846 (1958).
314. J. W. Eckert and M. J. Kolbezen, Phytopathology, 60, 545 (1970).
315. J. A. Seberry and R. A. Baldwin, Australian J. Exptl. Agr. Animal Husbandry, 8, 440 (1968).
316. A. Vanderweyen, R. Huet, and A. Ledergerber, Awamia, 14, 19 (1965).
317. Y. Gutter, Israel J. Agr. Res., 17, 167 (1967).
318. A. A. McCornack, Proc. Fla. State Hortic. Soc., 86, 284 (1974).
319. A. A. McCornack and E. F. Hopkins, Proc. Fla. State Hortic. Soc., 77, 267 (1965).
320. C. F. Pierson, Plant Disease Reptr., 50, 913 (1966).
321. R. O. Magie, Proc. Fla. State Hortic. Soc., 76, 458 (1964).
322. E. T. Sims and J. D. Ridley, HortScience, 2, 169 (1967).
323. P. F. McDonnell, Plant Disease Reptr., 55, 771 (1971).
324. G. C. Pratella and G. Tonini, Frutticoltura, 32, 43 (1970).
325. W. Grierson and F. W. Hayward, Proc. Amer. Soc. Hortic. Sci. (Trop. Reg.), 13, 124 (1971).
326. W. Grierson, Proc. Fla. State Hortic. Soc., 81, 238 (1970).
327. D. C. Graham, G. A. Hamilton, M. J. Nash, and J. H. Lennard, Potato Res., 16, 234 (1973).
328. D. C. Graham and G. A. Hamilton, Nature, 227, 297 (1970).
329. R. G. Tomkins and S. A. Trout, G.B., Dept. Sci. Ind. Res., Food Invest. Board, Rept., 1931, 1932, p. 117.
330. K. Prasad and G. J. Stadelbacher, Plant Disease Reptr., 57, 795 (1973).
331. K. Prasad and G. J. Stadelbacher, Phytopathology, 64, 948 (1974).
332. G. J. Stadelbacher and K. Prasad, J. Amer. Soc. Hortic. Sci., 99, 364 (1974).
333. G. L. Rygg, U.S., Dept. Agr., Handb., 482, 1975, p. 56.
334. F. Wesley, B. Rourke, and O. Darbishire, J. Food Sci., 30, 1037 (1965).
335. F. J. Bandelin, J. Amer. Pharm. Assoc., 47, 691 (1958).
336. E. S. Beneke and F. W. Fabian, Food Technol., 9, 486 (1955).
337. G. R. DiMarco and B. H. Davis, Plant Disease Reptr., 41, 284 (1957).
338. D. B. Sauer and R. Burroughs, Trans. Amer. Soc. Agr. Eng., 17, 557 (1974).
339. D. B. Sauer, T. O. Hodges, R. Burroughs, and H. L. Converse, Trans. Amer. Soc. Agr. Eng., 18, 1162 (1975).

340. G. M. Jones, D. N. Mourat, J. I. Elliot, and E. T. Moran, Jr., Can. J. Animal Sci., 54, 499 (1974).
341. K. S. Dhanraj, J. S. Chohan, M. S. Sunar, and S. Lal, Indian Phytopathol., 26, 63 (1973).
342. W. L. Smith, Jr. and J. T. Worthington, III, Plant Disease Reptr., 49, 619 (1965).
343. R. G. Tomkins, G. B., Dept. Sci. Ind. Res., Food Invest. Board, Rept., 1936, 1937, p. 149.
344. K. J. Scott, B. B. Beattie, and E. A. Roberts, Australian J. Sci., 29, 22 (1966).
345. K. J. Scott and E. A. Roberts, Australian J. Exptl. Agr. Animal Husbandry, 5, 296 (1965).
346. H. H. Gill, Dow Chemical Company, Analytical Report, Midland, Michigan, December 16, 1957.
347. J. N. Sharma, U.S. Pat. 2,054,392 (1936).
348. J. E. Van der Plank and J. M. Rattray, S. African Dept. Agr. Forest. Ann. Rept. Low Temp. Res. Lab., Capetown, 1938-1939, 1940, p. 93.
349. F. W. Hayward and A. A. McCornack, Proc. Fla. State Hortic. Soc., 81, 285 (1969).
350. A. Rajzman and A. Apelbaum, Pestic. Sci., 1, 59 (1970).
351. J. K. Long, D. Leggo, and J. A. Seberry, Australia, N. S. Wales Dept. Agr. CSIRO Div. Food Preserv., Citrus Wastage Res. Lab., Bull., 168, 1965, p. 7.
352. L. J. Kushman, W. R. Wright, J. Kaufman, and R. E. Hardenburg, U.S., Dept. Agr., Market. Res. Rept., 698, 1965, p. 12.
353. W. J. Martin, Plant Disease Reptr., 48, 606 (1974).
354. F. B. Cates and L. O. van Blaricom, Amer. Potato J., 38, 175 (1961).
355. A. D. Hoadley, Plant Disease Reptr., 47, 900 (1963).
356. R. L. Baichwal, R. M. Baxter, S. I. Kandel, and G. C. Walker, Can. J. Biochem. Physiol., 38, 245 (1960).
357. D. Leggo, J. Seberry, and J. K. Long, Australian J. Exptl. Agr. Animal Husbandry, 4, 247 (1964).
358. M. A. Slade and R. D. Gerwe, U.S. Pat. 2,578,752 (1951).
359. K. J. Scott and E. A. Roberts, Australian J. Exptl. Agr. Animal Husbandry, 7, 283 (1967).
360. A. Rajzman, Residue Rev., 8, 1 (1965).
361. B. C. Heiberg and G. B. Ramsey, Phytopathology, 36, 887 (1946).
362. P. R. Harding, Jr., Plant Disease Reptr., 48, 43 (1964).
363. F. W. Hayward and G. J. Edwards, Proc. Fla. State Hortic. Soc., 76, 318 (1964).
364. G. L. Rygg, A. W. Wells, S. M. Norman, and E. P. Atrops, U.S., Dept. Agr., Market. Res. Rept., 646, 1964, p. 22.
365. S. M. Norman, C. C. Craft, and D. C. Fouse, U.S., Dept. Agr., Market. Res. Rept., 896, 1971, p. 6.

366. S. W. Souci and G. Maier-Haarländer, Residue Rev., 16, 103 (1966).

367. P. R. Harding, Jr., Plant Disease Reptr., 46, 100 (1962).

368. J. S. Vandemark and E. G. Sharvelle, Science, 115, 149 (1952).

369. W. L. Smith, Jr., M. H. Miller, and T. T. McClure, Phytopathology, 46, 261 (1956).

370. C. E. D. Smith and W. A. Wynne, U.S. Pat. 2,853,415 (1958).

371. L. Chiarappa, J. W. Eckert, and M. J. Kolbezen, Amer. J. Enol. Viticult., 13, 83 (1962).

372. R. A. Cappellini and A. W. Stretch, Plant Disease Reptr., 46, 31 (1962).

373. J. M. Ogawa, S. D. Lyda, and D. J. Weber, Plant Disease Reptr., 45, 636 (1961).

374. J. M. Ogawa, J. L. Sandeno, and J. M. Mathre, Plant Disease Reptr., 47, 129 (1963).

375. J. M. Ogawa, R. H. Ramsey, and C. J. Moore, Phytopathology, 53, 97 (1963).

376. J. M. Wells and J. M. Harvey, Blue Anchor, 48, 17 (1971).

377. J. G. Stone, N. L. Wade, and B. B. Beattie, Agr. Gaz. N. S. Wales, 82, 375 (1971).

378. W. J. Martin, Plant Disease Reptr., 55, 523 (1971).

379. C. L. Burton, Trop. Agr. (Trinidad), 47, 303 (1970).

380. J. A. Freeman, Can. Plant Disease Surv., 45, 107 (1965).

381. W. L. Smith, Jr. and J. T. Worthington, III, Plant Disease Reptr., 49, 619 (1965).

382. E. Cohen and C. Dennis, Ann. Appl. Biol., 80, 237 (1975).

383. J. W. Eckert and B. Bretschneider, unpublished data, 1976.

384. V. L. Miller, C. J. Gould, E. Csonka, and R. L. Jensen, J. Agr. Food Chem., 21, 931 (1973).

385. C. Kowala, K. S. Murray, J. M. Swan, and B. O. West, Australian J. Chem., 24, 1369 (1971).

386. W. K. Lowen, Proc. Intern. Citrus Congr., 1st (in press).

387. J. A. Gardiner, private communication, 1976.

388. P. M. Upham and C. J. Delp, Phytopathology, 63, 814 (1973).

389. G. Crivelli, Freddo, 20, 25 (1966).

390. G. E. Brown, A. A. McCornack, and J. J. Smoot, Plant Disease Reptr., 51, 95 (1967).

391. Y. Gutter, Israel J. Agr. Res., 20, 91 (1970).

392. P. R. Harding, Jr., Plant Disease Reptr., 52, 623 (1968).

393. Y. Gutter, Israel J. Agr. Res., 20, 135 (1970).

394. A. A. McCornack and G. E. Brown, Proc. Fla. State Hortic. Soc. 82, 235 (1970).

395. A. Vanderweyen and H. de Trogoff, Awamia, 33, 113 (1971).

396. B. L. Wild, L. E. Rippon, and J. A. Seberry, Australian J. Exptl. Agr. Animal Husbandry, 15, 108 (1975).

397. N. D. Bondad, Philipp. J. Sci., 103, 21 (1974).

398. Y. Gutter, Plant Disease Reptr., 54, 325 (1970).
399. Y. Gutter and U. Yanko, Israel J. Agr. Res., 21, 105 (1971).
400. A. A. McCornack, Proc. Fla. State Hortic. Soc., 87, 230 (1975).
401. J. Cuillé and L. Bur-Ravault, Fruits, 24, 421 (1969).
402. J. W. Eckert and M. J. Kolbezen, Proc. Brit. Insect. Fungic. Conf., 6th, 3, 683 (1971).
403. Y. Gutter, Phytopathology, 65, 498 (1975).
404. P. du T. Pelser, Citrus Sub-Trop. Fruit J., 486, 15 (1974).
405. P. R. Harding, Jr., Plant Disease Reptr., 56, 256 (1972).
406. J. J. Smoot and G. E. Brown, Plant Disease Reptr., 58, 933 (1974).
407. I. F. Muirhead, Australian J. Exptl. Agr. Animal Husbandry, 14, 698 (1974).
408. B. L. Wild and L. E. Rippon, Phytopathology, 65, 1176 (1975).
409. M. Schiffmann-Nadel, E. Chalutz, J. Waks, and F. S. Lattar, HortScience, 7, 394 (1972).
410. M. Schiffmann-Nadel, E. Chalutz, J. Waks, and M. Dagan, J. Amer. Soc. Hortic. Sci., 100, 270 (1975).
411. T. I. Kokkalos, HortScience, 9, 456 (1974).
412. W. F. Wardowski, L. G. Albrigo, W. Grierson, C. R. Barmore, and T. A. Wheaton, HortScience, 10, 381 (1975).
413. R. N. Allen, Australian J. Exptl. Agr. Animal Husbandry, 10, 490 (1970).
414. D. M. Bailey, D. F. Cutts, L. Donegan, C. A. Phillips, and R. Pope, J. Food Technol., 5, 89 (1970).
415. J. Cuillé and L. Bur-Ravault, Fruits, 23, 351 (1968).
416. P. Frossard, Fruits, 25, 265 (1970).
417. L. E. Rippon and M. Glennie-Holmes, Agr. Gaz. N. S. Wales, 84, 229 (1973).
418. C. J. Jacobs, H. T. Brodrick, H. D. Swarts, and N. J. Mulder, Plant Disease Reptr., 57, 173 (1973).
419. H. S. Sohi, Phytopathol. Mediterranée, 12, 114 (1973).
420. Y. J. Fripp and E. B. Dettmann, Australian J. Exptl. Agr. Animal Husbandry, 9, 9 (1969).
421. J. L. Maas and I. C. MacSwann, Plant Disease Reptr., 54, 887 (1970).
422. T. O. C. Ndubizu and G. D. Blanpied, Plant Disease Reptr., 55, 791 (1971).
423. K. J. Scott and E. A. Roberts, Australian J. Exptl. Agr. Animal Husbandry, 10, 235 (1970).
424. S. R. M. Valdebenito and A. Pinto de Torres, Agr. Tec. Mex., 32, 148 (1972).
425. M. K. Roy, Plant Disease Reptr., 59, 61 (1975).
426. D. H. Spalding, Plant Disease Reptr., 54, 655 (1970).
427. R. Ben-Arie, Phytopathology, 65, 1187 (1975).
428. P. F. McDonnell, Plant Disease Reptr., 54, 83 (1970).

429. A. L. Jones and C. Burton, Plant Disease Reptr., 57, 62 (1973).
430. N. L. Wade and P. G. Gipps, Australian J. Exptl. Agr. Animal Husbandry, 13, 600 (1973).
431. J. L. Maas and W. L. Smith, Jr., Plant Disease Reptr., 56, 296 (1972).
432. J. L. Maas, Plant Disease Reptr., 55, 883 (1971).
433. J. A. Ekundayo and T. M. Daniel, Trans. Brit. Mycol. Soc., 61, 27 (1973).
434. S. K. Ogundana, Intern. Biodeterior. Bull., 8, 75 (1972).
435. R. Barkai-Golan and N. Aharoni, Israel J. Agr. Res., 22, 157 (1972).
436. D. H. Spalding and W. F. Reeder, Plant Disease Reptr., 58, 59 (1974).
437. R. L. Gabrielson, R. C. Maxwell, and O. O. Kienholz, Phytopathology, 62, 758 (1972).
438. P. Frossard and E. Laville, Fruits, 28, 617 (1973).
439. P. J. Griffee and J. A. Pinegar, Trop. Sci., 16, 107 (1974).
440. P. du T. Pelser, Citrus Sub-Trop. Fruit J., 474, 7 (1973).
441. L. E. Rippon and B. L. Wild, Australian J. Exptl. Agr. Animal Husbandry, 13, 724 (1973).
442. W. Koffman and P. F. Kable, Plant Disease Reptr., 59, 586 (1975).
443. J. W. Eckert and M. J. Kolbezen, Proc. Intern. Citrus Congr., 1st (in press).
444. J. W. Vonk and A. Kaars Sijpesteijn, Pestic. Sci., 2, 160 (1971).
445. P. Frossard, E. Laville, and J. Motillon, Fruits, 28, 195 (1973).
446. A. L. Jones, Plant Disease Reptr., 59, 127 (1975).
447. E. Laville and C. Piedallu, Fruits, 28, 33 (1973).
448. A. Vanderweyen, M. Besri, and R. Pineau, Awamia, 38, 45 (1971).
449. E. Laville, Fruits, 28, 545 (1973).
450. E. Laville, Meded. Fac. Landbouwwet., Rijksuniv. Gent, 39 (2, Pt. 2), 1121 (1974).
451. A. L. Jones, C. L. Burton, and B. R. Tennes, Mich., Agr. Expt. Sta., Res. Rept., 209, 1973, p. 10.
452. P. R. Harding, Jr., Plant Disease Reptr., 60, 543 (1976).

Chapter 10

FUNGICIDAL CONTROL OF PLANT DISEASES
IN THE TROPICS

R. H. Stover

Division of Tropical Research
United Fruit Co.
La Lima, Honduras

I. INTRODUCTION

The tropics encompass that vast, warm belt circling the globe between the latitudes of Cancer and Capricorn or about 23° N and S of the equator. It includes about one quarter of the earth's land mass. Except at extreme altitudes where few crops are grown and frequent frosts occur, mean minimum monthly temperatures are above 10°C. All of the most important crops in the tropics are grown below altitudes where frost occurs. Crops grown commercially where frost does occur, such as wheat and potatoes, or where temperatures often fall below 8°C are not strictly tropical. Within this global belt, all the important crops are grown in the humid tropics and not the arid regions, almost all of which are confined to Africa, Australia, and small portions of the Pacific coast of Chile, Peru, and Northern Mexico. In most areas where important tropical crops are grown, there are clearly defined rainy and dry seasons. These seasons, as well as minimum temperatures, determine the activity of plant pathogens and the timing of fungicide applications. All of the important pathogens are most active during rainy weather and at temperatures above 15°C.

With the exception of tea and coffee, all the important exports crops are grown at low elevations (below 600 m). These crops include pineapples, mangoes, papayas, bananas, abaca, cotton, cacao, rubber, African oil palm, coconuts, sugar cane, spices (pepper, cloves, cinnamon, ginger), tobacco, and peanuts. Citrus, widely grown at both low and intermediate elevations, is not an important tropical export crop. It is grown mostly for local consumption with little attention to disease control. The major citrus-growing areas, often called the subtropics, lie outside the true tropics. The major tropical food crops are grown at all elevations except where frost damage may occur. These include rice, corn, beans and other pulses, sorghum, plantains and bananas, cassava (Manihot utilissima), taro (Colocasia esculenta), malanga (species of Xanthosoma), yams, and sweet potatoes.

II. USE OF FUNGICIDE IN THE TROPICS

Most of the crop area in the tropics receives no fungicide treatment. Of this area, the major portion is devoted to the peasant food crops listed above where fungicides are not used. Other large areas devoted to sugar cane, coconuts, and African oil palm do not require fungicide treatment or treatment is uneconomic. Even with the tropical crops (see Table 1) requiring fungicide treatments in some locations, the majority of the areas devoted to coffee, tea, cacao, and rubber receive no treatment.

Of the crops widely grown in temperate as well as tropical areas, potatoes, tomatoes, muskmelons, watermelons, cucumbers, and tobacco are receiving increased amounts of fungicides in the tropics. There has been

TABLE 1

Important Tropical Diseases Controlled with Fungicides in the Field

Crop	Disease	Causal organism	Fungicides used
Bananas	Leaf spot	Mycosphaerella fijiensis and M. musicola	Oil, maneb or benomyl + oil
	Rust	Uredo musae	Maneb + oil emulsion
	Fruit spots		Maneb
	Pitting disease	Pyricularia grisea	
	Brown spot	Cercospora hayi	
	Diamond spot	Cercospora hayi	
		Fusarium solani	
Mangoes	Anthracnose	Colletotrichum gloeosporioides	Benomyl, maneb, copper (post bloom)
Papayas	Anthracnose	C. gloeosporioides	Maneb, benomyl
	Phytophthora rot	Phytophthora parasitica	Maneb, captafol, captan
Avocados	Cercospora spot	Cercospora purpurea	Benomyl, copper
	Anthracnose	C. gloeosporioides	Benomyl, copper
	Scab	Sphaceloma purseae	Benomyl
Coffee	Rust	Hemilia vastatrix	Copper, captafol, pyrocarbolid + oil
	Coffee berry disease	Colletotrichum coffeanum	Captafol, copper, benomyl
	American leaf spot	Mycena citricolor	Lead arsenate
	Cercospora leaf spot	Cercospora coffeicola	Dithiocarbamate fungicides, benomyl, copper

TABLE 1 (Cont.)

Crop	Disease	Causal organism	Fungicides used
Tea	Blister blight	Exobasidium vexans	Copper
Cacao	Black pod	Phytophthora palmivora	Copper
	Die-back	P. palmivora, C. gloeosporioides	Copper
	Witches broom	Marasmius perniciosus	Copper
	Monilia pod rot	Monilia roreri	Copper, dithiocarbamate fungicides
Rubber	Secondary leaf fall	Oidium heveae, C. gloeosporioides	Sulfur, chlorothalonil
	Abnormal leaf fall	P. palmivora, P. botryosa	Copper + oil
	Pink disease	Corticium salmonicolor	Copper (immature trees), tridimorph
	White root rot	Rigidoporus lignosus	PCNB (pentachloronitrobenzene)
	Panel moldy rot	Ceratocystis fimbriata	Benomyl, cycloheximide (Actidione 0.5%), captafol
	Black stripe	P. palmivora, P. botryosa	Captafol (Difolatan 2%), cyclohexi-mide (Actidione 0.5%)
Peanuts	Cercospora leaf spot	Mycosphaerella berkeleyii, Mycosphaerella arachidicola	Benomyl, maneb
		Puccinia arachidis	Maneb

an increase in the processing of vegetables, especially tomatoes, for local consumption in tropical countries. Also, muskmelons, watermelons, tomatoes, cucumbers, and other vegetables are grown for export to temperate areas during periods when these crops cannot be grown there. The rapid expansion in the use of refrigerated containers and container ships has contributed greatly to the export of fruit, vegetables, and ornamental plants from the tropics. Export and processing companies give technical assistance to the local growers in applying the modern technology developed to produce these crops in the United States and Europe. This includes use of fungicides for disease control and production of quality produce. Improved varieties of seed, especially hybrid corn, sorghum, peanuts, and cotton are now being produced in the tropics. This has led to increased use of fungicide and insecticide seed dressings. Since the materials used and the principles of disease control with fungicides are similar for these crops and their diseases in both temperate and tropical areas, they will not be described here. Rather, emphasis will be given to those crops listed in Table 1 that are grown largely in the tropics and are attacked by destructive tropical fungus pathogens. Control of these pathogens with fungicides results in an economic response in yield and quality in some but not all locations.

III. TROPICAL CROPS REQUIRING
FUNGICIDE TREATMENT

All of the important tropical export crops have field diseases that often require fungicide treatment for maximum yield and quality, except pineapples, sugar cane, coconuts, African oil palm, cotton, and abaca. Sugar cane seed pieces are often treated with a mercurial dip or spray for control of pineapple disease caused by Ceratocystis fimbriata [1]. African oil palm seedling nurseries often require applications of maneb (manganese ethylenebisdithiocarbamate including formulations with added zinc) or benomyl (methyl-1-(butylcarbamoyl)-2-benzimidazolecarbamate) for control of anthracnose (Colletotrichum gloeosporioides and other fungi) and leaf spots caused by Cercospora elaeidis. The latter pathogen can cause severe defoliation in young field palms in West Africa but fungicidal control is not practiced.

A. Bananas

Bananas of export quality cannot be grown without controlling leaf spot disease caused by Mycosphaerella fijiensis, M. fijiensis var. difformis, and M. musicola [2]. The strains causing black leaf streak and black Sigatoka are more virulent than Sigatoka and were not easily controlled until the advent of systemic fungicides of the benzimidazole group. However, in areas such as the Philippines where rust is also present, continued use of benomyl for black leaf streak control resulted in a buildup of the rust pathogen. It was necessary to apply maneb after every one to three cycles of benomyl to prevent an outbreak of rust.

The great advances in leaf spot control since the discovery of the therapeutic action of agricultural spray oil in 1954 have been described [3, 4]. Low volumes (either 9-10 liters of oil or 23 liters of a fungicide oil-in-water emulsion per ha) are applied by aircraft at intervals varying from two weeks to more than one month. In Sigatoka areas no more than 16 cycles of spray are applied annually and in the drier areas of the Antilles as few as six cycles of benomyl in oil are adequate. Where black leaf streak and black Sigatoka are present, however, 25-30 cycles of spray are applied. Maneb is applied with the spray oil in the form of an emulsion (9-10 liters of oil and 14 liters of water plus emulsifier per hectare) or in oil alone; the benzimidazole compounds are suspended in oil alone (9-15 liters/ha). This gave better control over leaf spot than oil spray alone. In a few areas, however, such as the Ivory Coast and Ecuador, oil alone gives adequate but not complete disease control.

With the large-scale planting of the fusarial wilt resistant Cavendish varieties in the early 1960s, and the conversion from ground-applied Bordeaux misture to aerial-applied low-volume sprays, fruit spots (Table 1) appeared in epidemic proportions throughout Central America, the Philippines, and in some parts of Ecuador and Colombia. Pitting disease, the most serious of the various spots, was especially troublesome because of latent infections that developed on fruit in transit [4]. An intensive research program on the biology of the pathogens and methods of control was undertaken [4]. Maneb fungicides were superior to either copper compounds or benomyl. They were applied weekly to the emerging hands of fruit until all the hands were exposed, a period of about 2 weeks. Then a perforated polyethylene bag was placed over the stem of fruit. This bag, used to protect fruit from insect injury and scarring, reduced weathering of the fungicide deposit from rain and sprinkler irrigation water. Maneb was found to maintain most of its activity throughout the 3 months before harvest. Therefore, it greatly reduced infections occurring just before harvest. Spray was applied to the fruit with a hand operated knapsack sprayer and a long lance from three different angles to insure an even deposit over and between the fingers. Baggers followed immediately behind the sprayers to protect those stems of fruit in which all hands had emerged from the washing effect of afternoon showers that would reduce the fungicide deposit. This combination of weekly spraying and prompt bagging has reduced fruit losses from 40 to a maximum of 3% during the peak fruit-spot season.

Bananas are subject to a number of in-transit rots of the crown, necks, and fingers [4], and all fruit exported is treated with a benzimidazole spray or dip just before packing. Either benomyl or thiabendazole [2-(4 thiazolyl) benzimidazole] formulations at 200-400 ppm active ingredient are most commonly used.

B. Small Fruit

The most important tropical fruits for export and local markets other than bananas and plantains, are pineapples, mangoes, papayas, and avocados. Citrus, mostly oranges, is important in local markets but seldom receives any fungicide. Pineapples usually require no fungicides in the field. Crowns for planting are dipped in a mixture of benomyl and captan [N-(trichloro-methyltio)-4-cyclohexene-1,2-dicarboximide] where Phytophthora parasitica and Thielaviopsis are present. Fruit is treated by dipping, following harvest to control basal rot caused by Ceratocystis fimbriata (Thielaviopsis paradoxa). Benomyl has replaced sodium o-phenylphenate for basal rot control.

The most serious disease of mangoes and the most difficult to control is anthracnose. Copper was found of limited value for bloom sprays in Florida, but good control was obtained with 227 g benomyl in 378 liters of water [5]. Maneb was also effective during bloom. Spray was applied weekly during the bloom period followed by copper sprays monthly after bloom and fruit set. Copper was superior to maneb in controlling latent anthracnose on fruit. Latent anthracnose in harvested fruit is controlled by a hot-water dip (55°C for 5 min). The addition of either benomyl or thiabendazole improved control [6].

Papaya fruit in Hawaii is attacked by anthracnose and Phytophthora. The latter also attacks seedlings and the trunk of mature plants. Hunter and Buddenhagen [7, 8] worked out the field biology and control of Phytophthora rot. Copper sprays increased stem-end rot due to Colletotrichum but maneb controlled the disease. It is applied with a high-pressure ground sprayer to the young or the upper half of the acropetally developing fruit cluster. Spray is applied weekly when there is more than 63 mm of rain and every 2 weeks with less rain. Aerial application was ineffective because of inadequate penetration of the canopy and hence poor fruit coverage. Postharvest diseases of papayas, caused by Ascochyta caricae, Botryodiplodia theobromae, Fusarium solani, and Phomopsis sp. [9] are usually controlled by a hot water dip (48-49°C for 20 min).

Avocado fruit is affected by Cercospora spot, anthracnose, and scab (see Table 1). Benomyl and copper compounds are used for control [5]. Spray is applied as bloom buds begin to open, near the end of the bloom, and 3 to 4 weeks after all fruit is set.

C. Coffee

Fungicidal control of two destructive diseases in Kenya, i.e., rust and coffee berry disease, has been extensively studied since 1956. The principles of rust control by copper fungicides and the importance of timing of the sprays had been worked out by 1962 [10, 11]. By 1965, Firman and Wallis [12, 13] had shown that low-volume copper sprays (93 liters/ha) applied with

knapsack mistblowers would provide economic control of rust. With six
carefully timed sprays per year, containing 1.1-2.8 kg of 50% copper per
hectare, a yield response was obtained equivalent to 5.6 kg even though rust
control was poor [12]. Beginning in the 1960s, coffee berry disease began
to increase in importance and an integrated control program for both diseases was developed [14-16]. At first it was believed that the bark was the
most important source of inoculum for early season berry infections.
Early sprays designed to reduce this inoculum, at first successful, soon
failed. Griffiths et al. [15] showed the major source of inoculum was produced on infected berries throughout the fruiting period. They worked out
a spray schedule starting at the commencement of the rains and continuing
at intervals throughout the fruiting seasons. They also showed that timing
was more important for the degree of control achieved than the fungicide
used, such as maneb or captafol [cis-N-(1,1,2,2-tetrachloroethylthio)-4-
cyclohexene-1,2 dicarboximide] with cuprous oxide 0.7%. Depending on
fungicide persistence, eight to eleven sprays applied at 3-to-4-week intervals may be needed in 1 year. In this schedule, at least one prerains and
one subsequent spray 3 weeks later is necessary for rust control [14].
Further advances in application methods have been made by the use of lower volumes of spray [17] and of specially designed multirow sprayers and
aircraft [18]. The former is applied with 1,100 liters/ha and the latter with
96 liters/ha.

     Of the systemic fungicides, benomyl and other benzimidazoles have been
very effective against coffee berry disease [47] whereas pyracarbolid
(2-methyl-5,6-dihydro-4-H-pyran-3-carboxylic acid anilide) was effective
against rust [48]. The latter was applied in an oil-based formulation and
was superior to 0.7% cuprous oxide.

     With the discovery of coffee rust in Brazil in January 1970 [19], renewed
studies on disease control with fungicides under Brazilian conditions are
being carried out. Spacing and pruning were never designed for spray
apparatus, however, and considerable change in cultural practices will be
required before efficient spray techniques can be applied in Brazil. Copper
compounds applied by mistblower or aircraft have been tried. A blend of
oils with or without added fungicide has also given good control when low
volumes (20 liters/ha) were applied by mistblowers [49].

     There are two important foliage diseases of coffee in Central and South
America caused by Cercospora coffeicola and Mycena citricolor ("ojo de
gallo"). The former is particularly severe in nurseries and in the field
where shade is deficient. It is controlled by spraying with dithiocarbamate
or copper fungicides every 2 weeks or benomyl every 3 to 4 weeks. Mycena
is controlled with one application of lead arsenate at the beginning and end
of the dry season; knapsack sprayers are used. Usually only the larger
growers spray for control of these diseases.

D. Tea

The most serious tea disease in Ceylon and India is blister blight. Kerr and his associates [20], working in Ceylon, studied the epidemiology of the disease and developed a highly precise method of determining just when to apply copper fungicides. They are applied by knapsack sprayers at low volume every 10 days throughout the monsoon [21]. Blister blight is not present in Africa or South America.

E. Cacao

Black pod is the most serious worldwide disease affecting cacao. Although it can be reduced with copper fungicides in some areas, most of the cacao acreage is not sprayed. For example, in the large cacao producing state of Bahia in Brazil a campaign was launched in 1972 for black-pod control based on use of fungicide and thinning of excess shade with arboricides [22]. Even with a government subsidy for fungicides to make their use more economic for the farmer, only 12,547 ha out of about 120,000 were treated. A total of 110,000 kg of fungicide were used, the largest quantity in the previous 12 years. The economics of spraying depends on number of pods present on the tree and world cacao prices, which fluctuate greatly [23, 24]. Spray is applied by mistblowers, now preferred in most areas, or knapsack sprayers with hand or pneumatic pumps. The interval between cycles is 3 to 4 weeks. Since pods can double in their surface area in 10 days, intervals between spraying should be shortened during the period of greatest pod development. Spray is directed at the pods rather than the entire canopy in West Africa [25, 26], whereas canopy spraying is recommended in Central and South America [27, 28] where dieback is also a problem. This is caused by P. palmivora and C. gloeosporioides which attack new flushes of leaf growth. Also in areas where witches broom is present canopy spraying improved yields [27]. Tollenaar [27] pointed out that canopy spraying increased flowering and fruit set and achieved production increases far beyond those to be expected from control of black pod only. However, results of spraying in different areas have varied from poor in West Africa [26] to excellent in Ecuador, Costa Rica, and West Cameroon [27, 28]. A rate of 0.9-1.4 kg of metallic copper spread over 300 trees with a mistblower was recommended in a high-rainfall area of Costa Rica [28]. The volume of water required would vary from 95 liters upward, depending on the efficiency of the spray operator. In West Africa, a minimum density of 1.0-1.5 μg Cu/cm^2 is required over the entire pod for good control [25]. In all areas Bordeaux mixture has been better than other copper formulations or non-copper compounds. Bordeaux residues have been shown to be more resistant to rain-induced weathering [25]. Other compounds approaching the effectiveness

of Bordeaux mixture were cupric hydroxide (Kocide) and cuprous oxide
(Perenox) in Costa Rica [28, 29].

Cacao seedlings in nurseries are often attacked by C. gloeosporioides.
Captafol (Difolatan) and maneb (Dithane M-45) at 0.9 kg per 378 liters of
water reduced spots from about ten per leaf to one or two [29].

F. Rubber

With the increased use of monoclonal plantings of high-yielding latex clones
in Malaysia, foliage diseases of rubber have increased in importance. One
of the two most important diseases is secondary leaf fall (SLF). Caused by
Oidium, SLF is most widespread, whereas the foliage disease caused by
Colletotrichum is locally severe in wet climates. Colletotrichum is more
severe in Liberia than Malaysia. These fungi attack the flush of new leaves
following the annual dry-season defoliation. Oidium is controlled by applying sulfur dust (8-9 kg/ha) four to six times at intervals of 5 to 7 days. The
first treatment is applied when leaves are starting to unfold and continues
until 80% of the leaves are past the susceptible stage [30]. No economic
fungicide control has been developed for Colletotrichum SLF although
chlorothalonil (Daconil) (2,4,5,6-tetrachloroisophthalonitrile) and benomyl
are effective. Even sulfur applications for Oidium are not always economical. A method of avoiding SLF was described by Rao [31]. He sprayed
the trees with a cacodylic acid defoliant, thus manipulating the wintering or
defoliation period. Mature refoliation generally coincides with a period of
dry weather which is not conducive for either C. gloeosporioides or
O. heveae, thus avoiding secondary leaf fall by both. About 10,000 acres
are now defoliated by aircraft spray annually for control of SLF.

The second most important foliage disease of rubber, especially in India
and Ceylon, is abnormal leaf fall caused by Phytophthora. Peries [32] in
Ceylon studied its epidemiology. Pods became infected first providing
inoculum for leaf petiole infection and abscission. He worked out a disease
forecasting system to be used during the 10-week period of weather suitable
for infection in Ceylon. In India, methods of applying copper by aircraft
were studied to replace the costly and cumbersome application of Bordeaux
mixture with ground rigs [33]. A formulation of 1.8 kg metallic copper is
applied in oil at 37-47 liters/ha and has replaced ground-applied Bordeaux
mixture. The cycles applied depend on local disease forecasts.

Helminthosporium heveae is primarily a disease of nursery seedlings.
It is readily controlled with dithiocarbamate fungicides at spray intervals of
no longer than 1 week during rainy weather [34].

White root rot is a serious disease in rubber plantations where debris
from previous jungle or old rubber was not thoroughly removed. Fungicides
can protect trees exposed to infection from the pathogen present in woody
debris from previous trees. Pentachloronitrobenzene (PCNB) is the active

ingredient of a collar- and root-protectant formulated with a nonaqueous base, making it highly persistent [30]. It acts by preventing growth over the root surface of mycelium or rhizomorphs of the fungus, without which the pathogen cannot penetrate the root. This compound is purely prophylactic and is ineffective if applied over an already infected root which must be removed before treatment.

Of the two panel diseases of rubber (Table 1) black stripe is the most widespread and important [30, 35]. Moldy rot is readily controlled with benomyl [36]. Phytophthora invades the cambium via the tapping cut preventing healthy regeneration of tapped bark. Once established, eradication is difficult, so frequent application of fungicide is required in areas prone to infection. A fungicide is applied to infected trees after every tapping during rainy weather. To protect adjacent healthy trees, a weekly application is made. Mercurial fungicides are used, but the less hazardous captafol is also effective.

Pink disease is a serious stem disease in parts of Malaysia during rainy weather and no clone is highly resistant [30]. Trees must be treated at the first signs of infection or the disease is difficult to eradicate. Bordeaux mixture has given effective control but cannot be used in mature rubber because of danger of latex contamination with copper. Fortunately, the disease is less severe on mature trees. Sticking agents have been studied to give the formulation the greater tenacity of paint with the aim of obtaining a quick-drying formulation that will seal off infected bark and make possible a single treatment. The Rubber Research Institute of Malaya now recommends a formulation of tridemorph (2% Calixin, N-tridecyl-2,6-dimethylmorpholine) in 75% natural rubber latex concentrate [46]. It is applied as a single treatment by brushing the affected parts and is superior to Bordeaux mixture.

G. Peanuts

Of the major oil crops grown in the tropics, peanuts rank in importance with African oil palm and coconuts. Although the planting of peasant crops is timed to avoid the rainiest period of the year and hence losses from leaf spot diseases, fungicidal treatment is necessary for optimum yields. None of the high-yielding varieties are resistant to either leaf spot or rust. Rust is not present in the African tropics but is serious in the American tropics. Under rainy conditions in Central America, Cercospora leaf spots can appear within 3 weeks of planting and half the leaves can be infected within 5 weeks of planting. Rust usually appears several weeks later than Cercospora. Good control of leaf spot and rust was obtained in Central America with weekly applications of maneb at 2.2 kg/ha. Biweekly applications of benomyl (280 g/ha) gave good control of Cercospora leaf spot but not rust. Rust incidence and severity were increased in plots treated with benomyl

in Central America. Hence a combined maneb-benomyl program is necessary for the control of both diseases. Good control has been obtained with
140 g of Benlate and 1.1 kg of maneb/ha applied weekly or every 10 days.
The addition of adjuvants did not enhance control. Application is by tractor-
powered boom sprayers, beginning about 35 days after planting. After
three cycles about 10 days apart, aircraft are used to apply spray every 2
to 3 weeks. The application is repeated if heavy rain occurs within 24 hr
of spraying. About 275 liters/ha are applied with ground rigs and 65-70
liters/ha by aircraft. Helicopters, because of downdraft forcing the spray
into the foliage, are reported to give better control than fixed-wing aircraft.

Smith and Crosby [50] found that application of benomyl (Benlate) in an
oil-water emulsion gave better control of <u>Cercospora</u> leaf spot in the southern
United States than when applied in water.

IV. FORMULATIONS

The duration and intensity of rainfall is much higher in most areas of the
humid tropics than in temperate areas. From 75-150 mm in 24 hr is not
uncommon during the peak rainy season. For this reason, copper formulations still play a role in disease control in tea, coffee, cacao, and rubber
(see Table 1). Copper formulations are more resistant to weathering during periods of high rainfall. Bordeaux mixture, the most tenacious of the
copper formulations, has been replaced in most areas by cuprous oxide
formulations such as Perenox and the newer cupric hydroxide (Kocide).
With Bordeaux mixture there were always problems of maintaining fresh
lime in a wet climate and cumbersome mixing requirements; once mixed,
it had to be used immediately or stabilizers added. The dithiocarbamates
of the maneb group have replaced coppers in control of many tropical diseases. In the mid-1970s the systemic benzimidazole fungicides replaced
maneb in many instances. Thus far problems of pathogen tolerance to
these compounds have not arisen.

The discovery of the therapeutic effect of petroleum oil on banana leaf
spot by Guyot and Cuille [37] in 1954 opened the way to the replacement of
Bordeaux in banana leaf spot control, a general trend toward low-volume
sprays in the tropics, often in oil formulations, and the use of aircraft to
apply them. Calpouzos [38] summarized information on oil use up to 1968
and there have been no major developments since then.

At first, copper compounds were added to the oil for control of banana
leaf spot but these proved phytotoxic and were replaced by maneb and more
recently by the benzimidazole fungicides [4]. Oil itself can also be phytotoxic and spray oils must meet rigid refining specifications and be applied
correctly [4, 38]. In fact, the sensitivity of most crops to oil injury, especially young leaves, has precluded its more widespread use. Oil and oil
emulsions with fungicide added are used in the control of abnormal leaf fall

of rubber in India [33] and banana leaf spot worldwide [3, 4]. A blend of oils without added fungicide has also been reported to control coffee rust in Brazil [49]. An oil emulsion improved control of Cercospora leaf spot of peanuts with benomyl [50].

In searching for an effective systemic fungicide for rusts in the tropics, Edgington et al. [39] found that oil dispersible carboxanildo-type compounds (e.g., carboxin) penetrated the cuticle of coffee leaves. The waxy upper surface of coffee leaves has no stomata. They suggested that oil emulsions dissolve some of the cuticular wax, making the cuticle more permeable. Therefore, use of emulsions rather than increasing the water solubility of fungicides, with resulting loss in adherence, should improve cuticular penetration of systemics. This may be the case in control of banana leaf spot with benzimidazole-type systemics. Control is much superior when the fungicide is used with oil or in an oil emulsion.

Since the early 1960s there has been an increasing trend toward low-volume spray formulations (less than 9.5 liters/ha) and the use of aircraft to apply them [40, 41]. In addition, the development of gasoline-powered mistblowers light enough to be carried on the worker's back has contributed to the widespread use of medium-volume formulations (93-280 liters/ha).

Many studies have been made in the tropics on the use of adjuvants and stickers to improve coverage and adherence. Apart from the nonionic wetters used in formulations on hard-to-wet waxy leaves and emulsifiers in oil-based sprays, none have been shown to consistently improve coverage or deposition. Increased adherence under artificial and natural rainfall is easily demonstrated with many sticker-type adjuvants but there are few instances where they have been shown to consistently improve disease control. In some instances, disease control was decreased because of excessive adherence. Redistribution of toxicants in rain and dew to foliage that did not receive a spray deposit is an essential aspect of disease control in the tropics. A balance must be sought between a too rapid washing of the spray droplet and a controlled release of toxicant in leaf drip or splash to other areas of the plant. In this way good fungicide coverage on the upper surface of upper canopy foliage can control lower surface infections in both the upper and lower canopy. This has been demonstrated in coffee rust [13] and banana leaf spot [4].

V. APPLICATION EQUIPMENT

Adequate coverage, initially measured visually and ultimately in disease control, is an essential criterion of a successful applicator. The best equipment is that which can place an adequate deposit of toxicant on 90% of the target area cheaply. For these purposes, aircraft, gasoline-powered mistblowers carried on the worker's back, and hand-operated knapsack sprayers are all widely used in the tropics. For rubber, special equipment has been designed for panel diseases, and applying sulfur dust for control of Oidium [42]. Ground-operated large power sprayers are not generally used in tropical bush or tree crops because of the uneven terrain, lack of good roads, and large volumes of water required to operate them. Also, they cannot complete with aircraft or mistblowers in costs of application and improved disease control through better timing of sprays. An aircraft can spray up to 162 ha per day even when spraying is confined to the first 2 or 3 hr following dawn. Then, turbulence and heat build-up as the day advances and prevent good coverage. Hence, spraying must be confined to the early morning hours in most locations, but even then, it cannot be done every day because of excessive turbulence or rain. Helicopters and fixed-wing planes are equally effective in placing the fungicide where required. It is a matter of costs, and usually the airplane is more economical. Also, the development of the rotary atomizers for use on aeroplanes has contributed greatly to improve coverage and application efficiency (e.g., wider swaths).

Apart from the airplane, the development of many models of gasoline-powered mistblowers that can be easily carried on the back is the most important advance in spray equipment for the tropics during the past decade. The capacity of the sprayer is from 11 to 20 liters and therefore it is less efficient when volumes applied are in excess of 280 liters/ha. The mistblower has, therefore, like the aircraft, contributed to the use of lower volumes of spray.

VI. EPIDEMIOLOGICAL CONSIDERATIONS: WHEN TO SPRAY

Timing of spraying without overspraying is the key to efficient disease control with fungicides. To improve timing, detailed studies on the biology of the pathogens and epidemiology of the diseases in relation to weather have been carried out on coffee [11, 13, 15], tea [20], bananas [3, 4], and rubber [32], all important plantation crops.

After determining the best time to spray, based on epidemiology and experimentation, a simple method must be devised that can be used by plantation technicians or small farm owners. A good example of this transfer of research knowledge to the farmer is the work done by Kerr [20] and his associates on blister blight of tea. Data on the current level of infection on tea brought into the factory and the mean daily sunshine for the previous 7 days is transferred to a simple disc calculating device. Using this, the planter can forecast disease incidence 2 to 3 weeks later. Less precise methods have been devised based on rainfall patterns for coffee [11, 15]; rainfall, humidity, and temperatures for bananas [4]; and recently, evaporation for bananas [43]. Where complicated disease assessment and weather calculations are required, however, these methods will not be used by the average farmer. Rainfall, measured either directly or indirectly (sun-shine, evaporation), is still the most important measuring stick of when and how much to spray. Also, when temperatures fall below 15°C, tropical pathogens are not active. Since the first rains following well-defined dry seasons can be predicted fairly accurately from historical weather records, these often serve as a guide to when to start spraying.

To avoid overspraying, one must establish some standard of disease control based on studies of the influence of disease on yields and quality. During periods favorable for infection, fungicides can be used to hold the level of disease below the threshold where economic loss occurs. Then the subsequent dry-season period, unfavorable for infection, can be used as an aid in "cleaning up" any residual disease before the start of the next rainy season. It is usually uneconomical to attempt to maintain very low levels of diseases with fungicides during tropical rainy seasons highly favorable for infection and highly unfavorable for spraying and spray deposits.

VII. ECONOMICAL CONSIDERATIONS

Whether or not fungicides can be used often depends on yield and current prices for the crop rather than the severity of disease. Thus the costs of spraying cacao [24] and coffee [19] cannot be recuperated except on farms with above average yields. As a result, diseases often force the poorer, less efficient producer to either improve his cultural practices or change crops. Banana leaf spot, black pod of cacao, coffee rust, and blister blight of tea undoubtedly forced many small growers to change crop either because they did not have the technology to control the diseases or their production was too low and control costs uneconomically high.

In Fiji, the once profitable banana export market has been lost because of failure to control the black-leaf streak strain of the banana leaf spot pathogen. The technology for control is available but the economics of air-craft spraying when the fields are small (less than 0.5 ha) and spread over a large mountainous terrain are dubious. Thus far, attempts to control the disease by mistblowers have failed. Spray does not penetrate the upper canopy of the tall, closely spaced plants in sufficient quantity to insure good coverage on the most susceptible, uppermost leaves. This is one of the reasons why plantation agriculture has been far more successful in the tropics than small farm holdings. Modern technology can be more readily applied more economically to large compact areas of a single crop than to small scattered holdings with variable cultural practices. Only when the latter are grouped into compact holdings via cooperatives and government sponsored programs of assistance can they compete in efficiency with large plantations.

With the current worldwide inflationary trend, skyrocketing prices, and scarcity of agricultural chemicals, including fungicides, changes in cultural practices in tropical crops must continue. Disease control with expensive chemicals will only be feasible in areas of high yields and low costs. Ground-applied sprays will become more uneconomical and the trend toward aircraft must continue. This will entail changes in planting patterns, varieties, and even the geographic areas devoted to the crop.

VIII. CONCLUSION

The first symposium on fungicide use in the tropics was published in 1955 by the American Chemical Society [44]. More recently Fulton [45] summarized some of the newer developments. Perusing the papers in the 1955 symposium, one readily grasps the tremendous progress made in 20 years. Copper compounds alone dominated the scene and high-volume sprays were the rule. Aircraft spraying was occasionally tried but did not really get off the ground until after 1955. Mistblower application also did not make progress until after 1955. The copper compounds have since been superceded by the dithiocarbamates in most but not all tropical diseases. Then, in the 1970s, the benzimidazole systemics arrived. Where once 26 applications per year of high-volume Bordeaux mixture was required to control Sigatoka leaf spot of bananas, it is now controlled with from six to sixteen low-volume sprays (9.5 liters/ha) based on oil, maneb, or a benzimidazole fungicide. Great advances have been made in our knowledge of the epidemiology of tropical diseases, especially of bananas, coffee, tea, and rubber. This has led to disease forecasting and more efficient timing of sprays. As a result, there is not only less total volume of spray applied but less overspraying. With the continuing rise in costs of fungicides, application equipment, and labor, additional changes are certain to come in how fungicides are used in the tropics in the next 20 years.

ACKNOWLEDGMENT

I wish to thank I. W. Buddenhagen, J. A. Salas, B. Sripathi Rao, and W. E. Bolton for information and suggestions.

REFERENCES

1. C. A. Wismer, in Sugar Cane Diseases of the World (J. P. Martin, E. V. Abbott, and C. G. Hughes, eds.), Vol. 1, Elsevier, Amsterdam, 1961, Chapter 10.
2. J. L. Mulder and R. H. Stover, Trans. Brit. Mycol. Soc., 67, 77 (1976).
3. D. S. Meredith, Phytopathology, Paper 11, Commonwealth Mycological Institute, Kew, England, 1970, p. 147.

4. R. H. Stover, Banana, Plantain and Abaca Diseases, Commonwealth Mycological Institute, Kew, England, 1972, p. 316.
5. R. T. McMillan, Jr., Proc. Florida State Hort. Soc., 84, 290 (1972).
6. D. H. Spalding and W. F. Reeder, Plant Disease Reptr., 56, 751 (1972).
7. J. E. Hunter and I. W. Buddenhagen, Ann. Appl. Biol., 63, 53 (1969).
8. J. E. Hunter and I. W. Buddenhagen, Trop. Agr. Trinidad, 49, 61 (1972).
9. J. E. Hunter, I. W. Buddenhagen, and E. S. Kojima, Plant Disease Reptr., 53, 279 (1969).
10. K. R. Bock, Trans. Brit. Mycol. Soc., 45, 301 (1962).
11. F. J. Nutman and F. M. Roberts, PANS, 16, 606 (1970).
12. I. D. Firman and J. A. N. Wallis, Ann. Appl. Biol., 55, 123 (1965).
13. I. D. Firman, Trop. Agr. Trinidad, 42, 111 (1965).
14. E. Griffiths, Kenya Coffee, 33, 393 (1968).
15. E. Griffiths, J. N. Gibbs, and J. M. Wallers, Ann. Appl. Biol., 67, 45 (1971).
16. S. K. Mulinge and E. Griffiths, Trans. Brit. Mycol. Soc., 62, 495 (1974).
17. J. L. Pereira, Turrialba, 22, 409 (1972).
18. J. L. Pereira, Intern. Pest Control, 6, 10 (1971).
19. I. D. Firman, F.A.O. Plant Protec. Bull., 20, 121 (1972).
20. A. Kerr and R. F. Rodrigo, Trans. Brit. Mycol. Soc., 50, 609 (1967).
21. C. R. Harter, World Crops, 25, 245 (1973).
22. M. D. Machado, Cacau Actualidades, 9, 1 (1972).
23. C. A. Thorold, Ann. Appl. Biol., 47, 708 (1959).
24. A. G. Newhall, Cacao, 12, 27 (1967).
25. E. C. Hislop and P. O. Park, Ann. Appl. Biol., 50, 67 (1962).
26. E. C. Hislop and P. O. Park, Ann. Appl. Biol., 50, 72 (1962).
27. D. Tollenaar, Trop. Agr. Trinidad, 36, 177 (1959).
28. A. G. Newhall, A. Paredes, and L. G. Salazar, Cacao, 13, 3 (1968).
29. A. G. Newhall, F. Diaz, and L. G. Salazar, Cacao, 11, 21 (1966).
30. R. L. Wastie, Rubber Research Institute of Malaya, Planters' Bull., 104, 199 (1969).
31. B. Sripathi Rao, Malaysian Crop Protection Conference, Rubber Research Institute of Malaya, 1970, p. 8.
32. O. S. Peries, J. Rubber Res. Inst. Malaya, 21, 73 (1969).
33. P. De Jong, Agr. Aviation, 3, 126 (1961).
34. R. L. Wastie, Exptl. Agr., 5, 41 (1969).
35. Rubber Research Institute of Malaya, Planters' Bull., 109 (1970).
36. Rubber Research Institute of Malaya, Planters' Bull., 118 (1972).
37. H. Guyot and J. Cuille, Fruits d'Outre Mer, 9, 289 (1954).
38. L. Calpouzos, in Oils (D. C. Torgeson, ed.), Vol. 2, Academic Press, New York, 1969, Chap. 8.
39. L. Edgington, H. Buchenauer, and F. Grossmann, Pestic. Sci., 4, 747 (1973).

40. R. H. Fulton, Ann. Rev. Phytopathology, 3, 175 (1965).
41. G. A. Brandes, Ann. Rev. Phytopathology, 9, 363 (1971).
42. Rubber Research Institute of Malaya, Planters' Bull., 123 (1972).
43. J. Ganry and J. P. Meyers, Fruits d'Outre Mer, 27, 767 (1972).
44. American Chemical Society, Advan. Chem. Ser., 13, 102 (1955).
45. R. H. Fulton, Phytopathology, 63, 1441 (1973).
46. Rubber Research Institute of Malaya, Planters' Bull., 130 (1974).
47. S. K. Mulinge, C. J. Baker, and J. L. Pereira, Phytopathology, 64, 147 (1974).
48. C. J. Baker, Kenya Coffee, 38, 185 (1973).
49. L. Calpouzos, Ann. Proc. Amer. Phytopathol. Soc., 1, 120 (1974).
50. D. H. Smith and F. L. Crosby, Phytopathology, 62, 1029 (1972).

Chapter 11

FUNGICIDES IN INDUSTRY

Charles C. Yeager

Ventron Corporation
Beverly, Massachusetts

I. INTRODUCTION

"After Moses had announced Joshua as his successor before all the congre-
gation, he disclosed to him that the entire course of his own life was run,
and that he would now depart to his fathers. As his inheritance he gave to
Joshua a book of prophecy, which Joshua was to anoint with cedar oil, and
in an earthen vessel lay upon the spot that from the creation of the world
God had created for it, so that His name might be invoked" [1]. These
instructions from Moses to Joshua appeared in the Pseudepigrapha Assumption of Moses, Chapter 1, and are one of the earliest, if not the earliest,
references to the preservation of materials. The use of cedar oil as a preservative was in common practice by the ancients and may account for the
excellent condition of the Dead Sea Scrolls when they were discovered in the
Caves near Khirbet Qumran in 1947, "the spot that from the creation of the
world God had created for it".

Egypt developed a long time ago a highly sophisticated preservation
process for human and animal bodies, but there are few if any references
to the preservation of inanimate objects. In their embalming processes the
Egyptians used such chemicals as bitumen (in which Tutankhamen was embedded), myrrh, cassia, cedar oil and natron [3].

We do not find further reference to preservation of materials until the
ancient Greek writings of Pliny the Elder reported that the law books of
Numa Pompilius were in excellent condition when his tomb was opened 350
years after his death, because they had been preserved in cedar oil [2].

The wide use of cedar oil by the ancients probably stemmed from the
practice of "anointing." It is doubtful if well defined testing procedures
had developed the practice. Its fragrance, availability, and oily character
probably inferred a long-term mystical property of preservation.

Through Roman history we find no reference to a requirement for preserving materials. It is doubtful that a dynamic civilization, constantly on
the move, would have stopped long enough to protect its artifacts.

During the long, sparsely recorded period of the dark ages and later
during the industrial revolution, mankind, in too much of a state of flux and
just trying to survive, gave little thought to the use of preservatives. The
early alchemists were too busy trying to find ways for converting base materials into precious metals to waste time in trying to protect their more
prosaic possessions. It was not until the early 19th century, when the
spread of humanity was outstripping its ability to produce, that a rapidly
developing chemical industry was investigating ways to prolong the usefulness of materiel. Still, we find few references to preservation processes
until the 1920s and 1930s, when, for the first time, procedures appeared in
print involving the protection of materials such as textiles, leather, paper,
glues, and adhesives.

Early uses of fungicides were haphazard, being based on empirical
rather than experimental information, and even today, with a sanitation-
minded public demanding product protection, some leading educational and

research institutions are prone to ignore the importance of experimental as
compared with empirical data. Before 1926, zinc chloride, usually in 0.8%
concentration, was the principal fungicide used for textiles, even though it
was known to be only marginally effective. Other compounds used then
were salicylic acid, sodium fluoride, sodium silicofluoride, phenol, cresylic
acid, and formaldehyde. Corrigan, in 1920, recommended the use of heat,
light, and such antiseptics as copper sulfate, potassium nitrate, zinc sulfate, zinc chloride, boric acid, formaldehyde, and phenols for the destruction of fungal growth on fabrics [4].

World War II brought the urgent need for sophisticated, effective preservatives into sharp focus. For the first time a major war was fought
under extreme tropical conditions. War materiel was deteriorating faster
than supply lines could replenish and it was necessary to engineer crash
programs to develop preservative systems and processes. Now, for the
first time in history, scientists developed experimentally sound preservation processes in a cooperative effort. Fungicide systems were designed
for specific jobs, and test methods and specifications were promulgated.
Man began to realize that he would last on this earth only as long as his raw
materials and manufactured goods were available.

II. TEXTILE FUNGICIDES

Textiles, in particular those made from natural fibers, such as cotton,
wood, manila, jute, and sisal, have been known for hundreds of years to
deteriorate. Fabrics and rope rapidly lose their tensile strength when
exposed to high humidity and in contact with the soil. Prior to the 1920s
little was done to protect such fabrics against deterioration. Most often it
was easier and more economical to replace the item than to protect it.
Thus, there was no incentive to develop preservatives. As populations
increased and manufactured goods were in greater demand, requirements
for longer-lasting fabrics emerged.

Though Corrigan claimed that the proper conditions coupled with an antiseptic treatment would protect fabrics against destruction, little attention
was given to determining factors which constitute the fitness of a fungicide,
such as the degree of resistance to leaching and weathering, complex-ion
exchange and buffering systems in soil, or compatibility with other compounds in the materials to be protected. In the early 1930s Fargher et al.
[5] developed screening methods by which thousands of chemical compounds
were examined to determine their fitness as fungicides for the preservation
of textiles. Again, little attempt was made to determine their resistance
to environmental influences. Because of this, many so-called experimentally
excellent fungicides failed miserably when World War II and its prodigious
demands required the treatment of millions of yards of fabric.

During the war many governmental agencies established special research
facilities to study the effects of different fungicides on all items subject to

microorganism attack. Although many of those organizations developed their own particular fungicide preferences and variations, and test methods became as numerous, order slowly emerged. Many of the compounds heretofore believed to be excellent were discarded, and others with the characteristics necessary to prolong fabric life were established as the standards by government and industry.

It is significant to note that with the exception of certain specialized areas, most of the products and practices recommended and specified during or shortly after World War II are still in use today.

Before a fungicide can be considered completely satisfactory for use as a textile finish it must have the following characteristics: (1) good stability; (2) high efficacy; (3) low toxicity; (4) inoffensive odor; (5) no color change; (6) good handle; and (7) no chemical effects on the fabric.

1. Stability is determined by resistance to heat, light, water, and oxidizing agents. Very few chemical compounds possess all of these characteristics. The fungicide must be stable as a compound and when applied to the fabric. It must retain its stability through long periods of storage and the life of the finished goods.

2. The efficacy may be either fungistatic or fungicidal. Many products possess good fungistatic qualities, that is, they prevent the germination of fungus spores. A few are fungicidal and they may destroy the spores before they can germinate and attack the fabric; these fungicides can kill the growing organism.
   The compound must be effective in relatively low concentrations, so as to enable the user to retain low-weight add-on, and keep the cost within reasonable limits.

3. It must be nontoxic or of an extremely low order of toxicity. This is one of the most essential requirements of a good fungicide. The fungicide industry today is controlled by the Environmental Protection Agency in the United States and similar agencies in other parts of the world. The regulations promulgated by these agencies require that extensive toxicity tests be conducted before a fungicide may be manufactured and sold. Human patch tests must be performed before a fungicide may be used on clothing items.

4. The fungicide must not impart an unpleasant odor to the fabric, particularly if used in wearing apparel.

5. If color is of importance to the sale of the product, the fungicide must not discolor the fabric before or during exposure to weathering conditions.

6. The fungicide must not significantly change the handle of the fabric, particularly if it is to be used in wearing apparel.

7. The fungicide must have no adverse chemical effects on the fabric. Many fungicides though effective, seriously affect the tensile strength through oxidation or photooxidation such as in the case of certain chlorinated phenols.

A. Organometallic Compounds

Various inorganic and organic salts had been used for a long time to control micro- and macroorganisms. Even in ancient times it was well known that lead and copper clad ship bottoms resisted the growth of barnacles and algae. Bordeaux mixture (copper sulfate and lime) had been used to prevent fungal diseases on crops. Thus it was understandable that man would turn to the products with which he was most familiar to control fungal deterioration. He soon found that the inorganic salts were too soluble to withstand weathering conditions, and that certain organometallic derivatives were less soluble and more resistant to weathering.

It had long been known that certain metals are more effective than others. Mercury and silver were found to be the most effective, followed in order by copper, zinc, tin, and aluminum. The remaining metals were relatively ineffective. It was only fairly recently that arsenic compounds were added to the list. Since mercury and silver compounds were either toxic or expensive, copper salts became the preferred antimicrobials.

Copper naphthenate was adopted in the beginning of World War II as the leading fungicide. Most of the articles to be protected were olive drab or khaki in color. Its greenish-blue color did little to change the color of the finished product. The stiffening effect obtained through the treating process was of little consequence, since the fabrics treated were used in tents, tarpaulins, and other heavy-duty items.

Although copper naphthenate imparted a rather unpleasant odor to the fabric, it performed so well in prolonging the life of the product, that the odor was ignored. It was low in human toxicity, yet most effective in controlling the various cellulytic organisms that so rapidly deteriorated cotton fabrics.

Copper naphthenate is derived from naphthenic acids of petroleum and a mixture of salts of fatty acids. The principal sources of naphthenic acids for the production of copper naphthenate are California and Venezuela crude oils. It is soluble in many organic solvents such as mineral spirits, kerosene, xylol, and other inexpensive petroleum oils.

Copper naphthenate was shown to prevent rotting of cotton fabric in soil contact, at lower concentrations than other copper soaps such as copper tallate and copper oleate [6]. This was apparently due to the lower solubility of the naphthenic acid derivative. Naphthenic acid itself is fungitoxic, whereas tall-oil fatty acids and oleic acid are ineffective or susceptible [7].

A fabric treated with 0.1-0.5% metallic copper would resist deterioration even after exposed to Gulf Coast weather for 2 years [8].

Copper-tolerant organisms, such as the Penicillia and Aspergilli, could solubilize many copper compounds, but not copper naphthenate. This was undoubtedly due to the fungitoxicity of the naphthenic acids. Soil burial tests of cotton ducks treated with a mixture of copper or zinc naphthenate and phenylmercuric naphthenate showed that the mercurial reduced the loss of copper, preventing breakdown of the fabric, even though it was found that the mercurial itself gave little protection in soil contact [9].

Copper naphthenate, because of its excellent leach resistance, could be used without water repellents. It was a common practice, though, for the military specifications to require that the finished fabric be mildew resistant, water repellent, and fire resistant. Millions of yards of FWMR (fire, water, mildew resistant) fabric were purchased by the military toward the end and after World War II.

These standards usually required an add-on of from 0.6 to 0.8% copper as metal, and quality-control specifications called for the analysis of the treated fabric for copper content.

Copper salts have always been known to accelerate photooxidation of cellulose. Sunlight alone will deteriorate cotton, and by measuring cupraammonium fluidity, one can distinguish between microbiological deterioration and photochemical effects. The fluidity is higher in the case of the latter [10]. Some copper-accelerated oxidation has been noticed in the case of copper naphthenate treated fabrics, although it is not as severe as that obtained with other copper salts and its effectiveness in controlling microbiological deterioration outweighs this minor weakness in the presence of sunlight [11].

Copper naphthenate is still used today for the treatment of heavy fabrics, burlap, and wood.

The drawbacks of copper naphthenate, such as color, odor, stiffening effect, and its photochemical effect on cellulosics, encouraged the search for other less objectionable copper salts. In 1947, a product known as copper 8-quinolinolate (copper 8-hydroxyquinoline or copper oxinate) was announced as an outstanding textile and paint preservative [11].

For over 50 years, 8-hydroxyquinoline was well known and used for extracting heavy metals from solution by chelating them to form highly insoluble salts. Furthermore, 8-hydroxyquinoline was known as an antibacterial agent. The copper salt of 8-hydroxyquinoline was found to provide greater and more durable protection to cellulosics than any previously known copper salt. Its solubility in water is only 0.8 ppm [11]. It is less toxic than other salts and does not stiffen fabrics. Because of its low solubility in water, it does not tender fabrics photochemically as do other copper salts, and the amount required to effect maximum protection is less than that required for other compounds.

Copper 8-quinolinolate is not soluble in ordinary solvents and early recommendations for its application required a two-bath process or a water or solvent dispersion. The two-bath process required first the dipping of a fabric in a solution of 8-hydroxyquinoline, followed by a bath of copper acetate. This formed the copper salt in the fabric. It involved a costly processing technique which rarely produced an even or controlled application. The dispersion process required excellent agitation and often resulted in a spotty treatment and severe "mark-off" or dusting. In 1951, Kalberg developed a solubilization process in which the copper 8-quinolinolate was dissolved in a mixture of a metal octoate and nonyl phenol [12, 13]. The resultant solubilized copper 8-quinolinolate was readily diluted with mineral spirits or could be emulsified to give a stable dispersion in water. For the first time, copper 8-quinolinolate could be easily and evenly applied to any type of textile, permitting good penetration and excellent control of the application process.

Many theories have been advanced to explain the superior fungitoxicity of copper 8-quinolinolate. As a general rule, fungitoxicity is not a well understood process. It is known that many chemicals have the ability to penetrate a cell wall and membrane. Only a few of these will kill or retard fungal growth. Copper 8-quinolinolate, because of its superior fungitoxicity, was studied at great length by many biochemists. It was first proposed that, because of the strong chelating properties of 8-hydroxyquinoline, trace essential metals within the cell were complexed, thus removing those elements essential for growth [14]. This was reinforced by the discovery that any modification of the structure that prevented chelation from taking place, would destroy fungitoxicity [16]. Others suggested that, since the parent compound is already saturated with metal, chelation could not take place [15]. The theory was even advanced that since amino acids in the cell were stronger chelators than 8-hydroxyquinoline, the latter served only as a mechanism to get the fungistat into the cell. Once there, the amino acids chelated the copper and the cell was poisoned, the so-called "shaped charge" theory. A modification of this theory advanced the idea that the iron and copper chelates were the cause of toxicity and not the 8-hydroxyquinoline. The statement is made that "the toxic effect is undoubtedly initiated by chelation, and is completed by subsequent poisoning effect of the metal, either in combination with oxine (8-hydroxyquinoline) or as free ferrous ions liberated after transport through the cell wall in the chelated condition. This poisoning apparently takes effect through catalysis of the oxidation of an essential enzyme or metabolite. This destructive oxidation can be prevented by lowering the oxidation-reduction potential" [16]. The fact that only a few of the more toxic metals in the form of the metal 8-hydroxyquinolate are toxic to microorganisms seems to emphasize that the 8-hydroxyquinoline is merely a mechanism for getting the metal through the complex cell wall and membrane barriers after which it combines with enzyme systems or metabolites and poisons the cell.

Other copper salts, such as the tallate, stearate, resinate and oleate, will provide protection against cellulose-destroying fungi, but none are durable enough or have a use-cost ratio favorable enough to warrant wide use.

Aside from copper naphthenate and copper 8-quinolinolate, the only copper salts that remain economically useful today are the cupraammonium complexes which are still used in the treatment of low-cost sand bags and burlap. These are basic copper compounds dissolved in aqueous ammonia to give complexes stable in aqueous solution. When the ammonia is driven off during the drying, the treated fabric becomes resistant to leaching and weathering. Great care must be taken in the application of copper compounds to prevent stiffening and a glazed finish. The treatment provides excellent protection in soil burial against cellulytic organisms, but little protection against surface-growing organisms such as the Aspergilli and Penicillia [17].

Several other organometallic chemicals are still used in limited quantities. Whereas zinc naphthenate was once widely used as a so-called colorless textile preservative, its high add-on requirement, stiffness and strong odor were too objectionable. A synergistic mixture of zinc dimethyldithiocarbamate and zinc 2-mercaptobenzothiazole has limited use in some fabrics to be subsequently coated with polyvinylchloride. The low solubility and high stability of this mixture provide good weathering properties with good fungitoxicity. Unfortunately the treatment may be safely used on white material only. For dark colored coated fabrics the mixture has a tendency to "bloom" on the surface of the coating, interfering with the aesthetic quality of the goods.

Phenyl mercuric compounds have long been used as inexpensive treatments for fabrics. These aryl mercurials, because of their great activity against fungi are used at extremely low levels, thus reducing the high toxicity factor. Unfortunately, all mercurial pesticides because of their potential hazard to the environment may soon be banned for use where there is any danger of effluent contamination. Textile manufacturers have already begun to change their use of these highly effective, fungitoxic derivatives to products less toxic to humans, but perhaps not quite so toxic to fungi.

B. Nonmetallic Compounds

One of the first effective fungicides to emerge during the latter stages of World War II was "dichlorophene," 2,2''-methylenebis(4-chlorophenol). More familiarly known as "Compound G-4", dichlorophene is a condensation product of formaldehyde and p-chlorophenol. It was soon found to have excellent protective properties when applied to cellulosic fabrics, imparted little color change to the substrate and only a slight antiseptic odor. Compound G-4 has low solubility in water and is applied to textiles in a variety of ways, in an alcohol solution, as an emulsion, or in a two-bath process, in which the G-4 is first dissolved in a caustic bath, applied to the fabric, and then acidified by passing the fabric through an acetic acid or an acidwax water-repellent bath. The G-4 is thus deposited within the fabric structure. Water-repellent systems are usually desirable to prevent mechanical leaching during the weathering process. Compound G-4 is nonirritating to the skin at use levels and is active against a wide range of organisms, both fungi and bacteria. It is still the preferred nonmetallic preservative in U.S. military specifications.

FUNGICIDES IN INDUSTRY

Pentachlorophenol is one of the most effective fungicides used in the preservation of cellulosics. It was widely used in the early stages of World War II on tentage and tarpaulins. Unfortunately, it is also a powerful skin irritant. Since pentachlorophenol crystallizes readily, the complaints from military personnel became so numerous that it was necessary to shift to other less objectionable products for military equipment. Pentachlorophenol was found to tender cotton fabrics through photooxidation. It was suggested that actinic exposure would accelerate decomposition, releasing chlorine to readily form hydrogen chloride when combined with moisture. Extensive research has been undertaken to take advantage of the powerful preservative forces without releasing unpleasant side effects. In 1949, the Hercules Company announced the development of a new product, dehydroabietyl amine (Rosin Amine D). This product readily combined with pentachlorophenol to produce a derivative without the high toxicity or tendering properties. Considerable effort was expended by industry, the military, and the National Research Council in Canada, to prove the value of this compound for textile preservation. But it soon became obvious that Rosin Amine D Pentachlorophenate was an excellent and stable preservative as long as it was not exposed to sunlight. Apparently actinic exposure would reverse the reaction process so that Rosin Amine D and pentachlorophenol were reformed. The pentachlorophenol was soon leached out leaving Rosin Amine D, a syrupy liquid which produced a sticky fabric surface to which sand and dirt adhered. Rosin Amine D pentachlorophenate is still employed today on fabrics used in mines, in rubber formulations, and as a preservative for latex systems.

Meanwhile, in Europe a different approach was being taken. Since the hydroxyl site on the pentachlorophenol molecule is quite reactive, it was found that condensation with a fatty acid would produce an ester with activity similar to the parent compound itself. Lauryl pentachlorophenate or pentachlorophenyl laurate was introduced to the textile industry and soon became the leading fungicide for fabric preservation in England and Europe. The name is a misnomer since only a small percentage of lauric acid is employed. The U.S. Environmental Protection Agency approved the name fatty acid (C_6-C_{20}) esters of pentachlorophenol for the product.

The ester linkage is quite stable at ordinary textile finishing conditions. The compound imparts little or no change in color, has good weathering properties, and low toxicity. It is the preferred treatment for textiles in Great Britain and most of the Commonwealth countries. A late comer to the United States, it has not enjoyed the widespread usage as in other parts of the world. This is probably due to its lack of acceptance for use on military equipage. This fine fungicide was introduced into the United States after the military had decided that 2,2'=methylenebis(4-chlorophenol) for light fabrics and copper 8-quinolinolate for dark textiles were quite satisfactory to provide adequate protection of their materiel. It is currently being developed as a treatment for camping tents, awnings, and tarpaulins and will be an excellent substitute for phenyl mercurics if the latter are banned by the government.

Dodecylguanidine terephthalate was developed within the last ten years and appears to have many of the characteristics required for the treatment of textiles. This product produces a finish with interesting properties. It is colorless and exceedingly stable when subjected to UV exposure. Although dodecylguanidine terephthalate is an eye irritant, it is relatively low in dermal and oral toxicity. As in the case of quaternary compounds, it has back-wetting properties and requires exceptionally effective additives to permit the development of satisfactory water-repellent properties in the finished fabric.

As in all industries, the ideal solution to a fungus resistant treatment for susceptible textiles requires that the substrate be funginert, i.e., nonsusceptible to fungal attack. During the 1960s much time and effort was expended on cellulose modification, e.g., by acetylation. Cellulose is .modified by acetylation making it nonsusceptible to the enzyme systems of microorganisms and immune to fungal deterioration. Modification was tried using many chemical systems, e.g., methylolmelamine, urea-formaldehyde, and indeed many other reactive compounds, in attempting to satisfy enough of the hydroxyl groups on the cellulose molecule to afford funginertness. Too often excessive cross-linkage occurred. Unfortunately, in modifying cellulose many of its desirable characteristics were lost. The handle of the fabric almost always suffered, tensile and tear strength were seriously affected, and costs skyrocketed.

In recent years with the expansion of the synthetic market, it has been found unprofitable to continue efforts on the modification of cellulose. More effort is being expended in the development of add-on chemical treatments for 100% cotton and cotton-polyester blends.

Wool is far less susceptible to fungal growth than is cotton. Since it is a proteinaceous fiber it is highly susceptible to bacterial attack. Woolen fabrics, although not decomposed by fungi, will support the growth of surface organisms which spot and stain, causing uneven dyeing and an unsightly appearance. Pure scrubbed wool is nonsusceptible to fungal growth, but there are always oils, soaps, debris, and various conditioning agents added to wool that will support the growth of surface-fungi [18]. Since wool is more hygroscopic than cotton, mildew will occur rapidly, often while the wool is being processed [44].

The preservation of wool was one of the earliest processes to be studied [19]. Some of the fungicides recommended were sodium fluoride, salicylanilide, sodium fluosilicate, and β-naphthol. It was soon found that good housekeeping and proper processing would obviate most of the mildew problems. Wool shoddy, in which reclaimed wool fibers are used, although a prime candidate for mildew attack, is processed today under carefully controlled conditions so as to prevent mildew from occurring. The wool industry is not considered a candidate for fungicides, although it has at various times shown interest in antibacterial compounds and, of course, is vitally interested in moth repellents.

Synthetic fibers are, as a general rule, considered funginert, that is, they do not support fungal growth, and are not particularly susceptible to surface growth. The enzyme systems of fungi are not specific enough to attack synthetic molecules. Numerous attempts have been made to prove that certain organisms will attack synthetic fibers such as Nylon and Dacron, but no one yet has shown this to be the case. It is conceivable as synthetic fabrics are in wider use, organisms will adapt their enzyme systems to utilize the carbon sources that so far have been unavailable to them. At the present time fungicides are used only on those fabrics that are blends of synthetic and natural fibers.

III. PLASTICS FUNGICIDES

Most synthetic polymers are relatively funginert. If only the polymer was to be considered, there would be little need for a fungicide. In the development of a useable polymer system many ingredients are used which must be protected.

Flexible vinyl chloride systems, for instance, include plasticizers, stabilizers, mold release agents, etc., and there are many references in the literature to the susceptibility of plasticizers. A typical table lists 20 general types of plasticizers, some susceptible, some resistant and others doubtful [45]. So-called resistant types will support fungal growth under certain conditions. "Doubtful" indicates that the plasticizer under one type of test will not support growth, whereas in other instances it will [46]. Probably no plasticizer is funginert. All, under the right conditions, will support fungal growth and any plasticized polymeric system will be susceptible. Stabilizers, which are essential to permit high-temperature processing, frequently contain fatty acids. These fatty acids will usually support fungal growth and further add to the susceptibility of the total system. It is important therefore, that, if resistance to fungal growth is required, a fungicide be used. The ability of the fungicide to perform is affected by a variety of natural and chemical phenomena. It is vital that we understand these factors in selecting the ideal control agent. The selection process is further complicated by the requirement that any suitable control agent must be effective against bacteria and algae as well as fungi. These organism types all affect the stability of the complex vinyl system.

Fungi are those organisms, usually filamentous in nature, that not only attack the plasticizer system, but most always cause unsightly discoloration and destroy the aesthetic qualities of a system. This group of organisms may destroy a fabric component adjacent to the vinyl through enzymatic action.

Bacteria, on the other hand, are less spectacular, but far more insidious in their effects. In most instances their presence cannot be visually determined, and yet they may rapidly affect the physical qualities of the system by destruction of the plasticizer and in some instances by the creation of foul odor. This group includes some of the most virulent human pathogens.

Algae are of concern only in systems exposed to excessive moisture conditions. Although there are numerous reports on the presence of algae growth on roof and siding in subtropical areas, the most serious problem is found in such systems as swimming pool, ditch, and pond liners. Algae themselves are not normally considered to seriously affect the physical properties of a vinyl system, but they are always unsightly and provide an excellent substrate for adherence of slime bacteria, which in turn may attack the plasticizer system. Before anyone may recommend a particular solution to the problem of deterioration of vinyl products, an adequate test procedure is first needed.

In the past it was sufficient merely to adapt existing test methods, developed for such materials as fabrics, leather, and rubber. Most of these were based on standard soil burial, zone of inhibition, and qualitative surface-examination procedures. The current laboratory testing procedures for flexible vinyls are empirical rather than definitive. They are most suitable as screening techniques, but should be correlated with outdoor exposure to provide evidence of the processes going on under service conditions.

The ASTM Committee G-3, on Deterioration of Non-Metallic Materials, has promulgated two test procedures in Tentative Recommended Practice for Determining Resistance of Plastics to Bacteria, ASTM D 22-67T [47]. Both of these methods outline the only pathways available at the present time and are quite adequate for systems as they come from the calendar or coating process. They fail, at least in part, in determining the resistance of systems that may be subjected to excessive heating, weathering, or leaching. This is not a criticism of these methods, but merely an attempt to point out that all factors should be considered in the testing of any system.

Klausmeier and Jones [20] in several publications have demonstrated the susceptibility of many of the plasticizers used in vinyl formulations. There have been numerous attempts to select plasticizers that will not only give the proper physical characteristics to the systems, but at the same time resist the attack by microorganisms. Some of these attempts have been moderately successful as proved by the fact that a Federal Specification, CCC-W-408 is widely used. This is said with some reservation, since recent work has demonstrated that, although this so-called resistant system will pass the usual laboratory tests without significant change in properties, profound changes occur when exposed to service conditions. The nature of these changes is due not only to plasticizer deterioration, but also to changes in the chemical structure of the total vinyl system. These formulations which easily pass the present laboratory tests are far from satisfactory in the field.

A polyvinylchloride (PVC) system is a complex of many components all contributing to the required physical and chemical properties of the end product. It is important to know the part played by each component of the total system in the overall deteriorating process. If all materials other than the plasticizers could be ruled out it would be easy to find a compatible, durable antimicrobial that would protect the plasticizers. As long as the plasticizers remained in the system they would be protected by the fungicide against deterioration. It is generally believed that if a good stable antimicrobial is added to a vinyl system without disturbing the original physical properties of the formulation, it will protect the system against all invasions of microbial entities. Or, if a completely nonsusceptible plasticizer were used it would not be attacked by microorganisms. Present empirical testing methods would then be sufficient for any situation which might arise.

It is not difficult to preserve good flexibility and maintain elongation properties through the use of an antimicrobial agent. Vinyl systems today may be weathered for long periods of time with only minor changes in flexibility or elongation. The use of nonfugitive and nonspewing plasticizers has accomplished this. Those same systems, though, containing sufficient levels of new sophisticated antimicrobials to preserve this flexibility, rapidly lose their ability to resist growth of surface organisms as service life is extended. There are two probable reasons for this enigma.

First of all, as better spew-free systems are developed, the mobility of the plasticizer is reduced. Once the natural surface film present in all vinyl systems is removed, the ability to control surface contaminants is lost.

Migration of the plasticizer and its accompanying antimicrobial may be attained in many ways. The aim of the different variations is to keep the plasticizer coming uniformly to the surface, in minute quantities to be sure, but moving always to the surface. The ideal system, therefore, is one that maintains its flexibility without sacrificing the control of surface organisms. A number of screening tests can be used, including the standard petri dish in any of its various modifications. The most important factor of this test is proper preparation. A candidate may be first laundered in any standard laundering device for a number of cycles in a given detergent or soap formulation with its accompanying rinsing, or leached in running water for a specific period of time. After drying the sample may be placed in a petri dish and inoculated with one or more organisms, usually associated with surface fouling of vinyls. The number of cycles used before total failure should indicate the degree of mobility of the plasticizer to the surface. The gradual disappearance of the protected area is an excellent indicator of this mobility. The same end result may be obtained through a program of natural exposure where frequent rainfall will gradually remove the surface film of plasticizer and antimicrobial, or through a Weather-Ometer exposure with accompanying water spray. Therefore, in the case of a system where the emphasis has been placed on nonspewing effects, the

ability of the antimicrobial to control surface-growing organisms may be
lost as the active principle is removed from the surface by leaching and
not replaced by migration from within the vinyl.

The second serious stumbling block is faced in attempting to coordinate
laboratory and field testing. The PVC systems are complex polymers
which under conditions of excessive sunlight or heat, liberate HCl as the
molecule is split. Cross-linking of the remaining components may then
produce encapsulation of the plasticizers and accompanying antimicrobial.
Such systems may retain all of their elasticity and tensile strength, but will
support the growth of surface organisms. There are numerous instances
where such vinyl products are admirably protected while in service, but
fail miserably in the laboratory. It is important, therefore, that laboratory
tests and field exposure be utilized to permit sound prediction of the performance of vinyl systems under service conditions.

This is most dramatically expressed in the phenomenon called "pink
staining" [21]. In this case, the causative organism in the vinyl system is
not necessary to produce the stain. The discoloration may be due to a
migration of the stain from a nearby substrate on which the organism is
growing. The vinyl must contain on its surface an antimicrobial which will
produce a zone of inhibition which will be produced throughout the life of
the vinyl. This principle applies to any organism growing outside the vinyl
system.

It is believed that methods currently employed for determining the
effects of microorganisms on plasticizer systems are adequate. The present ASTM procedures are used as standards for this purpose. Much has
been accomplished in determining effectiveness of vinyl systems against
surface growers. In 1962, the author outlined the requirement for testing
vinyl films or coated fabrics against a "pink staining" producing organism,
Streptoverticillium rubrireticuli [22]. A testing procedure was developed
by the Ford Motor Company and published as a company standard [23].
This is currently used by most of the automotive companies and their sup-
pliers. The procedure is based upon the presence in the vinyl system of a
suitable antimicrobial which produces a zone of inhibition necessary to
prevent formation of the red pigment immediately adjacent to the sample
tested. This phenomenon also calls for the protection of any material which
may be in contact with the vinyl, since the extreme mobility of the red pigment will often result in staining, even though the vinyl itself is unable to
support growth of the organism.

An equally important area of interest is concerned with the control of
algal organisms. With the spread of the swimming pools, the buildup of
algae growth on the vinyl liners has created serious problems. The algae
not only are able to utilize certain types of vinyl additives, but serve as
excellent traps for bacterial slime development. Since bacteria are able to
grow on most types of plasticizer, vinyl liners quickly lose their flexibility.
The same problem arises in the case of irrigation-ditch liners. The vinyl

industry has become vitally interested in combating this problem and is currently developing a suitable testing procedure. Algae can be easily cultured and are quite predictable. The fact that they will grow only in the presence of sunlight forces us to adopt techniques with sunlight as an integral ingredient. A rather simple screening procedure has been developed. Strips of vinyl film containing various amounts of biologically active compounds are fastened by means of paraffin wax or biologically inert cement to glass microscope slides and immersed in large (150 ml) culture tubes. The tubes are then fitted with aeration devices, filled with standard algal culture media and suspended in glass water tanks. The culture media are seeded with actively growing cultures of Oscillatoria sp., and treated films containing an active biocide should remain free from growth for considerable periods of time [24].

Since sunlight has been an important ingredient of these tests, and chemical changes normally associated with service exposure must be considered, this laboratory procedure closely approximates the conditions under which the pool or ditch liners will be exposed.

At this point it is wise to consider the problems immediately ahead. More effort must be spent on the chemistry of flexible vinyl systems. Effects of heat, light, water, and other external factors on the mobility of plasticizers through the system must be understood. The progressive polymerization of the surface of the resin system resulting in encapsulation must be prevented. The plasticizer lost in weathering must be replenished without causing excessive spewing. It is important that we know the effects of formulation changes on microbiological stasis. Finally, and most important, we must utilize both laboratory and field testing procedures to reasonably predict service performance.

There are only five stable antimicrobials on the market today. One of these, copper 8-quinolinolate, is rarely used in any system other than those for military equipment. It has an intense yellowish-green color which seriously affects the standard pigment system.

The remaining four compounds are used in a variety of systems and end items. Each one has its own characteristics. Formulations containing 10,10'-oxybisphenoxarsine produce spectacular zones of inhibition against almost all microorganisms. Its maximum value can be obtained if used with a properly selected heat stabilizer and a UV screening agent [48].

This organic arsenical is the most effective of all plastics fungicides and the backbone of the flexible vinyl industry. Its low cost and great effectiveness made 10,10'-oxybisphenoxarsine the fungicide employed by much of the vinyl industry for such uses as tents, awnings, shower curtains, hospital sheeting, marine fabrics, baby pants, wall coverings, swimming pool liners, etc.

A new vinyl additive is 2,3,5,6-tetrachloro-4-(methylsulfonyl)pyridine. This product, developed originally for the preservation of exterior paints, appears to be useful as a vinyl fungicide. It is colorless, stable, effective at reasonably low levels, but expensive. In addition, the lack of long-term weathering data indicate that more work is needed before this product is accepted as a suitable plastics preservative.

The final two compounds are the analogs N-(trichloromethylthio)phthalimide and N-(trichloromethylthio)tetrahydrophthalimide. These chemicals have excellent activity against fungal and bacterial organisms, but are reported to affect the heat stability of various vinyl formulations. Kaplan and Wendt [25] suggest that these compounds "failed to entirely suppress surface growth during plate tests of PVC film."

Antimicrobials should have the following characteristics:

1. Highly effective against a wide variety of microorganisms, both bacterial and fungal
2. Durable at high temperatures and withstand weathering
3. Priced so as to keep the product within economic limits
4. Relatively colorless
5. Essentially nontoxic
6. Compatible with all vinyl formulations without affecting heat, light, or chemical stability.

The polyurethanes, like the flexible vinyls, range from highly resistant to susceptible. They are used in many industrial and commercial products such as fabric coatings, adhesives, and flexible and vinyl foams. Poly-ester-derived polyurethanes are the most susceptible, probably because their diols contain various fatty acids. The polyether-derived products may be completely or almost completely resistant. It is claimed that the attack by enzymes can occur only if there is a sufficiently long unbranched carbon chain between the urethane linkages of the polymer and that three adjacent methylene groups are required for appreciable attack to occur in the polyether types [26]. Since many of the most useful commercial polyurethanes are polyester derived, the least expensive being castor-oil based, it became obvious that antifungal agents were essential. Little work has been done as yet in resolving this problem. The most successful compound is 10,10'-oxybisphenoxarsine at levels of 300-500 ppm. This highly effective fungicide is dissolved in a low-molecular-weight, unbranched alkane diol such as 1,4-butanediol, and then added to the monomer prior to reaction with the diisocyanate. The nature of the reaction rules out many antifungal additives because of interference or discoloration.

There are but few other plastics that are candidates for fungicides. Polymers such as Nylon, Dacron, polystyrene, polyethylene, etc., are generally conceded to be funginert, although from time to time there are suggestions that fungal growth is observed on a system normally thought to be nonsusceptible. This phenomenon is believed to be due to the presence of debris or close proximity of a susceptible substrate. For example, glass is considered funginert, and yet during World War II, millions of dollars were spent in the treatment of optical equipment to prevent fungal damage attributed to debris, skin oils, or lubricants on which the fungi were

actually growing. The metabolic products given off by fungi growing on
debris in contact with the glass, caused severe etching to occur. The same
phenomenon probably occurs in the case of various plastics, which, although
funginert, are contaminated or marred by surface-growing fungi.

IV. PAINT FUNGICIDES

Before 1940 there was little effort expended in the preservation of paint.
The most commonly used practice involved the selection of certain pigments
which provided some protection of paint systems. The susceptibility of
films, therefore, was determined by the type and volume concentration of
the pigment system. It was known that zinc oxide, and in some instances,
Paris Green, provided protection, although at the same time accelerated
blistering or other objectionable side effects.

In 1948, Goll and Coffey [27] reported the widespread growth of Pullularia
pullulans as the predominant organism affecting exterior paint films. Rothwell [28] confirmed this and added Cladosporium sp. as predominating in
certain geographical areas.

Studies by others on interior paints used in food-processing plants [29]
proved that Pullularia pullulans and Cladosporium sp. were not as prevalent
as were various Aspergilli and Penicillia. It was later found that Pullularia
pullulans and Cladosporium sp. grow much better on weathered rather than
on unweathered films.

On paint films two forms of fungi are found, i.e., clusters of spherical
dark colored spores or thread-like, mycelial structures. The form of
growth is determined by the condition of the film and the particular environmental conditions at time of observation. When weather conditions are
proper for growth the mycelial forms predominate. During dry weather
clusters of spores enable the organism to survive. Both forms disfigure
the film to the point where replacement is necessary. Most observers easily
confuse fungal attack with a dirty surface, but, whereas dirt can usually be
removed, fungal growth cannot. Repainting can only temporarily solve the
problem as the organism may erupt through the new coating.

The only successful method for combating fungal growth on paint systems
requires the addition of a suitable fungistat. There has been more work on
paint systems than on any other substrate. As a result, more chemical
agents are recommended for the preservation of a paint system than for any
other system.

The following list includes most of the commercially important fungistats
used today in the manufacture of exterior paint systems.

Inorganic pigments:
  1. Zinc oxide
  2. Cuprous oxide
  3. Barium metaborate

Organic additives:
1. Phenylmercuric oleate
2. Di(phenylmercury) dodecenylsuccinate
3. Phenylmercuric acetate
4. Bistributyltin oxide
5. 2,3,4,6-Tetrachlorophenol
6. 2,3,5,6-Tetrachloro-4-(methylsulfonyl)pyridine
7. N-(trichloromethylthio)phthalimide
8. N-(trichloromethylthio)4-cyclohexene-1,2-dicarboximide
9. 2,4,6-Dichloro-6(O-chloroaniline)-S-triazine
10. 1,2-Bis(N-propylsulfonyl)ethene
11. 2-(4-Thiazolyl)benzimidazole

Others, used in interior paints only, include chlorinated phenolics, copper 8-quinolinolate, quaternary ammonium compounds, and phenylmercurials. It is quite conceivable that the use of phenylmercurials will soon be outlawed by the U.S. Environmental Protection Agency if they are successful in their effort to ban all mercurials.

The extensive market for paint fungicides has created almost as many tests as there are additives. Ranging from testing the paint film on paper, string, metal, and wood, to the ultimate field-exposure tests, all claim to be the most effective and reproducible in proving the worth of a chemically treated system. The most recent and seemingly most practical is the ASTM Environmental Chamber Test [30] which avoids the artificiality of other procedures. It is obvious that such a laboratory procedure is essential as a screening tool, before resorting to the more time-consuming and demanding field-exposure trial.

It is expected that many more chemical agents will be recommended for paint protection as manufacturers continue to look for substitutes for the phenylmercurials. The substantial paint market will continue to demand the greatest effort by researchers who are looking for new and more effective fungicides.

V. PAPER FUNGICIDES

There has been little demand for paper fungicides other than for the control of slime during manufacture. Slime-control agents are largely concerned with bacterial inhibition and much literature is available on the subject. A well-organized control program in a paper mill considers the use of therapeutic agents that control both bacteria and fungi, but there is no direct effort to distinguish the effects of one type of organism as compared with another.

The paper industry has been inclined to approach the use of fungicides on finished papers with circumspection; perhaps because entry into the field of mildew-resistant papers involves the development of an entirely new sales promotional program.

Chemical additives such as the chlorinated phenolics and certain sulfonyl pyridines have been used in substantial quantities in paper used as wrappers by the soap industry. Mildew resistance has been claimed, indicating that it is undoubtedly true that such papers are more resistant to fungal attack than those containing no fungicide.

Efforts are increasing to use paper as a substitute for various textiles. When these uses call for weather resistance or soil contact, more than "an element of resistance" is essential. More durable and effective fungicides must be used. Paper to be used in sandbags, grain bins, tarpaulins, barrier materials, dam liners, etc., requires the following characteristics: It must be able to completely resist the growth of surface-defacing organisms as well as cellulytic types, and to withstand long exposure in contact with soil and long-term weathering.

The most comprehensive work in the development of durable fungistatic resistance was undertaken in 1955 by a leading paper mill in Wisconsin [31]. Five fungicides were applied to Kraft paper on an experimental fourdrinier paper machine at Forest Products Laboratories, Madison. The resultant machine-made Kraft paper was then subjected to baking for 2 hr at 121°C, leaching in running water for 24 hr, weathering for 100 hr in a Weather-Ometer and finally, burying in a well-composted soil for 14 days. The only fungicide that continually passed all the tests described was copper 8-quinolinolate. The paper developed during these extensive procedures was subsequently marketed as a mulch for agricultural uses. A large mill in southern United States has produced sophisticated versions of agricultural mulch for areas of the country where several crops a year are produced. The market for other types of fungus-resistant papers is still limited. As interest in the use of paper as a textile substitute develops, other fungicides will probably be examined.

VI. RUBBER FUNGICIDES

There are few references in the literature to the use of fungicides for the protection of rubber. Rubber elastomers are not highly susceptible to fungal attack. In the formulation of various rubber recipes, additives are used to provide plasticity and good mold release. As in vinyl film production, these additives will support fungal growth and lead to deterioration problems.

Many rubber constructions utilize supporting structures such as cotton fibers or fabrics. These structures are often highly susceptible, and upon deterioration, due to fungal attack, will affect the strength of the system.

Conveyor belts are usually made of numerous plies of fabric embedded in rubber. The loss in tensile strength of the supporting fabric will quickly cause serious failure in the total system.

There are two methods for overcoming the susceptibility of supporting fabrics. An effective fabric treatment can be applied prior to the use of this support medium in the finished rubber product. Almost any good fabric fungicide may be used whether resistant to environmental conditions or not. Once embedded in rubber it is screened from rain and sunlight and cannot be readily leached out.

Copper compounds cannot be used in any rubber system because of their deleterious effect on rubber elastomers. Products such as 2,2'-methylenebis(4-chlorophenol) are excellent when applied to fabric for use in conveyor belting.

The least expensive method for protecting rubber-fabric structures involves the addition of a suitable fungicide to the rubber recipe itself. In this case, the number of fungicides suitable is limited. Rubber is cured for considerable periods of time at 176°C or higher. Fugitive fungicides or those unable to stand high-temperature processing are lost or decomposed.

The fungicide most widely used in all rubber constructions is dehydroabietylamine pentachlorophenate. Combined in a mineral-oil plasticizer it provides long-lasting protection to the entire system. It has a low order of toxicity and is highly stable to the curing temperatures. It provides the protection of pentachlorophenol without its fugitive characteristics and it is probably the least expensive of all fungicides available for this use. This interesting fungicide has no deleterious effect on any of the many rubber ingredients and its addition requires no extra steps to the manufacturing process.

The manufacture of fire hose requires a somewhat different approach. In this case the supporting fabric system is on the outside of the rubber liner. A fungicide must be used that will withstand weathering and frequent wetting and drying. The product most commonly used is 2,2'-methylenebis-(4-chlorophenol). To produce a highly weather-resistant system, the fungicide is used with a good water repellent and applied to the jacket before the rubber liner is extruded therein. Here again, the fungicide, a straight organic product, has no deleterious effect on the rubber liner.

VII. FUNGICIDES FOR ADHESIVES AND EMULSION POLYMERS

Although adhesives were probably known to prehistoric man, it is only within the last 40 years that techniques have been developed to prolong their useful life [32].

In 1934, Herrmann [33] published his important paper "Mold Resistance of Casein Glues." He listed a large number of chemical compounds tested as preservatives. Among the successful additives were nitrobenzene, salol, sodium salicylate, thymol (especially in crude casein), oil of turpentine, tricresol, formalin, borax, boric acid, and sodium fluoride. For optimum preservation he recommended 2,2'-methylenebis(4-chlorophenol) 0.1 to 0.3%, phenol 0.6%, formaldehyde 2.0%, sodium o-phenylphenate

0.5 to 1.0%, p-chlorometacresol 0.1 to 0.15%, and the p-hydroxybenzoates 0.14% in adhesive. These compounds are still used today.

In 1936, von Artus [34] published his work on glue conservation. Both the Herrmann and von Artus papers were mentioned in 1938 by Hadert [35], whose article "Chemistry, Manufacture and Commercial Application of Casein, Including Formulae and Recipes," was the first American publication to mention the preservation of adhesives.

It is curious to note that work on the preservation of adhesives has never kept pace with similar efforts in other fields. Adhesives are usually inexpensive and the addition of any compound, regardless of its cost, has an important effect on the cost of the adhesive itself. It is not surprising then that we still use such additives as formalin, phenol, and sodium benzoate in many of our adhesive systems.

The deterioration of an adhesive is caused usually by bacterial contamination rather than fungal attack. Adhesive systems such as starch, dextrine, PV acetate, and gum arabic are rarely attacked by fungal organisms. On the other hand, animal-glue- and casein-based products are highly susceptible to fungal attack. Dried adhesive films of all types can support fungal growth.

Many of the additives recommended by Herrmann are not good fungicides or have such a high vapor pressure that they are lost during drying. Fungicides such as 2,2'-methylenebis(4-chlorophenol) and sodium o-phenylphenate are useful in the preservation of the liquid adhesive as well as the film. Both are good fungicides.

Since adhesives used in food packaging are under the control of the U.S. Food and Drug Administration, only a limited number of products are suitable to prevent fungal growth. The ones most commonly used are 2,2-methylenebis(4-chlorophenol) and copper 8-quinolinolate. The latter is particularly effective when used as fungicide in animal and casein glues. Since the market is limited it is unlikely that any extensive work would be undertaken to qualify new fungicides.

VIII.   MISCELLANEOUS FUNGICIDE APPLICATIONS

A. Leather

A proteinaceous material, leather, like wood, is highly susceptible to bacterial deterioration in its raw or skin state. After processing which includes salting, pickling, tanning, and finishing, the leather has been so changed that it is no longer destroyed by bacteria. The presence of various susceptible finishing aids increase its susceptibility to fungal organisms, which, although unable to destroy the substrate, will aesthetically be unacceptable for use. While fungal growth does not cause any appreciable deterioration of the hide material itself, it may cause stiffening and cracking as the leather oils and softeners are destroyed [36].

Extensive work was undertaken during World War II in attempting to find suitable preservatives for leather which came to a halt as the vinyl and polyurethane industries developed.

First suggested for the treatment of leather in 1934 [37], p-nitrophenol, is the only fungicide still used as a preservative. Chrome-tanned leather, less susceptible than vegetable tanned, requires the addition of 0.1%. The more susceptible vegetable-tanned product requires from 0.5 to 0.75% [38]. The British Leather Manufacturers Research Association still prefer β-naphthol for leather preservation [39], although it is necessary to use 0.82% as compared with the maximum of 0.75% for p-nitrophenol. Other compounds that were recommended included p-chlorometaxylenol, salicylanilide, tetrachlorophenol, and cresylic acid. Few of these are used today.

B. Fountain Etch Fluids

One of the most interesting uses for fungicides is found in the printing industry. A fountain etch fluid is used in the flooding of plates to permit selective adherence of inks. These fluids, consisting of a number of inorganic salts and organic dispersing agents, are recirculated to be used over and over again. As they recirculate they pick up pigments and debris which are screened out. The various additives are susceptible to fungal growth. The mycelia of the organisms build up on the screens until blockage occurs. To clean the screens requires expensive shutdowns. It was found that blockage can be prevented by the use of preservatives which are recirculated with the fountain etch fluid. The high level of inorganic salts prevent the use of emulsion systems and requires that the preservative not be inactivated by nonionic surfactants. The fungicide must also be effective against a wide spectrum of microorganisms.

There are few products which meet all of the requirements of this demanding use, namely water-soluble, nonoily, wide-spectrum activity, stable in the presence of nonionic surfactants, nondiscoloring and inexpensive. The product found to be uniquely suitable is 2,6-dimethyl-m-dioxan-4-ol-acetate, low in toxicity and chemically degradable, so that an exhausted fountain etch fluid may be discarded without contaminating water systems or sewers.

C. Food Plant Sanitation

1. Food Processing Areas

One of the most insidious hazards of food plants is fungus contamination. Not only are fungi apt to affect the appearance of the products to the point of making them unsaleable, but they may, by destroying the products, cause severe monetary loss. Fungi serve as food for mites which further add to the loss of revenue.

Efforts to control the mold problem in food plants date back to 1927, when leading meat processors instituted fungal control procedures in sliced bacon, fresh sausage, smoked meats, and margarine departments [40]. Fungal growth on walls, ceilings, and equipment was mechanically removed after which the contaminated areas were washed with a solution of sodium hypochlorite. These washdowns were expensive and time consuming and only of short value. Frequent shutdowns of processing areas reduced production and increased costs. The U.S. Department of Agriculture, through their inspection program, forced the shutdown of badly contaminated processing areas in their attempt to protect the integrity of the products processed and the subsequent health of the consumer.

Painting was instituted wherever feasible, but even the paint was soon badly contaminated. It was obvious that a treatment was needed that would, at least, reduce the number of cleanups that had become burdensome for the producer.

The development of a solubilized version of copper 8-quinolinolate in 1947 [13], made available for the first time a fungicide that, in various forms, could be utilized in such divergent areas as cold rooms, and damp, painted, and warm areas and even on tile walls. It could be applied as a penetrant on wood, sealer on metal and concrete, or emulsion on the grout between tiles. This versatile fungicide could not be easily removed during cleaning operations and was so effective against fungi that it reduced cleanups from once every 3 months to once in 2 years. Spore counts in processing areas were reduced by 90%.

The low toxicity and solubility of the fungicide was instrumental in insuring its acceptance by the U.S. Department of Agriculture, Meat and Poultry Inspection Division. It is still the only durable product used for this purpose today in food-processing plants.

2. Harvesting Containers

As in the case of food processing areas, harvesting containers such as lug boxes, pallet bins, baskets, pea-viner aprons, fish trays, and the like, are contributors to the fungal contamination of food products. The treatment of these containers with a low toxicity, nonleaching preservative extends their service life from 50 to 100% and maintains them in a sanitary condition. The preservative, if combined with a suitable water repellent, makes the wood water and fruit-juice repellent, and prevents warping, cracking, splitting, and mildewing.

The product most widely accepted for this purpose is again copper 8-quinolinolate, in solubilized form. It does not impart taste or odor to food products and it is acceptable to the U.S. Food and Drug Administration as a preservative for wooden equipment used to package, transport, and hold raw agricultural products [43, 44]. Field tests conducted by brewers, bottlers, sugar refiners, meat packers, and other food processors demonstrated its properties for extending the service life of wooden pallets, cases, and bulk bins.

Wood pallets treated with copper 8-quinolinolate and subjected to wide variations in moisture conditions did not shrink and swell as did the untreated pallets. They maintained their dimensional stability, important to automatic palletizing systems. Nail popping, a problem encountered when pallets are subjected to alternate shrinking and swelling, did not occur with treated pallets [41, 42].

Drosophila (fruit fly) growth and propagation appear to be inhibited on treated wooden containers used in harvesting blackberries and tomatoes. Tests by leading producers showed a drop of from 48 pupae on untreated containers to two on treated ones [43]. The larvae utilize fungi growing on fruit debris as a principal food source, and, although copper 8-quinolinolate is not an insecticide, its control of fungi removes the food source and prevents the development of the insect larvae.

IX. CONCLUSION

There are many other industrial areas where fungicides are required. Indeed, most nonmetallic products being manufactured today, require at some point, the addition of a fungicide to prolong their useful life. The fungicide industry has had its ups and downs. During times of war, fungicides cannot keep up with government demands. During depression or recession it is difficult to sell fungicides to any industry. It is inconceivable that the need for fungicides will ever cease. As industry moves toward dependence upon synthetic systems, requirements for industrial fungicides will shift from one substrate to another.

The foregoing discussion on fungicides for industry presents their current status. Work is constantly in progress to find less expensive, more stable, and more efficient chemicals. In 10 years industry may well have found the perfect fungicide, but, more than likely, it will still be using those products that have served it so well for 20 years.

REFERENCES

1. L. Ginzberg, The Legends of the Jews, The Jewish Publication Society of America, New York, 1942, p. 401.
2. G. A. Bravo and J. Trupka, 100,000 Jahre Leder, Birkhauser Verlag Basel and Stuttgart, 1970, p. 216.
3. A. Lucas, Ancient Egyptian Materials and Industries, 4th ed., Edward Arnold Publishers, London, 1962.
4. J. F. Corrigan, J. Soc. Dyers Colourists, 36, 198 (1920).
5. R. G. Fargher, L. D. Galloway, and M. E. Probert, Shirley Inst. Mem., 9, 34 (1930).
6. P. B. Marsh, G. A. Greathouse, K. Bollenbacher, and M. L. Butler, Ind. Eng. Chem., 36, 176 (1944).

7. S. S. Block, Ind. Eng. Chem., 41, 1783 (1949).

8. C. H. Bayley and M. W. Weatherburn, Can. J. Res., 25, 209 (1947).

9. J. D. Dean and R. K. Worner, Amer. Dyestuff Reptr., 36, 405 and 423 (1947).

10. L. Shanor and The Sub-Committee on Textiles and Cordage, National Defense Research Committee, Office of Scientific Research and Development Report No. 4513, Washington, 1945.

11. P. G. Benignus, Ind. Eng. Chem., 40, 1426 (1948).

12. V. N. Kalberg, U.S. Pat. 2,561,379 (1951), to Scientific Oil Compounding Co., Inc.

13. V. N. Kalberg, U.S. Pat. 2,561,380 (1951), to Scientific Oil Compounding Co., Inc.

14. G. A. Zentmyer, Phytopathology, 33, 1121 (1943).

15. A. Albert, S. D. Rubbo, R. J. Goldacre, and B. C. Balfour, Brit. J. Exptl. Pathol., 28, 69 (1947).

16. A. Albert, Selective Toxicity, Wiley, New York, 1951, 109.

17. J. E. Goodavage, Amer. Dyestuff Reptr., 32, 265 (1943).

18. R. Burgess, J. Textile Inst., 20, 333 (1929).

19. R. Burgess, J. Soc. Dyers Colourists, 50, 138 (1934).

20. R. E. Klausmeier and W. A. Jones, Develop. Ind. Microbiol., 2, 55 (1961).

21. T. A. Gerard and C. F. Koda, Modern Plastics, 36, 148 (1959).

22. C. C. Yeager, Plastics World, 20, 14 (1962).

23. Ford Motor Co., Resistance of Vinyls to Mildew, (Pink Staining), Company Specification, Qual. Lab. Dearborn, Mich., and Chem. Eng., Phys. Test Methods, 1965.

24. ASTM G.29-72T, Annual Book of ASTM Standards, 1974.

25. A. M. Kaplan and T. M. Wendt, Polymer Eng. Sci., 10, 4 (1970).

26. R. T. Darby and A. M. Kaplan, Appl. Microbiol., 16, 900 (1968).

27. M. Goll and G. Coffey, Paint, Oil, Chem. Rev., 14, 111 (1948).

28. F. M. Rothwell, Off. Dig. Fed. Paint Varnish Product. Clubs, 30, 368 (1958).

29. P. H. Krumperman, Amer. Paint J., 42, 72 (1958).

30. R. T. Ross, ASTM Paint Testing Manual, 13th ed., 1972, p. 369.

31. C. C. Yeager and R. T. Seith, TAPPI, 39, 557 (1956).

32. C. C. Yeager, Adhesives Age, June 1962.

33. A. Herrmann, Gelatine, Leim, "Klebstoffe," No. 10, 224 (1934).

34. F. von Artus, Gelatine, Leim, "Klebstoffe," 7-8, 129 (1936).

35. R. Hadert, Chemistry, Manufacture and Commercial Applications of Casein Including Formulae and Recipes, Chemical Publication Co., Inc., New York, 1938.

36. J. R. Kanagy, A. M. Charles, E. Abrams, and R. F. Tener, J. Amer. Leather Chemists Assoc., 41, 198 (1946).

37. L. D. Jordan, Leather Trades Rev., 67, 197 (1934).

38. J. F. Richardson, Hide Leather, 1940, p. 99.

39. British Leather Manufacturer's Research Association, <u>Mold Resistant</u> <u>Leathers for Use in Tropical Conditions, Monthly Digest</u>, Feb. 1945.
40. J. H. Richardson and V. J. Del Guidice, <u>Mod. Sanitation</u>, <u>4</u> (1951).
41. J. O. Blew, <u>U.S. Forest Products Laboratory Report No. 2166</u>, 1959.
42. R. S. Kurtenacker, T. C. Scheffer, and J. O. Blew, <u>U.S. Forest</u> <u>Products Laboratory Report No. 2054</u>, 1956.
43. W. H. Sardo, Jr., <u>Food Processing</u>, May 1944.
44. S. S. Block, <u>Applied Microbiology</u>, <u>1</u>, 287 (1953).
45. <u>Encyclopedia of Polymer Science and Technology</u>, Vol. 2, Wiley-Interscience, New York, 1963, p. 391.
46. A. David Baskin and Arthur M. Kaplan, <u>Appl. Microbiol.</u>, <u>4</u>, 288 (1956).
47. <u>ASTM</u>, Part 41, American Society for Testing Materials, Philadelphia, 1974, p. 756.
48. C. C. Yeager, <u>Biodeterioration of Materials</u>, Applied Science Publishers, Ltd., England, 1968, p. 151.

Chapter 12

FUNGICIDES IN WOOD PRESERVATION

Michael P. Levi

School of Forest Resources
North Carolina State University
Raleigh, North Carolina

I. INTRODUCTION

In 1973, approximately 100 million gal (380 million liters) of creosote, 40 million lb (18 million kg) of pentachlorophenol and 12 million lb (5.5 million kg) of copper-containing inorganic salt-type biocides were used in the pressure treatment of wood products in the United States [1]. In addition, several million pounds of pentachlorophenol-based biocides were applied by nonpressure techniques for the prevention of decay, sap stain and mold growth, and insect attack of wood. When these figures are expanded to include total world usage, it will be quickly recognized that the wood preservation industry represents one of the major consumers of fungicides in both the agricultural and nonagricultural commodity areas.

Despite the extensive use of preservatives for the prevention of fungal attack of wood, these organisms cause, at a conservative estimate, at least $1 billion worth of damage to wood products annually in the United States. This does not include the cost of damage caused to living trees by fungi, or the damage caused to wood products by other wood-destroying organisms such as termites, wood-destroying beetles, and marine borers. Fungal damage can occur when the tree is cut, at the various stages of primary and secondary manufacture, or during use of finished products. As repair costs and the demands on our forest resources continue to increase, more attention must be given to the prevention of fungal damage to wood products, including wider use of fungicides for control purposes.

The history of wood preservation is long, beginning approximately 4,000 years ago when man first coated wood with oils to increase its durability. For detailed discussion of the history of wood preservation the reader is referred to Graham [2] and Van Groenou et al. [3]. The methods and materials employed by the wood preservation industry change relatively slowly. The major fungicidal preservatives in use today, creosote, pentachlorophenol, and chromated copper arsenates (CCAs), were patented in 1834, the late 1920s, and 1933, respectively. Common pressure processes were patented between 1838 (full-cell process) and 1906 (Lowry empty-cell).

There are many reasons why the wood preservation industry has changed its methods so slowly. Among these are the relatively high cost of equipment which delays major changes in process technology; the long period required to verify the effectiveness of potential wood preservatives which, in some instances, will be required to perform for 30 or more years under high-hazard conditions; and the effective performance and relatively low cost of materials and techniques now available. This chapter considers the fungicides and application methods currently used in wood preservation and possible future developments. Fungicides and application methods which are of historic interest only are not considered.

II. CHARACTERISTICS OF WOOD-INHABITING FUNGI

Wood-inhabiting fungi play a vital role in the carbon cycle of the earth by decomposing millions of tons of carbohydrate residues into carbon dioxide and water. Without the action of wood-inhabiting fungi, life as we know it on earth would disappear for lack of carbon dioxide, soil fertility and aeration would decrease, and masses of lignified cellulosic material would be deposited. However, when wood-inhabiting fungi attack living trees or wood products they cease to be useful organisms and can cause major economic losses.

Hundreds of different fungal species have been isolated from wood products. Their effects on the wood range from complete destruction to deep staining, surface staining, and, in some instances, no apparent effect. It has not been possible in many cases to relate the effects of the fungi to their taxonomic class. Thus, certain Ascomycetes cause only discoloration of wood, whereas others metabolize cell-wall components of the wood. The situation is further complicated by the fact that the behavior of a single fungal species sometimes depends on environment and type of wood under attack. This variability in behavior of fungal classes is one of the reasons why wood-inhabiting fungi are characterized by their effects on wood rather than by their taxonomic classification. For a more detailed discussion of the physiological, chemical, and microscopic characteristics of wood-inhabiting fungi, see Ref. 4-6.

A. Surface Molds

Hyphae of surface molds grow primarily in the sapwood of both coniferous and deciduous tree species. Although the hyphae penetrate deep into the wood, attack by surface molds is characterized by surface discoloration caused by masses of colored spores. Common colors are green, black, orange, pink, and yellow. The discoloration can generally be removed by planing or sanding.

Surface molds primarily attack starch, sugars, fats, and proteins in the parenchyma tissues of the sapwood. They have little effect on strength properties of wood other than impact strength. Permeability is sometimes increased [7]. This can be beneficial, aiding in the treatment of impermeable wood species, or detrimental where it causes excessive absorption of preservatives or wood finishes.

Some of the more common genera of surface molds are Trichoderma, Gliocladium, and Penicillium, which all cause green discoloration, and Aspergillus and Alternaria which cause black discoloration. Surface molds rapidly attack wood which has not been dried. This is particularly true in warm, moist environments. Round wood, timbers, and lumber should all be dipped in, or sprayed with, a fungicidal solution as soon after conversion as possible if molding is to be avoided. Dried wood with a moisture content greater than 20% is also susceptible to attack by surface molds. Surface molds are the only wood-inhabiting fungi which can grow on wood below a moisture content of 25-30% (the fiber saturation point).

B. Sap Stain Fungi

There are many similarities in the effects of surface molds and sap stain fungi on wood. Both discolor sapwood, have little effect on strength properties other than impact strength, sometimes increase permeability, and utilize storage rather than structural components of the wood material. The primary difference is that sap stain fungi not only penetrate, but also discolor wood deeply. The color is either present in the hyphae, or caused by diffusible pigments produced by them. Another difference is that stain fungi normally discolor wood only when its moisture content is above fiber saturation point.

Sap stain fungi color wood blue, brown, black, or occasionally red. Common genera include Ceratocystis, Diplodia, Graphium, Aureobasidium, and Cytospora. Stain is found most frequently in logs and lumber prior to seasoning. However, stain also occurs in wood after drying where it remains moist for long periods or is rewetted regularly.

Control of surface molds and sap stain fungi is of greatest economic importance where the appearance of the finished product is important, for example, veneer logs or paneling. These fungi are acceptable on construction lumber, poles, and posts providing that decay fungi have not also become established in the products. As it is often difficult to determine whether or not decay fungi are present, ideally all wood products should be kept free of surface molds and sap stain fungi as well as decay fungi during manufacture and use.

C. Decay Fungi

Decay fungi are responsible for a large proportion of the economic losses caused by wood-inhabiting fungi. They attack not only the storage materials in wood, but also the cell-wall components and the living tree and moist or wet wood products at all stages of utilization.

Decay fungi are generally divided into two groups, brown-rot and white-rot fungi. Most of these belong to the class Basidiomycetes. Wood attacked by brown-rot fungi darkens in color as decay proceeds, becomes brittle or powdery, cracks across the grain, rapidly loses strength, and shrinks abnormally. The fungi metabolize cellulose and hemicelluloses but only modify lignin, increasing its alkali solubility and decreasing its methoxyl content.

White-rot fungi utilize both the carbohydrate fractions and the lignin. The relative rates of removal of the different components of the wood depend on the species of attacking fungus [8]. As decay proceeds, the wood becomes bleached in appearance and loses strength but does not shrink abnormally until decay is advanced. The wood frequently becomes stringy or fibrous. Decayed areas are sometimes enclosed by dark "zone lines" which may provide the first indication of decay.

The type of decay which occurs in wood depends on location and species. Wood from deciduous trees in above-ground use is generally attacked by white-rot fungi, and wood from conifers by brown-rot fungi. The reasons for this specificity are not known. For wood in ground contact these preferences do not exist, as both types of fungi occur in deciduous and coniferous woods. In the United States, the genera of decay fungi most frequently isolated from wood products are Coniophora, Lentinus, Lenzites, Polyporus, and Poria. The genera and species isolated will depend upon location and type of wood product.

Decay fungi metabolize only wood above the fiber saturation point, the optimum moisture content range for decay being 40 to 80%. Most fungi will grow only on wood above the fiber saturation point. However, a few species are able to transport water to dry wood through specialized structures called rhizomorphs, and thus grow on, and then decay wood that normally would be too dry to decay. The two major species able to do this are Merulius lacrymans, common in much of Europe, and Poria incrassata, which occurs primarily in North America. These two fungi often cause rapid and extensive damage because of their ability to conduct water.

D. Soft-rot Fungi

The importance of soft-rot fungi in the decay of wood was first recognized in 1954 [9]. These fungi, which belong to the classes Ascomycetes and Fungi Imperfecti, are able to metabolize the structural components of wood in addition to the storage materials. Thus, they resemble the decay fungi to some extent. However, they are more limited in the damage they cause and are generally grouped separately from decay fungi.

Soft-rot attack usually occurs in the outer layers of wood, although occasionally it may penetrate deeply into wood that has been treated with low concentrations of preservative chemicals. The hyphae of soft-rot fungi grow within the secondary wall layers of wood fibers and tracheids forming cylindrical cavities with pointed tips. This distinguishes soft-rot fungi from decay fungi which grow primarily in the lumena. The surface of wood attacked by soft-rot fungi can be easily scraped off when the wood is moist. When the wood dries, its surface is dark and covered by numerous cracks both along and across the grain.

The effects of soft-rot fungi on the various components of wood cells are not well known, Chaetomium globosum being the only fungus which has been studied in detail [10]. Chaetomium is similar in its behavior to brown-rot fungi, metabolizing cellulose and hemicellulose but only modifying lignin. However, it does not markedly increase the alkali solubility of decayed wood, as is the case with brown-rot fungi. These results support the hypothesis that Chaetomium has a highly localized effect on the cell wall whereas brown-rot fungi have a more general effect.

Soft-rot fungi occur in wood which is too moist to support growth of decay-fungi, for example, cooling-tower fills, and also in wood subjected to extremes of temperature or dessication too great for decay fungi to survive. Thus, soft-rot fungi are associated with the breakdown of wood surfaces exposed to natural weathering [4]. Laboratory efforts to decay coniferous woods with pure cultures of soft-rot fungi have been largely unsuccessful. However, these fungi decay both coniferous and deciduous woods in the field.

III. NONFUNGICIDAL CONTROL TECHNIQUES

Wood-inhabiting fungi have five basic requirements, namely food, oxygen, moisture, warmth, and a suitable pH. Modification of the environment to eliminate any one of these will control decay. Application of fungicidal wood preservatives to eliminate a suitable food source is often regarded as the most effective and commonly used control method. However, other methods are available. Control of moisture, the most important of these, is not only the oldest known but also the most widely used.

Control of the environment so that wood is kept below the fiber saturation point is the basis for prevention of decay of wood used in buildings. A general discussion of construction methods recommended to prevent decay is given by Scheffer and Verrall [11]. Wood used for construction of a building should have a moisture content below 20% (this gives a safety margin below the fiber saturation point) and should not be decayed (molded or stained wood can be used providing it is dry and there is no evidence of decay). When dry wood has been installed in a building it must be kept dry if decay is to be avoided. Wood close to the ground should be protected from soil moisture by installation of moisture barriers below the sills, under concrete slab floors, and on soil under basementless houses. Crawl spaces and attics should be well ventilated. Water should drain away from houses, not under them. Construction which brings soil close to untreated wood should be avoided. There should be at least 6 inches (15 cm) vertical clearance between wood siding and soil, and 8 inches (20 cm) between soil and structural members of the building. This is particularly important with earth-filled porches, carports, and patios, common sources of Poria incrassata when the correct clearance is not achieved.

Regular inspection and maintenance of the building must be kept up to prevent damage resulting from leaks in gutters and roofs, around showers, bathtubs, sinks, washing machines, and finally in water pipes. Vents must be kept clear of debris, and paint and caulking around windows and doors must be kept in good condition. If these simple rules of construction are followed, wood will give many years of service with a minimum of maintenance, and the oldest, most economical control method in wood preservation will continue to be successful.

If wood cannot be kept dry in buildings, chemically treated wood should be used if decay is to be avoided. The type of treatment will depend upon the severity of exposure. Wood in the ground should be pressure treated, whereas window frames will perform well with only a dip, brush, or spray treatment.

Control of moisture by rapid drying of wood is also one of the most commonly used methods for preventing decay in unseasoned lumber. The lumber should be stacked so that air can circulate freely over each piece of wood. The stack should be covered to protect it from rain and should be at least 8 inches (20 cm) off the ground. In warm, humid climates it is advisable to dip the green lumber in a fungicide immediately after cutting to prevent growth of surface molds and sap stain fungi. Alternatively, the wood can be kiln dried which achieves in hours what takes weeks with air-drying.

Decay of wood can be prevented not only by removing water from the wood, but also by saturating it with water so that there is insufficient oxygen present to support fungal growth. Ponding, i.e., the storage of logs in water, has been used for many years in different parts of the world. Its success in preventing attack by surface molds, sap stain, and decay fungi has led to the development of continuous water spraying where ponding is not practical [12]. Wood piling continually below the water table is also protected from attack by fungi because of the lack of oxygen. However, wood immersed in water is susceptible to attack by bacteria. These can increase the permeability of the wood and occasionally produce unpleasant odors due to the formation of organic acids. Whenever these effects will be economically objectionable water storage is not recommended.

An alternative method for controlling decay by excluding oxygen has been developed for the protection of pulpwood, which can also be used for storage of storm-damaged timber. This patented method involves completely encapsulating the wood pile in a plastic casing. Natural respiration of living cells in the wood quickly removes oxygen, and providing that the plastic is sealed properly, oxygen levels remain too low for fungal growth for at least two years [13].

Modification of temperature and pH have limited application for the control of wood-inhabiting fungi. Temperatures of approximately 90° C are reached in the kiln drying of lumber. These will kill most organisms within the wood. Similarly, heat sterilization techniques are used for the control of <u>Merulius lacrymans</u>, the heat sensitive, water-conducting brown-rot fungus commonly found destroying wood in buildings in Europe. Alteration of pH has been implicated in the protective action of ammonia treatment of wood against attack by brown-rot fungi [14].

Elimination of a suitable food source can be achieved by several methods other than addition of fungicides. Decay can be avoided by the use of the heartwood of naturally durable woods such as redwood, cypress, cedar, or black locust. Natural durability is generally associated with the deposition of fungicidal extractives as the heartwood is formed. Further information on natural durability can be found in a review by Scheffer and Cowling [15].

Concern has been expressed over the use of persistent fungicides in wood preservation. This has promoted interest in the prevention of decay by alteration of the wood so that fungi cannot metabolize the cell wall constituents. Esterification of the hydroxyl groups in cellulose has been used successfully. Acetylation with acetic anhydride in an inert solvent gives protection against decay fungi with no decrease in the impact strength of the wood [16], but the high cost of the process has prevented its general use. Numerous other studies have been reported on modification of cellulose to impart decay resistance and dimensional stability, but these have been either more expensive or technically less successful than acetylation. Modification with alkylene oxides including propylene oxide and epichlorohydrin has recently been reported [17]. Information is not yet available on either the cost of these treatments or their effects on the strength of wood.

An alternate to modifying wall constituents is the removal of minor constituents in the wood which are essential for the growth of wood-decay fungi. Baechler [18] suggested several methods for achieving this, including destruction of thiamine with alkali and deactivation of essential trace elements, for example by chelation. It is unlikely that these methods could be widely applied because of the difficulty of excluding trace elements from treated timber after it is placed in service.

Research on the use of microorganisms to prevent attack of wood by decay fungi indicates that such techniques may be more valuable for the short-term protection of unseasoned timber than for long-term protection of wood products. Spraying freshly cut ends of birch logs with a suspension of <u>Trichoderma viride</u> almost completely prevented colonization by wood-destroying fungi over a 6-month period [19]. Biological control of decay in poles has been reviewed by Ricard [20].

IV. FUNGICIDES IN WOOD PRESERVATION

A. Requirements

The development of new biocides is highly speculative with a very low success rate. The manufacturer faces several problems of both a technical and commercial nature. It has been estimated that on the average, only one

out of 4,000 chemicals tested as potential biocides is commercially accepted
[21]. In 1968 it was estimated that research and development costs for a
new pesticide are in the range of $1 to 2 million [22]. These costs are now
considerably higher (see Chapter 4). The requirements for an acceptable
wood preservative are in some respects more stringent than those for other
biocides, particularly those concerning permanence. Thus, it is not surprising that introduction of new wood preservatives is relatively rare.

The major requirements for a wood preservative formulation are:

1. It must be toxic to the wood-destroying organisms to which the
   treated wood will be exposed.
2. It must be sufficiently permanent in treated wood to give satisfactory
   service life.
3. It should have little or no effect on the strength properties of treated
   wood.
4. It should have good penetration properties into permeable wood
   species.
5. It should be readily available and relatively inexpensive.
6. Treated wood should be, (a) safe to handle and nontoxic to animals
   and plants; (b) noncorrosive to fastenings and other materials which
   come in contact with it; and (c) for certain end uses, paintable, odorless, colorless, water-repellent, or compatible with adhesives.

Considerable effort has been devoted to discovering a single all-purpose
preservative formulation which satisfies all of these requirements. This
has not been, and probably never will be achieved. It is likely that in the
future more effort will be devoted to the development of preservatives for
specific end-uses.

The range of organisms against which the preservative should be effective is totally dependent on the location in which the treated wood is to be
used. Generally, acceptable preservatives must be toxic to brown- and
white-rot decay fungi. Effectiveness against soft-rot fungi is necessary
where wood will be used in ground contact. Protection against surface
molds and sap stain fungi has been considered of minor importance for preservation of wood in use. However, as more wood is being used in situations where appearance is important, for example, siding, decks, etc.,
increasing attention is being given to combining mold and stain control with
decay control in a treatment that has little or no effect on the color of
treated wood.

Insects against which protection is required include termites, Lyctid,
Anobiid, and Cerambycid beetles. The economic importance of these insects varies from one country to another. Established preservatives such
as creosote, pentachlorophenol, and chromated copper arsenate have a
broad spectrum of activity against both fungi and insects. Some newer preservatives, for example, tributyltin oxide, have a more limited spectrum
of activity. They must be used in combination with other biocides where
protection against insects as well as fungi is required.

Permanence is regarded as an undesirable feature for many agricultural biocides. However, wood preservatives are expected to maintain their effectiveness for periods exceeding 50 years in some instances. Thus permanence in the environment where the treated wood will be used is essential. Preservatives to be used for the treatment of wood in ground contact must be nonvolatile and resistant to water leaching. They should not react chemically with the wood to produce inert compounds. Where wood is used away from the ground, resistance to water-leaching is not essential unless the treated wood will come in contact with foodstuffs or will be used in high-humidity situations where condensation may lead to loss of preservative.

Permanence of preservatives in wood is achieved in one of two ways. Some preservatives, for example creosote, are relatively insoluble and nonvolatile when they are introduced into the wood. Other preservatives are not permanent when introduced into the wood, but they react with the wood either chemically or physically to become insoluble or less volatile. Chromated copper arsenate is an example of a preservative which becomes relatively insoluble after impregnation. Both types of preservative are modified by the environment over long periods of time. Modification can be physical, e.g., slow loss of creosote due to evaporation; chemical, e.g., leaching of chromated copper arsenate in acidic environments; or biological, e.g., microbiological breakdown of pentachlorophenol, creosote, and chromated copper arsenate (see Section IV.D). Thus, permanence is a relative, not an absolute term. A satisfactory preservative is one which is sufficiently permanent to provide protection against wood-destroying organisms for an economically acceptable time in a particular environment.

Most preservatives in commercial use today have little effect on the strength properties of treated wood when used at normal concentrations. It has been suggested that high concentrations of chromated copper arsenate used for the treatment of marine piling increase the brittleness of piles. However, it is more likely that this is due to the kiln schedule used for the drying of the piles after treatment. Creosote and pentachlorophenol do not markedly affect the strength properties of treated wood. The only preservatives which have a serious effect on the strength of wood are the inorganic water-soluble fire retardants. Strength losses are not excessive under normal conditions, but can become high under very humid conditions. When building with lumber treated with a fire retardant, allowable stresses are reduced by 10% from those for untreated wood [23].

A preservative must be formulated so that it will penetrate deeply into wood if it is to be effective. Surface treatments are not sufficient under medium to high-hazard conditions because fungi and insects can enter through checks in the wood which develop after treatment. In addition, untreated wood can be exposed by physical removal of the protected layer. The heartwood, and in some instances the sapwood, of many wood species is resistant to penetration by oil- and water-soluble preservatives. This considerably reduces the proportion of our wood resources which can be satisfactorily treated. Development of preservative systems which will penetrate deeply into impermeable wood remains one of the greatest challenges in wood preservation. Possible solutions to this problem include development of gas-phase preservative systems, and physical, chemical, or biological pretreatments which increase the permeability of resistant wood species.

Safety in use, and nonharmful effects on the environment are essential prerequisites for any acceptable wood preservative today. Hazards in the handling and use of biocides can exist at the treatment plant and in the treated product in service. Most research effort has been concerned with reducing the hazard at the treatment plant because this is where it is higher. Recommendations for prevention of pollution have been summarized by Thompson [24]. Experience indicates that when used according to industry specifications, pressure-treated wood does not pose a threat to the environment. This is not surprising, in view of the permanence requirements for wood preservatives. Wood freshly treated with pentachlorophenol or creosote can damage plant tissue adjacent to the wood. This effect diminishes rapidly as volatile fractions of the preservative evaporate. Fixed waterborne preservatives such as chromated copper arsenate and ammoniacal copper arsenate have no effect on plant foliage close to treated products. Analysis of grape plants adjacent to posts treated with chromated copper arsenate failed to detect any increase in the copper, chromium, or arsenic content of the leaf or fruit tissue [25]. As a result of the current concern about the environment, efforts are now being made by industry to collect existing information on the effects of treated wood on plants and animals and determine the need for additional research [117, 118].

Acceptable wood preservatives must not corrode materials which come in contact with them. The combination creosote/pentachlorophenol is used rarely because of its corrosiveness to the standard equipment used in treatment plants. Water-soluble fire retardants used in moist environments can cause severe corrosion of metal fastenings. Currently used oilborne fungicidal preservatives generally do not corrode fastenings or other materials which come in contact with the treated wood, although they can soften some plastic and rubber materials. Waterborne preservatives, such as chromated copper arsenate and ammoniacal copper arsenate, are noncorrosive once they have reacted with the wood to become water insoluble.

Requirements for wood to be used inside buildings are more stringent than for wood to be used outside. Thus, creosote or pentachlorophenol in heavy oil, although excellent for heavy-duty uses such as piling and bridge timbers, are unsuitable for treatment of wood to be used inside, primarily because of potential odor problems. Where a colorless treatment is required, pentachlorophenol in recoverable solvent or light oil is most satisfactory. Waterborne preservatives provide a paintable, odorless, slightly colored treatment. Where water repellency is important, creosote, pentachlorophenol with water repellents, or chromated copper arsenate with water repellents should be specified. Compatibility between adhesives and preservative-treated wood is important in some situations. It is difficult to make general recommendations concerning gluing of treated wood because of the great range in types of glue and preservative treatment now available. Whenever compatibility is of conern, advice should be sought from both the manufacturer of the glue and of the treated wood.

B. Evaluation of Biocidal Activity

Effectiveness against wood-destroying fungi is the most important requirement for a potential wood preservative, and also the most time consuming to evaluate. Rapid laboratory methods are available to determine the biocidal activity of fungicides. However, accurate prediction of performance in the field on the basis of rapid laboratory tests cannot be guaranteed.

Laboratory and field test methods for evaluating potential wood preservatives are designed to determine the threshold concentration required to prevent decay. The threshold concentration is defined as the lowest concentration of preservative required to prevent decay under a given set of conditions. It is dependent on a number of variables including species of fungus, species of wood, duration of test, and size of wood specimen. Thus, the primary value of threshold concentrations is for the comparison of new with established preservatives under similar test conditions. Thresholds should not be compared where very different test conditions are used. Alternate systems for determining the effectiveness of preservatives which are less dependent on test method have been suggested [26, 27], but have not found general acceptance.

Several types of test are used for determining fungicidal activity of potential preservatives. These range from laboratory tests requiring only a few days to complete, to field or service tests requiring several years.

Agar-plate toxicity tests are often used for the initial screening of chemicals. Varied concentrations of the test compound are incorporated in nutrient agar and their effects on the growth of pure cultures of economically important wood-decay fungi are noted. This method is rapid and requires only small amounts of chemical. The major disadvantage is that its results may be misleading because the substrate is unlike wood in chemical and physical characteristics. Consequently, a variety of tests using veneer-size samples of wood have been developed for rapid screening of potential preservatives. These include methods which measure changes in strength [28] or weight [29] properties of exposed wood. These methods are more realistic than agar-plate tests but caution must still be exercized in interpreting results because of the small size of the treated specimens in relation to the amount of fungus. This can lead to the rejection of preservatives which would perform satisfactorily in larger samples of wood [30].

The laboratory test which has become accepted as the standard technique for evaluating fungicides is a 3-month, weight-loss method. In Europe, blocks of wood $5 \times 2.5 \times 1.5$ cm are exposed to pure cultures of wood-decay fungi growing on malt-agar [31, 32]. In North America, 1.9-cm cubes of wood are exposed to pure cultures of fungi growing on soil and an untreated wood feeder strip [33]. In both methods, the dry weights of the blocks before and after exposure to fungus are recorded. The threshold is determined as the retention of preservative required to prevent decay. These methods sometimes give erratic results and are quite lengthy and need improvement. The most promising new technique for evluating potential preservatives is respirometry [34, 35]. Carbon dioxide production or oxygen consumption is measured in flasks containing fungally infected blocks of wood treated with different amounts of preservative. The results obtained by this method compare favorably with those obtained by conventional weight-loss methods. In addition, results are obtained in approximately one-third of the time [36].

A variety of accelerated weathering tests for use in conjunction with fungicidal tests have been developed to estimate in the laboratory the probable permanence of preservatives. These include hot-water leaching (United Kingdom), cold-water leaching (United States and Germany), and high-temperature exposure (United States and Germany). These methods can be used singly or in combination. Efforts are now being made to standardize weathering tests and determine their relationship to service conditions.

Laboratory tests for the evaluation of effectiveness of preservatives against soft-rot fungi and surface molds and sap stain fungi are not as well developed as those for decay fungi. Soft-rot fungi can be tested in pure culture on nutrient agar, soil, or vermiculite [37, 38]. Alternately, preservatives can be exposed to natural populations of soft-rot and other fungi using unsterilized soil low in organic matter [39]. Surface molds and sap stain fungi are normally tested in mixed culture [40].

The choice of decay fungi for use in laboratory evaluation of potential preservatives will greatly affect the relevance of results to service conditions. Test fungi are chosen for their ability to cause rapid decay in untreated wood, ease of maintenance in the laboratory, stability in culture, tolerance to one or more types of preservative, and economic importance. Among the most widely used test fungi are Lentinus lepideus, a brown-rot fungus tolerant to creosote-type preservatives; Gloeophyllum (Lenzites) trabeum, a brown-rot fungus tolerant to arsenate and pentachlorophenol; Coniophora cerebella (puteana), Poria monticola, and Poria vaillantii brown-rot fungi tolerant to copper; and Polystictus (Polyporus) versicolor, a white-rot fungus tolerant to pentachlorophenol and tributyltin oxide.

Careful selection of test fungi and methods for laboratory testing enable the researcher to evaluate potential preservatives in a relatively short time.

However, field tests are essential before a preservative can be introduced commercially. This is necessary because of the impossibility of reproducing in the laboratory the microorganism complex found in the field. For example, tests with pure cultures of decay fungi cannot anticipate the possibility of preservative metabolism or detoxification by nondecay fungi or bacteria.

The types of field tests in current use are as diverse as the types of laboratory tests. Small stake tests are the most widely used of the field tests [41, 42]. Cross sections of stakes range from approximately 2 × 2 cm to 5 × 10 cm, and the stakes are generally at least 45 cm long. When a preservative is being evaluated, the stake should be completely permeable and of low natural resistance. Normally only sapwood is used. The stakes are saturated with a range of concentrations of preservative chosen to bridge the threshold concentration obtained in laboratory tests. Untreated stakes and stakes treated with known concentrations of established preservatives, such as creosote or chromated copper arsenate, should always be installed with test stakes as controls. When preservative solvent has evaporated from the stakes they are inserted in the ground.

Stakes are normally inspected visually once a year after careful removal from the ground. They are rated according to the degree of deterioration. Decay grades are 10, sound; 9, slight surface decay; 7, moderate decay; 4, heavy decay; and 0, failure. Rating is sometimes complicated by the presence of termite damage which should be evaluated separately. This procedure is quite subjective. Whenever possible, the same person should evaluate the whole of the test plot at any particular inspection. Ideally, in addition to visual inspection, treated stakes should be removed at regular intervals for chemical analysis to determine the permanence of the preservative and also for characterization of invading microorganisms. Unfortunately, this is rarely done. At least ten replicates per variable are normally installed, the ratings for these are averaged, and then the performance of the preservative under test is compared with the performance of the various control stakes. Thus, the performance of the test preservative is compared to that of established preservatives under the same conditions. These data are then used in the derivation of national specifications for the treatment of wood products to be used in different environments.

Field tests with larger specimens, for example, posts, poles, and crossties, will normally follow successful completion of stake tests. The final step in field evaluation involves regular monitoring of treated wood products in service to confirm that recommended preservative specifications are satisfactory.

Most field tests are concerned with the performance of wood in ground contact. However, attempts are now being made in Europe and North America to develop standard methods for evaluation of preservatives for the treatment of wood to be used above ground. Methods currently in use include exposure of rail-post units [43] or stacked boards above ground to

natural infection and decay. Visual inspection is normally used to determine the performance of the preservative.

Evaluation of the fungicidal performance of potential preservatives is
not yet an exact science. Some of the variables contributing to this evaluation have been discussed. For a more comprehensive review of decay test
methods, see Refs. 44 and 45.

C. Types of Fungicide

Traditionally, fungicides used in wood preservation have been of two basic
types, oilborne and waterborne, distinguished by the solvent in which they
are applied. Oilborne preservatives can be further divided into distillates
of coal tar (creosotes), and solutions of oil-soluble fungicides in organic
solvents. Fumigants, such as Vapam and chloropicrin, which recently
have been discovered to be highly effective for the prevention and control of
decay of wood in use [46], do not fit into these categories and can be considered a third type of fungicide. It appears that this group of chemicals
will have an increasingly important role in wood preservation. However,
since so little is known about their mode of action, they will not be discussed further in this chapter.

The primary factors determining the effectiveness of fungicides in
treated wood are preservative retention and penetration. It is usually possible to obtain satisfactory penetration in permeable wood species with
both oilborne and waterborne preservatives by varying treatment schedules.
Thus, selection of preservative does not depend primarily on biocidal activity but upon ancillary properties such as color, odor, effect of treatment
on the dimensions of the wood, or water repellency.

For more detailed information on the chemical and physical properties
of preservatives and factors influencing the effectiveness of preservative
systems, the reader is referred to articles by Hartford [47] and Arsenault
[48]. Reviews have appeared recently on creosote [49], organic solvent-
based wood preservatives [50], tributyltin oxide [51], boron compounds [52,
53], fluorine compounds [54], and chromated copper arsenates [55].

1. Oilborne Fungicides

a. Creosote. Creosote was patented for the preservation of wood in 1838
by Bethell. It contains over 50 biocidal components which make it effective
against almost all wood-destroying and wood-inhabiting microorganisms.
It is also effective against wood-destroying insects and marine borers, with
the exception of Limnoria tripunctata. It is used for the preservation of
poles, posts, piling, and sawn timber at retentions ranging from 6 to 20 lb/
cubic foot (96 to 320 kg/m^3), depending on end use. However, it should not
be used where odorless, uncolored, or paintable timber is required.

Creosote (or coal-tar creosote) is a distillate of coal tar produced by carbonization of bituminous coal. In the distillation the first fractions obtained contain so-called light oils and the residue left at the end of the process is pitch. The liquid fraction recovered between these materials is creosote. At ordinary temperatures some of the components of creosote solidify, thus the preservative is impregnated into wood at elevated temperatures.

The three major groups of constituents in creosote are (1) tar acids, including phenol, cresols, xylenols, and naphthols; (2) tar bases, including pyridines, quinolines, and acridines; and (3) hydrocarbons, including naphthalene, acenaphthene, phenanthrene, anthracene, and fluorene. Fungicidal constituents are found in each of these groups [56]. Evidence is not available to support the hypothesis that one of the groups is primarily responsible for the fungicidal activity of creosote.

Creosote is sometimes used in solution with coal tar or petroleum, primarily to reduce the cost of the preservative. Coal tar or petroleum can be added to creosote up to a 1:1 ratio. Creosote/coal tar can be used where cleanliness of the treated product is not critical as the preservative has a tendency to bleed to the surface of treated wood. Addition of either coal tar or petroleum reduces the surface checking of treated wood. Creosote/petroleum is a less effective preservative than creosote alone.

b. <u>Pentachlorophenol</u>. Polychlorinated phenols were introduced for wood preservation in the late 1920s. Pentachlorophenol (penta) was introduced a few years later in the United States. It is the second most widely used wood preservative after creosote. Pentachlorophenol and its water-soluble sodium salt are toxic to decay fungi, sap stain fungi, and surface molds. They are used to prevent attack on freshly felled logs, unseasoned timber, and wood in service. Preservative retentions in pressure-treated wood range from 0.4 to 0.6 lb/cubic foot (6.4 to 9.6 kg/m^3). A 5% solution is normally used for the surface treatment of dry lumber, for example, joinery.

The persistence, and hence effectiveness of penta is largely dependent upon the solvent system and treatment method used to apply the preservative. Solvent systems include heavy (nonvolatile) oils, light (volatile) oils (where a water repellent is often incorporated in the preservative formulation), and recoverable solvents, such as liquid butane and methylene chloride. It appears that when penta is deposited in the wood in the crystalline state after impregnation in recoverable solvents, or is kept in solution in a nonvolatile oil, it maintains its effectiveness for many years. When applied in a volatile oil, it migrates to the surface of the timber which reduces its permanence unless suitable antiblooming agents are added to the solvent.

Pentachlorophenol is normally manufactured by direct chlorination of phenol, or a mixture of phenol and lower chlorinated phenols, using aluminum chloride as catalyst. Other processes include reduction of hexachlorophenol and hydrolysis of hexachlorobenzene.

Several derivatives of pentachlorophenol have been developed in an attempt to increase its permanence and overcome its irritant properties. These include copper pentachlorophenate, dehydroabietylamine-pentachlorophenate (Rosin amine D pentachlorophenate), and pentachlorophenyl laurate. They have been used successfully as textile preservatives, but have not found acceptance as wood preservatives.

c. Tributyltin Oxide. Tributyltin oxide is a relatively new addition to the list of compounds used in wood preservation. Reports on the effectiveness of organotin compounds as preservatives first appeared in 1958 [57]. When cost, permanence, and mammalian toxicity are all considered, tributyltin oxide is the most effective of the organotins for wood preservation. It is effective against a wide range of wood-decay fungi. However, because of variable stake test results it is not recommended for preservation of wood in ground contact. It is used primarily for the protection of wood in buildings. It has several advantages over pentachlorophenol for this use. These include lower mammalian toxicity, lower irritant properties, better paintability, and greater solubility in light petroleum solvents so that antiblooming agents and cosolvents are not required. Where insect hazard is high, tributyltin oxide is used in combination with a contact insecticide, for example, lindane. It is recommended for use at a retention of 0.06 lb/cubic foot (0.64 kg/m^3).

d. Copper Naphthenate. Naphthenic acids, produced as byproducts in petroleum refining, react with a number of metal salts to form gummy or waxy naphthenates. Copper naphthenate is the most important of these compounds used in wood preservation. It was introduced commercially in Denmark in 1911 and its use spread until after World War II when it partially replaced creosote, which was then in short supply. However, as creosote supplies increased, usage of copper naphthenate decreased because of its relatively high cost. It is now used primarily for dip, spray, or brush treatment of lumber, boat timbers, and wood used in horticultural products. It is toxic to a wide range of wood-destroying organisms including fungi, insects, and marine borers. It is normally applied in solutions containing 1-3% copper metal.

Copper naphthenate solutions give treated wood a bright green color. Zinc naphthenate, which is light colored, is occasionally used in place of the copper salt where color is important. However, zinc naphthenate is a less effective biocide than copper naphthenate.

e. Copper-8-quinolinolate. Copper-8-quinolinolate is the copper chelate of 8-hydroxyquinoline. It is effective against a wide range of decay fungi, is odorless, and has very low human toxicity. These properties have led to its specification for treatment of specialty items such as railroad car floors, greenhouse flats, and containers where foodstuffs come in contact with treated wood. Because of its high cost, its use is generally limited to treatment of specialty items.

2. Waterborne Fungicides

The usage of waterborne fungicides is increasing more rapidly than that of the various types of oilborne preservative. Major reasons for this are the cheapness and ready availability of the solvent, i. e., water, and the development of waterborne preservatives which do not leach out of wood when it is used in moist environments. Disadvantages of waterborne preservatives are the swelling effect of the solvent, the general need for redrying after treatment, and the difficulty of providing protection against weathering.

Various metal salts formulated alone or in combination with each other and with arsenic, boron, and fluorine compounds have been used in wood preservation. Today only copper, chromium, and to a lesser extent, zinc are in general usage. Other metals, such as mercury and lead, have been discarded because of high human toxicity. Silver, cadmium, and nickel are too expensive.

a. Borax and Boric Acid. Taken separately, borax and boric acid have quite low solubilities in water, but when used together, their solubility increases considerably. This mixture is used in Scandinavia, Australia, New Zealand, and Canada for the diffusion treatment of green lumber. The mixture is effective against a wide range of surface molds and sap stain and decay fungi, but because of its relatively high solubility in water, it is not recommended for treatment of wood to be used in ground contact. The preservative is also effective against wood-destroying insects other than termites.

b. Acid Copper Chromate, Chromated Copper Arsenate, and Ammoniacal Copper Arsenate. These three classes of preservative are the most important of the waterborne fungicides used in North America. Acid copper chromate (ACC) was patented in 1928. The chromate serves two purposes. It reduces corrosiveness of the copper, and it precipitates the soluble copper as insoluble copper chromate to produce a nonleachable preservative in the wood. It is applied by vacuum/pressure treatment. Wood treated with ACC is susceptible to decay by copper-tolerant fungi. Therefore the preservative is not recommended for the treatment of structural timber, piles, or poles in ground contact.

Chromated copper arsenate was patented in 1933. Introduction of arsenic broadens the spectrum of activity of the preservative so that it is effective against decay fungi, wood-destroying insects, and marine borers, except pholads. However, it does not prevent growth of surface molds. Several formulations are available commercially. They differ slightly in the relative proportions of the active component and raw materials used for formulation. The preservatives are available in liquid concentrates, slurries, and dry powders. These compounds are recommended for treatment of wood to be used in ground contact, above ground, and in marine situations where pholads are not a major hazard. Retentions specified in the United States range from 0.25 lbs/cubic foot (4.0 kg/m^3) for wood above ground to 2.5 lb/cubic foot (40 kg/m^3) in the outer 1/2 inch of wood to be used in salt water. After impregnation, the components of CCA react with each other and with the wood. There is a rapid increase in pH as a result of ion-exchange and absorption reactions with the wood. Some copper reacts with carboxyl groups in the wood. In addition, a series of water-insoluble complexes are formed between copper, chromium, and arsenic and the wood. Final products are basic copper arsenate, tertiary chrome arsenate, and chromic acid [58]. The resultant CCA complex is distributed throughout the cell walls and also forms a protective layer on the lumen walls [59].

Ammoniacal copper arsenate (ACA) was first patented in 1928. It contains copper and arsenic salts dissolved in an ammoniacal solution. After vacuum/pressure impregnation into dry wood, the ammonia is allowed to evaporate, leaving an insoluble precipitate of copper arsenate in the wood. Original formulations were prepared with arsenite which is rapidly oxidized in the treating solution to arsenate (this explains why the preservative is often referred to as ammoniacal copper arsenite). Recently, the Canadian Forestry Service has applied for patent coverage for ACA (and ammoniacal zinc arsenate) formulations prepared directly with arsenate [60]. The form of arsenic in the preservative solution and treated wood product is important because arsenite has a much higher mammalina toxicity than arsenate; ACA can be used interchangeably with CCA. It has been suggested that deeper penetration in impermeable wood can be obtained with ACA than with CCA. However, supporting evidence has yet to be published. It is more difficult to handle ACA at the treatment plant than CCA because of the ammoniacal solvent.

Wood treated with ACC, CCA, and ACA is odorless and clean, and when dry can be painted or stained. Thus these preservatives are widely used in buildings and other situations where people and animals will come in contact with the treated wood.

c. Copper Chromium Fluoride and Copper Chromium Boron. These preservatives contain a toxic component which remains soluble after impregnation into wood (fluoride and boron, respectively). This is claimed to give improved penetration in impermeable timbers. However, this advantage is offset by the leachability of the soluble components in wet conditions. Stake tests with copper chromium boron at Saucier, Mississippi, indicate that very high retentions of preservative will be required to prevent decay and termite attack [61]. Thus the preservative may be limited to use in above-ground situations.

d. Fluor-Chrome Arsenate Phenols. These are the oldest of the multisalt waterborne preservatives in use in the United States, first patented in 1918.

They have a broad spectrum of activity against fungi and insects and have been used widely for vacuum/pressure treatment of lumber in buildings, but are not recommended for use in ground contact or other moist situations because the preservative components are partially soluble in water. Usage of fluor-chrome arsenate-phenols is rapidly decreasing as they are being replaced by the more permanent copper-containing multisalt preservatives.

e. Other Waterborne Fungicides. In Germany, fluoride compounds are widely used for the preservation of wood. Copper-chromium-fluoride and fluor-chrome arsenate phenols have been briefly mentioned. Other fluoride compounds in use include potassium and ammonium hydrogen fluoride, and magnesium silicofluoride. These compounds are used for the treatment of wood in buildings, remedial treatment of wood in use, and for protection of particle and fiber board. The hydrogen fluoride salts in particular have excellent penetrating properties because of the diffisuion of hydrogen fluoride formed in the wood. The hazards associated with diffusion of hydrogen fluoride out of the wood do not appear to have been considered.

The primary concern in wood preservation is prevention of decay. However, control of the growth of surface molds and sap stain fungi is sometimes essential, particularly on unseasoned wood and seasoned lumber to be used in situations where appearance is important. Sodium pentachlorophenate, sometimes used in combination with borax and an insecticide, has long been regarded as the most effective fungicide available for this use. Recently, concern has been expressed about the hazards involved in its use. An intensive search has been made for alternate compounds and two formulations appear promising at the moment. Field tests in New Zealand have shown that a 0.2% aqueous suspension of captafol (Difolatan) is effective in controlling mold and stain growth [62]. In Canada, laboratory tests with ammoniacal zinc oxide indicate that a 5% zinc oxide in ammonium carbonate/ammonium hydroxide solution is effective [63]. Field trials of this treatment are being undertaken.

D. Mode of Action of Fungicides

Wood-decay fungi metabolize timber by secreting enzymes or other chemicals which act either in contact with, or at some distance from the hyphae. Carbohydrates, and in some cases lignin in the cell wall are depolymerized either hydrolytically or oxidatively to smaller molecules which are further metabolized by intracellular enzymes [5]. Thus white-rot fungi can destroy almost 100% of the wood substance and brown- and soft-rot fungi more than 60%.

Fungicides acceptable for use as wood preservatives can prevent decay in two ways. They can disrupt normal intracellular metabolic processes by killing or inhibiting growth of the fungi. Alternately, they can inhibit

production of, or inactivate extracellular enzymes responsible for breakdown of the wood cell walls, so that in the absence of an alternate food
source the fungi die. Most of the methods currently used to evaluate wood
preservatives do not distinguish between these two modes of action since
they measure only the effect of the fungus on treated wood, not the effect of
the treated wood on the fungus.

The ability of fungi to grow in wood without decaying it is one of the reasons for recommending the use of wood as a substrate even in the preliminary stages of preservative evaluation. When agar is used as substrate,
the extracellular enzymes which metabolize cellulose and lignin are not
required. Thus, the test would not detect as effective a fungicide which
acts primarily against these enzymes. When wood is used as substrate,
however, the fungicide would prevent decay.

Another important reason for using wood as substrate is to take into
account any effects wood may have on the solubility, chemical form, and
fungicidal effect of the preservative. For example, the chemical form of
soluble chromium, copper, and arsenic salts in agar and wood is different.
This may affect their fungicidal activity.

Wood-inhabiting fungi can be divided into those which decay wood and
those which discolor it. The sap stain fungi and surface molds discolor
wood but do not metabolize the cell-wall components. Thus, the role of
extracellular enzymes is less critical than for decay fungi. To be effective
against sap stain fungi and surface molds, a fungicide must disrupt intracellular activity. This distinction may explain why fungicides such as the
chromated copper arsenates which prevent deterioration by decay fungi do
not prevent discoloration by molds.

The majority of wood preservatives are water insoluble, whereas decay
and staining fungi require the presence of moisture for growth. In some
instances, this raises the problem of mobilization of fungicide to the site of
action on the fungus. Where the site of action is external to the fungus, for
example, if extracellular enzymes are denatured, mobilization is not necessary because the fungicide is present at the site of action after impregnation
into the wood. However, where the primary mode of action is intracellular
it is necessary to explain how the fungicide reaches the fungus as this could
have a significant effect on the long-term performance of the formulation.

The two concepts which appear to be of the greatest significance in explaining the action of wood preservatives are first, that fungal metabolites
mobilize the fungicide, an idea which has received considerable experimental support, particularly for insoluble copper fungicides [64-66]; and
second, Overton [67] developed the lipoid theory of cell permeability which
states that movement of a substance in and out of a cell is determined by the
coefficient of distribution between the lipoid membranes and aqueous cytoplasm. There is considerable dispute about the validity of this hypothesis
[68], but it probably explains in part how some of the oilborne wood preservatives work.

Until recently, little progress has been made in elucidating the mode of action of wood preservatives. However, development of techniques for localizing the position of fungicides within cells has enabled some progress to be made, although much of the information available is still only preliminary and many basic questions remain to be answered.

1. Creosote

Some work has been published on the mode of action of individual compounds found in creosote against nonwood decay fungi. To the author's knowledge, no research has been done on the mobilization of creosote by decay fungi, or its penetration into, and effect on, hyphal cells.

Phenols (found in tar acids) are general biocides effective against a broad spectrum of microorganisms. Their fungicidal activity is probably due to combination with wall membranes or to penetration of undissociated lipid-soluble chemicals into fungal cells where they react with essential metabolites, for example, by denaturation of protein.

Heterocyclic nitrogen compounds (found in tar bases) may interfere with normal metabolic processes, such as membrane transport and growth. However, little is known about the mode of action of the types of nitrogen compounds found in creosote, other than that they are general metabolic suppressants. Unsubstituted aliphatic and aromatic hydrocarbons appear to act similarly to heterocyclic nitrogen compounds.

In addition to the fungicidal activity of creosote, some of its effectiveness is due to its water-repellent properties. Retardation of moisture uptake will inhibit the growth of fungi in treated wood. The importance of the contribution made by this type of action will be difficult to determine.

2. Pentachlorophenol

Pentachlorophenol acts primarily by uncoupling oxidative phosphorylation [69]. Thus it must be transported through the fungal cell wall before it becomes effective. Phosphate metabolism in the cell is altered so that sugar phosphates and nucleic acids decompose to inorganic phosphate and phosphorus compounds, some of which are lost from the cell. Uncoupling leads to an increased consumption of oxygen as the cell tries to produce the phosphorylated compounds needed for continued growth and cell death ensures. Pentachlorophenol may also inhibit the action of extracellular enzymes in situ in the wood cell wall [70]. Electronmicroscope examination of pentachlorophenol-treated wood decayed by brown-rot fungi showed that pentachlorophenol was not present in decayed areas of the cell wall. It was not determined whether or not the chemical had been removed by the fungi [70].

3. Tributyltin Oxide

Tributyltin oxide inhibits oxidative phosphorylation in the mitochondria of animal tissue [71], but this action has not yet been demonstrated in fungi.

Bravery [72] observed that extracellular enzymes secreted by <u>Polystictus</u> <u>versicolor</u> removed deposits of tributyltin oxide from treated wood at certain concentrations without affecting their subsequent extracellular action. At higher concentrations, hyphal ultrastructure was destroyed, indicating absorption of lethal amounts of preservative.

It has also been suggested that tributyltin oxide prevents attack of wood by reacting with cellulose so that it cannot be metabolized by extracellular enzymes of wood-decay fungi [73]. However, treatment of ball-milled wood with toxic concentrations of tributyltin oxide does not inhibit its digestion by cellulase enzymes, indicating that substrate protection is not an important mechanism of action for tributyltin oxide [74].

4. Copper Naphthenate

Both the copper ion and naphthenic acid portions of this compound contribute to its fungicidal effectiveness [75]. The compound is soluble in lipids which may also be an important factor in its mode of action. Mobilization of copper naphthenate by decay fungi has not been studied in detail. Many fungi discolor wood treated with the preservative before causing decay. This suggests that it is readily dissociated or absorbed by hyphae.

The mode of action of naphthenic acids is not known, but the action of copper, particularly its effect on the germination of spores, has been studied extensively. Most heavy metals, including copper, can form complexes with ligands containing sulfur, nitrogen, or oxygen as electron donors. Thus copper can react with most of the essential compounds in the cell. The reactions which occur depend upon the accessibility of the cell constituents to the metal toxicant. If exposure is for only a few minutes, copper does not penetrate into fungal cells and its effect is fungistatic. For extended exposures, copper penetrates into the cells and acts fungicidally. If sufficient quantities of copper are available, the fungus is killed by nonspecific denaturation of enzymes and other proteins.

5. Copper-8-quinolinolate

Experiments on the mode of action of copper-8-quinolinolate have been reviewed by Lukens [76]. It is suggested that being lipid soluble, it permeates cellular membranes. It then dissociates to the half-chelate, which may act by inserting copper at metal-binding sites on enzymes and other proteins or other cellular constituents where metals normally function. Alternately, 8-hydroxyquinoline may compete with coenzymes for metal-binding sites on enzymes. In both cases, the fungicide effectively blocks the normal metabolic processes of living cells and either inhibits growth or kills the fungus.

6. Acid Copper Chromate, Chromated Copper Arsenate,
   and Ammoniacal Copper Arsenate

These preservatives all rely on copper as the main fungicidal component, although arsenic plays an essential role in controlling some copper-tolerant fungi. They are neither lipid nor water soluble, so the first problem that must be solved in explaining their mode of action is that of migration to the site of action. It has been shown [66] that several brown-rot fungi solubilize chromated copper arsenate in treated wood even when sufficient preservative is present to prevent decay. The preservative components are not only solubilized but also absorbed by Poria monticola. After exposure to the fungus, some areas of the cell walls of treated wood contain no preservative and quite large concentrations of preservative are found in adjacent fungal hyphae. With most of the fungi examined, the effect of the preservative was primarily fungistatic. The fungi could be subcultured from wood treated with toxic amounts of preservative. Thus it appears that after invasion of treated wood by decay fungi, some of the preservative is solubilized and absorbed by the fungus. If sufficient preservative is present, growth of the fungus is arrested before decay occurs. In addition to this intracellular action, evidence was also obtained to suggest that prevention of decay may be due in part to inhibition of action or production of extracellular enzymes. This study by Chou et al. [66] clearly indicated that in explaining the mode of action of fungicides it is essential to consider not only the effect of preservatives on fungi, but also the effect of fungi on preservatives.

E. Tolerance

Selection of test fungi is one of the first decisions in evaluating a potential wood preservative. The choice of species and strain determines the threshold concentration obtained, because preservative tolerance varies widely with both species and strain of test fungus. Tolerant fungi are defined as those which require above-average amounts of fungicide to prevent decay. Thus tolerance is a relative, not an absolute term.

The practical significance of tolerance observed in the laboratory depends on the frequency with which the tolerant fungi occur in nature and also on whether or not they exhibit the same degree of tolerance under field conditions. It has been demonstrated in the laboratory that fungi are able to solubilize and assimilate preservatives such as chromated copper arsenate. Where such a mechanism operates, the toxicity of a given amount of preservative is dependent on the amount of fungus to which the preservative is exposed. Thus the tolerance of fungi in the laboratory will be considerably higher than that in the field where the ratio of preservative to fungus is much higher. This has been found to be the case with chromated copper arsenates [77]. Preservative cost also affects the practical significance of tolerance. This determines the amount of fungicide which can be introduced into the wood economically.

Some of the more important of the tolerant fungi have been listed in Section IV.B. A more comprehensive list is given by Levi [78]. The widespread occurrence of creosote-tolerant <u>Lentinus lepideus</u> has led to the premature failure of many creosoted poles and posts which contained sufficient preservative to control decay fungi other than <u>Lentinus</u>. High, and consequently uneconomic retentions of copper, zinc, and their chromated salts must be used to prevent decay of timber in ground contact by tolerant species of <u>Poria</u>, <u>Coniophora cerebella</u>, and <u>Gloeophyllum trabeum</u>.

Several mechanisms have been suggested to account for differences in the tolerance of fungi to preservatives [79], including the following:

1. Destruction of the fungicide
2. Production of a metabolite which inactivates the fungicide
3. Decreased permeability of the cell to the fungicide
4. Formation of, or increase in activity of, metabolic pathways bypassing the inhibited reaction
5. Increased production of the enzyme inhibited by the fungicide
6. Decreased requirement for a product of the inhibited metabolic system
7. Formation of an altered enzyme, with decreased affinity for the fungicide or with increased relative affinity for the substrate.

Little is known about the mechanism of tolerance of decay fungi to wood preservatives, but there is evidence for the operation of some of these mechanisms.

Extracellular oxidases produced by white-rot fungi detoxify phenols by oxidation [80]. This work was carried out with extracts from fungi and has yet to be reproduced in vivo. Such a mechanism would be an example of tolerance by destruction of the fungicide.

Increased production of metabolites which inactivate the fungicide may explain copper tolerance in decay fungi. A reduction of the pH of the wood is closely associated with copper tolerance. It has been suggested that copper is prevented from chelating with amino acids at low pH, so that enzymes will not be deactivated [81]. Alternatively, extracellular precipitation of copper as metabolically inert copper oxalate [82] or copper sulfide [29] may account for tolerance in some fungi. Such precipitation does not occur in wood treated with chromated copper arsenate.

It is clear from this limited discussion that a considerable amount of research will be necessary before preservative tolerance can be understood. Much of this research will be concerned with the mode of action of preservatives.

Two factors, other than preservative tolerance by wood-decay fungi, may be involved in the deterioration of properly treated wood in service. The first is the effect of timber species on fungicide performance. The second is the detoxification of fungicides by nonwood-decay microorganisms. Greaves [83] has shown that the distribution of chromated copper arsenate in treated wood varies with the species of timber treated. In certain hardwoods the preservative is located primarily in vessels leaving the fibers untreated. This may explain the greater susceptibility to decay of some hardwoods treated with chromated copper arsenate than softwoods treated with the same amount of preservative.

Laboratory experiments have shown that bacteria are able to absorb components of chromated copper arsenate from treated wood [84]. Fusarium oxysporum is able to convert dinitrophenol to less toxic compounds, possibly allowing entrance of the nontolerant decay fungus Coprinus micaceus into wood treated with fluor-chrome arsenate dinitrophenol [85, 86], and Trichoderma viride is able to remove large quantities of pentachlorophenol from treated wood [87].

The importance of these detoxification mechanisms in the field is not known. Similarly, the role of organism synergism in determining preservative effectiveness has yet to be investigated. These are areas where research needs to be initiated.

V. APPLICATION METHODS

The performance of preservative-treated wood is dependent not only on the properties of the preservative, but also on its retention and penetration in the treated product. This is dependent on the application method. Where wood is exposed to the ground or other high-hazard conditions, the application method used must guarantee deep penetration of chemical into the wood.

Several methods are available which, if used properly, will impregnate chemicals deeply into permeable wood species. If not used properly, these same methods will not provide a satisfactory treated wood product. The ease of controlling the quality of treatment obtained with the different methods is highly variable. Where treatment is carried out by unskilled or semiskilled operators poor quality control may result; whereas the greater the degree of automation, the greater the possibility of quality control. Thus in most situations, pressure treatment of dry lumber, poles, posts, and timbers is the most reliable method for commercial preservation. With this method, the quantity of preservative impregnated into the wood can be easily measured and controlled. In addition, large volumes of material can be treated relatively rapidly and economically. Because of these advantages, pressure treatment has established itself as the most widely used method around the world for long-term preservation of wood. Other methods are available for obtaining deep penetration. Some of these can be used for the preservation of wood which cannot be satisfactorily treated by pressure impregnation, such as double-diffusion treatment of unseasoned poles. Others require simpler and less expensive equipment so they can be used where pressure treatment would not be economically feasible.

A number of factors can affect the results obtained with a particular application method, including (1) species and nature of wood being treated, i.e., the sapwood of most (but not all) wood species is relatively permeable to both waterborne and oilborne preservatives, whereas the heartwood of many species is impermeable; (2) moisture content of wood being treated, i.e., diffusion techniques are unsatisfactory on dry wood, whereas pressure impregnation is most effective on dry wood; and (3) nature of the solvent system, i.e., preservatives impregnated with solvents which penetrate into the cell wall behave differently when deposited by other solvents in the lumen [88], and solvents also affect the permanence and distribution of preservatives such as pentachlorophenol. For these reasons, an application method should always be designed to impregnate a given amount of preservative into a given zone of wood. Such "results-type" specifications are being adopted widely by consumers of treated wood products. They are technically superior to "process-type" specifications which do not concern themselves with the quality of the treated wood product.

A. Unseasoned Wood

1. Diffusion Treatment

Diffusion treatment is one of the simplest methods available for obtaining deep preservative penetration into round or sawn timber. An important prerequisite for successful use of any diffusion technique is that the wood must be unseasoned and its moisture content should be as high as possible since the rate of diffusion decreases rapidly as the fiber saturation point is approached.

Diffusion methods currently in use rely upon single or multicomponent fungicides applied in either one or two stages. End-diffusion treatment of freshly cut undebarked posts with zinc chloride or chromated zinc chloride can considerably increase the service life of posts. Sweetgum, red oak, pine, post oak, and elm posts 4-6 inches (10-15 cm) in diameter were still serviceable after 20 years in service in Mississippi when treated to an average retention of 1 lb of zinc chloride per cubic foot [89].

Borate diffusion treatment of freshly cut lumber is used in Australia, New Zealand, and Canada to protect wood from fungal and insect attack during shipment, storage, and use above ground. Lumber is generally dipped for a few minutes in a hot concentrated solution of a highly soluble borate, corresponding approximately to disodium octaborate tetrahydrate. The lumber is then bulk piled and covered to reduce the rate of drying. The borate diffuses slowly throughout the wood, providing the moisture content remains above approximately 40% [90]. Diffusion time can be reduced from 60 to 80 days for a momentary dip to 7 to 9 days by using a hot and cold bath treatment with borate and storing the wood at 49°C [91]. A multisalt preservative for the diffusion treatment of building timbers is widely used in Papua, New Guinea. The borofluoride chrome-arsenic preservative was developed by the Commonwealth Scientific and Industrial Research Organisation,

Australia, to replace borate treatment which was not sufficiently effective against termites [92]. Timber fresh from the saw is given a momentary dip in the preservative solution. The dipped timber is immediately removed to diffusion storage where it is enclosed in an air tight cover for at least 21 days. Experience has indicated that a single-treatment schedule can be satisfactorily used for different species and sizes of lumber. A detailed description of the development and use of the process has been given by Levy et al. [93], who feel the method has great potential in developing countries where both economic considerations and available timber species rule out the use of pressure treatment.

A major disadvantage of the diffusion techniques described is that the fungicidal chemicals not only diffuse into the wood, but also diffuse out when the product is placed in the ground or other moist environments. Double diffusion has been developed to overcome this problem [94]. Unseasoned poles or posts are first soaked in copper sulfate solution for several hours, rinsed with water, and then transferred to a sodium arsenate solution (other chemicals which have been used for the second stage include borate, chromate, fluoride, and phosphate). The two salts react in the wood to produce a relatively insoluble compound, toxic to fungi and insects. Work on the treatment of several Alaskan species by double-diffusion indicates that the process has an important role in the preservation of wood where economical considerations or available timber species make pressure treatment unsuitable [95].

2. Sap Displacement

The sap-displacement technique was first patented in 1838 by Boucherie. A container of preservative solution, such as copper sulfate, was attached to a standing or freshly cut tree with bark, branches, and leaves still attached. Preservative solution was drawn into the tree to replace sap lost by evaporation. A slightly modified form of this method is still used for the treatment of freshly cut, undebarked poles. The poles are placed on skids with the butt ends slightly elevated. Water-tight caps are attached to the butt ends and preservative solution run in from containers elevated 25-30 feet (7 1/2-9 m) above ground. Treatment is limited to the sapwood but takes only a few days. The time required for satisfactory treatment of poles has been considerably reduced by the application of vacuum [96], vacuum and low pressure [97], or high pressure to the poles [98]. A recent survey shows that commercial operation of sap displacement treatments is limited to Switzerland, Argentina, France, and Denmark [99]. However, other countries are interested in the application of sap displacement as a relatively cheap and simple method for treatment of posts by farmers.

The major roles of sap-displacement techniques appear to be similar to those of diffusion techniques; namely in the protection of species impermeable to pressure treatment, or in areas where pressure treatment is not economically feasible.

3. Superficial Treatment

Freshly cut logs and lumber can be protected from decay and discoloration during drying, transport, and storage by superficial treatment with fungicides. If such treatments are to be effective they must be applied as soon after cutting as possible, certainly within 24 hr. If this is not done, the surface of the product may remain clean but sap stain fungi will develop in the interior. The methods of applying preservative include brushing, spraying, and dipping. Spraying can be done with portable spray equipment or by passing material to be treated on a moving belt through a continuous spray device. Dipping can also be achieved by passing logs or lumber on a moving belt through a trough of preservative solution. Alternatively, batches of wood can be immersed in a tank of preservative momentarily and then lifted out and allowed to drain. In addition to treatment as soon after cutting as possible, it is also important to cover all exposed surfaces of the wood with preservative solution and maintain the concentration of the preservative solution high enough to prevent fungal growth.

Sodium pentachlorophenate, sometimes in combination with borate and organomercurial compounds, is the fungicide most widely used for superficial treatment of wood. Considerable efforts have been made to discover alternatives for these compounds, but none have achieved wide acceptance. The most promising replacement compound at the moment is captafol (Difolatan) [62], but this requires further testing before it can be recommended for general use. Insecticides such as lindane (γ-benzenehexachloride), are frequently incorporated with fungicides for the treatment of unseasoned wood in order to prevent attack by ambrosia, and bark and longhorn beetles.

B. Seasoned Wood

1. Pressure Treatment

When it is necessary to treat large quantities of timber for use in areas where the product will be exposed to a high risk of fungal or insect attack, pressure treatment is generally the most economical method for preserving the wood. It provides a method for controlling the retention and penetration of preservative. Pressure treatment is the most rapid method for obtaining deep penetration into permeable wood species and it generally gives more uniform penetration patterns than other treatment methods. Finally, higher retention of preservative can be obtained than with other treatments. The major disadvantages of pressure treatment are the relatively high cost of the equipment required and its immobility, which often necessitates high transportation costs for both material to be treated and for treated material. These disadvantages have been partly offset by the development of portable pressure-treatment plants. However, these have not been widely accepted.

Several different types of pressure process have been developed. They all use basically similar equipment. The heart of a pressure treatment system is the retort or treatment cylinder. These range in size from 6 to 9 ft (2-3 m) in diameter and up to 180 ft (60 m) in length. They are usually built to withstand pressures up to 250 psi, although pressures up to 1,000 psi are being used for the treatment of some refractory species. Other essentials include storage tanks, a system for measuring preservative usage, pressure and vacuum pumps, and a method for moving wood in and out of the treatment cylinder. A more detailed discussion of treating plants and equipment is given in Refs. 100 and 101.

Pressure processes can be divided into two main groups, that is, full-cell and empty-cell treatments. The full-cell or Bethell process is designed to impregnate wood with as much liquid, to as great a depth, as possible. It is used frequently for treatment with aqueous preservative solutions and also with creosote and similar preservatives when a high retention is required, for example, in marine piling. Empty-cell processes are designed to give deep penetration with minimal solvent retention so that the cells are coated with preservative, rather than filled. These processes are used routinely with organic-solvent preservatives, creosote, and aqueous preservatives used in the treatment of highly permeable timber species.

a. <u>Full-Cell Process</u> (Bethell). In a typical full-cell treatment cycle for dry wood, the charge is loaded into the retort and a vacuum of at least 22 inches (55 cm) of mercury is drawn to remove air from the wood. Preservative is then introduced into the retort while the vacuum is maintained. When the treatment cylinder is completely filled with preservative solution, pressure is applied to force the solution deep into the wood. Pressure is applied until absorption of the solution virtually ceases and is then released slowly. Finally, the preservative is returned to storage. Pressures of approximately 150 psi are normally used. With southern pine, it is common for a final vacuum to be applied to reduce surface "wetness" of the treated material. Total cycle time usually ranges from 4 to 8 hr.

Frequently unseasoned poles, piling, and railway ties are treated by the full-cell process. Before the unseasoned wood is treated, it must be dried. This can be achieved by steam conditioning, followed by a 1-3 hr vacuum period; by vapor drying, where the vapors of high-boiling organic solvents are condensed on the surfaces of unseasoned wood; or by Boultonizing, where wood is heated in creosote or an oil-solvent preservative while under vacuum. These preconditioning techniques are described more fully by Hunt and Garratt [100]. Once excess moisture has been removed from the wood, the treatment cycle is the same as that for dry wood.

b. <u>Empty-Cell Processes</u> (Lowry and Rueping). There are two major types of empty-cell process. In the Lowry process, preservative solution is introduced into the retort at atmospheric pressure. A pressure of approximately 150 psi is applied when the cylinder is completely filled with preservative. When the desired absorption is reached, pressure is released, the preservative is returned to storage, and a final vacuum of at least 22 inches (55 cm) is drawn. The Rueping process is very similar, except that pressure is applied to the wood before preservation solution is introduced into the treatment cylinder. The cylinder is filled with preservative while the original air pressure is maintained constant. Additional pressure is then applied and the remainder of the cycle is similar to the Lowry process.

The major advantage of the empty-cell processes over the full-cell process is that similar penetrations can be obtained with much smaller absorptions of preservative solution. This is particularly critical with organic-solvent preservatives. Air trapped in the wood when preservative solution is run into the retort is used to expel preservative from the wood when the final vacuum is drawn. In the Rueping process, up to 50% of preservative absorbed may be recovered with the final vacuum. In the Lowry process, recovery is generally closer to 20%.

c. Recoverable-Solvent Processes (Cellon and Dow). Empty-cell processes were developed to obtain deep penetration with a limited absorption. In recent years, an alternate method for achieving this has been developed. This relies on the use of volatile solvents which can be recovered from the wood at the end of the treatment cycles. Liquid petroleum gas (butane) and methylene chloride are the two solvents now in commercial use in the Cellon and Dow Processes respectively.

In the Cellon process [102], the wood is completely covered with the preservative solution, i.e., pentachlorophenol in butane. Pressure can be generated by heating the treatment solution or by applying pressure. Penetration is more rapid and deeper than with conventional preservatives because of the low viscosity of the solution. When the required absorption has been obtained, preservative solution is returned to the storage tank and a vacuum is drawn on the wood. Solvent evaporates from the wood, is condensed, and then returned to storage. Either empty- or full-cell cycles can be used.

The Dow process [103] uses an empty-cell cycle at ambient temperature with methylene chloride as solvent system and pentachlorophenol as preservative. Solvent is recovered from the wood by steam distillation immediately after the pressure cycle is over and preservative has been returned to the storage tank. Water is introduced into the bottom of the retort and boiled over steam coils. A mixture of water vapor and methylene chloride vapor from the wood is passed through a heat exchanger. The condensed vapors separate and the methylene chloride layer is returned to storage.

Wood treated by either the Cellon or Dow processes comes out of the treatment cylinder virtually free of solvent. Consequently, the treated products, which show no discoloration, can be easily glued, stained, or painted.

Several other pressure processes have been developed, some of which have been adopted commercially. Description of the more important of these can be found in Refs. 100 and 101.

2. Vacuum Treatment

Single- and double-vacuum treatment techniques are used quite widely for the preservation of millwork, where the depth of penetration and absorption obtained with conventional pressure treatments would be unnecessarily high.

Vacuum treatment is essentially a very mild pressure treatment in which atmospheric pressure is used to push preservative into the wood. In the simplest form of vacuum treatment, dry, fully machined lumber is loaded into a treatment cylinder and a vacuum of up to 25 inches (63 cm) of mercury is drawn. Preservative solution is run into the cylinder, the vacuum released to atmosphere and the wood removed after a predetermined time. A final vacuum can be incorporated into the cycle to allow for removal of excess preservative solution and greater control over preservative retention and distribution [104]. The double-vacuum process is used widely in some parts of Europe. The time required for the complete double-vacuum treatment is often less than 1 hr.

3. Hot and Cold Bath or Thermal Treatment

Hot and cold bath treatment is the most effective nonpressure process available for treatment of dry wood when long service is required. It has been used quite widely for the treatment of poles and posts where pressure treatment is not feasible. The process has been used for the butt and full-length treatment of wood species with narrow sap bands such as cedar, where pressure treatment does not produce deep penetration. In addition, the process has been used for do-it-yourself treatment of small numbers of posts.

Dry timber is immersed first in hot preservative solution and then in relatively cool preservative solution. The hot preservative causes the air in the wood to expand. Immersion in cool preservative then causes the air to contract creating a partial vacuum which draws preservative into the wood. Creosote, pentachlorophenol, and single-component waterborne salts have all been used in this process, although creosote and pentachlorophenol are generally more satisfactory. Timber is normally held in the hot bath for 6 or more hours and in the cold bath for at least 2 hr. However, this is dependent on a number of variables including wood species, type of product, and moisture content. Transfer to the cold bath can be achieved by allowing the hot bath to cool, by replacing the hot preservative with cold preservative, or by placing the heated timber in a second bath containing cold preservative. In the latter methods, transfer must be rapid if maximum penetration is to be obtained. Sometimes a second hot bath immersion is used after the cold bath to reduce the likelihood of preservative bleeding in creosote treated products. Working temperatures for creosote are generally in the range 99 to 104°C for the hot bath and 38 to 65°C for the cold bath. A variety of adaptations of the hot and cold bath process are discussed by Hunt and Garratt [100].

4. Cold-Soak Treatment

Cold-soak treatment is probably the simplest and least expensive method available for treatment of small numbers of posts where long service life is necessary. The method utilizes very simple equipment. However, if it is to be successful, posts must be air-dry and free of bark before treatment. This method is popular with farmers.

Pentachlorophenol (5%) or copper naphthenate (1% copper metal) in fuel oil are commonly used preservatives. Cold soaking in creosote is less satisfactory, unless the preservative is warmed to make it less viscous. Once the wood to be treated has been debarked and dried, it is totally immersed in preservative solution until the desired amount of preservative has been absorbed, or until no more preservative is absorbed, whichever occurs first. Immersion periods normally range from 2 days to 1 week or more. Pine is the most satisfactory species for treatment by this method. As a class, hardwoods do not respond well. In some species, for example sweetgum and aspen, end-grain penetration and total absorption are good; but radial penetration is poor. Consequently, service life is poor.

5. Superficial Treatment

The performance of treated wood products is dependent on the amount of preservative present in the wood and also on its penetration. Thus where wood is exposed to moderate to high-hazard biodeterioration conditions it is essential to use treatment methods which provide deep preservative penetration. However, where wood is exposed to moderate to low-hazard conditions, for example siding, exterior trim, and other items exposed to irregular wetting, superficial treatments can provide good protection at minimal cost.

Superficial treatments include brush, spray, dip, and short-soak methods. Several papers have appeared which discuss the effects of wood properties, treatment variables, and preservative formulations on penetration and decay protection obtained with different superficial treatments [43, 105, 106].

For satisfactory treatment, wood must be air-dry and should also be machined to final dimensions. Pentachlorophenol (5%) in light oil, with or without the addition of water repellents and color additives to assist recognition, is the most widely used preservative for superficial treatment. Others include creosote, copper naphthenate, zinc naphthenate, and tributyltin oxide, usually combined with an insecticide.

Dip and short-soak treatments generally provide better coverage and end-grain penetration than brush and spray methods. If brush or spray treatment is used, at least two liberal coats should be applied; the second one after the first has dried or soaked into the wood. Brush or spray treatment should be used whenever untreated or poorly treated surfaces are exposed when pressure-treated wood is cut.

Dipping in preservative requires immersion from a few seconds to several minutes. A dip of a few seconds is normally recommended where short-term protection is required, as during storage and transport of framing lumber to a building site. Where longer protection is required, for example in millwork, a 3-min dip is recommended by the National Woodwork Manufacturers' Association as a minimum. However, even with a 3-min dip lateral penetration in the wood is slight and the treated wood will not perform well in areas of moderate to severe decay hazard.

Short soaks have found little commercial application because of the risk of excessive absorption in some softwoods. This causes problems in handling and finishing the treated products. In addition, the improvement in service life obtained by lengthening the soak period does not compensate for the increased cost and time involved. Generally, a vacuum or vacuum/pressure treatment is preferable where a dip treatment will not give sufficient protection.

C. Reconstituted Wood Products

Plywood, particle board, flake board, and hardboard are all being used in areas where there is a high risk of fungal and insect attack. The resistance of these reconstituted wood products to biodeterioration is often different from that of the wood from which they are made. New methods are being developed for their preservative treatment. Although, in some instances they can be treated with preservatives in the same way as solid wood products.

Most attention has probably been given to the preservation of plywood. This can be treated in the veneer form or after assembly by conventional pressure techniques. A considerable amount of work has been carried out on the development of glue-line additives for prevention of fungal deterioration. Effective glue-line additives have not yet been developed [107].

When plywood is treated after manufacture, glue lines sometimes retard preservative penetration from the faces. This depends on the wood species, thickness of the glue lines, and type of glue. However, lathe cracks in veneers and the presence of gaps in core veneers generally lead to acceptable penetration. A simple method of protecting plywood which requires little specialized equipment consists of applying a concentrated solution of chromated copper arsenate to the surfaces of green veneers as they are peeled from logs. Providing only thin veneers are treated in this manner, satisfactory penetration of preservative is obtained before the preservative becomes fixed in the wood.

Literature on the biodeterioration of particle board has recently been reviewed [108]. Although particle boards are generally less susceptible to decay than solid wood, they are attacked by fungi in moist situations. Preservative chemicals are effective when applied to either wood chips or the adhesvie [109]. However, it is more usual to incorporate them in the adhesive [110, 111]. Pentachlorophenol or its sodium salt are the most widely used preservatives for particle board. Where the treated product is also exposed to termite attack, insecticides such as inorganic arsenicals or lindane are added. There is still a need for the development of more effective preservative compounds for particle board [112].

Little information is available on the preservative treatment of flake board, hardboard and fiber board, although the need for protection in moist situations has been demonstrated [113]. It would appear that technology for the preservation of these products will be similar to that for particle board.

D. Wood In Use

Preservative treatment of wood before it is placed in service is the most economical method for preventing biodeterioration. However, remedial or in-use treatment is sometimes necessary and economically justifiable.

Groundline inspection and treatment of poles by utility companies is now a regular part of line maintenance. Several different methods are used commercially. They can be divided into two basic groups, those involving surface application of chemicals and those involving injection of chemicals into the poles. A comprehensive review of methods for treating standing poles is given by Hunt and Garratt [100].

Surface treatment involves removal of earth from around the pole to a depth of about 2 ft (60 cm), followed by application of the preservative, usually in the form of a thick gel or grease, from about 4 inches (10 cm) above ground level to 18 inches (45 cm) below. The treated area is then wrapped with a waterproof paper or polyethylene bandage to prevent excessive loss of preservative into the soil. Generally, pentachlorophenol, creosote, and sodium fluoride, either alone or in combination, are the major components of commercial formulations. Surface treatment will not control decay present in the interior of poles. However, its effectiveness in preventing surface decay is demonstrated by results obtained in U.S. Forest Service tests in Mississippi. Untreated southern pine pole stubs were groundline treated with either a paste reported to contain 10% pentachlorophenol or with a formulation containing sodium fluoride, dinitrophenol, potassium bichromate, and coal tar fortified with pentachlorophenol and then wrapped with a polyethylene-kraft-paper bandage. After 13 years service all pole stubs were still serviceable [114]. Untreated controls had an average life of 3.3 years.

When control of internal decay is critical, for example, in a species such as Douglas fir which has a narrow sapband and susceptible to decay, heartwood injection of chemicals into the poles is necessary. A most promising technique for controlling internal decay, which uses fumigant chemicals, has recently been described [46]. Holes are drilled into poles within 2 ft (60 cm) above groundline and Vapam, Vorlex, or chloropicrin is introduced into the holes. Bioassay indicates that the chemicals move both laterally and longitudinally in the sapwood and heartwood of the poles sterilizing the wood. It appears that retreatment will not be necessary for at least 5 years. Wrapping of the poles after treatment does not improve performance. If preliminary results are confirmed, fumigation will provide a very simple and relatively cheap method for protecting standing poles. It also has great potential for use in the control of internal decay in other wood structures, for example, laminated beams, columns, and above-water sections of marine piles.

Control of wood decay in buildings is usually achieved by elimination of the source of moisture. Where moisture cannot be controlled, it is usual to replace decayed wood with pressure-treated wood. Where this is not possible, surface application of conventional preservative solutions will usually have little effect on the rate of decay. This is because they do not penetrate deeply into the wood and cannot be applied adequately in joints and other inaccessible areas. However, deep penetration can be obtained with mayonnaise-like emulsion preservatives containing pentachlorophenol (e.g., Woodtreat TC) [115, 116]. The preservative is applied either as a continuous layer on the surface of the wood or in beads. As the emulsion breaks, pentachlorophenol is carried into the wood in the oil solvent and water evaporates from the surface. Penetration of 1-1/2 to 2 inches (3-5 cm) can be obtained in sapwood. Best results are obtained when the preservative is applied to dry unfinished surfaces. Mayonnaise-type preservatives have been used successfully for control of decay in subfloor areas of buildings, attic timbers, and exposed laminated beams.

Finally, brush or spray application of oilborne preservatives, preferably containing a water repellent, to wood exposed to irregular wetting, such as in column bases, steps, rails, and decks, will reduce the rate of decay but not control it. Such treatments do not reduce the rate of internal decay.

REFERENCES

1. T. G. Gill and R. B. Phelps, Amer. Wood Preservers' Assoc. Proc., 70, 331 (1974).
2. R. D. Graham, in Wood Deterioration and Its Prevention by Preservative Treatments (D. D. Nicholas, ed.), Vol. 1, Syracuse University Press, 1973, Syracuse, N.Y., pp. 1-30.
3. H. B. Van Groenou, H. W. L. Rischen, and J. Van Den Berge, Wood Preservation During the Last 50 Years, A. W. Sijthoff's Uitgevers-maatschappij N.V., Leiden, Netherlands, 1951.

4. T. C. Scheffer, in Wood Deterioration and Its Prevention by Preservative Treatments (D. D. Nicholas, ed.), Vol. 1, Syracuse University Press, 1973, Syracuse, N.Y., pp. 31-106.
5. T. K. Kirk, in Wood Deterioration and Its Prevention by Preservative Treatments (D. D. Nicholas, ed.), Vol. 1, Syracuse University Press, 1973, Syracuse, N.Y., pp. 149-182.
6. W. W. Wilcox, in Wood Deterioration and Its Prevention by Preservative Treatments (D. D. Nicholas, ed.), Vol. 1, Syracuse University Press, 1973, pp. 107-148.
7. R. M. Lindgren, Amer. Wood Preservers' Assoc. Proc., 48, 158 (1952).
8. T. K. Kirk and W. E. Moore, Wood and Fiber, 4, 72 (1972).
9. J. G. Savory, Amer. Appl. Biol., 41, 336 (1954).
10. M. P. Levi and R. D. Preston, Holzforschung, 19, 183 (1965).
11. T. C. Scheffer and A. F. Verrall, U.S. Dept. Agr. Forest Serv. Res. Paper FPL 190, 1973.
12. J. R. Hansbrough, Forest Prod. J., 3, 33 (1953).
13. R. R. Chase and J. C. McKee, U.S. Pat. 3,431,061 (1969), to Union Bag-Camp Paper Corporation.
14. T. L. Highley, Phytopathology, 63, 57 (1973).
15. T. C. Scheffer and E. B. Cowling, Amer. Rev. Phytopathol., 4, 147 (1966).
16. I. S. Goldstein, E. B. Jeroski, A. E. Lund, J. F. Wilson, and J. W. Weaver, Forest Prod. J., 11, 363 (1961).
17. R. M. Rowell and D. I. Gutzmer, Wood Sci., 7 (1975).
18. R. H. Baechler, Forest Prod. J., 9, 166 (1959).
19. J. K. Shields, Bi-m Res. Notes, Dept. Forestry Can., 24, 9 (1968).
20. J. Ricard, International Research Group on Wood Preservation, Document No. IRG/WP/135, Princes Risborough Laboratory, Princes Risborough, Aylesbury, Bucks., England, 1975.
21. G. A. Thomas, in Biodeterioration of Materials, Microbiological and Allied Aspects (A. H. Walters and J. J. Elphick, eds.), American Elsevier, New York, 1968, pp. 506-516.
22. J. M. Wincester and D. Yeo, Chem. Ind., 4, 106 (1968).
23. National Forest Products Association, National Design Specifications for Stress-Grade Lumber and Its Fastenings, Washington, D.C., 1968.
24. W. S. Thompson, in Wood Deterioration and Its Prevention by Preservative Treatments (D. D. Nicholas, ed.), Vol. 2, Syracuse University Press, 1973, Syracuse, N.Y., pp. 345-395.
25. M. P. Levi, D. Huisingh, and W. B. Nesbitt, Forest Prod. J., 24, 97 (1974).
26. D. A. Belenkov, Ivuz, Lesnoi Zh., 2, 83 (1968). (Library translation No. 1816, Building Research Establishment, Watford, England.)
27. W. H. Hartford, Amer. Wood Preservers' Assoc. Proc., 70, 273 (1974).
28. B. A. Richardson, Intern. Pest Control, 10, 14 (1968).

29. M. P. Levi, J. Inst. Wood Sci., 23, 45 (1969).
30. M. P. Levi, Proc. Brit. Wood Preserving Assoc., 19, 113 (1969).
31. Brit. Standard 838, British Standards Institute, London (1961).
32. Ger. Standard DIN 52176, Bundesanstalt fur Materialsprufing, Berlin (1972).
33. Amer. Wood Preservers' Association Standard M10, Amer. Wood Preservers Association, Washington (1974).
34. E. A. Behr, Forest Prod. J., 22, 26 (1972).
35. R. S. Smith, Wood Sci., 2, 44 (1969).
36. R. S. Smith, Forest Prod. J., 25, 48 (1975).
37. C. G. Duncan, U.S. Forest Serv. Forest Prod. Lab. Rept. No. 48, 1965.
38. P. Kaune, Mater. Organismen, 2, 229 (1967).
39. J. G. Savory and A. F. Bravery, Mater. Organismen, 5, 59 (1970).
40. A. J. Cserjesi and J. W. Roff, Mater. Res. Std., 10, 18 (1970).
41. Anon., Holzforsch. Holzverwert., 24, 15 (1972).
42. Amer. Wood Preservers' Association Standard M7, Amer. Wood Preservers Association, Washington (1973).
43. A. F. Verrall, U.S. Dept. Agr. Forest Serv. Tech. Bull. No. 1334, 1965.
44. R. H. Colley, Bell Telephone System Tech. Pub. Monograph No. 2118, AT&T Co., New York, 1953.
45. E. A. Behr, in Wood Deterioration and Its Prevention by Preservative Treatments (D. D. Nicholas, ed.), Vol. 1, Syracuse University Press, 1973, Syracuse, N.Y., pp. 217–246.
46. R. D. Graham, Forest Prod. J., 23, 35 (1973).
47. W. H. Hartford, in Wood Deterioration and Its Prevention by Preservative Treatments (D. D. Nicholas, ed.), Vol. 2, Syracuse University Press, Syracuse, N.Y., pp. 1–120.
48. R. D. Arsenault, in Wood Deterioration and Its Prevention by Preservative Treatments (D. D. Nicholas, ed.), Vol. 2, Syracuse University Press, 1973, Syracuse, N.Y., pp. 121–278.
49. Anon., J. Inst. Wood Sci., 6, 17 (1972).
50. H. Alliot, Brit. Wood Preserving Assoc. News Sheet No. 127, 1973.
51. T. Hof, J. Inst. Wood Sci., 23, 19 (1969).
52. R. Cockcroft and J. F. Levy, J. Inst. Wood Sci., 33, 28 (1973).
53. R. Bunn, Tech. Paper, Forest Res. Inst. New Zealand Forest Serv. No. 60, 1974.
54. G. Becker, J. Inst. Wood Sci., 6, 51 (1973).
55. E. M. Wallace, Amer. Wood Preservers' Assoc. Proc., 60, 50 (1968).
56. B. Schulze and G. Becker, Holzforschung, 2, 97 (1948).
57. G. B. Fahlstrom, Amer. Wood Preservers' Assoc. Proc., 54, 178 (1958).
58. S. E. Dahlgren and W. H. Hartford, Holzforschung, 26, 142 (1972).
59. C. K. Chou, J. A. Chandler, and R. D. Preston, Wood Sci. Technol., 7, 151 (1973).
60. M. R. Clarke and J. R. Rak, Forestry Chron., 50, 114 (1974).

61. L. R. Gjovik and H. L. Davidson, U.S. Dept. Agr. Forest Serv. Res. Note FPL 102, 1973.
62. J. A. Butcher and J. Drysdale, Forest Prod. J., 24, 28 (1974).
63. J. K. Shields, R. L. Desai, and M. R. Clarke, Forest Prod. J., 24, 54 (1974).
64. S. E. A. McCallan and F. Wilcoxon, Contrib. Boyce Thompson Instr., 8, 151 (1936).
65. R. L. Wain and E. H. Wilkinson, Ann. Appl. Biol., 30, 379 (1943).
66. C. K. Chou, R. D. Preston, and M. P. Levi, Phytopathology, 64, 335 (1974).
67. E. Overton, Vierteljahresschr. Naturforsch. Ges. (Zuerich), 44, 88 (1899).
68. L. P. Miller, in Fungicides (D. C. Torgeson, ed.), Vol. 2, Academic Press, 1969, New York, pp. 1-59.
69. H. Lyr and H. Ziegler, Phytopathol. Z., 35, 146 (1959).
70. W. W. Wilcox, N. Parameswaran, and W. Liese, Holzforschung, 28, 211 (1974).
71. W. N. Aldridge, Biochem. J., 69, 367 (1958).
72. A. F. Bravery, 2nd Intern. Congr. Plant Pathol., Minneapolis, 1973, Abst. No. 0791, 1973.
73. B. A. Richardson, in Biodeterioration of Materials, Microbiological and Allied Aspects (A. H. Walters and J. J. Elphick, eds.), American Elsevier, New York, 1968, pp. 498-505.
74. M. P. Levi, unpublished work, 1974.
75. P. B. Marsh, G. A. Greathouse, K. Bollenbacher, and M. L. Butler, Ind. Eng. Chem., 36, 176 (1944).
76. R. J. Lukens, in Fungicides (D. C. Torgeson, ed.), Vol. 2, Academic Press, 1969, New York, pp. 395-445.
77. E. W. B. DaCosta and R. M. Kerruish, Forest Prod. J., 14, 106
78. M. P. Levi, in Wood Deterioration and Its Prevention by Preservative Treatments (D. D. Nicholas, ed.), Vol. 1, Syracuse University Press, 1973, Syracuse, N.Y., pp. 183-216.
79. J. Ashida, Phytopathol. Rev., 3, 153 (1965).
80. H. Lyr, Nature (London), 195, 289 (1962).
81. J. G. Horsfall, Principles of Fungicidal Action, Chronica Botanica, Waltham, Mass., 1956.
82. A. Rabanus, Mitt. Deutsch. Forstver., 23, 77 (1939).
83. H. Greaves, Holzforschung, 28, 193 (1974).
84. H. Greaves, Mater. Organismen, 8, 85 (1973).
85. C. Madhosingh, Can. J. Microbiol., 7, 553 (1961).
86. C. Madhosingh, Forest Prod. J., 11, 20 (1961).
87. H. H. Unligil, Forest Prod. J., 18, 45 (1968).
88. A. F. Bravery, Intern. Biodeterioration Bull., 6, 145 (1970).
89. E. R. Toole and W. S. Thompson, Treating fence posts by the end-diffusion method, Mississippi Forest Products Utilization Laboratory Information Series No. 18, 1974.

90. J. W. Roff, West. Forest Prod. Lab. Inform. Rept. VP-X-125, 1974.

91. A. J. McQuire and K. A. Goudie, New Zealand J. Forestry Sci., 2, 165 (1972).

92. CSIRO, Australian Pat. 246,298 (1963).

93. C. R. Levy, S. J. Colwell, and K. A. Garbutt, International Research Group on Wood Preservation, Document No. IRG/WP/310, Princes Risborough Laboratory, Princes Risborough, Aylesbury, Bucks., England, 1972.

94. R. H. Baechler and H. G. Roth, Forest Prod. J., 14, 171 (1964).

95. L. R. Gjovik, H. G. Roth, and H. L. Davidson, U.S. Dept. Agr. Forest Serv. Res. Paper FPL 182, 1972.

96. H. Gewecke, Holz Roh - Werkstoff, 15, 119 (1957).

97. C. G. W. Mason, New Zealand Pat. 141,860 (1965).

98. M. S. Hudson, U.S. Pat. 3,443,881 (1966).

99. F. B. Shorland and C. G. W. Mason, International Research Group on Wood Preservation, Document No. IRG/WP/329, Princes Risborough Laboratory, Princes Risborough, Aylesbury, Bucks, England (1974).

100. G. M. Hunt and G. A. Garratt, Wood Preservation, 3rd ed., McGraw-Hill, New York, 1967.

101. W. T. Henry, in Wood Deterioration and Its Prevention by Preservative Treatments (D. D. Nicholas, ed.), Vol. 2, Syracuse University Press, 1973, Syracuse, N.Y., pp. 279-298.

102. R. H. Bescher, U.S. Pat. 3,200,003 (1965).

103. S. R. Marouchoc, Amer. Wood Preservers' Assoc. Proc., 68, 148 (1972).

104. R. W. Watson, Proc. Brit. Wood Preserving Assoc., 20, 59 (1970).

105. T. C. Scheffer, A. F. Verrall, and G. Harvey, Mater. Organismen, 6, 27 (1971).

106. J. W. W. Morgan and D. F. Purslow, Holzforschung, 27, 153 (1973).

107. E. W. B. DaCosta, K. Hirst, and L. D. Osborne, Holzforschung, 26, 131 (1972).

108. E. R. Toole and H. M. Barnes, Forest Prod. J., 24, 55 (1974).

109. M. Petrovic, Glasnik Sumarskog Fakulteta, Univerzitet Beogradu, No. 41, 1973; Forestry Abstr., 35, 5667 (1974).

110. H. A. Huber, Forest Prod. J., 8, 357 (1958).

111. H. J. Deppe and M. Gersonde, Mater. Organismen, 4, 123 (1969).

112. G. Becker, Wood Sci. Technol., 6, 239 (1972).

113. E. A. Behr, Forest Prod. J., 22, 48 (1972).

114. L. R. Gjovik, Proc. Fifth Wood Pole Institute, Colorado State University, 1971, p. 20.

115. D. Boocock, Wood, 31, 41 (1966).

116. D. Boocock, Wood, 31, 55 (1966).

117. R. D. Arsenault, Amer. Wood Preservers' Assoc. Proc., 71, 126 (1975).

118. R. D. Arsenault, Amer. Wood Preservers' Assoc. Proc., 77, 122 (1976).

Chapter 13

FUNGICIDES IN MEDICINE

Smith Shadomy,[*] H. Jean Shadomy,[*] and Gerald E. Wagner[†]

Virginia Commonwealth University
Medical College of Virginia
Richmond, Virginia

I. INTRODUCTION

Since the 1950s, there has been a dramatic increase in the number of cases of human disease attributable to pathogenic fungi. The diseases caused by these organisms range from superficial dermatological conditions, which

---

[*]Departments of Medicine and Microbiology.
[†]Department of Clinical Pathology.

result in cosmetic problems of little serious consequence, to severe, debilitating, often fatal diseases. Most of this increase in the number of
reported mycoses is due, in part, to an increased awareness on the part of
the clinician as well as to improved diagnostic practices in the laboratory
[1]. However, a major portion of the increase must be attributed to the
prolonged lifespan of patients in whom opportunistic fungal infections are
becoming increasingly frequent [2]. To this group belong patients with
naturally induced immunosuppression, as in the case of malignant diseases
and diabetes, as well as those individuals with artificially induced immunosuppression such as caused by transplantation chemotherapy.

The medical mycologist is concerned with three other disease groups in
addition to the opportunistic fungal infections. These include the deep or
systemic mycoses, mycotic infections of subcutaneous tissues, and the
superficial mycoses or dermatophytoses. In order to better comprehend
the problems involved in the treatment of fungal infections, a brief description of these disease groups and of the more important human mycoses
follows.

II. THE HUMAN MYCOSES

A. Systemic Mycoses

The systemic, or deep, mycoses are serious, sometimes fatal diseases
involving deep tissues and one or more internal organs or organ systems.
They usually begin as mild or subacute pulmonary or subcutaneous infections which are acquired by inhalation of or traumatic contact with infective
spores produced by vegetative mycelium growing saprophytically. In most
instances, the initial infections remain localized. However, such infections
can disseminate to various organs and organ systems. The factors which
may or may not lead to disseminated disease are poorly understood [3].
They may involve nothing more complex than the number of infective spores
to which an individual is exposed. They also may involve much more complex medical and physiological factors which combine to predispose certain
individuals to develop systemic or disseminated fungal disease. These
include normal physiological factors such as age, race, and sex. They also
include less normal factors such as preexisting disease including hemato-
logical disorders (leukemia, Hodgkin's disease, aplastic anemia, etc.),
cancer, diabetes mellitus, and tuberculosis [4]. Specific local physiological
changes such as those often associated with the use of broad-spectrum antibiotics, corticosteroids, and other therapeutic agents as well as local lesions of the mucosa which offer a portal of entry to the infecting fungus also
are involved. Several of the more important of the systemic mycoses will
be considered here. These include blastomycosis (North American blastomycosis), paracoccidioidomycosis (South American blastomycosis),

coccidioidomycosis (San Joaquin Valley fever), cryptococcosis (European blastomycosis), histoplasmosis, and systemic candidiasis.

1. Blastomycosis

Blastomycosis (North American blastomycosis or Gilchrist's disease) is a chronic, granulomatous disease caused by Blastomyces dermatitidis. B. dermatitidis is dimorphic, i.e., it exists in one form during its vegetative growth phase and in a second during its pathogenic or tissue growth phase. Infections are most commonly seen in males between the ages of 20 and 40 years. The infection is endemic in the north central and eastern United States. Infection is usually pulmonary at onset but also may be manifested by either localized, cutaneous, and subcutaneous lesions or as a systemic infection. Cutaneous lesions, particularly of the facial area are common, and the pleura, lungs, and bones are often involved. The disease may simulate progressive pulmonary tuberculosis.

Diagnosis of blastomycosis is made on the basis of demonstration of the pathogenic yeast-like phase in tissue or sputum by examination and by culture. B. dermatitidis grows well on most bacteriological culture media and identification is based upon recognition of specific macroscopic and microscopic morphology. Serological tests are of little value in the diagnosis of blastomycosis. Protective vaccines are not available.

2. Paracoccidioidomycosis

Paracoccidioidomycosis (South American blastomycosis) is found exclusively in Central and South America. Most cases have been found in Brazil where the disease is endemic and is most often seen in males who work in rural regions. The etiological agent of this disease is Paracoccidioides brasiliensis. The disease begins as a primary, pulmonary infection. Following dissemination, lesions generally begin on the buccal mucosa and are eventually seen on the mouth, face, and nose. Frequently there is lymph node involvement; lymphatic dissemination to the lungs and viscera also may occur. The early granulomatous lesions often resemble those of blastomycosis, but lymphatic involvement with draining sinuses differentiates the two diseases.

Diagnosis is based upon demonstration of multiple budding yeast-like cells in pathological material, and by culture of the specific organism which in the vegetative phase closely resembles B. dermatitidis.

3. Coccidioidomycosis

This disease, caused by the dimorphic soil saprophyte Coccidioides immitis, affects a large majority of the inhabitants of the arid southwestern United States. It is endemic in the San Joaquin valley of California where the nonprogressive primary form of disease is referred to locally as "San Joaquin Valley fever." Roughly 60% of the cases are asymptomatic and are discernible by skin testing only; chronic cavitary disease may develop in another 5% [5]. The disseminated disease, which occurs rarely in male Caucasians and more frequently in negroes, Filipinos and pregnant, female Caucasians, may affect any part of the body and is associated with a high mortality rate. The granulomatous, disseminated form of the disease shows a distinct predilection for dark-skinned races. The etiological agent is found in soils of endemic areas which are usually characterized climatically as being semiarid. It is spread by airborne dust aerosols which contain arthrospores produced following fragmentation of mature vegetative hyphae.

Diagnosis of the primary pulmonary disease is difficult as it may resemble tuberculosis, influenza, or other pneumonias. Specific mycological diagnosis may be made by demonstrating the characteristic pathogenic phase in tissue or pathological material. As such, C. immitis appears as rounded, thick-walled spherule containing endospores in various stages of development. When cultured in the laboratory, C. immitis produces a characteristic colony which, when mature, reveals arthrospores upon microscopic examination. However, isolation and growth of C. immitis both require time and pose a hazard of major magnitude for laboratory personnel. Serological tests may be used to establish the diagnosis. Reliable skin test and serological test antigens are available. No vaccine is available.

4. Cryptococcosis

Cryptococcosis (torulosis or European blastomycosis) is caused by the yeast-like organism Cryptococcus neoformans. Unlike the preceding fungi, C. neoformans appears as a yeast both in pathologic material and in its saprophytic growth phase. Cryptococcosis may perhaps best be described as the prototype of the opportunistic fungal pathogen, that is, while it does produce disease in normal individuals, it is most commonly seen as a complication to other diseases processes. These include Hodgkin's disease, lymphomas, and immunosuppressed and transplant recipients. C. neoformans is a common saprophytic fungus in nature; it is most commonly isolated from pigeon manure. Initial infection follows inhalation of aerosols containing the organisms. Dissemination of the disease from the primary pulmonary lesion results in infection of internal organs and tissues including the kidneys, bone marrow, and most frequently, the central nervous system. Initial symptoms are most commonly associated with central nervous system disease. Diagnosis is based upon demonstration of the organisms in cerebrospinal fluid or other specimens. When mixed with india ink and examined microscopically, C. neoformans is seen to be an encapsulated, yeast-like budding cell. If undiagnosed or untreated, disseminated cryptococcosis is usually fatal.

5. Histoplasmosis

Histoplasmosis, caused by the dimorphic fungus Histoplasma capsulatum is extremely variable in its clinical manifestations. H. capsulatum grows in

soil contaminated with bird or bat manure. H. capsulatum is endemic in
Kentucky, Tennessee, and other parts of the central United States. It produces spores which easily become airborne when the soil is disturbed and
infection follows inhalation of these spores. In the normal individual, the
initial pulmonary infection is usually self-limiting and asymptomatic. In
individuals exposed to large numbers of spores, clinical disease may
develop, usually mimicing influenza. Such individuals usually recover without serious sequelae. It is estimated that about 200,000 persons are infected each year; of these, two-thirds have asymptomatic infections [6].

Disseminated histoplasmosis is rare, occurring in less than 1% of those
exposed. Infants and individuals with defective cellular immunity are particularly prone to this form of the disease. There is multiple-organ involvement in disseminated histoplasmosis. Chronic pulmonary histoplasmosis
may develop with underlying pulmonary disease. Epidemics of acute primary
histoplasmosis have been described [7]; these frequently followed group
activities in endemic sites where there has been a significant amount of
disturbance of soil or dust.

Diagnosis of histoplasmosis may be difficult as the primary pulmonary
form of the disease cannot be distinguished clinically from tuberculosis or
coccidioidomycosis. Diagnosis thus requires demonstration of the etiologic
agent in clinical material. This may be done culturally or histologically.
In culture, the vegetative phase of H. capsulatum produces characteristic
spores by which identification is made. In tissue, the organism has a characteristic appearance which can be demonstrated by appropriate staining.
Both a reliable skin test and serological test are available. Vaccines are
not available.

6. Candidiasis

Candidiasis is an acute or subacute fungus infection which includes mild,
localized infections, serious infections of specific organs, as well as severe,
systemic disease. Most infections are endogenous in origin and a number
of different species of Candida are involved [8]. These include Candida
albicans, Candida tropicalis, Candida parapsilosis, Candida krusei, and
others. Most are endogenous to man; C. albicans in the intestinal tract
and vagina, C. parapsilosis on the skin. Infections can include oral, vulvovaginal, pericardial, meningeal, and cutaneous candidiasis. Systemic candidiasis most often ensues following alterations of the host's normal de-
fenses or metabolism. This may be the result of a natural disease process
such as leukemia or the result of intervention on the part of the physician
with antibiotics, steroids, or immunosuppressive drugs. Urinary-tract
infection is common in systemic disease; infection of the kidney is generally
secondary to another disease. In such patients there also is usually a fungal septicemia which precedes systemic disease. The mortality rate in
systemic candidiases is about 75%.

Diagnosis of systemic candidiasis is difficult as the clinician and tech-
nician are dealing with organisms which are regarded to be part of the normal

flora. Apart from isolation of C. albicans or C. parapsilosis from blood
or cerebrospinal fluid, the recovery of yeasts from any clinical specimen
is difficult to interpret. Frequently, the diagnosis of systemic disease is
made only post mortem.

B. Cutaneous and Superficial Infections

Cutaneous and superficial fungal infections include those which involve skin,
hair, and nails. One group, the dermatophytoses, include true fungal infections of keratinized tissues. These present distinct clinical entities,
depending on the site of infection and the causative organism. Principle
symptoms of the dermatophytoses include itching, scaling of skin, and loss
of hair. Superficial fungal infections represent infections at sites so remote
to the host that there generally is no host response. At most, patients seek
treatment only for cosmetic reasons. These latter infections will not be
further discussed here.

Dermatophytic infections are produced by three genera of fungi. These
include species of Microsporum, species of Trichophyton, and Epidermophyton. Certain species of Microsporum and Trichophyton are able to invade
skin and nails as well as attack hair shafts while Epidermophyton lacks this
capability and infects skin only.

Dermatophytoses, also called ringworms or "tineas," are most commonly observed in children and infection generally occurs by direct contact,
although fomites and animal vectors occasionally are involved.

The dermatophytoses may be grouped according to the natural habitat of
the infecting fungi. Geophilic organisms are those naturally found in the
soil; zoophilic organisms are generally found on animals and man, while the
anthropophilic organisms are found exclusively on man. Dermatophytic
infections of man also are classified according the area of the body involved:
tinea capitis affects the scalp and hair, tinea barbae affects facial hair,
tinea corporis affects smooth skin, tinea cruris affects the groin, tinea
pedis affects the feet, tinea manuum affeects the hands, and tinea unguium
affects the nails.

1. Tinea Capitis

Seen most frequently in prepubertal children, tinea capitis, or ringworm of
the scalp, generally occurs in one of two forms. These include a noninflammatory form usually of human origin, and an inflammatory form of
animal origin. The noninflammatory form, usually caused by Microsporum
audouinii, is characterized by scaling oval areas which enlarge by coales-
cence into large greyish areas. Infected hairs break just above the scalp
to produce characteristic "grey-patch" lesions. Although relatively resistant to treatment, this infection will clear spontaneously at puberty. It
formerly occurred in epidemics being transmitted indirectly by fomites
such as caps and theater seats.

Inflammatory tinea capitis is caused by <u>Microsporum canis</u> and is usually acquired by direct contact with an infected animal. It is similar to the grey-patch form in that scaling patches appear; however, and unlike grey-patch, there frequently is an inflammatory reaction which may become quite severe. In its most severe form, the inflammatory reaction leads to the development of deep ulcerative, boggy, and inflamed lesions known as kerions. This infection may result in permanent hair loss from the infected areas. <u>Microsporum gypseum,</u> acquired directly from soil, can produce a similar type of infection.

"Black-dot" ringworm is a third form of tinea capitis which is seen both in children and adults. Caused by one of several species of <u>Trichophyton,</u> "black-dot" ringworm is characterized by multiple scattered lesions within which infected hairs have broken off at the surface of the scalp. The infection is transmitted directly from man to man. In the Americas, it is most commonly caused by <u>T. tonsurans</u>; <u>T. violaceum</u> is the predominant etiological agent in Europe and Asia.

Tinea favosa or favus is another form of tinea capitis most commonly seen in Europe. Caused by <u>Trichophyton schoenleinii</u>, favus does not clear spontaneously. The infection begins as small, punctated lesions which enlarge into cup-shaped crusts or scutula; scarring may be prominent. Hair loss may be permanent.

2. Tinea Barbae

Tinea barbae is an infection of bearded regions of the neck and face. It is most commonly seen in adult males who acquired their infection from animals. The two fungi producing tinea barbae are <u>Trichophyton verrucosum,</u> most commonly acquired from cattle, and <u>Trichophyton mentagrophytes</u> which may be acquired from several different animals. Thus tinea barbae is an occupational infection in dairy workers. In its more severe form, tinea barbae is characterized by highly inflammatory pustular lesions which may become secondarily infected with bacteria. The inflammatory lesions, similar to kerions, frequently lead to extensive scarring and permanent hair loss.

3. Tinea Corporis

Ringworm infection of the smooth skin in both children and adults is called tinea corporis. It may be caused by any one of a variety of dermatophytic fungi, usually one of the species of <u>Trichophyton</u> or <u>Microsporum</u>. The infection, which is limited to the stratum corneum, begins as small macules that enlarge into annular patches or rings which clear in the center. Reinfection of the cleared central area may occur, resulting in a series of circinate rings. Depending on the specific organism involved, the lesions seen in tinea corporis may be either dry and scaly or vesicular. The more acute, vesicular form of infection caused by <u>T. mentagrophytes</u> and <u>T. verrucosum</u> usually clears spontaneously. Chronic, scaly lesions, most commonly caused by <u>Trichophyton rubrum</u> both are of longer duration and more resistant to treatment.

4. Tinea Cruris

While commonly associated with ringworm infections of the groin, tinea cruris also may involve axillae, inframammary folds, as well as other intertriginous areas. It is most frequently caused by <u>Epidermophyton floccosum</u> and <u>T. rubrum.</u> The infection is characterized by elevated plaques set apart from noninfected skin by a raised, vesicular erythematous border. Infections of the groin caused by <u>E. floccosum</u> are frequently nearly symmetrical on each side, except for isolated satellite lesions. <u>T. rubrum</u> tends to produce a more extensive infection involving the buttocks and thighs. <u>C. albicans</u> can produce a similar infection of the groin and a specific etiological diagnosis is required in tinea cruris to insure appropriate therapy.

5. Tinea Pedis and Tinea Manuum

Like tinea corporis, tinea pedis or ringworm of the foot and tinea manuum or ringworm of the hand can be caused by a variety of dermatophytic fungi. Symptomatology and severity of the infections vary, according to the specific etiological agent and site of infection. <u>T. mentagrophytes</u> and <u>E. floccosum</u> can produce an acute form of tinea pedis characterized by peeling, pruritic macerated lesions between the toes. Infections are more common in men and during the warmer part of the year. Denuding of the dermis may occur. <u>T. rubrum</u> produces chronic, hyperkeratotic infections of the hands and feet, but these are rarely associated with marked inflammation and vesicle formation. In contrast, <u>T. mentagrophytes</u> produces an acute, highly inflammatory vesicular infection of both hands and feet which may extend to other skin surfaces. Infections of this type are more frequent in warm humid climates.

6. Tinea Unguium

Infection of the nails by true dermatophytic fungi is called tinea unguium and is to be differentiated from onychomycosis or infection of nails caused by nondermatophytic fungi and yeasts.

In tinea unguium, infection usually begins at the tip or, less commonly, at the side of the nail. The infection, as it develops, invades the subungual region beneath the plate which leads to an accumulation of debris favoring the growth of other fungi and bacteria. Paronychial infections involve inflammation of the nail root and often results in the formation of a distorted nail. The hallmarks which distinguish tinea unguium from onychomycosis include both the accumulation of subungual debris and distorted frayed nails. Any of the dermatophytes can cause tinea unguium; <u>T. rubrum</u> and <u>T. mentagrophytes</u> are the most common causes.

Onychomycosis, an infection which involves actual invasion and destruction of the nail plate, may be caused by a variety of fungi including Scopulariopsis brevicaulis and C. albicans. The latter is most commonly associated with paronychia; onycholysis or separation of the nail from the bed may result.

7. Diagnosis

Diagnosis of dermatophytic infections involves several different types of examination. The simplest of these is the direct examination of infected skin scales, nail scrapings, or hairs. In this examination, a small amount of material is mixed with 10% KOH on a microscopic slide. After a cover glass has been placed over the specimen it is examined microscopically. This examination may reveal the presence of true filamentous fungi or, possibly, yeasts. This type of information, i.e., filamentous dermatophytic fungus or yeast-like organisms, is sufficient to permit specific therapy. In tinea capitis, examination of the scalp and infected hair under filtered ultraviolet (Wood's light) is of value. Hair infected with either M. canis or M. audoinii has a characteristic brilliant greenish-yellow secondary fluorescence. Diagnosis by culture may be required in a limited number of cases. These include children with epidemic ringworm where identification of the possible source of infection is critical, as well as possible failures due to highly resistant organisms.

C. Subcutaneous Mycoses

The subcutaneous mycoses include those localized infections of mucocutaneous and subcutaneous tissues in which dissemination generally is rare. In contrast to the systemic mycoses, initial infection generally follows traumatic implantation of infectious particles into the skin. The source of infection usually is vegetative growth of the organism on plant material or in soil.

1. Sporotrichosis

Sporotrichosis is a chronic, subcutaneous mycosis in general involving only the lymphatic system. The etiological agent is Sporothrix schenckii and it is worldwide in distribution. S. schenckii is associated with living and dead vegetation and inoculation most commonly occurs following implantation of a thorn or splinter wound in the skin or subcutaneous tissue. Primary lesions resembling those of syphilis appear on the skin, followed by lymphadenopathy after about 7 to 10 days which then progresses to necrosis and ulceration of the affected lymph nodes. Otherwise, the patient remains asymptomatic except in the occasional cases of disseminated disease. Rarely, sporotrichosis may involve extracutaneous sites such as bones or joints; in such cases there is usually a history of cutaneous infection.

Disseminated sporotrichosis, also rare, takes the form of multiple subcutaneous nodules which eventually ulcerate to form chronic ulcers. Prognosis in this form of the disease is poor. Pulmonary sporotrichosis, resulting from primary pulmonary infection is becoming more common. Like
primary pulmonary histoplasmosis, clinically it resembles tuberculosis as
well as other primary, pulmonary mycotic infections. Beginning as an
acute pneumonitis, it eventually develops into an often fatal chronic, cavitary infection.

2. Chromomycosis

Chromomycosis may be caused by the morphologically similar dematiaceous fungi Fonsecaea pedrosi, Fonsecaea compactum, and Phialophora ver-
rucosa. The disease is seen most often in the tropics. The chronic benign,
slow-growing wart-like, sometimes ulcerative lesions, have a tendency to
grow along lymphatic channels. The lower extremities are most often
involved although other skin surfaces may become infected. It is believed
that infection follows direct implantation of the organism from soil or plant
material. Chromomycosis, limited to skin and subcutaneous tissues, first
appears as a papule which ulcerates to form classical verrucous lesions.
With time, these lesions develop into large, verrucous masses set above
the surface of the infected limb. Clinical diagnosis is based upon the
demonstration in tissue of distinctive, brown or yellow yeast-like bodies
which divide by septation rather than by budding. Mycological diagnosis
is based upon recovery of one or more of the several different fungi associated with this disease from clinical or biopsy material.

3. Lobomycosis

Lobomycosis, or keloid blastomycosis, is a chronic, localized infection of
subepidermal tissue. The etiological agent, Loboa loboi, has not been cultured in vitro. The disease, found in the tropical jungles of South America,
is characterized by localized keloidal, firm, nodular lesions. Dissemination or spread to subcutaneous tissue does not occur. Infected tissue contains clusters of the spheroidal yeast-like organisms. The epidemiology
and etiology of lobomycosis is still obscure. While the etiological agent,
L. Loboi, has not been isolated from clinical or natural material, it has
been demonstrated in pathological materials from dolphins. Rippon postu-
lates that it may represent an obligate parasite of some lower animal form
[9]. While disfiguring, lobomycosis causes little pain or mortality. Treatment is primarily surgical.

4. Mycetoma

Mycetoma, also known as madura foot or maduromycosis, is a localized,
deforming infection of the hand or foot, characterized by massive destruction of cutaneous and subcutaneous tissues, muscle, and bone. It has a

diverse etiology, that is, it can be caused by one of several different fungi (eumycotic mycetoma) and bacteria (actinomycotic mycetoma).

One organism, Actinomyces israelii, is an anaerobic member of the normal flora of human tonsillar crypts. Infection is man with A. israelii follows traumatic introduction of the organism into the mucous membrane of the oral cavity. This infection, known as cervico-facial actinomycosis is characterized by multiple, draining sinus tracts containing purulent material in which are found characteristic and diagnostic "sulfur granules." The infection is treated with surgery and antibacterial antibiotics. Nocardia brasiliensis, an aerobic soil saprophyte, has been associated with actinomycotic mycetomas of the foot. Such infection, which follows traumatic implantation of the organism, is characterized by a swollen foot studded with multiple, draining sinus-tract openings. Underlying tissues are destroyed and involvement of bone is common. Pus from the sinuses contain granules which are white in the case of infections caused by N. brasiliensis. Similar infections are caused by true fungi such as Allescheria boydii and Madurella grisea. Establishment of the true etiology of mycetoma is essential for selection of appropriate therapy.

5. Rhinosporidiosis

Rhinosporidiosis is a chronic infection of mucocutaneous tissue caused by the fungus Rhinosporidium seeberi. Most commonly, the nose and conjunctiva are infected although other mucocutaneous tissues may be involved. The disease is characterized by the formation of large, friable polyps, tumors, papillomas, and wart-like lesions. The disease is almost exclusively seen in natives of the Far and Middle East. The taxonomic status of R. seeberi is uncertain as the organism has not been propagated in vitro.

D. Opportunistic Fungal Infections

Opportunistic fungal infections are caused by a variety of organisms, most of which are ubiquitous in nature or part of the normal flora of man. These infections are most commonly seen in persons whose immunological or physiological integrity has been compromised by underlying disease or therapeutic manipulation. Any organ or tissue of the body may be involved. The route of entry of the fungus into the body most often is the respiratory or intestinal tracts. Commonly encountered opportunistic fungal infections are caused by Phycomycetes (Mucor, Rhizopus, and Absidia sp.); Aspergillus fumigatus, Aspergillus niger, and other aspergilli; Penicillium sp.; and yeasts including C. albicans, C. neoformans, as well as Geotrichum sp. A major enigma in the diagnosis of opportunistic fungal infection lies in the determination of the clinical significance of positive cultures for these organisms in view of their roles as members of normal flora or of their possible sources being from environmental contamination.

1. Phycomycosis

Phycomycosis (mucormycosis, zygomycosis) is an acute, fulminating infection most commonly seen in diabetics. The classical form of this disease,
rhinocerebral mucormycosis, occurs in uncontrolled acidotic diabetics and
presents as an infection of the paranasal sinuses. The mucous membranes
are marked by areas of necrosis which rapidly extends to the hard palate
and nasal turbinates. The infecting agent, usually Rhizopus oryzae, invades ophthalmic and internal carotid arteries resulting in thrombosis and
leading to infection of the eye and of the central nervous system. Death, the
usual outcome, occurs within 2 to 7 days. Treatment which is often
unsuccessful depends, first, upon control of the diabetic ketoacidosis,
second upon surgical removal of infected tissues, and finally, upon specific
chemotherapy.

2. Aspergillosis

The genus Aspergillus includes a number of organisms of low virulence
which are common to the environment and to which man is constantly exposed. As a result, a variety of clinical conditions are seen both in normal
individuals and in the compromised host. The more serious forms of infection, usually caused by Aspergillus fumigatus, include invasive disease in
which there is extensive growth of the organism in tissue and aspergillomas,
or fungus balls, in which there is localized growth of the organism in preformed cavities. Invasive aspergillosis, most frequently seen in patients
with leukemia, is usually pulmonary and is associated with extensive
necrosis and destruction of functional tissue. Aspergilloma, usually associated with preexisting cavitary disease due to histoplasmosis, tuberculosis,
or sarcoidosis, is characterized by the growth of fungus balls in preformed
cavities. These balls, consisting of masses of mycetoma, produce a
minimal amount of pathology and many patients are asymptomatic except
for recurrent hemoptysis.

3. Opportunistic Yeast Infections

Both C. albicans and C. neoformans, as well as other yeasts, are associated with serious, secondary infections in patients, where they may
gain access either from the respiratory or the gastrointestinal tract, or
along the pathway of an indwelling catheter or cannula. Transient septicemia resulting from infected catheters may lead to massive, invasive disease. Patients, particularly renal transplant recipients, with primary,
pulmonary lesions due to C. neoformans may develop both meningitis and
multiple organ disease as well as localized infections as a result of immunosuppressive therapy.

III. MEDICINAL FUNGICIDES

Before the advent of antifungal antibiotics, the only treatment for the systemic mycoses was supportive therapy. Dermatophytic infections were treated with a variety of topical preparations containing keratolytic and antimicrobial substances [10]. The first specific antimycotic therapeutic agent was potassium iodide, the use of which was limited primarily to sporotrichosis [10]. Even now, the agents available for therapeutic use in treatment of fungal diseases are still limited in number. Most of the compounds currently utilized are insoluble or only slightly soluble in aqueous solutions. Furthermore, their usefulness is limited both because of restricted antifungal spectra and because of high levels of toxicity.

A principal problem in the development of new antifungal agents relates to the similarities between host and pathogen. That is, both being eucaryotes with similar metabolisms, a chemotherapeutic agent effective against fungi also may be inhibitory or lethal to the host cell.

As with antibacterial agents, antifungal agents can be categorized into two general groups of compounds, namely, antibiotics developed from bacterial or fungal secondary metabolites, and synthetic compounds developed by the organic and biological chemists. These two categories will be discussed separately.

A. Antifungal Antibiotics

Species of streptomyces produce over 60 antifungal antibiotics, most of which are polyenes [11]. Although many polyenes have been shown to have potential clinical usefulness, at present only two are utilized extensively as chemotherapeutic agents in human mycotic infections. These are amphotericin B and nystatin. The polyenes consist of a large lactone ring with rigid lipophilic and flexible hydrophilic constituents [12]. Both amphotericin B and nystatin are only slightly soluble in water and most common organic solvents. They dissolve readily, however, in highly polar solvents such as dimethylsulfoxide.

The polyenes exert their biological activity by binding to sterols in cell membranes. Ergosterol, to which amphotericin B binds most specifically [13], is found in cell membranes of most fungi, algae, and protozoa. This binding of the polyene allows for leakage of cellular components such as glucose and potassium by altering membrane permeability. It has been postulated that eight molecules of amphotericin B are incorporated into an ergosterol-containing portion of the cell membrane in such a manner as to form a pore or tunnel [14]. Unfortunately, a similar but lesser reaction takes place between amphotericin B and cholesterol, which occurs in membranes of mammalian cells, thus accounting for part of the toxicity associated with the polyene compounds.

1. Amphotericin B

Amphotericin B is an amphoteric heptaene produced by Streptomyces nodosus [12]. Having a molecular weight of 960.1, it is insoluble in water and only sparingly soluble in dimethylformamide and dimethylsulfoxide. The in vitro and clinical antifungal spectra of amphotericin B include systemic pathogens, pathogenic yeasts and some opportunistic pathogens. Although also inhibitory for dermatophytic pathogens, the activity of amphotericin B against these organisms is not sufficient for the drug to be useful clinically. Amphotericin B is inactive at clinically significant concentrations against several important pathogens including A. boydii, many isolates of Aspergillus, and most of the dematiaceous fungi.

The poor solubility of amphotericin B in water posed problems in early clinical studies of the drug. However, these were eliminated with the development of the intravenous preparation which now is used. This preparation contains sodium desoxycholate and buffer in addition to amphotericin B. The former acts as a dispersing agent while the latter maintains a suitable pH. Specifically, the intravenous formulation contains 50 mg of amphotericin B, 41 mg of sodium desoxycholate, and 25.2 mg of sodium phosphate buffer [15]. It is prepared by first dissolving the dry powder in 10 ml of sterile water for injection; this solution is then diluted in 5% dextrose injection U.S.P. to give a final concentration of 0.1 mg/ml. It is critical that the dextrose solution have a pH above 4.2 and be free of electrolytes and bacteriostatic agents.

Amphotericin B normally is administered intravenously; however, intrathecal administration and direct injection into the bladder or kidney also may be used. For intravenous administration, the initial dose consists of 0.25 mg/kg body weight administered in intravenous glucose at a concentration not exceeding 10 mg/100 ml over a 2- to 6-hr period [15]. The dosage is increased by 5 to 10 mg increments over several days. A total dosage of 1.5-2.5 g of amphotericin B is administered over a 4- to 6-week period. Serum levels of the drug range from a maximum of about 3.0 μg/ml to a minimum of 0.4 μg/ml; cerebrospinal fluid levels are approximately 1/40th the serum concentration [16]. Amphotericin B is bound to serum lipoproteins and slowly excreted in the urine. Thus, measurable urine levels will be detected for several days or more following cessation of treatment.

Toxicity of amphotericin B is legendary [15]. The patient may experience a whole range of side effects including fever, headache, malaise, nausea, vomiting, anorexia, and depression. Of critical concern is the nephrotoxicity of amphotericin B. Temporary and permanent renal failure may result from the administration of this drug and, in fact, the course of therapy is often governed by the patient's renal status. Thus, blood urea nitrogen and serum creatinine clearance values must be closely monitored.

Amphotericin B has been used in the treatment of a wide variety of systemic, subcutaneous, and opportunistic fungal infections [15]. Susceptible fungi generally exhibit a minimal inhibitory concentration (MIC) of from 0.02 to 1.0 μg/ml. The minimal fungicidal concentration (MFC) of the drug for susceptible organisms rarely exceeds 2.0 μg/ml. B. dermatitidis is commonly very susceptible while many isolates of Aspergillus species are most likely to be resistant. Phycomycetes tend to be the most variable, ranging from distinctly sensitive to highly resistant [16].

Intravenous amphotericin B remains the drug of choice in patients with proven progressive, potentially fatal infections [16]. It should not be used in patients with apparent disease where the only evidence of infection is a positive skin test. Specifically, intravenous treatment with amphotericin B is indicated in cryptococcosis, blastomycosis, histoplasmosis, systemic candidiasis, coccidioidomycosis, phycomycosis caused by susceptible organisms, disseminated sporotrichosis, and aspergillosis [15]. It also may be of value in treatment of mucocutaneous leishmaniasis. Intrathecal therapy may be of value in severe life-threatening cases of fungal meningitis.

Amphotericin B is available in most countries in the intravenous form as well as in the form of topical creams or ointments [15]. The latter preparations are used primarily in the treatment of cutaneous and mucocutaneous candidiasis but are of little value in treatment of superficial dermatophytic infections or bacterial skin infections.

2. Nystatin

Nystatin is an amphoteric tetraene, $C_{46}H_{77}O_{19}N$, produced by Streptomyces noursei [12, 17]. Like amphotericin B, it is insoluble in water; it is sparingly soluble in alcohols but soluble in dimethylformamide and dimethylacetamide. It is very unstable. In vitro, it is active against a variety of pathogenic fungi. However, it is not absorbed from the human gastrointestinal tract and is highly toxic when administered intravenously; thus its clinical use is limited to topical applications [18].

In the United States, nystatin is prescribed in cream, powder, ointment, or suspension form for topical use in mucocutaneous candidiasis, and in tablet form for intestinal candidiasis [18]. The drug is used extensively in the treatment of vulvovaginal candidiasis. In this instance, the preparation consists of a vaginal tablet containing 100,000 units of drug together with lactose, stearic acid, ethyl cellulose, and starch [18]. Oral doses of nystatin for adults range from 500,000 to 1,000,000 units given three times per day; in children, the oral dose is 100,000 units given three or four times daily. Topical nystatin cream or ointment usually contains 100,000 units per gram and is applied once or twice daily. Side effects to nystatin are uncommon. These include nausea, vomiting, diarrhea, and are seen primarily with oral treatment.

Resistance to nystatin, as to amphotericin B, is uncommon [19-21]. Mutants, either naturally occurring or laboratory induced, of yeasts resistant to the polyene compounds have been described. Generally, the resistance to polyene compounds has been associated with changes in sterol content of the yeast cell membranes [22].

3. Primaricin

Primaricin is a polyene tetraene macrolide produced by Streptomyces natalensis. Having a molecular weight of 665.8, it is insoluble in water. Its principal clinical value appears to be in the treatment of cutaneous fungal infections; its most promising value appears to be in the treatment of my- cotic keratitis [23].

4. Griseofulvin

Griseofulvin is a phenolic benzofuran cyclohexene produced by Penicillium griseofulvum and other species of Penicillium [12]. Having a molecular weight of 352.8, it is insoluble in water, sparingly soluble in organic sol- vents but soluble in dimethylformamide and dimethylsulfoxide. Unlike the polyene antifungals it is a highly stable compound. Structurally, the drug resembles purine riboside and radioactive labeling studies have shown it to be incorporated into the ribonucleic acid of actively growing fungal cells. The antifungal activity of griseofulvin is directed exclusively against the dermatophytes. The actual mode of action of the drug is not clear, but it acts only on actively growing cells. It seems to interfere with protein, nucleic acid, and maleic acid synthesis as well as to cause destruction of organelle membranes. The drug has also been reported to prevent the penetration of fungal hyphae into host cells [16].

In vitro, griseofulvin is active against dermatophytic fungi but not against yeasts or dimorphic pathogens. In vitro resistance is rare. The drug is only poorly absorbed in the gastrointestinal tract. It is not active in vivo until deposited in keratinizing tissues. Thus there is a delay between onset of treatment and onset of therapeutic response. Griseofulvin is admin- istered only orally. Because of the poor absorption, only a small portion of the drug becomes biologically available. Absorption can be increased by several mechanisms. One is to provide the drug in the form of a "micro- nized" preparation, thus increasing surface area and, hopefully, absorption. The other is to administer the drug with a high-fat-content meal.

Griseofulvin is supplied in capsules, tablets, and suspension [24]. As noted, the drug is only poorly absorbed in the gastrointestinal tract. It is often administered with a high-fat meal which aids absorption. Most of the drug is excreted unchanged in the feces and less than 1% is detected in the urine. Only rarely are detectable amounts of griseofulvin found in other body fluids and there are indications that the drug is metabolized in the liver. However, the compound is detected is growing skin, hair, and nails. The drug is customarily administered in four equally divided doses at 6-hr intervals and the doses are variable: 10 mg/kg body weight per day for infants; 125 to 250 mg for children; 250 to 500 mg for children over 50 lb (22.5 kg); and 500 to 1,000 mg for adults [16].

Severe or prolonged toxic reactions to griseofulvin therapy are rare. Physiological side effects include mild headache, fatigue, lethargy, and vertigo. Occasional gastrointestinal side effects including nausea, vomiting, and diarrhea are encountered. Leukopenia, neturopenia, proteinuria, and urinary casts have also been occasionally reported [16].

Griseofulvin is indicated only in the treatment of true dermatophytic infections. Thus cultural or microscopic confirmation of the clinical diagnosis is essential. Treatment failures are rare and usually occur only in cases of nail infections and infections of soles of feet. In such cases prolonged treatment, several months or more, may be required.

B. Synthetic Antifungal Compounds

Synthetic antifungal compounds are available for treatment of most forms of fungal infection. However, the economics of pharmaceutical research and development have dictated that the major area of application of such compounds primarily be in the area of topical preparations useful in the treatment of dermatophytic infections. Availability of similar agents for treatment of systemic diseases is limited. Structures for several of the more commonly used synthetic antifungal compounds are shown in Fig. 1.

1. 5-Fluorocytosine

5-Fluorocytosine (5-FC) or flucytosine is a synthetically produced fluorinated pyrimidine, originally synthesized as a potential antileukemic agent. Its most important pharmacological properties are that it is well absorbed in the gastrointestinal tract and is of low toxicity [25]. However, the drug has a limited in vitro and clinical antifungal spectrum [26]. This includes susceptible strains of species of Candida, Cryptococcus neoformans, Torulopsis glabrata, and species of Cladosporium. The drug has been shown to be highly effective in the treatment of candidiasis, cryptococcosis, and toruloposis [27]; limited success in the treatment of aspergillosis [28] and chromomycosis [29] has been reported.

The mode of action of 5-FC is unique. It is taken up by the yeast cell via a relatively nonspecific permease enzyme system which recognizes 5-FC [30]. A similar enzyme is present in mammalian cells. Exogenous pyrimidines and purine may interfere with the uptake of the drug by this enzyme system. Once inside the cell, 5-FC is rapidly deaminated by the enzyme cytosine deaminase to 5-fluorouracil which is then further metabolized and eventually incorporated into the cellular ribonucleic acid as an analog of uracil [31]. The fluorouracil substituted ribonucleic acid, in turn, causes the production of nonfunctional proteins causing inhibition of growth and sometimes cell death [32]. The low toxicity of the drug in man is apparently due to the inability of most mammalian cells to deaminate 5-FC to 5-fluorouracil.

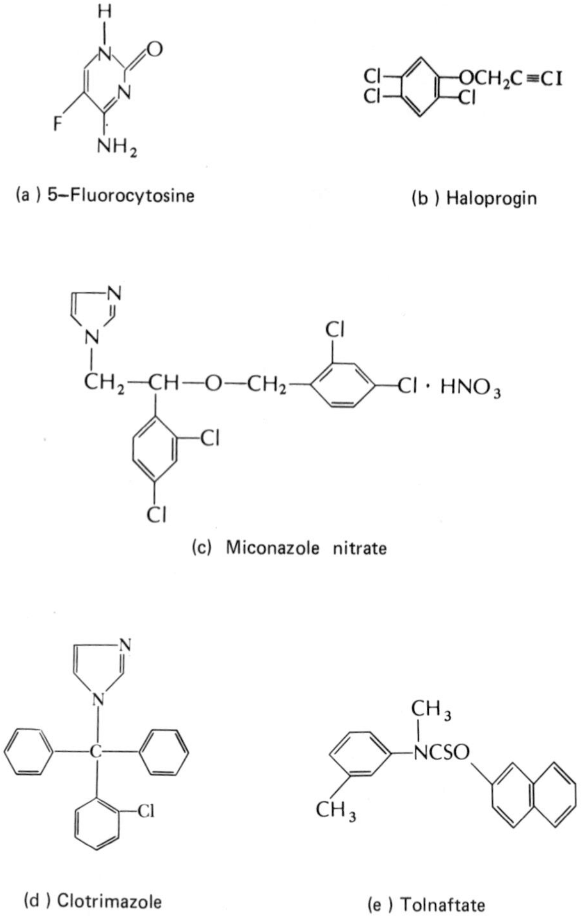

(a) 5-Fluorocytosine

(b) Haloprogin

(c) Miconazole nitrate

(d) Clotrimazole

(e) Tolnaftate

FIG. 1. Synthetic antifungal chemotherapeutic agents.

Natural resistance to 5-FC is common among the yeasts [26]. In addition, resistance is easily induced in most susceptible organisms. Resistance occurs at three levels within the cell. First, the cytosine permease system may be inoperative or totally lacking, thus preventing uptake of the drug. The second site of resistance is at the level of the enzyme cytosine deaminase. The lack of or mutation of this enzyme prevents the formation of the metabolically active 5-fluorouracil. Finally, the cell may lack the enzyme uridine kinase which is responsible for the phosphorylation of uridine to form the nucleotide which is eventually incoporated in the ribonucleic acid of the cell. The ease with which resistance develops makes necessary the use of in vitro susceptibility data for proper clinical use of the drug [33].

Despite the low solubility of 5-FC in water, the drug is well absorbed in the gastrointestinal tract and the usual route of administration is oral. The drug is detectable within the serum in 30 min and peak serum concentrations of 50 to 75 μg/ml are seen in 4 to 6 hr. The drug is administered at a dosage of 150 mg/kg body weight per day in four equally divided doses [16]. 5-Fluorocytosine readily crosses the blood-brain barrier and cerebrospinal fluid levels reach 60-80% of obtainable serum levels. Better than 90% of the compound is excreted in the urine within a 24-hr period. Since the drug is excreted solely through the kidneys, peak serum levels in excess of 100 to 200 μg/ml have been seen in patients with impaired renal function. It is in these patients that toxic side effects are most often observed and the therapeutic dosage of 5-FC must be reduced accordingly. Serum levels should be monitored in such patients if possible. Abdominal discomfort has been seen occasionally; rare incidences of leukopenia and thrombocytopenia have also been reported [16].

The drug is most often employed in the treatment of cryptococcosis and candidiasis caused by susceptible organisms. 5-Fluorocytosine has also been reported useful in the treatment of aspergillosis and chromomycosis and a 10% topical ointment has been applied to otomycosis due to Aspergillus sp. The major drawback in the use of the drug is the high incidence of naturally resistant yeasts and the ease which susceptible organisms may become resistant to treatment with 5-FC.

Amphotericin B and 5-FC have been shown to be synergistic in their action against yeasts [34], and recent reports have demonstrated this synergism to be possibly operational at the clinical level [35]. The synergistic interaction is felt to be twofold. First, amphotericin B potentiates the uptake of 5-FC through alterations in cell membrane permeability [34]. Second, amphotericin B also may act to suppress the emergence of 5-FC resistant clones during the treatment. The emergence of such clones was associated with most therapeutic failures with 5-FC in earlier studies of this drug. In both animal and clinical studies with combination therapy, the emergence of such clones was reduced or eliminated [35, 36].

2. Imidazole Derivatives

The imidazoles represent a group of compounds with previously demonstrated antiparasitic activity. As a group, they are broad-spectrum agents being active against fungi, bacteria, and protozoa. Several have recently become available for use as antifungal agents, including clotrimazole, miconazole, econazole, and thiabendazole. The site of biological activity of these compounds appears to be in the cell wall and plasma membrane. Electron microscopy studies of cells of C. albicans treated with miconazole reveals destructive changes in the cell wall [37]. Leakage of amino acids, proteins, and cations from imidazole-treated cells has also been observed [38]. Vacuole formation in treated cells suggests destruction of the cytoplasmic membrane. Exposure to imidazole derivatives may also enhance nucleoside transport in yeast cells [39]. The compounds are poorly soluble in water although they can be dissolved in dimethyl sulfoxide, pyridine, or ethylene glycol. Further dilution in aqueous solvent results in a turbid solution which maintains its activity.

Miconazole, 1-[2, 4-dichloro-β-(2,4-dichlorobenzyloxy)phenethyl] imidazole, is available in topical, vaginal, oral, and intravenous preparations both as the base and as a nitrate salt. Miconazole has been used extensively in the treatment of vulvovaginal candidiasis and found to be superior to nystatin [40]. A 10-day course of therapy generally produces an effective cure in over 80% of the patients treated [41]. Systemic use of miconazole has been limited to Europe and principally to cases of pulmonary candidiasis. Oral doses of 50 mg/kg body weight per day for several weeks are administered. Prophylactic use of 1 mg/day miconazole in renal transplant patients has also proven effective. Peak serum levels are in the range of 2 μg/ml with detectable amounts of the drug appearing in the serum within 15 min of administration. The drug seems to be well tolerated and few side effects have been noted [42].

In vitro, miconazole is active against a large number of fungi, including systemic pathogens, pathogenic yeasts, and dermatophytes. However, available data suggest that serum levels will not be sufficient for treatment of systemic disease. Clinically, miconazole appears to be of greatest potential value in treatment of vaginal and cutaneous infections. Its principal advantage is that it is active against both yeast and dermatophytic skin infections.

Clotrimazole, bis-phenyl-(2-chloro-phenyl)-1-imidazolyl methane, is available as topical and oral preparations. A 1% cream or suspension applied topically has proved dramatically effective in the treatment of mucocutaneous candidiasis and dermatophytic infections. No drug was detected in the blood of patients receiving heavy applications of the cream over large areas of skin [43]. Trials with topical clotrimazole in the treatment of candida vaginitis have been equally encouraging. Prophylactically, the drug has been used antepartum to prevent neonatal thrush [44].

Oral clotrimazole therapy has been employed in the treatment of urinary tract candidiasis, candida septicemia, and pulmonary aspergillosis. Dosages for both children and adults were in the range of 100 mg/kg body weight per day, administered in four equally divided doses. Peak serum levels ranged from 0.15 μg/ml in children to 0.35 μg/ml in adults [43]. The highest levels of the drug obtainable in sputa were 0.05 μg/ml while as much as 110 μg clotrimazole per gram wet weight was found in feces. Treatment was considered most effective in the candida infections. Fungal infection appeared to be cleared in the cases of pulmonary aspergillosis reported, but complete clinical follow-ups were not possible. Duration of therapy ranged from several weeks in the cases of candidiasis to several months in the cases of aspergillosis. Gastrointestinal intolerance was observed in approximately half the patients treated with oral clotrimazole; this was

severe enough to cause cessation of therapy in about 20% of the cases.
Transient mental disturbances were also noted in some patients receiving
clotrimazole therapy [44]. Unfortunately, clotrimazole has no potential
value in treatment of systemic disease.

Econazole and thiabendazole are still under study at this time. Both
have been reported to be effective in the treatment of various fungal infections of the eye. In addition, thiabendazole has been employed in the management of chromomycosis [45].

3. Haloprogin

Haloprogin, 3-iodo-2-propynyl 2,4,5-trichlorophenyl ether, is marketed
as a 1% cream for topical application. It is active against species of Candida
in vitro and against most dermatophytes clinically [46]. It has not been
successful in the treatment of either tinea capitis or tinea unguium. The
minimal inhibitory concentration (MIC) for Trichophyton rubrum is 0.78
μg/ml and for T. tonsurans 0.48 μg/ml. Most other dermatophytes exhibit
MICs within the same range of values. In vivo studies with rats showed the
drug to be absorbed through the skin, appearing in the urine. No detectable
amounts of haloprogin were found in other tissues. The drug appears to be
well tolerated in man although it does have associated side effects. These
include local irritation and enhancement of preexisting lesions as well as
sensitization in some individuals.

4. Tolnaftate

Tolnaftate, m-N-dimethylthiocarbanilic acid O-2-naphthyl ester, was one
of the first synthetic antifungal chemical compounds used topically. It is
available as a 1% cream, suspension, or powder. The drug shows a broad
spectrum of activity against the dermatophytes with MICs in the range of
0.0075 to 0.075 μg/ml for T. mentagrophytes. It is active clinically in
treatment of tinea pedis [47] but not against infection due to yeasts or in
the treatment of tinea capitis, tinea unguium, or in treatment of infections
of heavily keratinized tissue. Relapses in infections caused by T. rubrum
are common. Tolnaftate is generally applied twice daily for up to 2 weeks.
There is no significant percutaneous absorption. Toxicity is limited.

5. Miscellaneous Compounds

Numerous other compounds have been used with varying degrees of success
in the treatment of both systemic and dermatophytic fungal infections.
Potassium iodide was the first chemotherapeutic agent used in the treatment
of systemic mycoses and is still employed in the treatment of sporotrichosis
where it is the drug of choice in lymphatic infections. Stilbamidine and
hydroxystilbamidine were utilized in the treatment of blastomycosis for a
number of years, but are active only against the cutaneous form of the disease. Interest in these compounds has lessened as they have been associated

with unacceptably high relapse rates. Hamycin, candicidin, and natamycin are polyene antibiotics with activity comparable to that of nystatin and they have been used previously in the treatment of fungal infections. However, they have not been made available in sufficient amounts to permit full clinical evaluation. Saramycetin shows activity against H. capsulatum and various other systemic fungi, but is not economical due to high production costs. A 2.5% suspension of selenium sulfide is commonly employed in a shampoo base for topical use. Other nonprescription shampoos often contain sulfur, coal tar, or coal-tar derivatives as their active ingredients.

IV. LABORATORY STUDIES

In vitro susceptibility testing with antifungal agents presents several unique problems for the clinical mycology laboratory [48]. Most of the agents are insoluble or only slightly soluble in water and therefore suitable solvents which do not interfere with fungal growth must be found. One of the more commonly employed organic solvents is dimethylsulfoxide which may be inhibitory at low concentrations for some fungi. In addition, it may require 72 hr or longer for many of the filamentous fungi to develop visible growth and the possibility of drug deterioration is a major problem, particularly with the polyenes. Except in a few cases with some yeasts, disc susceptibility testing is neither practical nor reliable and is not utilized in the clinical mycology laboratory. In vitro susceptibility data are of value only in the management of systemic opportunistic and subcutaneous infections. They are not indicated in the management of dermatophytic infection.

At present, three methods for antifungal susceptibility testing are available [49]: the broth–dilution method, the semisolid agar–dilution, method, and the agar–dilution method. All three employ twofold serial dilutions of the compound being tested at concentrations of either 128 to 0.06 μg/ml or 100 to 0.05 μg/ml. The broth–dilution and the semisolid agar–dilution methods are performed in an identical manner except that the medium for the semisolid agar–dilution method includes the addition of 0.5% agar. In the agar–dilution technique the drug is diluted in an agar medium and dispensed into petri plates [50]. One of the more critical aspects of in vitro testing with antifungal compounds is the size of the test inoculum. This generally consists of 10^6 cells, spores, or mycelial fragments per ml of physiological saline. Incubation is customarily at 30°C until growth appears in drug-free control tubes. Incubation at 37°C is not advised both because of the instability of amphotericin B and because of the failure of some fungi to grow at this temperature. Following incubation, the minimal inhibitory concentration (MIC) of the drug is read. This is defined as the lowest concentration of drug which inhibits visible growth of the organism. In the broth- and semisolid agar-dilution methods, 0.05 ml is removed from all tubes showing no growth with a calibrated loop and subcultured to drug-free medium. Following a second incubation period, the minimal fungicidal concentration (MFC) of the drug is determined. This is defined as that concentration which killed the organisms as determined by the lack of a viable subculture.

The agar-dilution technique has the disadvantage of not allowing for convenient determination of the MFC. An agar-dilution method for use with 5-FC and filamentous fungi which permits determination of MFC values has been described [51]. In this procedure, membrane filter discs are placed on the surface of the agar and the filter is then inoculated with the test organism. Following incubation, the filters showing no growth are removed and washed by filtration to remove any excess drug. The filters are then placed in tubes of drug-free broth or on drug-free agar and incubated to determine the MFC.

REFERENCES

1. E. L. Randall, in Infectious Diseases Reviews (W. J. Holloway, ed.), Vol. 1, Futura Publishing Co., Mount Kisco, N.Y., 1972, pp. 133-138.
2. T. L. Fleisher, in The Diagnosis and Treatment of Fungal Infections (H. M. Robinson, Jr., ed.), Charles C. Thomas, Springfield, Ill., 1974, pp. 113-127.
3. R. D. Baker, in Systemic Mycoses (G. E. W. Wolstenhome and R. Porter, eds.), Little, Brown and Co., Boston, 1967, pp. 9-18.
4. P. D. Hart, E. Russell, Jr., and J. S. Remington, J. Infect. Diseases, 120, 169 (1969).
5. C. Smith, D. Pappagianis, H. Levine, and M. Saito, Bacterial Rev., 25, 310 (1961).
6. L. Ajello, Proc. Intern. Symp. Mycoses, Sci. Publ. 205, PAHO, Washington, D.C., 1970, pp. 3-10.
7. K. R. Wilcox, Ann. Internal Med., 49, 388-418 (1958).
8. M. Marples and D. A. Somerville, Trans. Roy. Soc. Trop. Med. Hyg., 62, 256 (1968).
9. J. W. Rippon, Medical Mycology, W. B. Saunders Co., Philadelphia, 1974, p. 279.
10. N. F. Conant, D. T. Smith, R. D. Baker, and J. L. Callaway, Manual of Clinical Mycology, 3rd ed., W. B. Saunders Co., Philadelphia, 1971, pp. 729-735.
11. D. Perlman, Progr. Ind. Microbiol., 6, 1 (1967).
12. M. W. Miller, Pfizer Handbook of Microbial Metabolites, McGraw-Hill Book Co., New York, 1961, pp. 123-130.
13. J. Kotler-Brajtburg, H. D. Price, G. Medoff, D. Schlessinger, and G. S. Kobayashi, Antimicrobial Agents Chemotherapy, 5, 377 (1974).
14. T. E. Andreoli, Ann. N.Y. Acad. Sci., 235, 448 (1974).
15. E. R. Squibb and Sons, Fungizone® Intravenous, Amphotericin B for Injection U.S.P., Package insert J2-419B, Princeton, N.J., rev. 1970.

16. J. P. Utz, in Antimicrobial Therapy (B. M. Kagan, ed.), W. B. Saunders Co., Philadelphia, 1974, pp. 147-154.
17. E. L. Hazen and R. Brown, Proc. Soc. Exptl. Biol. Med., 76, 93 (1951).
18. C. E. Baker, Physician's Desk Reference, 31st ed., Medical Economics Co., Oradell, N.J., 1977, pp. 1527-1528.
19. G. R. Gale, J. Bacteriol., 86, 151 (1963).
20. M. L. Littman and M. A. Pisano, Antibiotics Annual, 1957-1958, Medical Encyclopedia, New York, 1958, pp. 981-987.
21. R. A. Woods, J. Bacteriol., 108, 69 (1971).
22. P. J. Trocha, S. J. Jasne, and D. B. Sprinson, Biochem. Biophys. Res. Commun., 59, 666 (1974).
23. J. W. Rippon, Medical Mycology, W. B. Saunders Co., Philadelphia, 1974, pp. 531-544.
24. C. E. Baker, Physician's Desk Reference, 31st ed., Medical Economics Co., Oradell, N.J., 1977, p. 1483.
25. E. Grunberg, H. N. Prince, and J. P. Utz, in Proc. 5th Intern. Congr. Chemother. Vienna (K. H. Spitzy and H. Hoschek, eds.), 1967, p. 69.
26. S. Shadomy, Appl. Microbiol., 17, 871 (1969).
27. S. Scholer, Mykosen, 13, 179 (1970).
28. W. Atkinson and H. L. Israel, Amer. J. Med., 55, 496 (1973).
29. A. A. Mauceri, S. I. Cullen, A. C. Vandervelde, and J. E. Johnson, Arch. Dermatol., 109, 873 (1974).
30. R. Jund and F. LaCronte, J. Bacteriol., 102, 607 (1970).
31. R. Giege, J. H. Weil, Bull. Soc. Chim. Biol., 52, 135 (1970).
32. A. Polak and H. J. Scholer, Chemother., 21, 113 (1975).
33. C. E. Baker, Physician's Desk Reference, 31st ed., Medical Economics Co., Oradell, N.J., 1977, pp. 1303-1304.
34. G. Medoff, M. Comfort, and G. S. Kobayashi, Proc. Soc. Exptl. Biol. Med., 138, 571 (1971).
35. J. P. Utz, I. L. Garriques, M. A. Sande, J. F. Warner, G. L. Mandell, R. F. McGehee, R. J. Duma, and S. Shadomy, J. Infect. Diseases, 132, 368 (1975).
36. A. Polak and H. J. Scholer, 9th Intern. Congr. Chemother., Abstr. SM-105, 1975.
37. H. Yamaguchi, Y. Kanda, K. Iwata, and M. Osumi, J. Electron-microscopy (Tokyo), 22, 167 (1973).
38. K. H. S. Swamy, M. Sirsi, and G. R. Rao, Antimicrobial Agents Chemotherapy, 5, 420 (1974).
39. H. van den Bossche, Biochem. Pharmacol., 23, 887 (1974).
40. J. E. Davis, J. H. Frudenfeld, and J. L. Goddard, Obstet. Gynecol., 44, 403 (1974).
41. A. Kunicki, Arzneimittel-Forsch., 24, 534 (1974).
42. J. M. Van Cutsem and D. Thienpont, Chemother., 17, 392 (1972).
43. H. Weuta, Arzneimittel-Forsch., 24, 540 (1974).

44. R. J. Holt, Infection, 2, 95 (1974).
45. R. Y. Cartwright, J. Antimicrobial Chemother., 1, 141 (1975).
46. H. W. Herman, Arch. Dermatol., 106, 839 (1972).
47. J. C. Gentles, E. G. V. Evans, and G. R. Jones, Brit. Med. J., 2, 557 (1974).
48. S. Shadomy and J. P. Utz, in Principles and Techniques and Human Research and Therapeutics (F. G. McMahon and M. Finland, eds.), Vol. VII, Drugs Useful vs. Infectious Diseases, Futura Publishing Co., Mount Kisco, N.Y., 1975, pp. 41–57.
49. S. Shadomy and A. Espinel-Ingroff, in Manual of Clinical Microbiology, 2nd ed. (E. H. Lennette, E. H. Spaulding, and J. P. Truant, eds.), American Society for Microbiology, Washington, D.C., 1974, pp. 569–574.
50. H. M. Ericsson and J. C. Sherris, Acta Pathol. Microbiol. Scand., Suppl., 217, 65–67 (1971).
51. G. Wagner, S. Shadomy, L. Paxton, and A. Espinel-Ingroff, Antimicrobial Agr. Chemother., 8, 107 (1975).

Chapter 14

RESIDUE ANALYSIS

H. P. Burchfield[*] and Eleanor E. Storrs

Division of Life Sciences
Gulf South Research Institute
New Iberia, Louisiana

I. INTRODUCTION

With few exceptions, the analysis of fungicide residues does not differ basically from the measurement of other groups of pesticides such as herbicides and insecticides. Those differences which exist arise primarily from

---

[*]Present address: Research Associates, New Iberia, Louisiana

unique physicochemical properties of individual groups of compounds. The dithiocarbamates are not volatile but readily decompose to yield carbon disulfide which is the compound measured. The quinones are photolabile, and compounds containing active halogen react readily with substrates containing nucleophilic groups. Therefore, special precautions must be used in handling members of these classes of compounds to prevent losses during extraction and analysis. Otherwise, general techniques have been developed which can be applied with modifications to most pesticides including fungicides [1, 2]. These include, among others, extraction methods, cleanup procedures, and analysis of residues by gas-liquid chromatography (GLC), high-pressure liquid chromatography (HPLC), colorimetry, and in a few cases, fluorometry.

II. EXTRACTION AND CLEANUP

Fungicide residues can be extracted from soil, tissues, and other substrates with polar solvents such as acetone or acetonitrile, nonpolar solvents such as benzene, or mixed solvents such as benzene and acetone. The latter are most efficient for removal of fungicides but also can result in the extraction of large amounts of other substances. Usually, tissue samples are homogenized with the solvent in a blender, using 200 to 400 ml of solvent per 100 g wet weight of tissue. If nonpolar solvents are used, the homogenate may be centrifuged to break emulsions. If water-miscible solvents are used, the homogenate is filtered to remove tissue debris.

Extracts in nonpolar solvents are almost always dried over anhydrous Na_2SO_4, which is sometimes also used to dry extracts during filtration.

Following extraction, it is often necessary to evaporate the solvent to replace it with another one. Losses can occur through volatilization of compound at this step, so evaporation is carried out at low temperature under a stream of dry nitrogen, and the residue is removed from the heat source and redissolved immediately after evaporation of the last traces of the solvent. In the case of highly volatile fungicides, a small amount of a nonvolatile compound such as polyethylene glycol is sometimes added.

In most cases, interfering substances must be removed from the extract before analytic determination. This is usually accomplished by two processes, liquid-liquid partition and liquid-solid chromatography on a column or thin-layer plate. In the case of oily samples, liquid-liquid partition is always required and precedes cleanup by chromatography.

The polar solvent is usually acetonitrile [3], dimethylformamide [4], or dimethyl sulfoxide [5, 6]. The nonpolar solvent is hexane or petroleum ether. During partition, fungicides are extracted into the polar layer and lipids into the nonpolar layer, which is discarded.

At this point, the polar layer can be evaporated and the residue redissolved in hexane, or it can be diluted with water and extracted with a nonpolar solvent, usually hexane.

Some samples can be analyzed at this point, but further cleanup with an adsorbant is usually required, such as charcoal [7], aluminum oxide [8], and Florisil [9].

In a few cases, such as surface strippings of fruit and foliage, the extract may simply be shaken with a few grams of adsorbant and then filtered. However, in most cases column chromatography is required. Although the search for a better stationary phase continues, Florisil is still used in most methods. Some fungicides may be unstable on it; in these cases, alternate methods must be sought.

The residue is usually applied to the Florisil column in hexane solution and the column washed with additional hexane. Impurities are eluted by washing the column with a mixture of hexane and a polar solvent such as ethyl ether. The fungicide is then removed from the column by increasing the amount of polar solvent in the mixture.

A variant of the partition method may be used in the cleanup of basic fungicides such as dodine or acidic compounds, e.g., pentachlorophenol (PCP). Here advantage is taken of the fact that on partitioning of these fungicides between aqueous acid and a nonpolar solvent, dodine is extracted into the aqueous phase and PCP into the nonpolar phase. On partitioning between aqueous alkali and a nonpolar solvent the reverse occurs. In both cases, levels of interfering compounds are reduced by transferring the fungicide between organic and aqueous solutions by successive partitioning with acid and alkali.

Following partitioning and column chromatography, the sample is usually ready for residue analysis.

III. MEASUREMENT OF RESIDUES

In the past, fungicide residues were usually determined by colorimetry, the method employed being specific for each compound or group of closely related compounds. Now most of these methods have been replaced by gas-liquid chromatography (GLC). There are, however, cases where colorimetry is still useful. Dexon, for example, is too unstable to chromatograph in the gas phase, although there might be a possibility of developing a method based on GLC by reducing it to dimethylaniline in a stream of nitrogen. However, colorimetry can also be useful in the analysis of compounds that are suitable for GLC. This is the case if the sensitivity of the colorimetry method is sufficient for the purpose, measurement of possible metabolites is not required, and a very large number of samples must be analyzed. The advantage of colorimetry in this situation is that only a minute or less of instrument time per sample is required, while GLC may require half an hour or more. Since colorimetric procedures are rather specific, only GLC and HPLC will be discussed in this section.

A. Gas-Liquid Chromatography

The GLC of fungicides is usually similar to that of other pesticides. In general, glass-lined injection ports are used to minimize losses caused by compound breakdown. One of the most widely used columns is a 6-ft by 1/4-inch o.d. glass "U" tube [10]. These minimize compound breakdown, are easy to pack, can be installed without danger of breakage, and are long enough to yield a sufficient number of theoretical plates for most separations.

The solid support is usually an 80-100 mesh diatomaceous earth which has been acid washed and silanized. Commonly used commercial brands include Chromosorb W, Chromosorb P, Anakrom ABS, and Gas Chrom Q.

Liquid phases which are commonly used include silicones such as OV-1, OV-17, QF-1, and DC-300, and polyesters such as Reoplex (polypropylene glycol adipate), and diethylene glycol succinate (DEGS).

These can be used alone or as mixtures to obtain specific separations. The very low liquid-solid ratios ($< 1\%$) and the very high ratios ($> 20\%$) which were popular in the past are now rarely employed. Most packings now contain 2-10% of liquid phase. The carrier gas is usually nitrogen at 60 ml/min, unless the detection system requires some other choice.

With the exception of fumigants, most fungicides are chromatographed at 180-200°C. Fumigants are chromatographed at 50-70°C. Unless the compound is unstable, the injection port is usually maintained 20-40°C higher than the column temperature. Detector temperature is maintained 10-20°C higher than column temperature unless this affects its operation.

The most variable and useful feature in the gas chromatography of fungicides and other pesticides is the detector. The thermal detector is of no value in residue analysis because of low sensitivity, and the flame ionization detector is of little value since it is sensitive to all organic compounds and most extracts would require very extensive cleanup before its use.

1. Electron Capture Detector (ECD)

Despite its lack of linearity and selectivity the electron capture detector continues to be used more widely than any other detector for the analysis of chlorinated compounds because of its high sensitivity to these compounds and low cost. Although this detector has been in use since 1960 [11], improvements continue to be made in its design and operation. In 1963, Lovelock described a parallel plate design which, when operated with pulsed collector potential, substantially freed the device from anomalous responses and errors induced by a continuous electrical field [12]. The tritium foil which was previously most commonly used as a radioactive source in the electron capture detector imposed a temperature limitation because a significant loss of tritium occurs at temperatures in excess of 200°C. ^{63}Ni radioactive sources are now readily available. Commercially available electron capture detectors can be operated at temperatures up to 350°C. Simmonds et al. [13] describe an electron capture detector which can be operated at 400°C. It was necessary to abandon the conventional parallel plate design and adopt a coaxial design because of the severe demands on insulating materials using the former configuration. The linearity of the ^{63}Ni electron capture detector has been improved by using analog conversion of the detector signal to a linear function of sample concentration [14]. Response curves have been obtained using pulse intervals from 100 μsec to 2,000 μsec at various detector temperatures. At sufficiently long pulse intervals, the converted response is linear to as high as 98% of detector saturation, making possible a linear dynamic range of 1×10^5. Recent studies have been made on variations of the response of the ECD to carrier-gas-flow rate [15], temperature [16], and polarizing voltage and the nature of the radioactive source [17].

2. Electrochemical Detectors

Electrochemical detectors are used to measure the reduction or oxidation products of organic compounds eluted from chromatographic columns. In the oxidative mode of operation, oxygen is added to the carrier gas and the organic compounds are converted to CO_2, H_2O, HCl, SO_2, and variable amounts of nitrogen oxides. In the reductive mode of operation, the compounds are converted to CH_4, H_2O, hydrogen halides, H_2S, and PH_3. The reductive process is cleaner and permits measurement of all four heteroatoms (N, S, Cl, and P). Oxidation permits determination of S and halogens only.

An electrical conductive cell can be used to measure HX, SO_2, and NH_3 [18]. Recently, techniques have been developed which permit continued and dependable use of the detector at or near its lower limit of sensitivity of about 0.1 ng of organic nitrogen [19]. Primarily, this involves keeping the hydrogen gas clean and maintaining high purity of the deionized water between pH 7 and 7.5. Improvements in the detector system include the use of hydrogen carrier gas, nickel wire catalyst, and Teflon tubing inserts to reduce absorptive losses of NH_3.

A microcoulometer can be used to measure HX, SO_2, H_2S, NH_3, and PH_3 [20]. This method is somewhat more specific than measurement of electrical conductivity since three cells can be employed: an I^-/I_3^- cell for the detection of compounds which are oxidizable, an Ag/Ag^+ cell for the detection of compounds which precipitate silver ion, and a platinum black $-H^+$ cell for compounds which change the pH of the electrolyte. However, specificity of both detectors can be improved by the utilization of adsorption or chemical reactions to selectively remove certain compounds. For example, in the determination of NH_3, strontium hydroxide is employed to remove acid gases.

3. Flame Photometric Detector (FPD)

The flame photometric detector can be used to measure compounds containing phosphorus and sulfur with very good selectivity. The carrier gas is mixed with hydrogen and air and the mixture burned. The lower part of the flame is shielded so that the detector does not respond to short-lived radicals derived from hydrocarbons and oxygenated compounds. Longer-lived radicals containing sulfur and phosphorus appear in the upper part of the flame. The light they emit is detected with a photomultiplier. A 526-mμ interference filter is used in the measurement of phosphorus and a 394-mμ filter in the analysis of sulfur in order to isolate the lines emitted by these radicals from other radiation. This detector is insensitive to temperature and pressure fluctuations, and shock and small changes in the burner gas flow rates. A disadvantage in the earlier models is that injection of the solvent containing the compounds to be chromatographed extinguished the flame, so that it was necessary to relight it after each injection. The FPD manufactured by Tracor, Inc. (Austin, Texas) is equipped with a screw-type needle valve to divert the solvent vapor during injection.

4. Alkali Flame Ionization Detector (AFID)

The flame ionization detector (FID) yields enhanced responses to compounds containing heteroatoms in the presence of alkali metal salts [21]. Enhancement is most pronounced with organophosphates and is believed to be due to increased ionization in the flame rather than volatilization of the metal salt. In early versions of this detector, the metal salt was fused onto a metal ring or helix which served as a collector. The response of these detectors decreased rapidly with time owing to evaporation of the metal salt. Improved methods for supplying salts to the flame have since been developed. A disc of alkali metal salt with a hole bored in the center can be used [22]. The disc is affixed to the quartz glass jet so that the mixture of carrier gas and hydrogen pass through the opening and are burned at its surface. The phosphorus detector manufactured by Hewlett-Packard (Palo Alto, California) makes use of a cylindrical platinum collector electrode containing a potassium chloride crystal with a bore in the longitudinal axis. This detector is said to be sensitive to 10 pg of parathion or less and to discriminate between organophosphates and hydrocarbons by a factor of 10,000. The sensitivity of the AFID varies widely with the nature of the alkali metal salt, the gas flow rate, and the positioning of the electrodes. Also, authors vary in the terms used to define sensitivity. Therefore, it is difficult to compare the relative sensitivities of various AFID designs reported in the literature.

B. High-Pressure Liquid Chromatography

High-pressure liquid chromatography will be briefly mentioned here, since it may be useful for the analysis of compounds which are impossible to chromatograph in the gas phase, such as the dithiocarbamates and dexon, and it has already been applied successfully to benzimidazole fungicides (see Section IV.E.1).

Basically, high-pressure liquid chromatography uses a pump to force the developing liquid through the chromatographic column at great speed. The pressure used is normally 150-4,800 psig and it can be constant or programmed. The column used is usually 1-2.4 mm in diameter and 10-300 cm long. The most commonly used column is 1 mm × 300 cm. Stationary phases usually employed for HPLC include ion-exchange resins coated on microglass beads, rigid porous silica (Porasil), rigid cross-linked polystyrene, and porous layer beads. The requirements for good packing materials include large surface area, a thin layer of adsorbant, and well-dispersed, open-structured surfaces readily available to the mobile phase. Of course, the stationary phase must not dissolve in the mobile phase or be modified by solute molecules.

The mobile phase can be a nonpolar solvent such as hexane, or highly polar solvents such as mixtures of water and methanol. For example, chlorinated hydrocarbon insecticides can be separated with the former and chlorobenzenes with the latter. Flow rate is generally 30-60 ml/hr. Column temperature is usually maintained at ambient to 80°C. The time required for elution of solutes is in the same range as for GLC, i.e., from a few minutes to an hour.

The detector most commonly used is a UV photometer employing the 254 nm mercury line from a low-pressure source. A spectrophotofluorometer has been used in our laboratories. Sensitivity is excellent, but of course the compound being measured must be fluorescent or capable of being converted to fluorescent derivatives. Other detectors which have been used include the flame ionization detector (following evaporation of the mobile phase), heat of adsorption detector, electrolytic conductivity detector, dielectric constant monitor, and polarograph.

It is believed that the electrical conductivity and microcoulometric detectors described in the previous section of this chapter would give excellent results, providing the mobile phase is evaporated before reduction or oxidation of the fungicides.

The principal drawback of HPLC compared to GLC is that the detectors available are not as numerous, selective, or sensitive. The main advantage is that it can be used for the analysis of heat-labile compounds. While HPLC may have important specific applications, it is probable that GLC will continue to be used for the analysis of most fungicides and other pesticides.

IV. SPECIFIC METHODS

In this section methods for the analysis of specific fungicides and groups of fungicides are described. Because of space limitations, complete details of the methods are not given, and the reader who wishes to use these procedures in the laboratory must refer to the original literature. However, sufficient detail is included to illustrate the principles on which the methods are based, their degree of complexity, limitations, and sensitivities, and the recoveries obtained when known amounts of fungicides are added to various substrates. Where satisfactory methods of analysis are not available, suggestions are offered on how these might be developed.

Although this chapter is not concerned with metabolism, analytical methods for fungicide metabolites are described when available. Most metabolism studies use radioisotopes. Therefore, the number of "cold" methods available for fungicide metabolites is relatively small.

The organization of this section posed some problems since fungicides, unlike insecticides, do not fall into simple categories such as chlorinated hydrocarbons, organophosphates, and carbamates. We finally decided to classify them according to both chemical structure and chemical reactivity and list these groups alphabetically. Any confusion which this may cause can be remedied by use of the index.

A. Active Halogen Compounds

Compounds of this class contain chlorine atoms which are easily replaced by nucleophilic groups such as RS⁻ and R-NH₂ in substitution reactions with metabolic substrates. This group of fungicides includes the captan family, s-triazines, and the chloroquinones. Compounds of this class are frequently unstable in the presence of tissue macerates and extracts, soil, and light. In addition, they may decompose on solid phases used for cleanup of more stable pesticides by chromatography. Hence, special precautions should be used with them. In particular, tissue and soil extracts should be processed immediately, and quinones should not be exposed to light.

Their chemical reactions with organic bases to yield chromophores are often used for their determination by colorimetry. Chlorothalonil (Daconil) could probably be included in this group since it contains reactive halogen atoms. However, its fungicidal properties were discovered long after colorimetry was replaced by GLC for most residue analyses.

1. The Captan Group

The captan group of fungicides consists of compounds containing a sulfur atom linked to both an imide nitrogen and a carbon atom substituted with chlorine.

(1) captan   (2) folpet   (3) captafol

All of these compounds contain active halogen atoms and react rapidly
with substrates containing thiol groups. They can be analyzed colorimetrically or by gas chromatography.

a. **Captan and Folpet.** Captan (1) and folpet (Phaltan) (2) can be analyzed
by colorimetric methods based on their reaction with resorcinol [23] or
pyridine and tetraethylammonium hydroxide [24].

In the resorcinol method, the fungicides are stripped from the fruit surfaces with benzene or the tissues are chopped and extracted with benzene in
the presence of Na_2SO_4. If necessary, the extracts are treated with activated charcoal to reduce interference from background color. The solvent
is evaporated, resorcinol solution is added, and the mixture is heated on an
oil bath for 25 min at 140°C. The reaction mixture is diluted to a suitable
volume with acetic acid and the absorbance measured at 429 nm.

Accuracy is ± 10% at 1 ppm, ± 25% at 0.5 ppm, and ± 100% at 0.1 ppm.
The limit of sensitivity is 5 to 10 μg of captan or folpet.

In the pyridine-alkali method, soil is extracted with dichloromethane on
a mechanical shaker or the surface of fruit is stripped with benzene. The
extract is decolorized with activated charcoal if necessary. The solvent is
evaporated and pyridine-tetraethylammonium hydroxide reagent is added.
Absorbance is read at 427 mm 3 min after the addition of the reagent.
Recoveries of captan from apples, strawberries, beans, and potatoes
ranges from 93 to 101%. The molar absorbance is 1.3×10^4 for captan
and 2.3×10^4 for folpet calculated from the concentration of fungicides in
the final test solutions.

Captan and folpet can be chromatographed on 10% DC-200 on 110-120
mesh Anakrom ABS at 185°C [25], 3% SE-60 at 195 to 200°C [26], and 5%
QF-1 on 80-100 mesh Chromosorb W [27]. Captan and folpet can be analyzed by the same method used for captafol except that the operating temperature should be lower [28].

b. **Captafol (Difolatan).** Captafol (3) can be measured colorimetrically by
reaction with pyridine and base. An unstable yellow compound is formed
which can be stabilized by addition of ethyl cyanoacetate [29]. However,
the preferred method of analysis is electron capture gas chromatography [29].

The crop sample is macerated and extracted with benzene in the presence of Na_2SO_4. Emulsions, if formed, are broken by addition of Celite
filter aid or centrifugation. The benzene is then filtered and dried with
sodium sulfate. If the sample contains 5 ppm or more captafol, it can be
analyzed without cleanup following concentration of the extract.

For oily crops, the benzene is evaporated and the residue is partitioned
between hexane and acetonitrile. The acetonitrile layer is evaporated and
the residue is applied as a streak at the origin of a tlc plate coated with
silica gel. Captafol standard is applied at both sides of the streak to serve
as indicators. The plate is then developed with 2.5% ether in benzene. The
center of the plate where the extract was applied is covered, and the sides

containing the standard are sprayed with $KMnO_4$ in acetone. The captafol spots are yellow against a pink background. The section of silica gel between the captafol spots is scraped off and extracted with acetone. The acetone is evaporated, the residue is dissolved in a measured volume of benzene, and an aliquot of this solution is injected into the gas chromatograph.

In the analysis of oils, the sample is added directly to the top of a column containing Florisil and allowed to drain to the top of the stationary phase. The column is washed successively with hexane, benzene–hexane (1:1), and finally benzene. The captafol is eluted with benzene and the solvent evaporated. The residue is then ready for further cleanup by TLC as described above.

Captafol is chromatographed on a column containing 5% QF-1 on acid-washed, silanized, 60-80 mesh Chromosorb G at 190°C using an ECD.

2. s-Triazines

a. Dyrene. The only s-triazine derivative to be developed as a commercial fungicide is dyrene, 2,4-dichloro-6-o-chloroanilino-s-triazine (4). Other members of this group such as atrazine and simazine have received widespread use as herbicides.

(4)

Two colorimetric methods are available for the analysis of dyrene. The first measures total extractable dyrene and reaction products of dyrene with metabolites [30]; the second measures unreacted dyrene only [31]. Since dyrene has a very short half-life in the presence of macerated tissues, results obtained by the two methods can be quite different.

To measure total extractable dyrene (some of the dyrene reaction products are not extractable) the sample is homogenized with isopropyl alcohol in a blender. Water is added and the mixture is extracted with benzene. The benzene is evaporated and the residue is hydrolyzed with HCl in ethanol for 5 hr to liberate o-chloroaniline. Zinc dust is then added to reduce background color. The mixture is cooled and filtered, and $NaNO_2$ is added to diazotize the amino group of the o-chloroaniline. Sulfamic acid is added to destroy the excess nitrite. The diazonium salt derived from dyrene is then coupled with N-1-naphthylethylenediamine to form a colored product, the absorbance of which is read at 540 nm.

Dyrene can also be measured by reaction with pyridine and alkali. The reactions which are believed to take place are illustrated in Figure 1. Maximum color development is obtained when the dyrene residue is permitted to react with aqueous pyridine saturated with glycine for 20 min. Aqueous NaOH is then added and absorbance of the solution read at 440 nm exactly 2 min later.

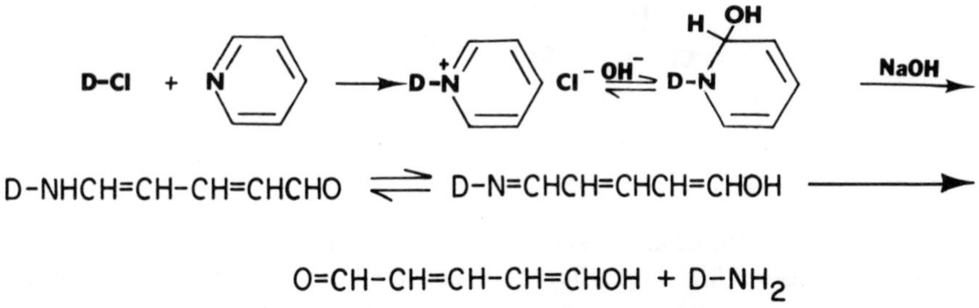

FIG. 1. Reaction of dyrene with pyridine and alkali; D-Cl symbolizes dyrene.

Both of these methods yield colored products with an absorbance of 0.1 at concentrations of 1 μg/ml in 1-cm cells. Recoveries from fortified soil and crop samples using the pyridine-alkali method are of the order of 90-100% if the samples are analyzed immediately after the addition of dyrene. However, it should be noted that dyrene has a half-life of only half a day in soil [32] and when added to animal chow is equally unstable, as determined by gas chromatography.

Dyrene residues can also be measured by electron capture gas chromatography (EC-GLC). Crop samples are extracted with acetonitrile in a Waring blender and the homogenates filtered [33]. Phosphate buffer is added to the filtrate which is then extracted with hexane. The hexane extract is washed and made up to volume. Aliquots of the extract are then chromatographed on a column containing 1.5% OV-17 and 1.95% QF-1 coated on Chromosorb W (HP) at a temperature of 190-195°C [34]. Recoveries of dyrene from tomatoes (10 ppm added) and strawberries (20 ppm added) averaged 90-93% when the standard used in making the calculations was dyrene added to plant extract. When the standards were injected directly into the gas chromatograph average recoveries were 103-112%, indicating that interfereing compounds were present.

3. Quinones

The two quinones used as agricultural fungicides are chloranil, tetrachloro-p-benzoquinone (5), which was used as a seed protectant, and dichlone (2,3-dichloro-1,4-naphthoquinone) (6), which is a seed and foliage protectant. Their residues can be measured by colorimetry, polarography, or gas chromatography. These compounds are unstable in the environment.

Dichlone has a half-life of only one day in moist soil [32], and both chloranil and dichlone are photolabile [35].

(5)   (6)

a. Chloranil. Chloranil on seed surfaces is extracted with acetone and the extract filtered [36]. Diethylamine is added and the intensity of the yellow-green color which is formed is read at 450 nm. Recovery of chloranil from seed is 90 to 118%. The absorbance of a solution containing 5 μg of chloranil per ml is 0.1.

Lane [37] developed a colorimetric method for chloranil on plant surfaces which is based on the sequence of reactions shown in Figure 2.

Chloranil is stripped from plant surfaces with benzene and the solution is treated with a mixture of adsorbants to remove interferences. Diphenyl-p-phenylenediamine reagent is added and the benzene is extracted with aqueous acetic-hydrochloric acid. The blue color which appears in the aqueous layer is read at 700 nm. Recoveries from broccoli, cabbage, cauliflower, and lettuce fortified with 0.06-0.50 ppm of chloranil averaged 100% with a high of 124 and a low of 80%. The limit of sensitivity is 0.03 ppm on these crops.

The gas chromatography of chloranil together with that of many other compounds has been investigated by several research groups [38-40].

b. Dichlone. A colorimetric method for dichlone on seed was developed which is based on extracting it with acetone and developing an orange color by addition of dimethylamine [41]. For the analysis of food crops, Lane [42] replaced the acetone with benzene to reduce background interference. Following surface stripping of the crops, the benzene is dried over anhydrous sodium sulfate, anhydrous dimethylamine is added, and the absorbance read at 495 nm. The method is applicable to all fruits and most vegetables. Recoveries of 18 fortified samples ranged from 85 to 100% with an average of 91%.

Newell et al. [43] describe a UV spectrophotometric method for the determination of dichlone in water, and Lane [44] has developed a colorimetric procedure for dichlone in milk which is based on causing it to react with pyridine and triethylamine in nitromethane solution.

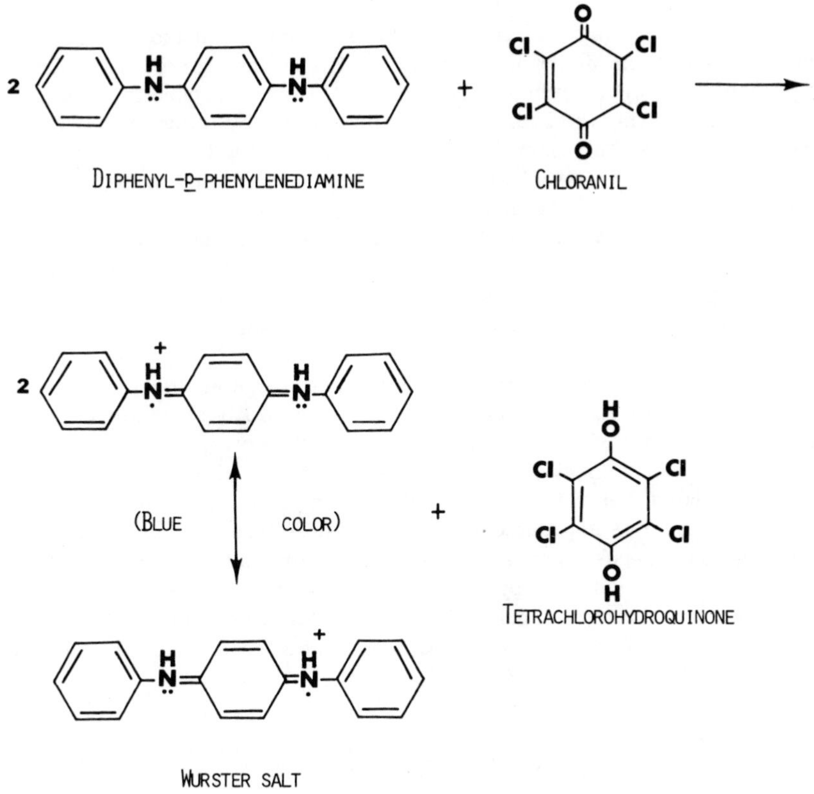

FIG. 2. Reaction sequence for colorimetric determination of chloranil.

B. Antibiotics

Antibiotics which have been used for the control of fungus diseases include cycloheximide (7), candicidin, and validacin.

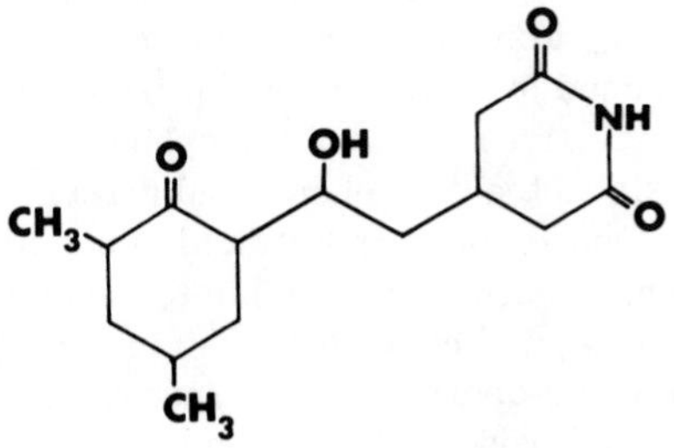

(7)

Compounds of this class are usually too polar to measure by GLC. However, some of them might be amenable to GLC following derivatization. It is probable that cycloheximide (7) could be chromatographed in the gas phase after masking the hydroxyl and imido groups by trimethylsilylation. The derivative could be measured selectively by using the electrical conductivity detector or the microcoulometer in the reduction mode for the detection of nitrogen as ammonia.

Another alternative would be to use high-pressure liquid chromatography using a UV photometer for detection. Sensitivity would be low since the carbonyl groups of this compound absorb weakly. However, this could be overcome by reacting the hydroxyl group with 2,4-dinitrobenzoyl chloride to form a compound with high UV absorbance.

In the absence of satisfactory chemical methods of analysis, it is necessary to measure residues of antibiotics by their ability to inhibit the growth of microorganisms such as Saccharomyces pastorianus.

For the measurement of cycloheximide, fruit samples are homogenized in a blendor and then refluxed with chloroform [45]. After cooling, the chloroform layer is separated and evaporated to dryness. The residue is suspended in aqueous ethanol and its capacity for inhibiting the growth of S. pastorianus is assessed by comparing diameters of zones of growth with those obtained in the presence of known amounts of antibiotic.

As little as 0.04 ppm of cycloheximide can be detected. Average recovery is 75% with a range of 62 to 87%.

C. Benzene Derivatives

The largest group of fungicides in this series are chlorine compounds. Unlike the compounds with active halogen, these are generally unreactive. Chlorothalonil is an exception, but is included in this group since colorimetric methods based on its reactivity with organic bases have not been developed.

In general, compounds in this group can be measured by GLC methods similar to those used for the chlorinated hydrocarbon insecticides. Following extraction and cleanup, the compounds are chromatographed and measured with an electron capture, microcoulometric, or electrical conductivity detector. In the case of pentachlorophenol (PCP), it is necessary to protect the hydroxy group by alkylation or acylation before GLC.

Several fungicides used for mildew control including dinobuton, dinacap and binapacryl are derivatives of 2,4-dinitrophenol. Their residues are measured by hydrolysis to the free phenol and determining its absorbance at 378 nm. However, these compounds can also be chromatographed in the gas phase. Therefore, it should be possible to develop methods for their residue analysis by using gas chromatography in combination with an electrical conductivity detector operated in the reduction mode. Ammonia would be produced through reduction of the nitro groups which can be measured with excellent sensitivity.

Dexon is not related to any of the fungicides included in this section except by virtue of possessing a benzene ring.

1. Compounds Containing Chlorine

a. Chloroneb. The fungicide chloroneb, 1,4-dichloro-2,5-dimethoxybenzene (8) is metabolized to 2,5-dichloro-4-methoxyphenol (9) by plants and animals [46, 47], and to a lesser extent to 2,5-dichlorohydroquinone (10) and 2,5-dichloroquinone (11) by cotton and bean plants [46]. In view of the known fungitoxicity and chemical reactivity of the chloroquinones it is interesting to speculate that chloroneb might be a fungicide precursor.

Chloroneb and (9) can be measured in soil and a variety of plant and animal tissues by microcoulometric gas chromatography (MC-GLC) [48]. A sample of the tissue is placed in a round-bottomed flask together with boiling chips, an antifoam agent, and 5N H_3PO_4. The mixture is steam distilled and the compounds are extracted continuously from the distillate with hexane in an apparatus described by Bleidner et al. [49]. The hexane is concentrated to a small volume and an aliquot injected into a gas chromatograph. Separation is accomplished on a column containing 10% DC-560 plus 0.2% Epon resin 1001 on high-performance Chromosorb W with the temperature programmed from 100 to 180°C. Recoveries of 90% can be obtained from a variety of samples at levels down to 0.2 ppm.

In order to liberate conjugates of (9) and possibly (10) in urine, the samples are acidified with concentrated HCl, heated under reflux for 4 hr, and

extracted continuously with ether in a liquid–liquid extractor [47]. Compounds (10) and (11) were not found in the urine of cows which had been fed
chloroneb.

b. Chlorothalonil (Daconil). Although chlorothalonil, 2,4,5,6-tetrachloro-
isophthalonitrile (12) contains active halogen atoms, its fungicidal properties were discovered long after gas chromatography had become an established method for residue analysis. Consequently, the development of
colorimetric methods was bypassed, and MC–GLC and EG–GLC are used
exclusively for its determination [50, 51].

The lability of these chlorine atoms must be taken into account in its
analysis, however, because in soil it is partially hydrolyzed to 4-hydroxy-
2,5,6-trichloroisophthalonitrile (13). Consequently, the initial analytical
procedure [50] was modified to include measurement of this compound [51].

(12)   (13)

The crop sample is chopped and extracted in a blender with acetone-
dilute H_2SO_4. The homogenate is filtered and the acetone evaporated. Then
HCl is added to the aqueous residue which is extracted with ether. The
ether is evaporated, and the residue is transferred to a Florisil column.
Chlorothalonil (12) is eluted first with 5% acetone in dichloromethane and
next compound (13) with 50% acetone in dichloromethane. The fraction containing (13) is treated with diazomethane to mask the hydroxyl group before
GLC.

The compounds are chromatographed separately on a column containing
20% DC–200 silicone oil on acid-washed 30–60 mesh Chromosorb P at a
temperature of 275°C.

Recoveries for both compounds are 80–100% at the 0.05 ppm level. Sensitivity is 0.02 ppm using the microcoulometric detector.

c. Dicloran. Dicloran, 2,6-dichloro-4-nitroaniline (14) can be measured
either colorimetrically [52-55] by MC–GLC [56, 57], or EC–GLC [58, 59].

In the colorimetric method described by Kilgore et al. [53], 400 g of
sample are chopped and extracted with 800 ml of benzene. The extract is

(14)

dried with Na_2SO_4 and filtered. An aliquot of the extract is chromatographed on a Florisil column to remove interferences. The eluate is
evaporated, the residue dissolved in acetone, and color developed by addition of aqueous KOH. Absorbance is read at 464 nm within 30 min after the
addition of alkali. Recoveries range between 75 and 100% with an average
of 86%. The method is sensitive to 0.05 ppm of dicloran.

In the gas chromatographic method [57], the sample is macerated with
benzene and the mixture centrifuged. The supernatant is separated and
evaporated to dryness. The residue is partitioned between hexane and
acetonitrile and the acetonitrile layer (which contains the dicloran) is
evaporated to dryness. The residue is transferred to a sedimentation tube
with benzene, solvent is evaporated, and the residue is then made up to the
desired volume. Extracts which contain chlorophyll require additional
cleanup on a Florisil column.

Chromatography is carried out on a column containing 5% DC-200 silicone oil coated on 90 mesh Anakrom ABS operated at 150-160°C. A microcoulometer is used for detection. Recoveries are 80% or better from samples fortified with 0.1-10 ppm of dicloran.

d. Pentachlorophenol (PCP). Gas chromatographic methods have been
developed for the analysis of PCP in soil, water, and fish [60], blood [61],
fats and oils [62], and blood, urine, tissue, and clothing [63]. The hydroxyl
group is masked by methylation or acetylation before chromatography, and
electron capture is used for detection.

In the analysis of plant tissues [64], the sample is chopped, acidified
with dilute H_2SO_4, and extracted with benzene. Anhydrous Na_2SO_4 is added
to take up water and the mixture is filtered. With some crops such as
apples no cleanup is required. When cleanup is needed, this is accomplished
by shaking the benzene twice with concentrated sulfuric acid followed by
washing with water. The solution is then dried over anhydrous Na_2SO_4 and
evaporated to dryness. Diazomethane reagent is added to the residue. When
methylation is complete as indicated by the persistence of a yellow color,
the solvent is evaporated and the residue is redissolved in a standard volume of benzene. An aliquot of this is injected into a gas chromatograph con-

taining a column packed with 5% Dow 11 silicone oil coated on acid–washed
60–80 mesh Chromosorb W at 180°C. Recoveries range from 76 to 90% and
as little as 0.01 ppm of compound can be detected.

   Chau and Coburn [65] mask the phenol group by acetylation rather than
methylation in the determination of PCP in natural and waste waters on the
grounds that esters of phenoxyacetic herbicides interfere in the measurement of the methyl ether of PCP by GLC.

   The preserved water samples are extracted three times with benzene to
remove PCP, which is then returned to the aqueous phase by extracting
three times with 0.1M K_2CO_3 solution. Chlorinated hydrocarbons and polychlorinated biphenyls remain in the aqueous phase. Acetic anhydride and
hexane are added to the carbonate solution and the mixture is stirred for
1 hr. The hexane layer is then removed and analyzed by EC-GLC. For low
levels of PCP (less than 0.5 ppb) the aqueous layer is extracted two more
times with hexane and the combined extracts reduced to a volume of 1 ml
before chromatography.

   Chromatography is carried out on two columns: one of these is packed
with 11% QF-1–OV-17 coated on 80–100 mesh acid-washed silanized Chromosorb Q; the second is packed with 3.6% OV-101 and 5.5% OV-220 coated
on 80-100 mesh acid-washed silanized Chromosorb W. Both columns are
operated at 200°C. Recoveries are of the order of 85 to 90% at levels of
10 to 1,000 ng/liter. As little as 0.01 ppb of PCP in 1 liter of water can
be detected.

e. Quintozene. Quintozene (15) can be determined by polarography [66, 67]
or colorimetry [68]. Gas chromatography is particularly useful for separating quintozene from the closely related fungicides and impurities such as
(16) and (17) [67, 69], and metabolites such as (19) and (20) [70, 71].

(15)   (16)   (17)

(18)   (19)   (20)

The colorimetric method of analysis for quintozene [68] is based on hydrolyzing it to nitrite and pentachlorophenol (PCP) with ethanolic KOH. The nitrite is used to diazotize procaine hydrochloride, and the diazonium salt is coupled with 1-naphthylamine to yield a magenta colored product. Absorbance is read at 525 nm.

Quintozene can be determined in crop samples by EC-GLC [72]. The sample is extracted with hexane, and the quintozene is partitioned into dimethylformamide. After further cleanup on an alumina column, the residue is chromatographed on three columns: 5% EGSS-X on 100-120 mesh Chromosorb W operated at 170 or 200°C; 1.3% Apiezon L and 0.15% of Epikote 1001 on 60-80 mesh silanized Chromosorb G at 200°C; and 1% Apiezon L and 0.15% of Epikote 1001 on 60-80 mesh silanized Chromosorb G.

For extracting quintozene and metabolites from animal tissues, the samples are homogenized with acetonitrile and Na_2SO_4. The acetonitrile is diluted with water and extracted with hexane to recover quintozene and its metabolites. The hexane layer is then dried over Na_2SO_4. Following concentration of the extract, an aliquot is chromatographed on a column containing 2% SE-30 coated on acid-washed, silanized, 80-100 mesh Chromosorb G operated at 170°C. The compounds elute in the order (17), (16), (18), (15), and (19); (15) and (18) are not completely resolved.

2. Dexon

Dexon (21) is a highly polar compound which is too unstable to chromatograph in the gas phase. However, it will couple with resorcinol (22) in a light-catalyzed reaction to yield product (23) which can be measured colorimetrically [73].

(21)   (22)   (23)

Dexon decomposes on exposure to light; therefore, during extraction it must be protected by the addition of aqueous Na_2SO_4. Its solubility in water makes purification by dialysis possible.

Crop samples (with the exception of cottonseed) are blended with aqueous 1% sodium sulfite solution. The slurry is transferred to a seamless cellulose dialysis tube and dialyzed in the dark against 1% aqueous Na_2SO_3 solution for 20 hr at room temperature. The volume of the diffusate is then

measured, following which it is transferred to a glass refrigerator tray set
in a larger tray containing ice. Resorcinol is added to the solution followed
by aqueous NaOH. The solution is then irradiated for 30 min followed by
addition of phosphate buffer and HCl. The dye (23) is extracted from the
aqueous solution with benzene and reextracted into aqueous KOH. Phosphate
buffer and KOH are added to the aqueous phase and the dye is again extracted
with benzene. The extract is dried over Na_2SO_4 and the absorbance is
read at 450 nm.

In the case of cottonseed, benzene is added during blending to dissolve the oil, and following dialysis the diffusate is extracted with benzene.

Sensitivity is of the order of 0.02-0.1 ppm depending on the crop. Recoveries range from 65 to 136%.

3. Dinitrophenol Derivatives

Several fungicides and acaricides are derivatives of dinitrophenol.

(24) dinobuton   (25) dinocap   (26) binapacryl

Their residues can be determined spectrophotometrically by hydrolysis and
measuring the intensely colored dinitrophenate ion [74].

The sample is extracted with hexane using Na_2SO_4 as a dehydrating
agent. The filtered extract is evaporated to dryness and the residue dissolved in benzene. The benzene solution is washed with acid $KMnO_4$, excess
reagent destroyed with H_2O_2 and the benzene layer separated and dried.
The solution is then added to a chromatographic column containing Grade-I
neutral alumina. Plant pigments are removed by washing the column with
a mixture of acetonitrile and propanol, and the nitrophenol is eluted with
1% n-butylamine in aqueous acetone. Absorbance is read at 378 nm. Recovery of dinobuton from apples, black currants, and cucumber is 83-109%
at levels of 0.2 to 2 ppm. This method can be applied to other esters of
dinitrophenol with minor modifications.

Dinocap can be measured in formulations by GLC [75]. The parent
compound is chromatographed on a column containing 60-80 mesh Gas
Chrom Q coated with 3% QF-1 silicone oil with the temperature programmed
from 100 to 230°C. The FID is used for detection, but use of the electrical
conductivity detector should render the method more specific in the analysis
of tissue extracts.

D. Fumigants

1. Haloalkanes

Volatile compounds including D-D (mixture of cis- and trans-1,3-dichloro-propene and 1,2-dichloropropane), ethylene dibromide, methyl bromide, and Nemagon (1,2-dichloro-3-chloropropane) are widely used as soil fumigants. In the past, these were measured by elemental analysis for chlorine or bromine. Methods employed include neutron activation and liberation of halogen atoms, followed by their titration or measurement by colorimetry [76]. Many of these methods have been replaced by GLC. However, these compounds are used primarily for the control of nematodes and wire worms, and are only secondarily fungicides.

An important exception is Lanstan (1-chloro-2-nitropropane) which is effective against a wide range of soil fungi. Lanstan residues can be analyzed by EC-GLC [77]. Crop samples are macerated in a blender and then extracted with benzene/methanol. The benzene layer is separated, washed with water, and dried over anhydrous Na_2SO_4. If necessary, the extract is shaken with Florisil to remove interferences. An aliquot is injected into a gas chromatograph packed with a column containing 4% XE-60 coated on 80 mesh Chromosorb W operated at a temperature of 95°C. The injection port is kept below 120°C to minimize interferences caused by pyrolysis of crop constituents. Sensitivity of the method is 10 ppb. Recoveries range between 75 and 105% in the 10- to 100-ppb level for 20 crops evaluated.

A metabolite of Lanstan, 2-nitropropanol, can be extracted from cotton-seed with ethyl acetate and measured by EC-GLC on a column packed with 5% Carbowax 20-M coated on 60-80 mesh Gas Chrom Q at 150°C [78].

2. Vapam

Vapam (27) is not volatile but decomposes in soil to yield methyl isothiocyanate (28) as shown below.

$$\underset{(27)}{CH_3\text{-}NH\text{-}\overset{\overset{S}{\|}}{C}\text{-}SNa} \longrightarrow \underset{(28)}{CH_3\text{-}N=C=S} \quad + \quad NaSH \xrightarrow{NH_3} \underset{(29)}{CH_3\text{-}NH\text{-}\overset{\overset{S}{\|}}{C}\text{-}NH_2}$$

Compound (28) is volatile, but it can be determined gravimetrically by first converting it to (29) [79].

In addition to methyl isocyanate, vapam also yields carbon disulfide, hydrogen sulfide, methylamine, and other products when it is mixed with soil [80]. These compounds can be separated on a column containing 4% Silicone Oil 550 and 1% Carbowax 400 coated on Fluoropak 80 which is operated at 70°C.

E. Heterocyclic Fungicides

Heterocyclic compounds used as fungicides include benzimidazole derivatives, and other compounds containing nitrogen or sulfur in the heterocyclic ring. Benzimidazoles are one of the few groups of pesticides which can be measured directly by fluorometry. In most other cases, substitution with halogen and other electronegative groups tends to quench fluorescence. Benzimidazoles can also be separated from impurities by high-pressure liquid chromatography. In the example given in this chapter, the column eluate is monitored with a UV photometer. Much greater selectivity and sensitivity could probably be achieved with a spectrophotofluorometric detector similar to the one used in the analysis of polynuclear arenes [81].

Since these compounds contain heteroatoms, they could also be measured by use of an electrical conductivity detector operated in the reduction mode for nitrogen and in the oxidation mode for sulfur. Sulfur could also be measured selectively with a microcoulometer or a flame photometric detector. At this time, however, these highly selective methods have been applied only to the analysis of carboxin.

1. Benzimidazole Derivatives

a. Benomyl. Benomyl (30) breaks down in aqueous solution and within plants to yield (31) [82]. This compound is then hydroxylated to yield (32) and (33) which are found as conjugates with glucuronide and sulfate in the urine of rats.

(30)   (31)

(32)   (33)

Fluorometric and colorimetric methods have been developed for the determination of benomyl which involve quantitative conversion of benomyl to (31) and hydrolysis of this latter compound to 2-aminobenzimidazole by a two-step procedure [83, 84]. A method has also been described for the direct measurement of (31) by fluorometry, following conversion of any benomyl present in the sample to (31) by quantitative catalytic decomposition on a magnesium oxide-Celite-alumina column during cleanup [85].

Compounds (30) through (33) can be determined in milk, urine, feces, plant and animal tissues, and soil by high-speed liquid chromatography [86, 87].

Details of sample preparation vary depending on its nature. In the analysis of milk [86], phosphoric acid is added to the sample and the mixture is heated on a steam bath for 1 hr to hydrolyze the conjugates to (32) and (33). The hydrolysate is then neutralized with NaOH solution and extracted with ethyl acetate, the phases separated, the organic layer dried, and the volume reduced in an evaporator. The concentrate is extracted with dilute HCl, and the aqueous phase backwashed. The HCl is neutralized with dilute NaOH and the compounds are extracted back into ethyl acetate. The organic layer is dried and the solvent evaporated. The residue is made up to standard volume in dilute aqueous phosphoric acid.

Chromatography is carried out in a $1,000 \times 2.1$ mm i.d. stainless steel column packed with a strong anion-exchange material (Zipax SAX), and the compounds measured with UV photometer operated at 254 nm.

Sensitivity is of the order of 0.01 ppm to 0.2 ppm, depending on the nature of the sample. Recoveries average 80% from milk and urine, and 50 to 80% from feces and tissues.

b. Thiabendazole. Thiabendazole (34) can be analyzed by spectrophotometry and gas chromatography [88, 89].

(34)

Merck [90] developed a fluorometric procedure for analysis in crops. A slurry of the tissue in water and sodium acetate buffer is extracted with ethyl acetate. The organic layer is extracted successively with dilute NaOH and distilled water to remove interferences. The thiabendazole is then extracted from the ethyl acetate with dilute HCl. The HCl extract is neutralized with NaOH, buffered with sodium acetate, and reextracted into ethyl acetate. The compound is finally extracted from the ethyl acetate into dilute HCl. Intensity of fluorescence is measured at an excitation wavelength of 302 nm and an emission wavelength of 360 nm.

2. Carboxin and Oxycarboxin

Carboxin (35) and oxycarboxin (36) residues in crops can be measured by a colorimetric procedure which is based on their hydrolysis to yield aniline which is then coupled with p-dimethylaminobenzaldehyde to yield a chromophore [91]. Alternatively, the aniline can be measured by MC-GLC [92].

The principal metabolite of carboxin is carboxin sulfoxide. Evidently, oxycarboxin is not reduced to the sulfoxide. The currently available methods do not distinguish between carboxin, oxycarboxin, and the sulfoxide since only the amount of aniline is measured in all three cases.

$$(35) \qquad (36)$$

In the colorimetric method, the sample is macerated and placed in a distillation flask together with aqueous sodium hydroxide and an antifoam agent. Zinc granules and $TiCl_3$ are added to provide a reducing environment in order to prevent loss of aniline. The mixture is heated until 150 ml of distillate is collected. Solid NaCl is dissolved in the distillate which is then extracted with hexane. The hexane layer is discarded, the pH of the aqueous phase is adjusted to 6.5 with H_3PO_4, and the aniline extracted into benzene. After the benzene is washed with water, the aniline is extracted from the benzene into aqueous acetic acid. Dimethylaminobenzaldehyde reagent is added, and after various adjustments, absorbance is read at 440 nm and corrected for background absorbance at 490 nm.

Recoveries from crops average 89% at the 0.2 to 20 ppm level. The minimum detectable level is 0.13 ppm in crops.

This method is not sensitive enough for the detection of carboxin and its sulfoxide in seeds at the 0.2 ppm level. This can be accomplished by measuring, by GLC, the aniline which is released on hydrolysis.

The seed is homogenized with methanol in a blender following which the solid and liquid phases are transferred to a Soxhlet extractor and extracted for 3 hr. The methanol is cooled and water and saturated NaCl solution are added. The solution is washed twice with hexane and the organic layers discarded. Water is added, the solution is extracted three times with

chloroform, and the chloroform is dried over Na_2SO_4. Most of the chloroform is removed by evaporation, toluene is added, and the solvent is again
reduced to a small volume.

Aqueous NaOH, zinc granules, and $TiCl_3$ are added and the mixture is
distilled until 100 ml has been collected. The pH is adjusted to 1.0 with
H_2SO_4, the solution is extracted with benzene, and the benzene is discarded.
The pH is then adjusted to 11 with NaOH and the aniline extracted into ether
which is dried over Na_2SO_4. The ether solution is added to a beaker together
with aqueous H_2SO_4 and the ether evaporated. The solution is then made
alkaline with NaOH and weighed. An aliquot of this solution is injected into
a gas chromatograph equipped with a microcoulometric detector.

The aniline is chromatographed on a column packed with 4% Carbowax
20M, 1% Igepal CO-880, and 1% KOH coated on 70-80 mesh Anakrom SD
with the temperature programmed from 100 to 200°C.

Recoveries range from 73 to 110% at the 0.2 ppm level. Modifications
of the method are required for the analysis of cotton and seed and seed oils.

3. Dichlozolin and Dimethachlone

Residues of dichlozolin (37) in grapes can be measured by EC-GLC [93].
Gas chromatography [94] has also been used to analyze formulations of dichlozoline and the related fungicide dimethachlone (38).

(37)   (38)

The fruit is extracted with acetone and the extract is diluted with water.
The dichlozoline is then extracted into chloroform and impurities removed
on a Florisil column.

Chromatography is performed using a column packed with 5% QF-1
coated on silanized acid-washed 80-100 mesh Chromosorb G at a temperature
of 170°C. The limit of detection is 0.05 ppm and recoveries average 95%.

These compounds can also be chromatographed on columns packed with
5% DEGS or 10% QF-1 coated on acid-washed Chromosorb W at a temperature of 185°C [94].

4. Drazoxolon

Drazoxolon exists primarily as (39) where hydrogen bonding occurs between
the carbonyl and NH groups [95]. However, in alkaline solution it tautomerizes
to (40).

(39)   (40)

The solubility of (40) in alkali and its absorbance are used in the cleanup and determination of drazoxolon by colorimetry [96].

Fruit samples are macerated with chloroform and the phases allowed to separate. The chloroform layer containing (39) is filtered through a bed of Hyflo Super-Cel supported by filter paper in a Büchner funnel. The chloroform is washed with dilute HCl and (40) is then extracted into dilute 0.1N NaOH. The aqueous layer is washed with chloroform and then acidified with HCl to shift the equilibrium in the direction of (39). The drazoxolone is extracted back into chloroform, the extract is filtered, and the absorbance is read at 400 nm. The method is sensitive to 0.05 ppm of compound. Recoveries range from 64 to 106%.

5. Morestan

Morestan (41) is determined by EC-GLC [97]. Because of the presence of nitrogen in the molecule it could also be measured with higher selectivity using the electrical conductivity detector or microcoulometer. The flame photometer could be used for selective measurement of sulfur.

(41)

For the extraction of moist crops, the sample is blended with acetone and water, and the homogenate extracted with Skelleysolve B. The organic layer is dried with Na_2SO_4 during filtration, and the solvent is evaporated. The residue is redissolved in a standard volume of Skelleysolve for chromatography.

For dry crops, the sample is blended with Skelleysolve at high speed, and the homogenate filtered into a separatory funnel. The extract is washed twice with aqueous NaOH followed by a wash with dilute HCl. The organic phase is dried during filtration with Na_2SO_4 and the solvent removed in a vacuum evaporator. The residue is dissolved in benzene and transferred to a Florisil column. The column is rinsed with benzene and the morestan eluted with acetone. The acetone is evaporated and the residue redissolved in Skelleysolve for chromatography.

Chromatography is performed on a column packed with 5% QF-1 coated on acid-washed, silanized, 70-80 mesh Chromosorb G at an operating temperature of 200°C. The method is sensitive to 0.1 ppm or less of morestan. Recoveries range from 82 to 107%.

6. Mylone

Mylone (42) is synthesized by the reaction of methylamine, carbon disulfide, and formaldehyde.

(42)

On heating in dilute acid the reaction is reversed and the carbon disulfide which is evolved is measured colorimetrically by a method similar to that used for the dithiocarbamates [98].

7. Parnon

Residues of parnon (43) in agricultural crops are measured by a procedure which is based on EC-GLC [99].

(43)

The crop sample is homogenized and blended twice with acetone. The homogenate is filtered and NaCl solution is added to the filtrate which is then extracted twice with hexane to remove the parnon. The hexane is extracted three times with aqueous HCl and the organic layer discarded. Na_2CO_3 solution is added to the aqueous phase until the pH is at least 10, and then the solution is saturated with salt. The parnon is extracted into chloroform, the solvent evaporated, and the residue made up to volume in benzene.

Chromatography is carried out on a column packed with XE60 coated on Gas Chrom Q at a temperature of 235°C.

8. Pipron

Pipron (44) and its principal metabolite, 3,4-dichlorobenzoic acid (45) can be measured in samples of high water content by a procedure based on EC-GLC [100]. The two compounds are obtained in different fractions during isolation; compound (45) is methylated before GLC.

(44)   (45)

The crop sample is blended twice with acetone, the homogenate filtered, and the acetone evaporated. The aqueous residue is transferred to a separatory funnel with Na_2CO_3 solution using enough to keep the pH above 10. The solution is then extracted twice with hexane and once with benzene. Pipron is partitioned into the organic layer and its metabolite into the aqueous layer.

Polyethylene glycol in dichloromethane is added to the organic layer and the solvent is evaporated. The residue is dissolved in benzene and the flask is washed with a coagulating solution containing NH_4Cl and H_3PO_4; then the solids are removed by vacuum filtration. The filter is washed with coagulating solution and the filtrate is extracted three times with chloroform. The combined organic layers are dried during filtration with Na_2SO_4, the solvent evaporated to near dryness, and the residue made up to standard volume. An aliquot of this solution is injected into a gas chromatograph.

Chromatography is carried out on a column packed with 2% ECNSS-M coated on 80-100 mesh Gas Chrom Q at a temperature of 170°C.

In the analysis of compound (45), the Na_2CO_3 solution is acidified with concentrated HCl to a pH of about 1 and extracted three times with chloroform. The combined extracts are evaporated to dryness.

Benzene is added to the residue and compound (45) is methylated with a freshly prepared reagent containing N-methylnitrosamine in ether which has been treated with concentrated NaOH in the cold. The solvent is then evaporated and the residue is made up to standard volume.

Chromatography is performed on a column containing 5% XE-60 coated on 80-100 mesh Gas Chrom Q at 140°C.

9. Terrazole

Terrazole (46) is metabolized in animals and plants to 3-carboxy-5-ethoxy-1,2,4-thiadiazole (47).

(46)   (47)

For the analysis of Terrazole, tissue is homogenized in a blender with freshly distilled hexane and anhydrous Na_2SO_4. An aliquot of the extract is analyzed by GLC without further treatment [101]. The column is packed with 3% SE-30 coated on acid-washed, silanized 80-100 mesh Chromosorb G at a temperature of 180°C.

The method is sensitive to 0.005 ppm of Terrazole in plant tissues and recovery at this level is greater than 70%.

F. Metal Compounds

1. Copper

Bivalent copper compounds used as fungicides include bordeaux mixture, burgundy mixture, and tribasic copper sulfate. A monovalent compound, cuprous oxide is also used.

These are protectant fungicides and are not absorbed or translocated to an appreciable extent. Therefore, residues are usually removed from fruit or foliage by surface stripping, and the copper measured colorimetrically.

In studies on bordeaux mixture [102], copper deposits were removed from leaf sections by agitating them with 1N HCl for 30 sec, followed by washing with water. The combined extracts were made up to volume, and aliquots were made alkaline by the addition of ammonium hydroxide. Aqueous sodium diethyldithiocarbamate solution was added, and absorbance read at 455 nm. Bis(2-hydroxyethyl) dithiocarbamate, known as cuprethol, has been used for the analysis of copper in water [103, 104].

2. Dithiocarbamates

The dithiocarbamates include metal salts of dimethyldithiocarbamic acid (48) and ethylenebisdithiocarbamic acid (49).

(48)   (49)

The most widely used fungicides in this group are the iron (ferbam), and zinc (ziram) salts of (48) and the manganese (maneb), zinc (zineb), and sodium salts of (49). The oxidation product of (48), bis(dimethylthiocarbamyl) disulfide (thiram) and a polymer of a mixed oxidation product of (49) and zineb (Polyram) are also included in this group.

a. Methods Based on CS_2 Evolution. When treated with aqueous sulfuric acid these compounds decompose to yield the corresponding metal ion, free base, and in all cases carbon disulfide [105]. Any of these products could be measured to determine residue levels but carbon sulfide evolution is the basis of most colorimetric procedures [106-108].

Compound (49) can yield (50) and hydrogen sulfide at high concentrations and if not properly wetted by the solvent [109] this would result in low recoveries.

(50)

To avoid this, the total amount of fungicides should not yield more than 100 μg of carbon disulfide [110] and aqueous tetrasodium (ethylenedinitrilo)-tetraacetate (Versene) is added to the reaction mixture to serve as a wetting and dispersing agent [108].

Carbon disulfide evolution is carried out in the apparatus shown in Figure 3, where A is a 500-ml flask which contains the sample, B is a dropping funnel from which sulfuric acid is added to the sample, C is a trap containing aqueous lead acetate to absorb hydrogen sulfide, and D is a trap which contains a reagent for trapping the carbon disulfide [110].

The crop sample is usually diced into 1/4 to 1/2-inch cubes before adding it to the reaction flask. However, if desired, ferbam, ziram, and thiram can be stripped from plant tissues with chloroform, maneb and zineb with aqueous versene, and nabam with water.

Water and boiling sulfuric acid are added to the sample and the mixture is refluxed for 30-45 min to liberate carbon disulfide. This is trapped in Viles reagent in D which is a solution of copper acetate, diethylamine, and

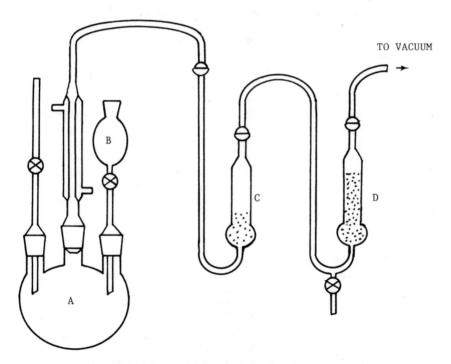

FIG. 3. Apparatus for evolving carbon disulfide from residual amounts of dithiocarbamates. Reprinted from Ref. 110, p. 1113, by courtesy of J. Assoc. Offic. Agr. Chemists.

triethanolamine in aqueous ethanol. The cupric diethyldithiocarbamate formed is measured colorimetrically at a wavelength of 380 nm.

This method has been applied to tissues of many plants and recoveries of 80-100% are reported [110]. Maximum sensitivity is on the order of 0.1 ppm.

Keppel [111] has suggested substitution of an aqueous solution of HCl and $SnCl_2$ for H_2SO_4 in the hydrolysis of the bisdithiocarbamates resulting in higher yields of ethylenediamine. The reasons for this are unknown.

McLeod and McCully [112] describe a head space gas procedure for screening food samples for dithiocarbamate residues. The sample is placed in a closed container and digested with $HCl-SnCl_2$ reagent for 30 min at 60°C. The head space gas is then sampled with a syringe and its CS_2 concentration measured by injecting it into a gas chromatograph equipped with a flame photometric detector and a filter to isolate the 394 nm sulfur line. The column contains 10% SE-30 coated on 80-100 mesh silanized Chromosorb W and is operated at 60°C.

b. Other Methods. Methods based on CS_2 evolution do not distinguish between the various dithiocarbamates. These compounds are sparingly volatile and gas chromatographic or other procedures have not yet been developed for the intact molecules. Therefore, in order to identify the dithiocarbamates in samples of unknown origin it is necessary to establish the identities of the amines and metal ions in addition to measuring the amount of CS_2 evolved.

Newsome [113] has recently described a method for the measurement of ethylenediamine produced by the hydrolysis of zineb, maneb, and Polyram by gas chromatography.

The crop sample is cut into small pieces and heated to reflux for 60 min with aqueous 1N HCl containing $SnCl_2$. The digest is filtered and the filtrate percolated through a column containing Dowex 50W-X8 ion-exchange resin to bind the ethylenediamine. The amine is then eluted with aqueous $NaHCO_3$, the solution acidified, and the water evaporated. The amine is converted to its bis(trifluoroacetate) by treatment with trifluoroacetic anhydride. The diacetate is chromatographed on a column containing 5% butanediol succinate on 100-120 mesh Chromosorb W at 180°C.

Residues of free ethylenediamine are determined by macerating the plant samples in dilute HCl, filtering the macerate, and applying the filtrate to the ion-exchange column without prior hydrolysis.

Recoveries of zineb, maneb, and Polyram when added to apples at levels of 0.16 to 1.4 ppm varied from 77 to 102%. Recoveries of zineb from lettuce and tomato at these levels was 84 to 101%. Sensitivity of the method is limited to 0.1 ppm by sample background.

Methods for the analysis of residues of dimethyldithiocarbamates by chromatography of derivatives have not been reported. However, such methods should be feasible since Onuska and Boos [114] have converted N,N-dialkyldithiocarbamates (48) to S-alkyl esters (51) by reaction with iodoalkanes or diazomethane.

(48)　　　　　　　　　　(51)

S-Methyl-N,N-dimethyl- and S-ethyl-N,N-dimethyldithiocarbamate, products which would be expected from ferbam and ziram, were chromatographed on a column containing 10% Apiezon L coated on 80–100 mesh Varaport 80 at a temperature of 250°C.

c. Ethylenethiourea. Any discussion of the dithiocarbamates must include mention of ethylenethiourea (52) since this compound is formed during the manufacture and storage of the bisdithiocarbamate fungicides, and may be generated from them after field application [115–120]. Moreover, this compound has been reported to be a carcinogen in mice [121].

Direct determination of ethylenethiourea by gas chromatography is not practical since the peaks tail because of its high polarity. Peak profiles can be improved by conversion to a butyl derivative by reaction with 1-bromobutane [122]. Enhanced sensitivity to the electron capture detector can be obtained by benzoylating the compound, followed by trifluoroacetylation [123]. However, this method did not give consistent and complete recoveries when evaluated by Nash [124], and an interfering peak was observed in the controls and reagents.

Nash recommends conversion of ethylenethiourea to 2-(benzylthiopentafluorobenzoyl)-2-imidazoline (54) by the following series of reactions.

(52)

(53)　　　　　　　　　　(54)

The reaction product (54) is chromatographed on columns containing 3% OV-17 on 100-120 mesh Gas Chrom Q; 3% OV-17; 3% OV-1; and 10% DC-200, all on 80-100 mesh Chromosorb W.

For the analysis of plant samples, 2-5 g of tissue are homogenized with methanol, filtered, and the filtrate refluxed with benzyl chloride. The mixture is diluted with aqueous HCl, extracted with chloroform, and the organic layer discarded. The aqueous phase is made alkaline and reextracted with chloroform to isolate the benzoyl derivative. After drying, the extract is treated with pentafluorobenzoyl chloride and pyridine for 15 min at 65°C. The reaction mixture is then added to a mini-Florisil column, the column washed with benzene, the product eluted with a mixture of ethyl ether and petroleum ether, and the volume reduced before EC-GLC.

Recoveries from soybean leaves, apples, and spinach at levels of 0.01 to 1.3 ppm of compound ranged from 93 to 120%. Maximum sensitivity of the method is 0.005 ppm. Although the results were not conclusive, it appeared that 1-3% of the parent fungicides (maneb, etc.) were degraded to ethylenethiourea during benzoylation.

3. Mercury

Methods for the analysis of mercury derivatives by chromatography have been reviewed by Fishbein [125]. In general, gas chromatographic methods using the ECD are most sensitive, the limit of detection of phenylmercuric chloride being about 10 ng.

Phenylmercury compounds are extracted from body fluids with benzene in the presence of 6 N HCl [126]. The benzene layer is washed with water, and the mercury compounds then extracted with a dilute aqueous solution of glutathione which serves to complex them. The glutathione solution is washed with benzene and then 6 N HCl is added. The compounds are extracted back into benzene and the volume reduced before injection of an aliquot of the solution into a chromatograph. Chromatography is carried out on 5% DEGS coated on 60-80 mesh Shimalite W at a temperature of 185°C. Alkyl-mercury chlorides can be separated on a column containing 25% DEGS on 60-80 mesh Shimalite at 160°C.

Mercury fungicides can also be chromatographed in the gas phase as their dithiozonates [127]. Crop samples (apples, potatoes, tomatoes), are macerated with a solution of cysteine in a mixture of 2-propanol and ammonium hydroxide at pH 8. After the debris settles, the supernatant is decanted and the tissue is extracted twice more. The combined extracts are centrifuged. The clear liquid is diluted with sodium sulfate solution, extracted with ether, and the extracts discarded. The organomercury compounds are extracted from the aqueous phase with a dilute solution of dithiozone in ether. Volume is reduced in a Kuderna-Danish evaporator and an aliquot of the solution is injected into a gas chromatograph.

Chromatography is conducted on 2% DEGS coated on acid-washed, silanized 60-80 mesh Chromosorb G at temperatures of 140 and 150°C for aliphatic and 170 and 180°C for aromatic mercury compounds. The latter tend to yield broad peaks; profiles can be improved by reducing the amount of stationary phase to 1%.

This method is sensitive to 0.5 ng of arylmercury compounds. Recoveries from fortified crop samples are 85-95% at levels of 0.01 to 5 ppm.

If an electron capture detector with a tritium foil is used, detector temperature should be kept at 180°C and the amount of mercury injected kept below 100 μg to avoid poisoning it.

Extracts prepared by this method can also be analyzed by thin-layer chromatography (TLC) on plates coated with silica gel or alumina. Hexane-acetone and petroleum ether-acetone in a variety of ratios are used for development. About 2 μg of each dithiozonate can be observed. Treatment of the plates with reagents is not required since these compounds are naturally yellow or red.

4. Tin Derivatives

Triphenyltin hydroxide can be determined colorimetrically [128]. The closely related miticide tricyclohexytin hydroxide (55) can be analyzed by gas chromatography [129], and it appears that this method could be used for the analysis of the aromatic analogue with minor modifications. Tricyclohexyltin hydroxide is degraded in the environment as shown below (where Cy is the cyclohexyl group) [130].

$$Cy_3SnOH \longrightarrow Cy_2SnO \longrightarrow CySnO_2H \longrightarrow SnO_2$$

$$\underline{(55)} \qquad \underline{(56)} \qquad \underline{(57)} \qquad \underline{(58)}$$

The reaction products are dicyclohexyltin oxide (56), cyclohexylstannoic acid (57), and stannic oxide (58).

a. Colorimetric Method. In the colorimetric method, triphenyltin hydroxide is extracted from peanuts with hexane. The hexane is extracted with a mixture of dimethyl sulfoxide (DMSO) and acetic acid to partition the tin derivative into the polar phase. The DMSO phase is diluted with aqueous ammonium sulfate and extracted with dichloromethane. The dichloromethane is washed with aqueous ethylenediaminetetraacetic acid (EDTA) solution to remove any inorganic tin which might be present. The dichloromethane is evaporated and the residue is wet-ashed with fuming nitric and perchloric acids in sulfuric acid. The inorganic tin in the ashed sample is reacted with phenylfluorone [131]. The absorbance of the product which is formed is read at 510 nm.

b. <u>GLC Method</u>. In the GLC procedure for tricyclohexyltin hydroxide, the crop sample (strawberries, apples, and grapes) is finely chopped and slurried with concentrated HBr to replace the hydroxyl group with bromine. The mixture is then extracted with benzene, centrifuged, and the benzene layer removed. This is repeated twice more, and the benzene layers combined, washed with water, and dried over Na_2SO_4. The volume of benzene is then reduced to about 1 ml and an aliquot is gas chromatographed.

If the residue level is high and interferences are negligible, the analysis is completed without further cleanup. However, if interferences are a problem, the extract is purified by liquid chromatography on silica gel. However, this results in loss of (56) and (57).

Gas chromatography is carried out using a column packed with 2% OV-225 coated on acid-washed, silanized, 100-120 mesh Chromosorb G at a temperature of 200°C for compounds (55) and (56). Compound (57) chromatographs best on 0.5% OV-225 coated on 100 mesh beads at 100°C. An electrical conductivity cell is used for detection.

Recoveries of 78-94% of (55) and (56) are obtained at the 0.1 ppm level from strawberries, apples, and grapes.

G. Organic Bases

Organic bases used as fungicides include glyodin (59), dodine (60), and guanoctine (61).

(59)   (60)

(61)

Compounds of this class form salts with anionic dyes such as bromocresol purple which can be extracted into organic solvents and measured colorimetrically [132, 133].

In the analysis for dodine (60), fruit is extracted with methanol-chloro-
form (2:1) and the pulp filtered and washed with additional solvent. A
measured volume of extract is evaporated to 50 ml. Following addition of
concentrated HCl, methanol and sodium chloride are then added, and the
solution is extracted repeatedly with CCl_4 to remove colored compounds.
The pH of the aqueous phase is adjusted to 5.5 and phosphate buffer and
bromocresol purple reagent are added. The solution is extracted once more
with CCl_4 which is discarded, and the colored complex is extracted into
chloroform along with some free bromocresol purple. The free and complexed dye are extracted into dilute alkali while the dodine remains in the
chloroform. The dodine is recomplexed in the aqueous phase by extracting
the chloroform with buffered reagent. Excess reagent is removed from the
chloroform by extracting it with buffer. The complexed bromocresol purple
is then extracted into dilute alkali, and the absorbance read at 590 nm. In
the analysis of glyodin, color is read in the chloroform phase at 410 nm [133].

Additional partitioning steps are required to reduce background interference in the analysis of foliage.

The limit of sensitivity is about 0.05 ppm for a 100 g sample of fruit.

H. Organophosphates

Organophosphates are primarily insecticides but some compounds of this
class are fungicides, including hinosan (62), conen (63), and pyrazophos (64).

(62)   (63)

(64)

The metabolic pathways proposed for hinosan in rice plants are shown
in Figure 4.

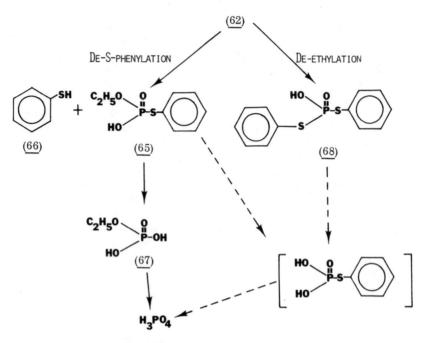

FIG. 4. Proposed metabolic pathway for hinosan in rice plants. Solid
lines indicate identified pathway, dotted lines assumed pathways.

Hinosan and its metabolites can be measured by flame photometric gas
chromatography following methylation of the polar compounds [134].

Plant tissue (rice) is homogenized for 5 min in a blender. Following
filtration, the tissue macerate is reextracted with methanol followed by tri-
chloroacetic acid. The combined extracts are evaporated to remove methanol, and the aqueous solution is extracted three times with hexane. The
hydrolysis products of hinosan remain in the aqueous phase, while hinosan
is partitioned into the organic phase. In the published procedure, the hinosan metabolites are separated by TLC in order to measure the radioactivity
of individual compounds. The TLC spots are scraped from the plate and
extracted with methanol. Following measurement of ^{32}P, the extracts are
combined, the solvent evaporated, and the residue methylated with diazomethane and ether. Both the methyl esters and hexane fraction containing
hinosan are chromatographed on a column packed with 10% DC-200 coated
on 80-100 mesh Gas Chrom Q at a column temperature of either 150 or 190°C.

Compounds (65) and (66) are found in the hexane extract, the other
metabolites being chromatographed as methyl esters.

Connen (63) isolated from formulations can be chromatographed on 2%
XE-60 coated on 60-80 mesh acid-washed, silanized Chromosorb W at
180°C [135].

REFERENCES

1. H. C. Barry, J. G. Hundley, and L. Y. Johnson, eds., Pesticide Analytical Manual, U.S. Food and Drug Administration, Washington, D.C., 1962.
2. H. P. Burchfield, D. E. Johnson, and E. E. Storrs, Guide to the Analysis of Pesticide Residues, No. 26/2-p 43/2, Vols. I and II, U.S. Public Health Service, Government Printing Office, Washington, D.C., 1965.
3. L. R. Jones and J. A. Riddick, Anal. Chem., 24, 569 (1952).
4. H. P. Burchfield and E. E. Storrs, Contrib. Boyce Thompson Inst., 17, 333 (1953).
5. M. J. Eidelman, J. Assoc. Offic. Agr. Chemists, 45, 672 (1962).
6. M. J. Eidelman, J. Assoc. Offic. Agr. Chemists, 46, 182 (1963).
7. E. W. Laws and D. J. Webley, Analyst, 86, 249 (1961).
8. M. J. de Faubert Maunder, H. Egan, E. W. Godly, E. W. Hammond, J. Roburn, and J. Thompson, Analyst, 89, 168 (1964).
9. P. A. Mills, J. Assoc. Offic. Agr. Chemists, 44, 171 (1961).
10. J. W. Rhoades, R. J. Wheeler, D. E. Johnson, and H. P. Burchfield, in Lectures on Gas Chromatography (L. R. Mattick and H. A. Szymanski, eds.), Plenum Press, New York, 1965, p. 33.
11. J. E. Lovelock and S. R. Lipsky, J. Amer. Chem. Soc., 82, 431 (1960).
12. J. E. Lovelock, Anal. Chem., 35, 474 (1963).
13. P. G. Simmonds, D. C. Fenimore, B. C. Pettitt, J. E. Lovelock, and A. Zlatkis, Anal. Chem., 39, 1428 (1967).
14. D. C. Fenimore and C. M. Davis, J. Chromatogr. Sci., 8, 519 (1970).
15. P. Devaux and G. Guiochon, J. Chromatogr. Sci., 7, 561 (1969).
16. B. C. Pettitt, P. G. Simmonds, and A. Zlatkis, J. Chromatogr. Sci., 7, 645 (1969).
17. P. Devaux and G. Guiochon, J. Chromatogr. Sci., 8, 502 (1970).
18. D. M. Coulson, J. Gas Chromatogr., 4, 285 (1966).
19. G. G. Patchett, J. Chromatogr. Sci., 8, 155 (1970).
20. H. P. Burchfield, J. W. Rhoades, and R. J. Wheeler, J. Agr. Food Chem., 13, 511-516 (1965).
21. A. Karman and L. Guiffrida, Nature, 201, 1204 (1964).
22. D. A. Craven, Anal. Chem., 42, 1679 (1970).
23. J. N. Ospenson, D. E. Pack, G. K. Kohn, H. P. Burchfield, and E. E. Storrs, in Analytical Methods for Pesticides, Plant Growth Regulators and Food Additives (G. Zweig, ed.), Vol. 3, Academic Press, New York, 1964, Chapter 2.
24. H. P. Burchfield and J. Schechtman, Contrib. Boyce Thompson Inst., 19, 411 (1958).
25. W. W. Kilgore, W. Winterlin, and R. White, J. Agr. Food Chem., 15, 1035 (1967).
26. A. Bevenue and J. N. Ogata, J. Chromatogr., 36, 529 (1968).

27. I. H. Pomerantz, L. J. Miller, and G. Kava, J. Assoc. Offic. Anal. Chemists, 53, 154 (1970).
28. G. Zweig and J. Sherma, in Analytical Methods for Pesticides, Plant Growth Regulators and Food Additives (G. Zweig, ed.), Vol. 6, Academic Press, New York, 1972, p. 549.
29. D. E. Pack in Analytical Methods for Pesticides, Plant Growth Regulators and Food Additives (G. Zweig, ed.), Vol. 5, Academic Press, New York, 1967, p. 299.
30. W. R. Meagher, C. A. Anderson, C. E. Gonter, S. B. Smith, and D. MacDougall, J. Agr. Food Chem., 8, 558 (1959).
31. H. P. Burchfield and E. E. Storrs, Contrib. Boyce Thompson Inst., 18, 319 (1956).
32. H. P. Burchfield, Contrib. Boyce Thompson Inst., 20, 205 (1959).
33. P. J. Wales and C. E. Mendoza, J. Assoc. Offic. Anal. Chemists, 53, 509-513 (1970).
34. C. E. Mendoza and J. B. Shields, J. Assoc. Offic. Anal. Chemists, 54, 986 (1971).
35. H. P. Burchfield, in Plant Pathology—An Advanced Treatise (J. G. Horsfall and A. E. Dimond, eds.), Vol. 3, Academic Press, New York, 1960, pp. 494-495.
36. H. P. Burchfield and G. L. McNew, Phytopathology, 38, 299 (1948).
37. J. R. Lane, J. Agr. Food Chem., 6, 667 (1958).
38. J. A. Burke and W. Holswade, J. Assoc. Offic. Anal. Chemists, 49, 374 (1966).
39. J. P. Barette and R. Payer, J. Assoc. Offic. Agr. Chemists, 47, 259 (1964).
40. D. M. Coulson, L. A. Cavanagh, and J. Stuart, J. Agr. Food Chem., 7, 250 (1959).
41. H. P. Burchfield and G. L. McNew, Phytopathology, 38, 665 (1948).
42. J. R. Lane, J. Agr. Food Chem., 6, 746 (1958).
43. J. E. Newell, R. J. Mazaike, and J. W. Cook, J. Agr. Food Chem., 6, 669 (1958).
44. J. R. Lane in Analytical Methods for Pesticides, Plant Growth Regulators and Food Additives (G. Zweig, ed.), Vol. 3, Academic Press, New York, 1964, pp. 148, 149.
45. G. C. Prescott, H. Emerson, and J. H. Ford, J. Agr. Food Chem., 4, 343 (1956).
46. R. C. Rhodes, H. L. Pease, and R. K. Brantley, J. Agr. Food Chem., 19, 745 (1971).
47. R. C. Rhodes and H. L. Pease, J. Agr. Food Chem., 19, 750 (1971).
48. H. L. Pease, J. Agr. Food Chem., 15, 917 (1967).
49. W. E. Bleidner, H. M. Baker, M. Levitsky, and W. K. Lowen, J. Agr. Food Chem., 2, 476 (1954).
50. D. E. Stallard and A. L. Wolf, Diamond Shamrock Corp., Report (1967).
51. H. N. Benedict, Diamond Shamrock Corp., Report (1974).

52. J. Roburn, J. Sci. Food Agri., 12, 766 (1961).
53. W. W. Kilgore, K. W. Cheng, and J. M. Ogawa, J. Agr. Food Chem., 10, 399 (1962).
54. K. Groves and K. S. Cough, J. Agr. Food Chem., 14, 668 (1966).
55. J. A. Heagy, J. Assoc. Offic. Anal. Chemists, 52, 797 (1969).
56. H. Beckman and A. Bevenue, J. Food Sci., 27, 602 (1962).
57. Pesticide Analytical Manual, Vol. II, Section 120-200, U.S. Food and Drug Administration, Washington, D. C., 1967.
58. K. W. Cheng and W. W. Kilgore, J. Food Sci., 31, 259 (1966).
59. H. V. Brewerton, P. J. Clark, and H. J. W. McGrath, New Zealand J. Sci., 10, 124 (1966).
60. A. Stark, J. Agr. Food Chem., 17, 871 (1969).
61. A. Bevenue, M. L. Emerson, L. J. Casarett, and W. L. Yauger, Jr., J. Chromatogr., 38, 467 (1968).
62. G. R. Higginbothan, J. Ress, and A. Rocke, J. Assoc. Offic. Anal. Chemists, 53, 673 (1970).
63. W. F. Barthel, A. Curley, C. L. Thrasher, V. A. Sedlak, and R. Armstrong, J. Assoc. Offic. Anal. Chemists, 52, 294 (1969).
64. K. W. Cheng and W. W. Kilgore, J. Food Sci., 31, 742 (1966).
65. A. S. Y. Chau and J. A. Coburn, J. Assoc. Offic. Anal. Chemists, 57, 389 (1974).
66. C. A. Bache and D. J. Lisk, J. Agr. Food Chem., 6, 747 (1958).
67. A. K. Klein and R. J. Gajan, J. Assoc. Offic. Agr. Chem., 44, 712 (1961).
68. H. J. Ackermann, H. A. Baltrush, H. H. Berges, D. O. Brookover, and B. B. Brown, J. Agr. Food Chem., 8, 459 (1960).
69. J. C. Caseley, Bull. Environ. Contam. Toxicol., 3, 180 (1968).
70. S. Gorbach and U. Wagner, J. Agr. Food Chem., 15, 654 (1957).
71. E. F. Kuchar, F. O. Geenty, W. G. Griffith, and R. J. Thomas, J. Agr. Food Chem., 17, 1237 (1969).
72. P. B. Baker and B. Flaherty, Analyst, 97, 378 (1972).
73. C. A. Anderson and J. M. Adams, J. Agr. Food Chem., 11, 474 (1963).
74. H. Crossley and V. P. Lynch, J. Sci. Food Agr., 19, 57 (1968).
75. C. P. Kurtz, H. Baum, and C. Swithenbank, J. Assoc. Offic. Anal. Chemists, 53, 887 (1970).
76. C. E. Castro, in Analytical Methods for Pesticides, Plant Growth Regulators and Food Additives (G. Zweig, ed.), Vol. 3, Academic Press, New York, 1964, Chaps. 16-18.
77. T. C. Cullen and R. P. Stanovick, J. Agr. Food Chem., 13, 118 (1965).
78. J. M. Devine, B. Fletcher and G. Zweig, J. Assoc. Offic. Anal. Chemists, 52, 1106 (1969).
79. R. A. Gray, in Analytical Methods for Pesticides, Plant Growth Regulators and Food Additives (G. Zweig, ed.), Vol. 3, Academic Press, New York, N.Y., 1964, Chap. 19.

80. N. J. Turner and M. E. Corden, Phytopathology, 53, 1388 (1963).
81. H. P. Burchfield, E. E. Green, R. J. Wheeler, and S. M. Billedeau, J. Chromatogr., 99, 697 (1974).
82. G. P. Clemons and H. D. Sisler, Phytopathology, 59, 705 (1969).
83. H. L. Pease and R. F. Holt, J. Assoc. Offic. Anal. Chemists, 54, 1399 (1971).
84. H. L. Pease and J. A. Gardiner, J. Agr. Food Chem., 17, 267 (1969).
85. N. Aharonson and A. Ben-Aziz, J. Assoc. Offic. Anal. Chemists, 56, 1330 (1973).
86. J. J. Kirkland, J. Agr. Food Chem., 21, 171 (1973).
87. J. J. Kirkland, R. F. Holt, and H. L. Pease, J. Agr. Food Chem., 21, 368 (1973).
88. R. Mestres, M. Campo, and J. Tourte, Ann. Fals. Expert. Chim., 63 (1970).
89. R. Mestres, M. Campo, and J. Tourte, Ann. Fals. Expert Chim., 63, 160 (1970).
90. Merck and Co., Inc., Method in Pesticide Analytical Manual, Vol. II, Food and Drug Administration, Washington, D.C.
91. J. R. Lane, J. Agr. Food Chem., 18, 409 (1970).
92. H. R. Sisken and J. E. Newell, J. Agr. Food Chem., 19, 738 (1971).
93. E. Lemperle and E. Kerner, Z. Anal. Chem., 256, 353 (1971).
94. A. Murano, T. Dor, and M. Nagase, Bunseki Kagaku, 20, 792 (1971); Gas Chromatogr. Abstr., 72-0137 (1972).
95. L. A. Summer and D. J. Shields, Chem. Ind., 1264 (1964).
96. S. H. Yuen in Analytical Methods for Pesticides, Plant Growth Regulators and Food Additives (G. Zweig, ed.), Vol. 7, Academic Press, New York, 1973, Chap. 40.
97. C. A. Anderson, Chemagro Corp., Technical Report No. 15,769, Chemagro Division of Baychem Corp., Kansas City, Missouri, 1966.
98. H. A. Stansbury, Jr., in Analytical Methods for Pesticides, Plant Growth Regulators and Food Additives (G. Zweig, ed.), Vol. 3, Academic Press, New York, p. 122.
99. E. Day, Eli Lilly & Co. Procedure No. 5801370, Greenfield, Indiana, 1966.
100. Eli Lilly & Co., Procedure No. 5801250, Greenfield, Indiana, 1966.
101. W. P. Griffith, Olin Corp. Method, New Haven, Conn., 1973.
102. H. P. Burchfield and A. Goenaga, Contrib. Boyce Thompson Inst., 19, 141 (1957).
103. L. G. Borchardt and J. P. Butler, Anal. Chem., 29, 414 (1957).
104. W. C. Woelfel, Anal. Chem., 20, 722 (1948).
105. F. J. Viles, Ind. Hyg. Toxicol., 22, 188 (1940).
106. W. K. Lowen, Anal. Chem., 23, 1846 (1951).
107. W. K. Lowen, J. Assoc. Offic. Agr. Chemists, 36, 484 (1953).
108. M. Levitsky and W. K. Lowen, J. Assoc. Offic. Agr. Chemists, 37, 555 (1954).

109. D. G. Clark, H. Baum, E. L. Stanley, and W. F. Hester, Anal. Chem., 23, 1842 (1951).
110. H. L. Pease, J. Assoc. Offic. Agr. Chemists, 40, 1113 (1957).
111. G. E. Keppel, J. Assoc. Offic. Anal. Chemists, 52, 162 (1969).
112. H. A. McLeod and K. A. McCully, J. Assoc. Offic. Anal. Chemists, 52, 1226 (1969).
113. W. H. Newsome, J. Agr. Food Chem., 22, 886 (1974).
114. F. I. Onuska and W. R. Boos, Anal. Chem., 45, 967 (1973).
115. L. E. Lopateki and W. Newton, Can. J. Bot., 37, 131 (1952).
116. G. Petrosini, Notiz. Mal. Pianti, 59, 59 (1962).
117. G. Petrosini, F. Tafuri, and M. Businelli, Notiz. Mal. Pianti, 65 (1963).
118. J. R. Iley and J. G. A. Fiskeli, Soil Crop. Sci. Soc. Florida, 23, 50 (1963).
119. R. A. Ludwig, G. D. Thorn, and D. M. Miller, Can. J. Bot., 32, 48 (1954).
120. J. W. Vonk and A. Kaars Sijpesteijn, Ann. Appl. Biol., 65, 489 (1970).
121. J. R. M. Innes, B. M. Ulland, M. G. Valerio, L. Petrucelli, L. Fishbein, E. R. Hart, A. J. Pallotta, R. R. Bates, H. L. Falk, J. J. Gart, M. Klein, I. Mitchell, and J. Peters, J. Natl. Cancer Inst., 42, 1101 (1969).
122. J. H. Only and G. Yip, J. Assoc. Offic. Anal. Chemists, 54, 165 (1971).
123. W. H. Newsome, J. Agr. Food Chem., 20, 967 (1972).
124. R. G. Nash, J. Assoc. Offic. Anal. Chemists, 57, 1015 (1974).
125. L. Fishbein, Chromatogr. Rev., 12, 167 (1970).
126. K. Sumino, Kobe J. Med. Sci., 14, 115 (1968).
127. J. O'G. Tatton and P. J. Wagstaffe, J. Chromatogr., 44, 284 (1969).
128. Thompson-Hayward Chemical Co., Method A - 136D, Kansas City, Kansas, 1970.
129. W. O. Gauer, J. N. Seiber, and D. G. Crosby, J. Agr. Food Chem., 22, 252 (1974).
130. M. E. Getzendaner and M. F. Corbin, J. Agr. Food Chem., 20, 881 (1972).
131. Thompson-Hayward Chemical Co., Method A-77B, Kansas City, Kansas, 1970.
132. E. F. Hillenbrand, W. W. Sutherland, and J. N. Hogsett, Anal. Chem., 23, 626 (1951).
133. W. A. Steller, K. Klotsas, E. J. Kuchab, and M. V. Norris, J. Agr. Food Chem., 8, 460 (1960).
134. K. Takase, E. Tan, and K. Ishizuka, Agr. Biol., Chem., 37, 1564 (1973).
135. A. Murano and M. Nagase, Bunseki Kagaka, 20, 732 (1971); Gas Chromatogr. Abstr., 72-0527 (1972).

Chapter 15

SAFE AND RESPONSIBLE USE OF FUNGICIDES

William C. von Meyer

Marketing Department—Agrochemicals
Rohm and Haas Company
London, England

René J. Lacoste[*]

International Division
Rohm and Haas Company
Philadelphia, Pennsylvania

---

*Present address: Department of Foreign Regulatory Affairs, Rohm and Haas Company, Philadelphia.

I. INTRODUCTION

"Safe" may be defined as meaning noninjurious or not harmful, while the word "responsible" indicates being answerable for actions, fulfilling obligations, or being able to choose between right and wrong. The title of this chapter encompasses all of these definitions, inasmuch as we are concerned with the right way to use fungicides and provide useful results without harm to the user. The enactment of legislation in many countries emphasizes the answerable aspect of responsibility by often including penalties for unsafe use of these materials. The concept of safety, originally focused on man, has expanded so that today we are concerned with man and his total environment.

Today it is recognized that chemicals may often be harmful or unsafe in different ways. Taking this into consideration, new as well as older fungicides used in industry and agriculture are being carefully examined by industrial and governmental organizations to assure that they do not have adverse effects, not only with regard to short-term exposures to the applier, but also with regard to repeated use, or during manufacture over a period of years, or by contamination of food, water, and the environment.

The concept of a safe or unsafe fungicide depends entirely on what is known about the chemical, how it is to be applied, and the way people and animals are contacted by the product or its metabolites. We must ask the obvious question, "safe to whom or what?" For example, a contact agricultural fungicide which is used in a paint applied to the pruning wounds of trees once a year involves a completely different type and degree of exposure to man and the environment as compared to a fungicide which is sprayed by airplane weekly over fields of potatoes, bananas, or rice. Certain operations are common to both cases, such as moving and storing the product between applications. In all cases, the degree of product safety must be a judgment decision arrived at by examining the toxicological properties of the chemical, its metabolites, and its intended use. Because of the frequent need for such decisions, both responsible manufacturers and public organizations employ large numbers of people to assess, monitor, and make such judgments which ultimately lead to an approved label for fungicide products. The purpose of this chapter is to discuss some of the general principles regarding the safe and responsible use of fungicides in agriculture and to provide an insight into the state of man's awareness of the need to use fungicides in a safe and responsible way.

The word fungicide, as used herein, means a chemical which kills a fungus or controls disease development by fungi in or on plants. We have also included in this definition those products which may not kill the fungus but render it inactive (fungistat).

Fungicidal molecules in certain instances (for example, mercurials) are quite toxic to the cells of animals [1]. However, many potent fungicides may have little acute toxic effect on mammalian cells or species as indicated by several papers in Torgeson's treatise [2]. The reason being that there are certain sensitive biochemical processes inherent in many fungi which either do not occur in animals or are not readily exposed in such a way as to be easily contacted by the fungicide. Therefore, the hazard exhibited by most fungicides is considerably less than that exhibited by many insecticides, which often are potent acetylcholineesterase inhibitors designed to affect enzymes common to the nerve tissues of many animals and insects. This is a fortunate circumstance, because the first fungicidal or fungistatic materials were discovered and used long before man realized the danger of chronic toxicity. The first materials used in early food preservation and plant disease control efforts were salt and sulfur, both relatively harmless.

II. THE NEED FOR SAFE HANDLING AND USE OF PESTICIDES

Concern over the possible hazards of fungicide use predates legislation to control their usage. Moreover, early legislation was aimed at insecticides which were based on compounds recognized as being toxic to man. Some time elapsed before legislation was enacted which specifically caused fungicides to be placed under regulatory control. In the mid-19th century, before the introduction of arsenic compounds as insecticides, pest control was accomplished with rather innocuous materials such as brine washes, saltpeter, alum, and lime. According to Bates [3], as early as 1864, a French Ordinance forbade the sale and use of arsenic compounds for the destruction of insects.

By 1903, the United Kingdom had established a maximum permissible arsenic content in fruit [3]. This latter law may be the earliest introduction of the concept of a tolerance for a chemical residue. It also created problems which were to serve much later as the basis for the formation of the Codex Alimentarius (see below), an international effort toward establishing food standards. For example, apples grown under New Zealand conditions retained arsenic residues below the permissible maximum. However, a heavier dosage was required for pest control in North America and South Africa, which frequently resulted in arsenic residues above the maximum permissible level in fruit from those countries. Such differences in regional needs for agricultural production created problems in the free and equitable trade of foods. These problems are currently being examined and, it is hoped, will be resolved by the Codex Committee.

The purpose of the United States Insecticide Act of 1910 was prevention of manufacture, sale, or transport of adulterated or misbranded Paris greens, lead arsenates, and other insecticides, as well as fungicides. This Act is of historical interest since it was the first U.S. Federal Statute involving pesticides and it specifically mentions fungicides. It addresses itself to protection of the product user and was passed in an effort to stop

unethical persons from merchandising ineffective or adulterated products.
Therein lies a fundamental difference in emphasis in the legislative control
of insecticides, fungicides, and other agricultural chemicals which have
collectively become known as pesticides.

It is of interest to review the development of national legislation in the
United States which reflects the different steps taken toward assuring the
safe and responsible use of fungicides and pesticides. In 1947, 37 years
after passage of the Insecticides Act of 1910, another significant step was
taken with the passage of the Federal Insecticides, Fungicide, and Rodenti-
cide Act (FIFRA); which was "An Act to regulate the marketing of economic
poisons and devices, and for other purposes." The important contributions
of this Act were embodied in the introduction of the concepts of product
registration and stricter control of product labelling. Product registration
required that the manufacturer furnish proof that the chemical would be
useful for the purpose stated on the label. Furthermore, it was required
that the label include adequate directions for the proper use of the product
to achieve the desired result without injury to living man and other vertebrate
animals, vegetation, and useful invertebrate animals.

While the Insecticides Act and the initial version of FIFRA established
legislative protection for the user of pesticides, concern for the safety of
consumers of products treated with pesticides was expressed along a different legislative route. In 1906 the Food and Drug Law was passed which
prohibited the interstate commerce of adulterated food or drugs. This law
established the concept of quantative tolerance levels for adulteration which
included such quality control as, "seven cherry maggots per tin of cherries."
An amended version, the Federal Food, Drug and Cosmetic Act of 1938
(FFDCA), established tolerances for chemicals 32 years later which were
essential to the production and storage of crops. The effectiveness of the
1938 amendment was diminished by the requirements that the government
prove the presence of a poisonous chemical in a food or drug before it could
initiate action to remove that product from the market. The 1954 Pesticide
Amendment (Miller Bill) remedied this inadequacy by placing the burden of
proof of safety on the manufacturer. It required that the manufacturer of a
pesticide develop analytical methods to determine chemical residues and to
submit results for crops treated in the field according to directions on the
label. This requirement was made a part of the registration process established under FIFRA. In the United States, the two Federal agencies
responsible for administering FIFRA and FFDCA are the United States
Department of Agriculture (USDA) and the Food and Drug Administration
(FDA), respectively. If the proposed directions for use of a product resulted
in finite residues in or on food or feed crops, a tolerance needed to be established by the FDA before the USDA would grant registration. This
merged the administrative processes of providing assurance of safety to the
user of pesticides and to the consumer of treated products. The final merger
of this process did not occur until after the enactment of an amendment to

FIFRA, the Federal Environmental Pesticide Control Act of 1972 (FEPCA), which placed responsibility for controlling the manufacture, sale, and use of pesticides within a single agency, the Environmental Protection Agency (EPA). The basic structure of FIFRA provided for penalties when adulterated or improperly labelled products were shipped interstate. There was no provision for direct control over actual use and interstate manufacture or sale. This was considered a significant gap in providing safety to users and to the environment. FEPCA added a use classification and regulation scheme to the labelling scheme of FIFRA. FEPCA also brought into the regulatory scheme a greater emphasis on environmental implications of pesticide use. Such legislation was to be expected from the increased environmental awareness after publication of Rachel Carson's Silent Spring in 1962 [4].

The development of the safe and responsible use of pesticides has been an evolutionary process with contributions being made by manufacturers, pesticide users, researchers, legislators, and consumers of the products which are protected and improved by these chemicals. The regulated and unregulated practices of today are not the same as those of 20 years ago. Furthermore, it is highly likely that these present practices will change as our knowledge and experience expands. The extent to which safe practices are observed by individuals or are regulated by governments varies considerably throughout the world. A field worker in a hot, humid climate is often less inclined to use adequate protective clothing and devices than one in a cooler environment. Potentially higher chemical residues in food are often of less concern to those who have an urgent need for chemical control of insects and diseases on a crop critical to their survival.

It is interesting to note that most countries which tend to have extensive legislative control of pesticides are in a relatively advanced stage of industrial development. Such countries include West Germany, the Netherlands, Japan, the United States, Canada, Sweden, and Switzerland. The least legislative control is encountered in most of Africa, South and Central America, and South Asia. The reasons for this situation are probably as complex as the diversity of the people and regions involved. Surely, the regional environmental, economic, and social circumstances bring different pressures to bear upon ways in which agriculture is conducted and controlled. One can readily perceive that, in some countries, the objective of providing an adequate food supply would take precedence over adequate safety precautions for the use of chemicals needed to protect that supply. That is not to say that people in those countries are oblivious to safety, but that their priorities are oriented in a different direction.

Most nations have some type of government control over the sale of pesticides within their borders. However, the degree of control varies considerably. It may consist of the simple procedure of requiring an application with minimal product information and payment of a fee for a license to sell. The information required about the product may not include toxicology. If toxicology is required, animal feeding studies often are not required. On the other hand, the government controls might be quite extensive and require the manufacturer to provide information about efficacy; chemical composition; complete toxicology, including feeding studies; metabolism studies in plants and animals; residue studies in plants and animals; and environmental impact, such as persistence and toxicology to nontarget animal species. Even in this latter case, effective control over actual use of the pesticide ceases without some sort of monitoring system. Control is accomplished by some authorities through periodic chemical inspection of food produce to determine if excessive pesticide residues are present. Excessive residues are taken as an indication that the pesticide has been misused. Recently in the United States, certification of pesticide appliers has been introduced as a requirement to permit use of products which are considered potentially hazardous.

As a result of pesticide residue monitoring consumers have become more aware of the presence of chemicals in foods. Nations with monitoring capability observed that the levels of residues in a given crop might depend on the particular country and environment in which that crop was produced. This has created problems, since the importing country could not determine whether or not residues in an imported crop reflected normal levels to be expected from proper use of the pesticide at dosages needed to control pests in the producing country. The uncertainty has been further aggravated in those instances where the producing country exercised little or no control over pesticide usage. These problems soon became the concern of various international groups.

In 1963, the Food and Agriculture Organization of the United Nations established the FAO Working Party on Official Control of Pesticides upon the recommendation of the 1962 FAO Conference on Pesticides in Agriculture. One of the tasks of this Working Party was to prepare a model scheme for the official control of pesticides and to provide guidance on labelling. Eight experts from Austria, Belgium, Italy, the United States, and the United Kingdom met in 1965 and 1966, and subsequently published a paper outlining such a scheme [5]. It was designed to accommodate countries with different needs and financial resources. Starting from the simplest scheme for countries with limited resources, the paper develops a complex scheme which can be adopted by the most highly developed countries. It was planned so that a country might enter into the process at a point related to its resources and needs, with the possibility of expanding as these grow. A unique feature of the paper is its effort to explain the reasons and philosophy behind the courses of action it recommends. A second paper was jointly published with the World Health Organization and proposed a model law for the control of pesticides [6]. It is appropriate to quote two paragraphs from the first paper [5] published by the FAO:

> It is necessary to warn against the implication that a 100 percent efficiency in safety and pest control will be attainable as a consequence of the operation of schemes to control pesticides and their use. Their

cost would be completely uneconomic in relation to the potential hazards involved, or to the possible economic gains from the more efficient control of the pest..... Overdetailed legislation or excessively exacting demands for information by the director of a scheme can result in "les lois inutiles affaiblissent les lois nécessaires" (Montesquieu). It is recognized that the potential demand for information on all phases of pesticides toxicology and use are practically unlimited. Therefore, it is important to first ensure that scientific manpower is conserved and the cost of developing pesticides kept within economic limits, if the world is to take proper advantage of the value of pesticides in increasing the food supply for a growing population; and second to give the director the authority to require only the submission of information which is necessary for the satisfactory evaluation of safety and efficacy.

Thus, while recognizing a need to control the use of pesticides, the FAO has interjected a note of caution that the desire to promote the safe use of these materials must be achieved realistically and without completely negating their beneficial effects.

The FAO and WHO were not the first and certainly are not the only organizations which are working on the promotion of safe and efficient use of pesticides. However, because of their extensive influence on a worldwide scale, their message may reach the major part of the agricultural community. In 1962, the Council of Europe published the first edition of a booklet which has since been twice revised [7]. It provides guidance for the type of experimental data which should be obtained to establish the safe use of a pesticide. The presentation of the subject in the Council of Europe publication is different than that of the FAO scheme and probably reflects the fact that it is primarily directed toward a region of relatively developed nations.

The pesticide industry has also played a role in the development of safe and efficient practices for the use of its products. In the United States, for example, the National Agricultural Chemicals Association (NACA) has literature, films, and publicity campaigns which are designed to promote such practices. NACA has joined with the equivalent industrial associations from other countries in an international group named Groupement International des Associations Nationales de Fabricants de Pesticides (GIFAP). Through GIFAP, the pesticide industry provides information input to various international groups such as FAO, WHO, the Council of Europe and the European Economic Community to enhance the efforts of these organizations to develop safe and efficient pesticide use through educational and legislative channels.

A rather unique cooperation between industry and government has developed in the area of agriculture and related business activity. The FAO has been engaged since its creation in helping developing countries improve their agricultural production. Most of this effort has been through government agencies with relatively little opportunity for direct participation of

industry. In 1966 the Industry Co-Operative Program (ICP) of Agro-Allied
Industries with FAO and other United Nations Organizations was formed.
This program is designed to provide a direct link between the world's
industry related to agriculture, forestry, and fisheries; the FAO and other
organizations of the UN system; and governments. Through ICP, member
companies participate in a unique worldwide joint effort to accelerate economic and social development. Industry participation is strictly limited to
those companies engaged in business related to agricultural activities in
the broadest sense in developing countries. In 1975, 97 companies, including
16 pesticide manufacturers, were listed as members of ICP. The objectives of the program are accomplished through industry working groups
which meet periodically with the FAO Secretariat to plan projects which are
suggested by governments or by industry. These projects emerge in various forms such as, high-level missions to countries to identify areas for
industrial partnership and the exchange of technical and economic information to assist in the development process. The Pesticides Working Group
of ICP has been particularly active in providing technical information by
sponsoring regional seminars on the "Safe, Effective and Efficient Use of
Pesticides in Agriculture and Public Health." Each seminar is of about 1
week duration and involves the participation of representatives of the sponsoring companies, invited experts, and regional government personnel
engaged in the control of pesticide use. Thus far, four such seminars have
been conducted in South America (Sao Paulo, Brazil, 1971); Central America
(Turrialba, Costa Rica, 1972); the Far East (Bangkok, Thailand, 1973);
and Africa (Nairobi, 1974). These seminars have stimulated the exchange
of philosophies regarding the use of pesticides and the special regional conditions which influence local practices. They have resulted in intergovern-
mental requests for further educational missions. They have also been
followed by some increase in local government activity in pesticide control
in those regions.

    Coincidental with increasing international concern over effective control
of pesticide usage, an international movement developed toward standardizing
chemical residue tolerances in food. This movement gained impetus in 1953
when the Sixth World Health Assembly expressed concern that the increasing
use of various chemical substances in the food industry was creating a new
public health problem. This meeting was followed by a joint FAO/WHO
meeting on food additives in 1955, and a FAO meeting on pesticides in 1959.
In 1961, FAO and WHO drafted a set of principles governing consumer safety
in relation to pesticide residues.

    In the meantime, the food industry became more aware of the possibility
that individual nations might attempt to circumvent trade agreements by
imposing arbitrary food standards in order to restrict imports in favor of
local agricultural interests. Thus, in 1962, the Codex Alimentarius Commission (CAC) was created to implement the joint FAO/WHO Food Standards
Program. The purpose of the program is to: "protect the health of consumers

and to ensure fair practices in the food trade; to promote coordination of all food standards work undertaken by international governmental and non-governmental organizations; to determine priorities and initiate and guide the preparation of draft standards through and with the aid of appropriate organizations; to finalize standards and, after acceptance by governments, publish them in a Codex Alimentarius either as regional or worldwide standards. "

Membership of the CAC comprises those member nations and associate members of FAO and WHO who have notified the Director General of FAO or of WHO of their wish to be considered as members. By March 1969, 65 countries had become members of CAC. This number increased to 100 by 1974.

The delegates to the Commission are high-level officials who have the authority to commit their respective governments to an international agreement. The groundwork for the Commission is conducted by committees which recommend specific food standards. The one concerned specifically with pesticides is known as the Codex Committee on Pesticide Residues (CCPR). It convenes approximately annually to consider recommendations for residue tolerances for agricultural commodities in the light of comments from governments relative to those recommendations. The CCPR is made up of persons of lower rank but who hold responsible positions in ministries of agriculture and/or health. The CCPR is aided in its work by the Joint Meeting of the FAO Working Party of Experts on Pesticide Residues and the WHO Expert Committee on Pesticide Residues (JMPR). This group is composed of scientific experts who are invited by FAO and WHO to convene about once a year to evaluate data provided by governments and industry and to recommend residue tolerances for pesticides which have been placed on the agenda by the CCPR. Although some degree of continuity is maintained, the experts who constitute the JMPR may not be the same from year to year. A series of ten steps are provided in the Codex procedure whereby the tolerance recommendations are discussed, modified, sent back to JMPR for reconsideration, sent to governments for comment, and advanced until the CCPR presents them to the Commission (CAC) for adoption.

In the Codex system, two influences operate toward a common goal, which is to facilitate the movement of food in international trade, while maintaining safety toward public health. The scientific influence is represented by the efforts of the JMPR, which is intended to be an apolitical body. The economic influence is exerted by the CCPR. At the conclusion of the Eighth Session of the CCPR in 1975, 267 tolerance recommendations for 32 different pesticide chemicals were awaiting governmental acceptance.

Thus, we see that international efforts toward the safe and responsible use of pesticides have proceeded via several routes involving both the public and the private sectors. A recent report identifies over 30 international organizations, governmental and nongovernmental, concerned with the issue [8].

III. PRINCIPLES OF RESPONSIBILITY AND SAFETY

The reader may wish to establish, to his own satisfaction, a set of principles governing the "safe and responsible" use of fungicides. We have developed the following suggestions which we consider most important.

A. Responsibility

1. Identify the causative organism, the degree of advancement of the disease, and source of inoculum.
2. Determine, if possible, what fungicides have been successfully used to control the fungus.
3. Assess the best control procedure considering:
   a. crop variety and quality desired,
   b. stage of disease,
   c. stage of field to be treated,
   d. weather forecast and harvest date,
   e. availability of labor and equipment,
   f. cost and return on investment required, and
   g. toxicity of the fungicide and its impact.
4. Establish a plan of handling the mixing, application, storage, and disposal of the fungicide product which minimizes personal contact:
   a. consider warehousing (where, how),
   b. consider delivery to the spraying rig,
   c. consider method of subdividing into units mixable into the carrier liquid (or duster),
   d. consider a holding tank for disposal of pesticide waste, and
   e. provide protective clothing and masks.
5. Keep an accurate record of the materials used, treatment procedure, costs, and results.

We have set forth five principles of responsibility. In controlling a plant disease, the person with the ultimate responsibility is generally the farm owner or producer of the crop. This man will often delegate various tasks to the managers of the farm or plantation and these persons in turn play a key role in pest and disease management.

The first principle of responsibility concerns the identification of the nature of the disease and its degree of advancement. Identification should be done preferably with the aid of a competent plant pathologist. The wrong fungicide may often be used because the cause of the crop damage has not been properly identified. Some common mistakes made are to use a fungicide when a bactericide is required, or to spray for a disease such as late blight of potato or tomato when actually a wilt pathogen is the culprit. In respect to establishing the stage of advancement of the disease, it is important to treat as soon as the disease has been recognized, using, if possible, a fungicide which will arrest or prevent incubating infections from developing into sporulating and necrotic lesions. In preventing outbreaks of disease, a grower should periodically inspect his crop. For crops covering large acreages, one should establish a program of inspecting various geographical areas of the farm or plantation and take careful note of the appearance of the plants.

Disease in plants usually manifests itself by a change in plant appearance. Once a change is noticed, one should quickly try to identify the cause. In this effort, several key questions need answering. Some of the important points are:

1. Is there a characteristic symptom on all the plants involved, e.g., wilting, color change, necrotic spots?
2. Are there visible structures of a pathogen associated with the change in the plant and can these be identified with a known pathogen?
3. Is the disease transmissable, and by what means?
4. Is there a cycle of the development of the disease apparent as related to weather, insects, irrigation, or other farm practice?

Once the disease and causative agent have been adequately described or identified, the next step is to determine if the problem is known or new. If the disease is known, one can then determine what fungicides have been successfully employed to control the pathogen. A list of potential treatments should be prepared and any changes in crop management required to use them should be noted. Once the list of potential treatments has been compiled, it should be carefully reviewed by the user and the following factors determined:

1. What materials will provide the best degree of control and also fit into the management scheme of the farm?
2. Considering the products and recommended treatment dose and schedule, labor and equipment, what is the estimated cost per hectare per year?
3. Considering the alternatige products, labor, and crop quality required, what is the expected return on investment for the control procedure? Return on investment may be defined as the net profit after tax divided by the total investment. The total investment should include labor, materials, and other overhead costs involved in producing the crop.

In considering the best control procedure, it is important to establish again the stage of development of the disease. For example, it may be possible to use an eradicant fungicide and prevent young lesions from further development. If the fungus pathogen has been sporulating for some time in a wide area under ideal conditions, one might need to treat the entire area with the appropriate fungicide under the assumption that a widespread outbreak will result unless action is taken. In this regard, an evaluation of weather conditions is of prime importance, including, in the case of fungi requiring moisture, an estimate of the amount of dew collecting daily on the plants. As an example, the Nile River delta in Egypt has little rainfall, but a heavy dew creates conditions ideal for the spread of cereal rusts and various windborne vegetable pathogens.

Often the farmer will not be able to enter the field to spray with ground equipment due to soft ground. Under these conditions, one must consider the advisability of aerial application of the fungicide. Yet, certain plant diseases may not be effectively controlled by aerial application. In the author's experience, <u>Mycosphaerella melonis</u>, incitant of gummy stem blight on watermelon, is one such pathogen. Thus the selection of the control procedure will obviously affect the farmer's cost and return on investment. It will affect directly the amount of labor used, safety of the workers, and other matters such as insurance.

The writers of this chapter have often been amazed at how little formal attention is given to the financial aspect of fungicide evaluation. Of the publications on plant disease control procedures and methods, over 98% have reported yields or percentage control without calculating the cost per unit of crop produced or the expected return on investment. While the fungicides themselves may often comprise only a small portion of the farmers' total investment, other costs (e.g., labor) may be substantial.

In the selection of the fungicide, it is advisable to consider the destiny of the crop and whether or not the product produces an acceptably safe residue as viewed by the government of the country receiving the crop. We discussed this in the preceding section. In 1975, significant amounts of citrus treated with postharvest fungicides were rejected at the dock in Japan, because the fungicides and their residues in citrus were not yet cleared by the Japanese Ministry of Agriculture and Fisheries.

In establishing a plan of handling the mixture, application, storage, and disposal of the waste, one should attempt to outline the handling procedure. This is particularly important for large farms where trips to and from storage and use areas constitute a significant expense and safety hazard. Therefore, a central warehouse is usually best, combined with delivery to the site of application by a covered trailer or, alternatively, one may arrange small storage sheds on various parts of the farm. Such storage areas should be weatherproof and appropriately locked when not in use. In the delivery of the fungicide to the application site, a scheme of handling water or other spray medium, such as spray oil, should be considered prior to the treatment. Reducing the handling of equipment will assure safety, reduce costs, and minimize trips to the warehouse. It is advisable to always work with another person. The measurement of the fungicide into the spray tank or duster should be done so as to avoid splash or inhalation of dust. Provision should be made for proper measurement of the material with scales and suitable containers.

Keeping an accurate record of the material used, its cost, the procedure, and results is a critical measure of the success of the program. Over a

CHART 1

Example Production Record

Crop	Location			Year	
Planted					
Harvest date(s)					

Pesticide schedule					
Treatment					
Date	Labor	Material	Amount/ha	Cost/kilo	

Weather/Water			
	Dates		Amount
Rain			
Irrigation			

Total cost						
	Fertilizer	Fungicide	Herbicide	Insecticide	Water	Harvest
Labor						
Material						
Equipment						
(depreciation)						
Total						

period of time such records will allow the comparison of products, measurement of interactions with the environment, and crop variety. This aspect of responsibility is the key to forward planning. A typical record appears in Chart 1.

In the case of certain perennial crops, yields may increase as a result of successive years of fungicide application. In general, this results from such factors as increased plant vigor, improved survival of buds, and increased root and stem growth. Improvement can often be detectable during the second and third years of fungicide treatment. However, the excessive continued use of copper fungicides, for example, has been known to reduce plant vigor. For these reasons, a record of treatments and yields is valuable.

In one case in the United Kingdom, as much as 1,500 ppm copper was detectable at the soil surface in an orchard which received copper treatment over a period of years [9].

B. Safety

Once a plan has been established and the appropriate fungicide has been selected, certain principles of safety concerning people, the product, application, and equipment should be followed. A farmer who has had an accident is the best judge of their importance. We have previously defined safe as being noninjurious or not harmful. Thus, the following proposed principles of safety are concerned with preventing harm or injury.

1. Use the product only as directed on the label.
2. Avoid contact with the chemical by wearing gloves, a mask, and proper clothing.
3. Store the product in a sealed container in a dry location.
4. Ensure that equipment is operable and calibrated to deliver the proper dosage prior to introducing the pesticide.
5. Apply the product under conditions so as to reduce drift.
6. Dispose of diluted unused material by procedures designed to hold the product while it degrades and dissipates.

In the development of a pesticide, attention has to be devoted to the development of the container label. An improper label may lead to an undesirable result in one or more of the following ways: disease control may be less than desired; undesirable chemical residues may appear in the crop at harvest; the crop may be injured; incompatibility with companion products may result; and someone may be injured in handling the product. A specific instance recalled is that of a tomato grower who believed that by tripling the recommended 2 lb/acre dose of maneb used for early blight control, he would achieve a similar tripling in disease control and yield. He thus raised the dose to 6 lb/acre on a weekly schedule and was astonished to find that at the normal harvest period he had massive vines and few fruit. While this was a painful lesson, it illustrates that using a product as directed is essential.

As we learn more about the toxicity and fate of residues of new as well as standard products, researchers have been able to establish the desired waiting period between the last spray and harvest so as to minimize the chemical residue at harvest and assure that a safe crop is harvested. The recommendation appearing on the label, such as "do not apply a certain number of days before harvest," is as critical as using the products during the growing season on the proper target pathogen.

Contact with a pesticide is possible during storage, mixing, application, and disposal. In all cases, one should perform any handling operation with the proper protective clothing and wear a mask. It is particularly important that all handling operations be done away from unprotected persons, food, and food containers. Concerning application, the current law in the United States [10] says, "No owner or lessee shall permit the application of a pesticide in such a manner as to directly or through drift expose workers or other persons except those knowingly involved in the application. The area being treated must be vacated by unprotected persons."

Opening of drums and bags of products should be done with the same care as spraying. Some rules to follow are:

1. Avoid unpacking in unventilated places.
2. Measure into containers used only for the product.
3. Wear gloves, a complete protective covering, and a face mask.
4. Where possible, make a dilute premix of dry products with water which can be subsequently measured as a liquid prior to placement in the spray tank.

During application of the product, similar personal precautions should be exercised, but in addition, the following factors must be considered:

1. Do not apply the product under conditions of possible spray drift.
2. Do not treat fields where workers will contact the spray or dust as applied unless they are properly clothed.
3. Consider the properties of the chemical and,
   a. know the symptoms of poisoning and
   b. establish a site of medical aid and means of reaching aid.

We do not propose here to discuss the details of the equipment used for the application of pesticides. The principles we have outlined encompass the use of all types of equipment. However, we do wish to point out that in studies of injuries on farms to date, equipment-related hazards have produced more known fatalities than pesticides. For example, figures from the Ministry of Agriculture in the United Kingdom, as cited by Barnes in his paper on the toxicology of agricultural chemicals [11], appear in Table 1 and clearly indicate that deaths involving handling tractors and accidents on farms, are infinitely higher than deaths due to pesticides.

Unused concentrated fungicide should be disposed of by incineration or burial under conditions of local ordinance requirements. Unused diluted spray material can best be disposed of by placing it in a holding tank, which should be emptied into a proper sewage treatment scheme or digester. Should this not be convenient, simply allowing evaporation and degradation in a holding tank may be sufficient in the case of organic fungicides. In no case, however, should fungicide materials be dumped near streams, ponds, or natural drains. Proper planning of the amount of material to be used should insure that little material remains. A general review of the problems and means of pesticide disposal for formulators and packagers of pesticides has recently been published. It provides valuable guidelines on specific schemes for large scale disposal problems [12].

TABLE 1

Accidents on Farms in the United Kingdom[a]

	1968	1969	1970	1971
Tractors				
Deaths	42	39	39	37
Injuries	311	451	378	395
Fall				
Deaths	14	22	15	14
Injuries	1705	1698	1368	1277
Pesticides				
Deaths	0	0	0	0
Injuries	18	10	23	15

[a]Courtesy of the Chief Safety Inspector, Ministry of Agriculture, Fisheries and Food, as cited by Barnes [11].

IV. RESEARCH ON PRODUCT SAFETY

The increasing awareness of the need for safety in the design and use of pesticides has had substantial impact on the development of new fungicides. Generally speaking, it is not difficult to synthesize compounds which are fungicidal, but it is difficult to prepare compounds which meet all the needs of modern agriculture, including the requirement for safety to man and his environment. Safety to man and environment cannot be assessed without long-term feeding studies of both the parent compound and its major metabolites. Thus, it is often 3 or 4 years after a chemical is defined as a fungicide before its properties are understood.

Part of the concern for safety encompasses a consideration of the impact of the chemical on the environment and whether or not the product or its residues biomagnify or have a negative effect on some critical link in nature's food chains. At present, as much as 50-70% of the research expended on a new fungicide may concern its fate and the effect of the product and its metabolites on the environment. In our experience, practically every compound which is new to the environment where it is used will have an effect on some component of that environment. We do not believe there is an organic compound which is completely inert in respect to the environment. A realistic goal is to minimize these effects and to achieve a balance of results which strongly favors the use of the fungicide. This is essentially the risk-benefit analysis.

The desire to achieve greater safety in pesticides has resulted in increasingly safer products, particularly insecticides. However, modern fungicides in some respects may also be considered safer than older products because industrial research objectives are being directed toward higher standards of safety as established by government agencies. Table 2 relates some interesting facts concerning the comparative acute oral toxicity of several fungicides and fungicide components. One may note that historically important materials, such as certain inorganic copper salts, are more acutely toxic when administered orally to rats than the organic fungicides captafol, maneb, or mancozeb. One should keep in mind, however, that acute toxicity is only one measure, or one aspect, of toxicity and that a comparison of the chronic toxicity of these materials would be needed to complete the picture.

In preparing this chapter, we found that there was little readily available long-term toxicology data on copper-based fungicides, probably because they were developed many years ago and are no longer proprietary products. Thus, industrial firms have not invested large sums to define their toxicology within today's concepts of safety. It is normally assumed that copper-based fungicides are safer to use and handle than other fungicides. However, in an acute sense and from the viewpoint of ingestion, their toxicity may be higher, as measured by LD_{50} values in rats, than several commonly used organic fungicides.

TABLE 2

Oral Toxicity of Some Widely Used Fungicides

	Acute Oral LD_{50} mg/kg			
	Rat	Ref.	Other species	Ref.
$CuSO_4$	300	[13]	200[a]	[14]
$CuCl_2$	140	[13]	200[a]	[14]
Captafol	2,500	[15]		
Maneb	6,750	[16]		
Mancozeb	>8,000	[17]		

[a]Fatal human dose.

Due to cost and our limited predictive ability it is impossible to evaluate fungicides on every possible important living species in the environment. Thus, there have been reports of the deleterious effects of widely used agricultural fungicides which were apparently missed in the precommercial testing of the products. These reports are as valuable to the manufacturer as to the public since the common goal is safety. In the case of thiophanate methyl and benomyl, for example, a harmful effect was noted on earthworm populations in apple orchards treated with the products [18]. In the case of captan, high toxicity to young fathead minnows (Pimephales promelas) and bluegills (Lepomis macrochirus) has been noted [19]. Information such as this is helpful in understanding the risk-benefit of these products when used in certain environments. One must, however, keep the impact of these agricultural fungicides in perspective. It might be relatively easy to search the literature and find incidents of negative effects of some molecule or a metabolite on some species. Consider, however, that the residue of the products may be occurring at a level which has no effect. Where such products do have an effect on the environment, it may be rather limited geographically and the balance sheet of utility may still favor use of the product.

To the author's knowledge, no fatalities or serious injuries have been attributable to agricultural fungicides when used as directed on the label. In contrast, in the United States, in 1973, it was estimated that there were 372,000 bicycle accidents resulting in various degrees of injury [20]. This is perhaps not a fair comparison, but it does help put fungicides in perspective in man's total environment. Of course, if research on fungicides for agriculture was not properly directed, one could have large-scale poisoning resulting from the use of these products.

Currently, each new fungicide is put through a strict sequence of toxicological observations during its development. Although government requirements vary throughout the world, the tendency has been to expand the number and complexity of toxicity tests needed to establish safety of a chemical. Table 3 shows some of the tests which are part of the registration requirements in many countries. Most of these are usually conducted with the product, but may also be done with metabolites if their residues at harvest are significant.

Generally speaking, many of the new fungicides are systemic (translocate) within the crop plant, and studies of their metabolism and fate in both plants and animals usually demonstrates several metabolites. For example, in the case of the new fungicide triforine, $\underline{N},\underline{N}'$-bis(1-formamido-2,2,2-trichloroethyl)piperazine, three metabolites appear to be formed in plants [21]. In the case of dimethirimol, 2-dimethylamino-4-hydroxy-5-n-butyl-6-methylpyrimidine, as many as 19 metabolites may be formed in the dog [22].

In recent years, research on fungicides has taken a different direction which, in turn, has and will continue to affect the aspect of safe and responsible use. From the early 1800s to the 1950s, research toward the development of fungicides stressed the control of killing fungi by direct contact with

TABLE 3

Toxicology Tests and Studies to Establish Fungicide Safety

Toxicity	Animal
Acute, single dose	
Oral	2 mammalian species
Dermal	Rabbit
Inhalation	Rat
Eye	Rabbit
Short-term, up to 90 days	
Oral	2 mammalian species
Skin irritancy	Rabbit
Inhalation	Rat
Allergic sensitization	Rabbit
Mutagenicity	
Long-term	
Chronic, 2-year feeding studies	Rat, dog
Reproduction, teratogenicity	Rat
Carcinogenicity	Rat, mouse
Metabolism	Rat
Neurotoxicity	

the pathogen when the fungus encountered a spray deposit of the chemical on the surface of the plant. The major shortcoming of such fungicides was recognized in their inability to redistribute to cover new plant growth and "hard to spray" plant parts. Thus research began on the development of systemic fungicides which enter the plant, move within the plant, and persist within the tissues for a certain period. Interestingly, several of the resultant products of this research either lack a direct toxic effect upon the pathogen (but may be acting through the formation of fungicidal metabolites) or induce host resistance through an effect on the biochemistry of the host-parasite interaction. Compounds largely systemic and apparently active by virtue of an active metabolite (Carbendazim), are the 2-substituted benzimidazoles [23]. The 4-n-alkyl-1,2,4-triazoles appear as a class of compounds active by their effect on the biochemistry of the host-parasite interaction, as in the case of control of brown rust in wheat. The most active member of this class, 4-n-butyl-1,2,4-triazole (Indar), does not affect the brown rust pathogen in vitro, is not active in crops other than wheat, and is only active against brown rust of wheat [24].

The development of "fungicides" that have more specific biochemical mechanisms of action and may act via metabolites, and which are transported within plants has already caused some changes in our outlook on the "responsible use" of fungicides because a new element of risk has developed, namely the occurrence of resistance by fungi to the fungicide. Systemic fungicides appear to have modes of action which are more specific than the multiple effects on the fungus which are caused by copper-based materials or the dithiocarbamates. The specificity of systemic fungicides is no doubt related to their effects on specific metabolic pathways which are genetically controlled and which, in turn, may be altered by a mutation or recombination of genes. To date, reports of the resistance of fungi to systemic fungicides have been mainly about benzimidazoles or other compounds which yield similar derivatives. Generally, resistance has also been reported when these fungicides were employed for control of pathogens of the ascomycete group or certain imperfect fungi which are near relatives of the ascus-forming fungi. These pathogens are characterized by high reproductive capacity, short incubation periods, coupled with frequent genetic and parasexual recombinations. The end result of the treatment of certain of these fungi with benzimidazole fungicides has been an occasional outbreak of resistance in the field of such diseases as apple scab, gray mold on grape, and powdery mildew on certain crops. We project, however, that benzimidazole fungicides will continue to enjoy a significant market when responsibly used alone or in admixture with contact fungicides.

The development of resistance is likely to occur more often in the future as fungicides are developed which have a highly specific locus of effect in the pathogen. The consequence of having such fungicides available is to increase the risk of error in the decision to use a specific spray program heavily dependent on this type of product. This can perhaps be overemphasized since the obvious advantages of systemic fungicides have permitted improved disease control of diseases which previously were difficult to control at all.

V. THE FUTURE

At the recent FAO/ICP Nairobi Pesticide Seminar (Oct. 30 to Nov. 8, 1974) sponsored by the Pesticide Working Group of the Industry Co-operative Program of the United Nations, M. Vandekar and J. F. Copplestone of the World Health Organization cited E. F. Edson and co-workers as follows:

In recent years there has been a comparative preoccupation and perhaps over-occupation with the possible long-term problems from extremely minute traces of pesticide residues in foods, humans and
wildlife..... The next hundred fatalities with present-day pesticides
will almost certainly involve negligence or ignorance, errors of judgment or deliberate carelessness.

We believe this statement is correct. The "EPA type" guidelines for
registration and use, and the resultant care given in the development of
modern fungicides to ensure that they can be safely handled, will assure that
the basic fungicide product is useable. The problem of reducing hazards
should now focus mainly on the education of those using the products, whether
they be fungicides or other pesticides. One can not help but cite the fact
that a large percentage of the fatalities resulting from pesticides in general
occurs among people who cannot read and make a decision, for example,
children. Some of the major calamities with insecticides have been as a
result of the contamination of foodstuffs due to outright stupidity [12]. In
some respects, the situation with pesticides parallels that of the automobile.
A manufacturer of an automobile does not intend his product to do harm to
others. However, accidental deaths involving autos are a major concern in
most developed countries. The problem, of course, relates to how the auto
is used. Similarly, in the case of pesticides, their safety depends on how
they are used.

Concerning research, the future will no doubt reveal many heretofore
unexpected results. In the area of pesticide interaction, we can expect to
find pesticide residues which produce more toxic effects as a result of additive or synergistic actions. Greater research efforts will be needed to
narrow the gap between toxic reactions to chemicals observed in experimental animals and their significance with respect to toxicity to man.

If we examine the safety considerations from the discovery of a fungicide
through registration, distribution, and finally to application, we can visualize this as a chain of events which currently has as its weakest link, the
end user. At the present time, the safety of a fungicide is a prime consideration of the manufacturers and distributors of pesticides. The laws of
many countries are designed so as to permit the sale of only those materials which have certain low toxicity levels. However, the end users are not
always as conscious as government bodies of the safe and responsible use of
fungicides. This chapter, in fact, will probably be read by government
specialists, persons specialized in developing fungicides, and the like, but
by relatively few end users.

It appears that governments and pesticide manufacturers have nearly
completed a so-called "Phase I" in establishing the toxicological criteria
for a safe and useful fungicide product. What we now require is more attention to a "Phase II," which we hope will consider education and assistance
to the end user in respect to applying principles of safety and responsibility.

Legislation may not be the answer to Phase II, but rather a program of advice to handlers and users of pesticides. The same people who cannot read a pesticide label are also those who cannot read a newspaper or a law. Lastly, in many underdeveloped areas, the sociology of change must be considered in implementing a pesticide safety program. If the person(s) responsible for implementing safety precautions in the storage and handling of pesticides lack status in the family or local social unit, we predict that the desired result will not be achieved. The key to safer use and responsible decisions in handling pesticides must first involve the development of responsibility in each community.

REFERENCES

1. A. Swenson and U. Ulfrason, Occupation. Health Rev., 15, 5 (1963).
2. D. C. Torgeson, ed., Fungicides: An Advanced Treatise, Vol. 2, Academic Press, New York, 1969.
3. J. A. R. Bates, Chem. Ind. (London), 1968, p. 1324.
4. R. Carson, Silent Spring, Houghton Mifflin Co., Boston, 1962.
5. Food and Agriculture Organization of the United Nations, AGP: CP 28, Rome, 1970.
6. Food and Agriculture Organization and World Health Organization, Joint Publication, PL-CP/21, Geneva, 1969.
7. Agricultural Pesticides, 3rd ed., Council of Europe, Strasbourg, 1973.
8. D. A. Kay, The International Regulation of Pesticide Residues in Food, A Report to the National Science Foundation under Grant GI 41472, Washington, D.C., 1975.
9. J. M. Hirst, H. H. Le Riche, and C. J. Bascomb, Plant Pathol., 10, 105 (1961).
10. Federal Register, 39, 92, Section 107.3, May 10, 1974.
11. J. M. Barnes, The Planter, 49, 460, 572 (1973).
12. T. M. Ferguson, R. R. Swank, and D. J. Becker, Environmental Protection Technology Series, EOA-660/2.74-094, Office of Research and Development, U.S. Environmental Protection Agency, Washington, D.C., 1975.
13. W. S. Spector, in Handbook of Toxicology, Vol. I (W. S. Spector, ed.), W. B. Saunders Co., Philadelphia, 1956, p. 76.
14. Food and Agriculture Organization (FAO), Nutrition Meetings Report Series No. 53A, 1974, Rome, p. 48.
15. R. J. Palazzolo, J. H. Kay, and J. C. Calandra, Data submitted by Chevron Corp., and cited in 1969 Evaluation of Some Pesticide Residues in Food, FAO, PL M/17/1, Rome, 1969.
16. J. W. Clayton, D. R. Hood, J. R. Barnes, and A. R. Borgman, Data submitted by the E. I. du Pont Co., and cited in 1967 Evaluation of Some Pesticide Residues in Food, FAO, PL M/11/1, Rome, 1967.

17. P. S. Larson, unpublished report by the Rohm and Haas Company, Philadelphia, 1965.
18. A. Stringer and M. A. Wright, _Pestic. Sci._, 4, 165 (1973).
19. R. D. Hermanutz, L. H. Mueller, and K. D. Kempfert, _J. Fisheries Res. Board Can._, 30, 1811 (1973).
20. _Business Week_, May 18, 1974, p. 60.
21. A. Fuchs, M. Vites-Verweij, and F. W. de Vries, _Phytopathol. Z._, 75, 111 (1972).
22. H. Bratt, J. W. Daniel, and I. H. Monks, _Food Cosmet. Toxicol._, 10, 489 (1972).
23. D. Peterson and L. V. Edgington, _Phytopathology_, 59, 1044 (1969).
24. W. C. von Meyer, S. A. Greenfield and M. C. Siedel, _Science_, 169, 977 (1970).

Chapter 16

FUNGICIDES: PROBLEMS AND PROSPECTS

Hugh D. Sisler

Department of Botany
University of Maryland
College Park, Maryland

I. INTRODUCTION

Although the use of chemicals may not be the preferred method of disease control, it is vitally important when alternative methods are lacking. In many respects the role of the fungicide in plant disease control programs closely parallels that of the antibiotic in human medicine. The chemicals, in either case, provide a means of disease control when natural resistance or other systems of control have failed.

The use of fungicides is a highly flexible control measure. It can be scheduled well in advance for anticipated problems or adopted quickly in an emergency. Disease situations requiring intervention with fungicides constitute a significant segment of the total disease problem, a situation which is likely to continue into the foreseeable future.

Progress in reducing disease losses will require further advances in plant breeding and biological control as well as in chemical control. Alternatives for the latter are currently receiving much emphasis particularly in cases where hazardous or persistent chemicals are involved. One approach to solving at least some of the problems of chemical control is to provide more effective compounds which pose essentially no hazard to man or the environment. In order to do so, the basis of chemical control may need to be shifted away from fungitoxicity and toward other types of action.

Currently, disease control with chemicals is based almost exclusively on compounds of intrinsic fungitoxicity, but there appears to be no fundamental reason why this should be. Theoretically, disease control could be accomplished by any mode of chemical action which impedes the pathogenic process. Such action might involve the host metabolism or secondary metabolism of the pathogen without affecting its saprophytic growth.

There is little doubt that past screening procedures favored selection of the intrinsically fungitoxic compound. However, the in vivo screening procedures which are now employed by industry should detect compounds which control disease irrespective of their mode of action. Consequently, it is reasonable to predict that modes of action other than that of direct fungitoxicity will be represented among the compounds available for chemical control in the future.

The following discussion presents some background and a few ideas concerning some problems and the directions of progress in the development of chemicals for the control of fungal diseases.

II. PROTECTIVE FUNGICIDES

Protective fungicides function essentially as toxic barriers on the surface of plants. Until very recently, surface protection was almost exclusively the means of chemical control. Dominance of the surface protectant for many years can be attributed in large measure to the early preponderant use of inorganics and to the use of in vitro fungitoxicity tests as the primary screening procedure for organics. Only future experience will reveal the magnitude of the possibilities for chemical control that were excluded by this screening procedure.

In vivo tests have recently yielded compounds with systemic activity which have displaced protectants in some cases. However, development of new types of fungicides is a slow process. Therefore, protectants developed a decade or more ago are still the mainstay of chemical control. Furthermore, low cost, dependability, improved methods of application and problems encountered in developing systemics render the position of the protectants relatively secure for the foreseeable future.

Although the loss of effectiveness through weathering and the lack of penetrativeness and mobility in plant tissues are evident weakness of protectants, they usually have several desirable characteristics. These include (1) a broad antifungal spectrum; (2) a mechanism of action unfavorable for development of fungal resistance; and (3) easily removable from food products.

Further search for protectants with antifungal spectra overlapping those of compounds already available is not a very promising endeavor. On the other hand, a hydrocarbon shortage may again stimulate the search for new protectants. Should vast amounts of sulfur become available through antipollution efforts, new investigations into ways of improving the performance of elemental sulfur and of inorganic sulfur compounds may be justified. Moreover, the fact that elemental sulfur is not regarded as a serious toxicological hazard would minimize problems related to its use on food products.

The possibility of utilizing other abundant inorganics as protectants might also be worthy of further consideration. Aluminum, for example, has long been regarded as nonfungitoxic. There is no reference to the fungitoxicity of this element by Horsfall [1] and only one brief mention by Lukens [2] in their textbooks on fungicides. However, aluminum ions have recently been shown to be quite toxic to ascospores of Neurospora tetrasperma [3] and to mycelia of Verticillium albo-atrum [4]. Lewis [5] showed that ions of aluminum as well as those of barium and iron inhibited growth of Aphonomyces euteiches by 95% at concentrations of 25 μg/ml or less. Aside from the effects that the barium ion has on growth, at subtoxic concentrations it is reported to inhibit an aspect of fungal metabolism [6] which may have a specific role in the pathogenicity of some organisms (see Section V).

Although there may still be merit for specific reasons to develop purely protectant fungicides, the pathway of general progress for this type of fungicide is extremely limited. On the other hand, there are excellent prospects for major advances with systemic compounds with conventional or unique modes of action.

III. SYSTEMIC FUNGICIDES

The term "systemic" is loosely used to designate chemicals displaying therapeutic or protective activity from within host tissues. In the narrow sense, only the applied fungitoxic compounds or their fungitoxic conversion

products (metabolites) would be included in this class. For the purposes of
this discussion, however, the term is used to include also compounds which
control disease by modifying the host or the pathogen without being directly
fungitoxic. The degree of mobility may vary according to the compound and
the site of application. Frequently the activity is limited to tissues near the
region of application, but in some cases it extends throughout much of the
plant. When applied to foliage, both surface protection and systemic activity
may be involved in disease control. The ability to eradicate established
infections and the extended time of retention are salient features of superiority of systemics over protectants.

The criteria for fungitoxic systemic activity are far more rigorous than
for fungitoxic surface protection. In addition to the requirement for low
phytotoxicity, a protectant need only be fungitoxic and relatively stable.
Poor penetration into host tissues as well as binding and degradation of
material that does penetrate each serve to prevent phytotoxicity. These
factors, however, tend to diminish the effectiveness of the systemic fungicide. A high degree of stability and biochemical specificity, therefore, are
necessary in systemic fungicides, although these criteria may apply more
rigidly to compounds entering readily into the symplast than to those remaining primarily in the apoplast.

The future of systemics would appear less exciting if success were anticipated only with compounds with innate fungitoxicity. Although the idea
is not new, the time is overdue for exploiting some of the numerous possibilities for disease control by chemically modifying the host or the pathogenicity of the fungus. What is needed now is the validation of one of these
possibilities by practical success.

With the exception of the benzimidazoles, most systemic fungicides have
a narrower antifungal spectrum than typical protectants. However, most
protectants do not control powdery mildews, and there are notable failures
of individual compounds to control other fungal pathogens. From an economic point of view, a narrow spectrum is undesirable because it limits the
usefulness of the compound. This factor may not be particularly detrimental
to the development of fungitoxic systemics, but it may prohibit the development of many systemics which are active by virtue of their ability to alter
the host susceptibility or suppress pathogenic factors in the fungus. A number of these compounds are likely to be active only against specific pathogens or in particular plant species. Unless the disease control is of considerable economic significance, the development of such compounds may
not be worthwhile. On the other hand, valuable compounds of this type for
the control of economically important diseases may never be discovered
unless specific host-pathogen combinations are used in the screening program.
Sbragia [7] has noted a criticism of the success in discovering a vast array
of bean rust fungicides, which point out the limitations, but also the potential
of the use of specific host-parasite combinations in screening for fungicides.

IV. MODIFICATION OF HOST SUSCEPTIBILITY

A. Increasing Effectiveness of Phytoalexin Systems

Host defense systems constitute foundations on which unusual types of chemical control might be based. The most clearly defined of these systems are
those which produce fungitoxic compounds (phytoalexins) in response to a
stimulus by the invading pathogen.
     Phytoalexins produced by higher plants have a broad antifungal spectrum
[8-10] and a mode of fungitoxicity which is evidently not easily averted by
the development of fungal resistance. A pathogen, in some cases, may,
however, represent a fungal strain with increased tolerance to a phytoalexin.
This is suggested by the observations that pathogens of a host are in general
more tolerant to its phytoalexin than nonpathogens [9, 10]. There are notable exceptions to this generalization [10], but these do not necessarily
invalidate the role of phytoalexins in disease resistance. A sensitive pathogen, for instance, may be a poor inducer or may repress production,
whereas an insensitive nonpathogen may lack certain factors required for
pathogenicity. Thus, failure of the host phytoalexin system to control a
pathogen may be due to inadequate phytoalexin production or to an insensitivity of the pathogen. Assuming that either inadequate production or metabolic destruction accounts for host susceptibility, at least two possibilities
exist for synergism between an applied chemical and the host system. The
first of these would involve intensification of phytoalexin production either
by direct action on the host or by indirect action on the pathogen to increase
production of inducer or decrease production of repressor. Chemical
action on the pathogen would be preferred, because phytoalexin production
would then be limited to the region of fungal invasion. Although generalized
phytoalexin production has been induced by chemical treatment of host tissue [11-13], little is known regarding possible phytotoxic consequences.
Practical success with this approach is possible only if fungitoxic levels of
phytoalexins can be attained in the tissue without injury to the host. Whether
or not this is feasible may well depend on the host species involved. At
least in the case of the phytoalexin, capsidol, late blight of tomato has been
controlled by foliar applications without injury to the host [14]. The exposure of host tissue to phytoalexin by this method of application, however,
is probably not analogous to production within host tissues.
     In cyanogenic plants, release of hydrogen cyanide (HCN) from cyano-
glucosides upon infection is thought to be a resistance mechanism [21].
Pathogens of cyanogenic plants have a low sensitivity to HCN which is evidently due to their capacity to convert HCN to formamide [22]. In several
instances, phytoalexin degradation by pathogens has been observed [15-20],
and this may explain host susceptibility. A second possibility for synergism of the host system, therefore, is chemical inhibition of phytoalexin

detoxication by the pathogen. Inhibition of insecticide metabolism has proved to be the basis of action of a number of insecticide synergists [23]. Frequently the synergist prevents oxidative detoxication of the insecticide by inhibiting microsomal mixed function oxidases in the insect. Some of these inhibitors, such as the 1-, 2-, and 4(5)-substituted imidazoles and 1,3-benzodioxoles [24], could be of value in restoring host resistance to fungal pathogens by inhibiting oxidative detoxication of the phytoalexin. Appropriate inhibitors of fungal enzymes involved in other types of phyto-alexin metabolism might likewise restore host resistance.

B. Other Effects on Host Systems

As noted previously, various possibilities exist for disease control by chemical modification of the host. These include desensitization to toxins, increasing wall resistance to enzymatic degradation, increasing capacity to inactivate toxin or to repress toxin and extracellular enzyme production by the pathogen.

The rate of lignification has been related to tissue resistance to fungal invasion in cucumber [25], potato [26], and reed canary grass [27]. In cucumber, there is evidence that nonfungitoxic levels of phenylthiourea increase peroxidase activity and the rate of lignification and thereby make the plants resistant to infection by Cladosporium cucumerinum [28].

Buchenauer and Grossmann [29] showed that fluorene derivatives restricted hyphal development, markedly reduced browning, and suppressed wilt symptoms in tomato plants inoculated with Fusarium oxysporum f. lycopersici. The basis of the therapeutic activity was not determined, but there was no correlation between the effects of the compounds in vivo and their fungitoxicity in vitro.

The possibility of stimulating host capacity for phytotoxin inactivation is suggested by studies with pyricularin, a toxin implicated in the pathogenesis of Pyricularia oryzae [30]. Chlorogenic or ferulic acid apparently detoxify pyricularin by forming inactive complexes. Addition of ferulic acid increases host resistance to infection by P. oryzae. Thus, chemicals which stimulate chlorogenic or ferulic acid production in rice plants would probably increase resistance to rice blast.

Host sensitivity to a pathotoxin is sometimes very specific and may be decreased by chemical treatment as Gardner and Scheffer [31] have demonstrated with oat tissue sensitive to Helminthosporium victoriae toxin.

Other cases of chemical modification of host susceptibility are cited in the literature. There are indeed many possibilities, but a case of practical success would greatly stimulate investigations of this approach to chemical control.

V. MODIFICATION OF PATHOGENICITY

Survival in nature involves reproduction in competition with other organisms. While most phytopathogenic fungi grow readily on a variety of substrates in pure culture, as saprophytes in nature, they are usually ecologically disadvantaged organisms of minor significance in comparison with many of the obligate saprophytes. Specific attributes of plant pathogens, however, confer on them a supreme ecological advantage on living plant substrates. In parasitizing higher plants, the pathogen must be able to penetrate and to cope with the natural resistance systems. Various enzymes and toxins, depending on the particular host-parasite combination, may be required for invasion of living plant substrates. Therefore, chemical suppression of one or more of the specific factors required for pathogenesis could result in disease control as effective as that attained by conventional inhibition of fungal growth and possibly with less environmental hazard.

A. Inhibition of Phytotoxin Production

Fungi are notorious for the production of a great variety of end products which are classified as secondary metabolites, because they play no obvious role in the economy of the organisms [32]. Some of these compounds may, however, be involved in the formation of pigments which protect spores from visible and UV irradiation and in the synthesis of resistant wall structures. Others may inhibit saprophytic competitors or aid in overcoming the resistance of living substrates. A variety of phytotoxic "secondary metabolites" are produced by phytopathogenic fungi in culture and in infected plant tissue. The potential importance of these compounds in plant disease is widely recognized, although much additional information is needed to assess the role of specific toxins [33] in a disease process. Toxins such as victorin are almost totally responsible for the disease symptoms produced by the pathogen, whereas other toxins may have little or nothing to do with the disease process [34].

Chemical inhibition of phytotoxin biosynthesis affords a potentially promising approach to disease control. While there appear to be no reported cases of practical disease control by this mechanism, there is evidence that some compounds may selectively inhibit pathways which lead to phytotoxin production. Consideration in this discussion will be limited to compounds which affect the polyketide (polyacetate) and terpenoid pathways.

1. Polyketide Synthesis

Isotope labelling data or structural considerations indicate that the following toxins are synthesized via the polyketide pathway: alternaric acid [35], diaporthin [32], skyrin [36], pyriculol [32], novarubin [37], and alternariol [38]. This pathway is also involved in the synthesis of aflatoxin by Aspergillus flavus [39]. Attempts to inhibit the production of this toxin have revealed some chemicals which appear to be selective inhibitors of the polyacetate pathway in fungi. Low concentrations of Ba^{++} were shown to inhibit aflatoxin production (and pigmentation) in Aspergillus flavus without affecting growth [6]. Similar results were obtained in Aspergillus parasiticus with low concentrations (10 ppm) of the insecticide dichlorovos (dimethyl-2,2-dichlorovinylphosphate) [40]. The polyacetate pathway appears to be the primary source of pigment production in many fungi [41] and, therefore, inhibition of pigment production may be presumptive evidence of inhibition in the pathway. In A. flavus high concentrations of dimethyl sulfoxide (DMSO) completely block conidial pigmentation [42] and markedly reduce aflatoxin production [43] while having little or no effect on fungal growth. High concentrations of DMSO also inhibit pigment production in perithecia of Nectria haematococca [44] and in conidia of Aspergillus niger [45].

Inhibition of pigment production may result from effects late in the pathway which do not block toxin production. Such may be the case with tricyclazole, a compound which markedly inhibits or modifies pigment production in several fungi without affecting growth [46]. Tricyclazole, 5-methyl-1,2,4-triazolo(3,4-b)benzothiazole, controls rice blast at concentrations which do not affect growth of the pathogen, Pyricularia oryzae. However, melanization of the fungal hyphae is blocked and a soluble brown pigment is produced instead, while production of the phytotoxin, pyriculol, is actually enhanced [46]. Although the fungicide may inhibit the polyketide pathway, its action in disease control does not appear to involve an inhibition of pyriculol production.

2. Terpenoid Synthesis

The terpenoid pathway leads to the synthesis of primary metabolites of the sterol group and several phytotoxins as well. The terpenoid toxins include the victoxinine moiety of victorin [47], ophiobolin A [48], helminthosporal [49], southern corn leaf blight toxin [50], and fusicoccin [32]. The gibberellins produced by Gibberella fujikuroi are also terpenoid derivatives. Hirol [51] suggested that gibberellins are special toxins associated with the pathogenicity of some Fusarium species.

Selective inhibition of gibberellin biosynthesis in G. fujikuroi has been demonstrated with a higher plant growth retardant. Kende et al. [52] showed that 2-chloroethyltrimethylammonium chloride (CCC) strongly inhibits gibberellic acid (GA_3) synthesis while having little or no effect on fungal growth. This indicates the potential for selective chemical suppression of toxin production in plant pathogens. The fungicide, triarimol, and the plant-growth retardant, ancymidol, likewise inhibit gibberellin production in cultures of G. fujikuroi (and in higher plant tissues) but in these cases growth of the fungus is severely curtailed [53]. Since gibberellins are evidently secondary metabolites in G. fujikuroi, the inhibition of growth is probably due to inhibition of ergosterol biosynthesis [54, 55].

B. Regulation of Extracellular Enzyme Production

Extracellular fungal carbohydrases which degrade host cell wall material are important factors in invasion of plant tissue by fungal pathogens [56]. Production of these enzymes is highly sensitive to metabolite regulation. In Verticillium albo-atrum and Fusarium oxysporum f. sp. lycopersici, enzymes synthesis is almost always induced specifically by the sugar or uronic acid unit predominant in the polymeric substrate of the enzymes [57]. Presence of the inducers in slight excess of the requirements for growth represses enzyme production.

Sands and Lukens [58] showed that production of endopolygalacturonase, cellulase, and β-glucosidase in Alternaria solani cultures are all markedly increased by adding adenosine phosphates and particularly by adding adenosine-5-phosphate (AMP). Although starvation (low sugar levels) usually favors enzyme production, AMP mimics the starvation effect in nonstarved mycelia with the result that high levels of enzymes are produced. Thus, lowering of AMP levels or increasing levels of specific substrates may regulate production of extracellular enzymes involved in pathogen invasion. Chemicals which counteract the effect of AMP or mimic the effect of specific sugars might prove useful in disease control where inducible extracellular enzymes are involved. Patil and Dimond [59] showed that glucose, monogalacturonic acid, and 2-deoxyglucose decreased wilt symptoms and polygalacturanase activity in tomato stems infected with F. oxysporum f. sp. lycopersici. The action was presumably due to repression of polygalacturonase synthesis in the pathogen. Chemical repression of enzyme production appears to be a more favorable approach than chemical inhibition of enzyme activity. In either case, penetration of the chemical into the host tissues would probably be required.

VI. RESISTANCE

Protective fungicides have met the challenge of fungal resistance over a relatively long period with a high degree of success. However, the relative obscurity of the resistance problem is not likely to continue as compounds with more specific modes of action come into use. In fact, fungal resistance looms as a major obstacle to the development of systemic fungicides. Already there is much evidence that some of the new systemics are poorly fitted to meet the challenge of resistance. Various aspects of the resistance problem are discussed in detail by Georgopoulos (see Vol. 2, Chapter 12); therefore, only selected aspects of the subject will be considered here.

A. Effect of Mode of Action on Resistance

The mode of action undoubtedly influences the extent to which a fungicide is victimized by resistance. While knowledge in this area is quite limited, the infrequency of resistance to protectants indicates that multisite action is the best safeguard against resistance. With either specific or multisite inhibitors, a lethal action at the cell surface is a favorable mode of action because it is not subject to resistance based on impermeability. On the other hand, compounds acting within the mitochondria or nucleus face the possibilities of resistance based on decreased permeability of two membrane systems.

While multisite inhibitors almost certainly have effects within the cytoplasm, the plasma membrane has been implicated in the action of compounds such as maneb [60], silver [61], copper [62], dialkyldithiocarbamates [63], dodine [64], and glyodin [65]. Compounds acting with a specific mode at the plasma membrane site are not likely to escape the resistance problem if polyene antibiotics are valid examples. These antibiotics bind to sterols in the plasma membrane and cause membrane damage [66]. Resistance based on the substitution of sterols in the plasma membrane to which the polyenes bind less readily have been described [67, 68]. A relatively small degree of increased resistance has even been described for the less-specific inhibitor, dodine [69, 70]. However, it is not known whether the increase in tolerance involves a loss in membrane sensitivity.

Mutations in the DNA of the nucleus or mitochondria which code for resistance must be translated before the cell is actually resistant. It is evident, therefore, that fungicides which prevent translation of the mutational event are less likely to allow the emergence of resistant strains than those which allow the translation to occur. Among the fungicides which prevent translation are those which inhibit protein or RNA synthesis and those interfering with energy-generating processes or causing rapid damage to the plasma membrane. Cycloheximide, for example, is an inhibitor of protein synthesis which should prevent the translation of resistance mutations. The results of Neuhäuser et al. [71] indicate that it does. They showed that the yield of cycloheximide-resistant mutants was 100-fold higher in UV irradiated conidia of <u>Neurospora crassa</u> preincubated in growth medium for 13 hr before treatment with cycloheximide than in conidia treated immediately after irradiation. Compounds which specifically inhibit DNA synthesis, mitosis, or cytokinesis would permit translation of resistance mutations and the ready emergence of resistant strains in their presence. The fact that the benzimidazole fungicides belong in this category [72-75] may partially explain resistance problems encountered by these compounds.

Unless a fungicide is a mutagen, the frequency of resistance mutations should be as high in the absence of the compound as in its presence. Thus in a large population of spores there may be mutants in which translation has occurred before the fungicide is encountered. Neither of the two

aforementioned types of inhibitors would escape this type of resistance mutation. On the other hand, some mutations may be expressed only after
adaptation in the presence of the fungicide. Compounds which directly or
indirectly prevent protein synthesis would prevent the multiplication of
strains carrying such mutations. These strains would be a practical problem, however, with compounds which do not prevent the adaptation.

B. Combatting the Resistance Problem

Once field resistance to a fungicide has developed, it may be difficult to
combat. The only practical solution in many cases will be the use of alternative fungicides. Therefore, stability to fungal resistance should receive
major consideration in the early stages of fungicide development.

Although success in dealing with an existing problem of resistance is
uncertain, several measures in addition to altering the application schedule
might be utilized. Resistance based on a change which lowers the affinity
of a fungicide for an affected site might be overcome by the use of an inhibitor
analogue with high affinity for the altered site. The resistant and wild-type
strains should exhibit a low or possibly even a negatively correlated cross
resistance to such an analogue (see Vol. 2, Chapters 10 and 12). Other
mutations leading to loss of affinity for the analogue are conceivable, but
these might restore sensitivity to the original compound. Moreover, there
are probably only a few possible alterations which do not lead to significant
loss of the survival value or pathogenicity of the fungus under field conditions.
There are obvious shortcomings to this approach. These include the development costs for the analogue which may approximate those of the original
compound. Furthermore, resistance in some fungal species may be based
on mechanisms which do not respond in a favorable way to the use of analogues.

Although some fungicides may enter cells via specific membrane transport systems, most organic fungicides probably enter by simple diffusion
as uncharged molecules through the lipid phases of the membrane. An appreciable penetration rate by simple diffusion for compounds of a molecular
weight of about 50 or more requires a high degree of lipid solubility [76].
This requirement implies a limited solubility in water, a property which is
characteristic of practically all organic fungicides. Either a decreased
permeability or a lowered effectiveness at the action site(s) may render a
compound ineffective because of extracellular aqueous solubility limitations.
Since the internal toxicant concentrations at equilibrium is proportional to
the amount of dissolved toxicant in the external solution, increasing the
solubility of the chemical in the external solution might aid in overcoming
resistance. Additives, such as dimethyl sulfoxide, surface-active agents,
or glycerol might increase the amount of dissolved toxicant.

In the foregoing discussion it has been assumed that only the chemical
which is dissolved is effective. Pinocytosis or contact dissolving of fungicide

particles by membrane lipids could also influence uptake. Possibly such phenomena account for the extension of fungicide effectiveness beyond the aqueous saturation point which is sometimes observed.

Metabolism of pesticides by mammals, insects, and higher plants is widely recognized as an important basis of tolerance. Fungi are known to metabolize fungicides (see Vol. 2, Chapter 3) but whether this capability serves as a basis of fungicide resistance remains to be proved. The rapid action and multiple sites of inhibition by many protective fungicides permit little opportunity for adaptative metabolic detoxication. The general failure of resistance to develop to these compounds over a long period of time is convincing evidence that neither this nor any other basis of resistance is of much significance. However, the possibility of fungicide metabolism serving as a basis of resistance will be more likely with specific metabolic inhibitors, particularly with those that do not inhibit protein synthesis.

In many situations, the ratio of fungicide available within easy diffusion distance is so massive relative to the fungal protoplasm present (e.g., a few hundred spores per milliliter of solution saturated with fungicide) that the rate of metabolic degradation would not detectably reduce the extracellular fungicide concentration. Metabolic degradation could, nevertheless, maintain intracellular levels below the toxic concentration and thus serve as a resistance mechanism. Only in the presence of a large number of fungal cells could a significant rate of reduction of extracellular fungicide be detected.

Should metabolic detoxication prove to be the basis of resistance to a fungicide, methods of combating it might include (1) the use of synergists to inhibit the enzymes involved and (2) alterations in molecular structure of the toxicant which prevent enzymatic degradation. The former approach has been used successfully with insecticides [23] and the latter to combat resistance to the penicillins [77].

Another approach to the problem of resistance to systemics is the use of an inert derivative from which a multisite inhibitor is released by the pathogen at the invasion site. In some respects this mechanism of chemical control would resemble that of the phytoalexin systems in higher plants. Cycloheximide acetate is an example of such an inert systemic derivative, although the toxin released (cycloheximide) is not a multisite inhibitor. Unfortunately, plant tissues release cycloheximide more readily than fungi and phytotoxicity results [78, 79]. The compound pyrazophos may prove to be a better example than cycloheximide acetate. Its fungal metabolic conversion product, 2-hydroxy-5-methyl-6-ethoxycarbonylpyrazolo(1,5a)pyrimidine (PP) is regarded as the toxic principle of the relatively inert parent compound, pyrazophos [80]. Piricularia oryzae is apparently sensitive to pyrazophos treatment because it converts the compound to PP. On the other hand, Pythium debaryanum and Saccharomyces cereviseae are not sensitive. Neither of these fungi convert pyrazophos to PP [80]. Cucumber plants are tolerant of pyrazophos but whether this is due to inability to convert the compound to PP is not known.

VII. TOXICOLOGICAL AND ENVIRONMENTAL HAZARDS

Many chemicals have come into widespread use without adequate knowledge of possible detrimental effects on man and the environment. In the case of pesticide chemicals, at least, some types of toxicological information has been required for a number of years. Nevertheless, there is justified concern that past procedures have not been adequate to reveal some of the potential hazards. New regulations, therefore, will almost certainly require information on mutagenic activity and persistence in the environment in addition to conventional data on acute toxicity. Environmental mutagens will be of particular interest, and eventually suitable test systems must be devised to facilitate detection of chemicals producing the following classes of mutation [81]: (1) changes in chromosome number; (2) changes resulting from chromosome breaks (including deletions, duplications, inversions, and translocations); and (3) single-gene mutations. Some of the possible detection procedures and various aspects of the environmental mutagen problem are discussed in a recent report by a committee of The Environmental Mutagen Society [81].

The clearance of a pesticide for practical use by any existing standards should be reviewed periodically, taking into consideration new information which tend to verify or dispute previous data on which clearance was based. Knowledge of the mode of action of a toxicant provides a basis for further assessment of its potential hazards to man. A fungicide presumably has a site (or sites) of action in mammalian cells corresponding to that in the fungal cell. Once the site of action in the fungus is known, it should be relatively simple to determine the extent to which the corresponding site in mammalian cells is affected. Aside from possible differences at the cellular and subcellular level that condition sensitivity in the two types of organisms, cells in the intact animal may be protected because of poor absorption from the intestinal tract or through metabolic detoxication. The validity, therefore, of toxin effects on isolated fungal cells (or isolated mammalian cells) might be legitimately questioned. On the other hand, the effect in isolated cell systems may have validity in predicting effects on cells of the digestive and excretory systems exposed before detoxication or excretion of the chemical.

In regard to safe mechanisms, the inhibition of fungal metabolic reactions which have no counterparts in mammalian cells such as those in certain aspects of secondary metabolism or cell-wall synthesis would be highly desirable. Selective inhibition of energy production or protein synthesis in fungal cells would ordinarily be viewed as acceptable since significant effects on these processes in the animal should be readily detectable in feeding tests, and minor effects should cause no obscure damage. Fungicides which inhibit mixed function oxidases or other enzymes involved in the degradation of pesticides could be harmful if they reduced the capacity of the animal to eliminate other toxins.

VIII. RECENT DEVELOPMENTS

Langcake and Wickins [82] presented evidence that 2,2-dichloro-3,3-dimethylcyclopropane carboxylic acid stimulates disease resistance mechanisms to P. oryzae in rice plants. Intrinsic fungitoxicity, which is low, did not correlate with the ability of the compound to prevent "blast" of rice. On the other hand, protective activity of several dichlorocyclopropanes correlated closely with the ability of the compounds to enhance peroxidase activity of the leaves [83]. The compounds intensified accumulation of melanin-like material around infection zones accompanied by the production of phytoalexin-like compounds [83, 84]. Dichlorocyclopropanes, therefore, may be interesting representatives of a new generation of chemicals which control disease by mechanisms other than that of direct fungitoxicity to the pathogen (see Section IV).

Some interesting results on cross resistance as a basis for combatting fungal resistance (see Section VI.B) have been reported by Uesugi and coworkers. Uesugi et al. [85] showed that a phosphorothiolate (PTL) resistant mutant of P. oryzae is more sensitive than the wild type to a number of phosphoramidate (PA) derivatives. A PA tolerant strain subsequently selected from the PTL tolerant population proved to have reverted to the PTL sensitivity of the wild type [86]. Thus, reciprocal negatively correlated cross resistance is an effective means of combatting resistance in this model system of PA and PTL fungicides.

Toxicity of PA was synergized by PTL in the wild type and in the PA tolerant strain but not in the PTL tolerant strain [86]. Synergism of PA most likely resulted from an interference with PA metabolism by PTL. Evidence suggests that PA sensitive strains have a decreased capacity to metabolize the fungicide. The basis of decreased sensitivity to PTL in the PA sensitive strains, however, remains a mystery.

REFERENCES

1. J. G. Horsfall, Principles of Fungicidal Action, Chronica Botanica, Waltham, Mass., 1956.
2. R. J. Lukens, Chemistry of Fungicidal Action, Springer Verlag, New York, 1971.
3. W. H. Ko and F. K. Hora, Soil Sci., 113, 42 (1972).
4. R. G. Orellana, C. D. Foy, and A. L. Fleming, Phytopathology, 65, 202 (1975).
5. J. A. Lewis, Phytopathology, 63, 989 (1973).
6. E. G. H. Lee, P. M. Townsley, and P. M. Walden, J. Food Sci., 31, 432 (1966).
7. R. J. Sbragia, Ann. Rev. Phytopathol., 13, 257 (1975).
8. K. Uehara, Bull. Hiroshima Agr. Coll., 1, 7 (1960).

9. I. A. M. Cruickshank and Dawn R. Perrin, Phytopathol. Z., 70, 209 (1971).

10. H. D. VanEtten, Phytopathology, 63, 1477 (1973).

11. R. E. Pierre and D. F. Bateman, Phytopathology, 57, 1154 (1967).

12. S. L. Hess and L. A. Hadwiger, Plant Physiol., 48, 197 (1971).

13. J. J. Reilly and W. L. Klarman, Phytopathology, 62, 113 (1975).

14. E. W. B. Ward, C. H. Unwin, and A. Stoessl, Phytopathology, 65, 168 (1975).

15. K. Uehara, Ann. Phytopathol. Soc. Japan, 29, 103 (1964).

16. A. deWit-Elshove, Neth. J. Plant Pathol., 75, 164 (1969).

17. V. J. Higgins and R. L. Millar, Phytopathology, 59, 1500 (1969).

18. J. W. Mansfield and D. A. Widdowson, Physiol. Plant Pathol., 3, 393 (1973).

19. H. D. VanEtten and D. A. Smith, Physiol. Plant Pathol., 5, 225 (1975).

20. V. J. Higgins, Physiol. Plant Pathol., 6, 5 (1975).

21. W. E. Fry and R. L. Millar, Phytopathology, 61, 494 (1971).

22. W. E. Fry and P. H. Evans, Proc. Amer. Phytopath. Soc., 2, 55 (1975).

23. R. D. O'Brien, Insecticides, Action and Metabolism, Academic Press, New York, 1967, p. 209.

24. C. F. Wilkinson, K. Hetnarski, and L. J. Hicks, Pestic. Biochem. Physiol., 4, 299 (1974).

25. T. Hijwegen, Neth. J. Plant Pathol., 69, 314 (1963).

26. J. Friend, S. B. Reynolds, and M. A. Aveyard, Physiol. Plant Pathol., 3, 495 (1973).

27. C. P. Vance and R. T. Sherwood, Proc. Amer. Phytopath. Soc., 2, 48 (1975).

28. A. Kaars Sijpesteijn, J. Sci. Food Agric., 20, 403 (1969).

29. H. Buchenauer and F. Grossmann, Nature, 227, 1267 (1970).

30. K. Tamari, N. Ogasawara, and J. Kaji, in The Rice Blast Disease, The Johns Hopkins Press, Baltimore, 1965, p. 55.

31. J. M. Gardner and R. P. Scheffer, Physiol. Plant Pathol., 3, 147 (1973).

32. W. B. Turner, Fungal Metabolites, Academic Press, London, 1971.

33. L. D. Owens, Science, 165, 18 (1969).

34. H. Wheeler and H. H. Luke, Ann. Rev. Microbiol., 17, 223 (1963).

35. W. B. Turner, J. Chem. Soc., 1961, p. 522.

36. S. Gatenbeck, Acta Chem. Scand., 14, 296 (1960).

37. H. Kern and S. Naef-Roth, Phytopathol. Z., 60, 316 (1967).

38. S. Gatenbeck and S. Hermodsson, Acta Chem. Scand., 19, 65 (1965).

39. M. Biollaz, G. Büchi, and G. Milne, J. Amer. Chem. Soc., 90, 5017 (1968).

40. D. P. H. Hsieh, J. Agr. Food Chem., 21, 468 (1973).

41. A. A. Bell, R. D. Stipanovic, and J. E. Puhalla, Tetrahedron, 32, 1353 (1976).

42. G. A. Bean, G. W. Rambo, and W. L. Klarman, Life Sci., 8, 1185 (1969).

43. G. A. Bean, W. L. Klarman, G. W. Rambo, and J. B. Sanford, Phytopathology, 61, 380 (1971).
44. C. R. Curtis and G. A. Bean, Phytopathology, 58, 884 (1968).
45. H. E. Carley, R. D. Watson, and D. M. Huber, Can. J. Bot., 45, 1451 (1967).
46. M. Chrysayi, unpublished work, 1976.
47. R. B. Pringle, in Phytotoxins in Plant Diseases (R. K. S. Wood, A. Ballio, and A. Graniti, eds.), Academic Press, London, 1972, pp. 141-145.
48. L. Canonica, H. Fiecchi, M. Galli Kienle, B. M. Ranzi, and A. Scala, Tetrahedron Letters, 1966, p. 3035.
49. P. deMayo, J. R. Robinson, E. Y. Spencer, and R. W. White, Experientia, 18, 359 (1962).
50. A. L. Karr, Jr., D. B. Karr, and G. A. Strobel, Plant Physiol., 53, 250 (1974).
51. I. Hirol, J. Fac. Agr., Tottori Univ., 5, 1 (1969).
52. H. Kende, H. Ninnemann and A. Lang, Naturwissenschaften, 50, 599 (1963).
53. J. B. Shive, Jr. and H. D. Sisler, Plant Physiol., 57, 640 (1976).
54. N. N. Ragsdale, Biochim. Biophys. Acta, 380, 81 (1975).
55. J. L. Sherald and H. D. Sisler, Pestic. Biochem. Physiol., 5, 477 (1975).
56. P. Albersheim, T. M. Jones, and P. D. English, Ann. Rev. Phytopathol., 7, 171 (1969).
57. R. M. Cooper and R. K. S. Wood, Physiol. Plant Pathol., 5, 135 (1975).
58. D. C. Sands and R. J. Lukens, Plant Physiol., 54, 666 (1974).
59. S. S. Patil and A. E. Dimond, Phytopathology, 58, 676 (1968).
60. A. L. Morehart and D. F. Crossan, Toxicol. Appl. Pharmacol., 4, 720 (1962).
61. L. P. Miller and S. E. A. McCallan, J. Agr. Food Chem., 5, 116 (1957).
62. R. Troger, Arch. Microbiol., 29, 430 (1958).
63. J. Goksøyr, Physiol. Plantarum, 8, 719 (1955).
64. I. F. Brown and H. D. Sisler, Phytopathology, 50, 830 (1960).
65. M. Kottke and H. D. Sisler, Phytopathology, 52, 959 (1962).
66. S. C. Kinsky, in Antibiotics I, Mechanism of Action (D. Gottlieb and P. D. Shaw, eds.), Springer-Verlag, Berlin, 1967, pp. 122-141.
67. R. A. Woods, J. Bacteriol., 108, 69 (1971).
68. E. G. Gollub, P. Trocha, P. W. Liu, and D. B. Sprinson, Biochem. Biophys. Res. Commun., 56, 471 (1974).
69. A. Kappas and S. G. Georgopoulos, Experientia, 24, 181 (1968).
70. M. Szkolnik and J. D. Gilpatrick, Plant Disease Reptr., 53, 861 (1969).
71. A. Neuhäuser, W. Klingmüller, and F. Kaudewitz, Mol. Gen. Genet., 106, 180 (1970).
72. G. P. Clemons and H. D. Sisler, Pestic. Biochem. Physiol., 1, 32 (1971).

73. R. S. Hammerschlag and H. D. Sisler, Pestic. Biochem. Physiol., 3, 42 (1973).
74. L. C. Davidse, Pestic. Biochem. Physiol., 3, 317 (1973).
75. D. V. Richmond and A. Phillips, Pestic. Biochem. Physiol., 5, 367 (1975).
76. R. Collander in Plant Physiology II (F. C. Steward, ed.), Academic Press, New York, 1959, pp. 3-102.
77. C. H. O'Callaghan and P. W. Muggleton, in Cephalosporins and Penicillins (E. H. Flynn, ed.), Academic Press, New York, 1972, pp. 438-495.
78. A. J. Lemin and W. E. Magee, Plant Disease Reprt., 41, 447 (1957).
79. H. D. Sisler, M. R. Siegel, and Nancy N. Ragsdale, Phytopathology, 57, 1191 (1967).
80. M. A. deWaard, Mededel. Landbouwhogeschool Wageningen, 74, 14 (1974).
81. Committee 17, Council of the Environmental Mutagen Soc., Science, 187, 503 (1975).
82. P. Langcake and S. G. A. Wickins, J. Gen. Microbiol., 88, 295 (1975).
83. P. Langcake and S. G. A. Wickins, Physiol. Plant Pathol., 7, 113 (1975).
84. P. Langcake, Abstracts, Symposium on Internal Therapy of Plants, Wageningen, The Netherlands (1976).
85. Y. Uesugi, M. Katagiri, and O. Noda, Agr. Biol. Chem., 38, 907 (1974).
86. Y. Uesugi and M. Katagiri, Neth. J. Plant Pathol. (Suppl.), 82, in press.

AUTHOR INDEX

Numbers in brackets are reference numbers and indicate that an author's work is referred to although his name is not cited in the text. Underlined numbers give the page on which the complete reference is listed.

A

Abrams, E., 391[36], 395
Abu-El-Haj, F. J., 132[91], 156
Ackermann, H. J., 480[68], 481[68], 503
Adam, D. B., 275[33], 339
Adams, J. E., 22[8], 43
Adams, J. M., 481[73], 503
Adams, L. E., 182[98], 220
Adams, M. J., 277[70,71], 340
Adams, P. B., 63[119], 68, 126[44, 46], 133[111], 141[185], 154, 157, 159
Adkisson, P. L., 41[166], 50
Adlung, K. G., 257[103], 266
Aelbers, E., 9[66], 16
Afifi, N. N., 151[247], 161
Agrios, G. N., 201, 222
Aharoni, N., 334[435], 352
Aharonson, N., 233[41], 248[72], 264, 265, 485[85], 504
Ahmed, E. M., 289[184], 343
Ahmedullah, M., 293[225], 313[225, 294], 345, 347
Ajello, L., 441[6], 459
Akamine, E. K., 288[179], 343

Al-Beldawi, A. S., 141[172,176, 177], 142[210], 159, 160
Albersheim, P., 539[56], 546
Albert, A., 377[15,16], 395
Albrigo, L. G., 277[84,85], 278[84], 282[84,85], 283[84, 85], 284[84,85], 287[84,85], 330[85], 331[412], 340, 351
Alconero, R., 140, 159
Aldred, W. H., 145[222], 160
Aldridge, W. N., 418[71], 435
Alexander, J. V., 126[40], 154
Alexandri, E., 11[80], 17
Allard, C., 8[53], 16, 165[5], 217
Allen, R. N., 331[413], 351
Allen, T. C., Jr., 59[86], 67
Alliot, H., 411[50], 434
Al-Rawi, N. Y., 31[85], 46
American Chemical Society, 368, 370
American Phytopathological Society, 55, 61[99], 64, 67, 228[5], 263
Amin-Zaki, L., 31[85], 46
Amsden, R. C., 97, 102[18], 105[18], 114
Anderson, C. A., 472[30], 481[73], 488[97], 502, 503, 504

SUBJECT INDEX